THE SECOND CREATION

MAKERS OF THE REVOLUTION IN TWENTIETH-CENTURY PHYSICS

Robert P. Crease and Charles C. Mann

RUTGERS UNIVERSITY PRESS

New Brunswick, New Jersey

To Sasha and Newell,
with lots of love
—*CCM*

To India, who changed my life
—*RPC*

First published by Macmillan Publishing Company. Revised edition published by Rutgers University Press, New Brunswick, New Jersey, 1996.

Portions of this book were first printed in *The Atlantic Monthly*, *Science 85*, and *The Sciences*.

Excerpts from *The Birth of Particle Physics*, edited by Laurie Brown and Lillian Hoddeson, are reproduced by permission of Cambridge University Press.

Library of Congress Cataloging-in-Publication Data
Crease, Robert P.
    The second creation : makers of the revolution in twentieth
-century physics / Robert P. Crease and Charles C. Mann.
        p.   cm.
    Includes bibliographical references and index.
    ISBN 0-8135-2177-7 (alk. paper)
    1. Physics—History.   2. Grand unified theories (Nuclear physics)—
History.   I. Mann, Charles C.   II. Title.
QC7.C74   1996
539'.09—dc20   95-47448

Printed in the United States of America

Contents

VI UNIFICATION

Foreword

Particle physics ranks among the most visionary intellectual activities of our times. It is also a remarkably open enterprise, full of scientists happy to explain as much about their work as a good listener is prepared to hear. Yet notwithstanding its importance and openness, few among the general public know much about it. Within living memory, physicists have written equations that can accurately predict the outcome of every fundamental process in nature, from the interactions of quarks to the pulsations of giant stars, and have begun to excavate the ancient foundations that link these equations to one another and to events that transpired during the first second of cosmic time. And yet few general readers have absorbed this astonishing news. It is as if the average citizen of Renaissance Florence had never seen a ceiling fresco.

Into the daunting vacuum separating the physicists from the wider public venture popularizers of science, eager to tell the exciting story of how our awakening species, from its perch on a small planet in the suburbs of one among billions of galaxies, has begun painting the big picture. Many fail at this important task. Some grapple in vain with admittedly strange concepts like quark confinement or chiral symmetry, and their accounts, while well-intentioned, are muddled. Some grasp the science but are defeated by the subtleties of the written word; their sentences rise and stall and flop back to earth like overloaded biplanes. Others, burdened with the baggage of senescent modes of thought, seek to make a priesthood of science, revelling in a grandiose obscurantism that cloaks their lucid subject in clouds of incense. But a few have done justice to this wonderful subject, writing clear, evocative, and accurate reports that convey the glory of particle physics without shrinking from its human faults and frailties.

Robert P. Crease and Charles C. Mann earned a place in this elect fraternity with *The Second Creation*, a work that combines first-rate storytelling with a sound and wide-ranging exposition of some of the most exotic science the twentieth century has to offer. The long-awaited publication in paperback of this estimable book brings to a fresh audience their distinctively witty and informative tale of how humankind sounded the depths of the atom and found there a reflection of the farflung universe. No reader who gives it the attention it deserves will go away unrewarded.

Timothy Ferris
Berkeley, California

v

Preface

ACROSS THE GLOBE, THEORETICAL PHYSICISTS ARE ENGAGED IN AN extraordinary enterprise: the attempt to construct, by sheer force of reason, what some refer to, only half-jokingly, as a Theory of Everything. More formally known as a unification theory, a Theory of Everything would be nothing less than a complete description of the foundations of matter, space, and time, a set of linked equations containing the elements of the cosmos. Physicists have always hoped for a unified picture of nature, and now and then individuals have actually sought one. What is remarkable about the present search is that the cream of the present generation of researchers has decided as a group that the time may be ripe for unification, and a task that fifteen years ago was relegated to the lunatic now preoccupies the science as a whole. *The Second Creation* is an account of how this great leap came to pass.

We have taken our title from one of the implications of the search for unification: Working back from the present, the new theories promise to explain the beginnings of the Universe, the awful, minute explosion of the Big Bang, as well as everything that ensued. Never before has science tackled such large questions; the prospect of discovering some answers is thrilling for those who find delight in the accomplishments of the human mind.

Unfortunately for nonscientists, the progress of physics has gone hand in hand with an increasing reliance on abstraction and mathematical representation, a remoteness that has obscured the intense play of style, personality, and thought that permeates this most human of creative practices. Although it is a history of science in this century, *The Second Creation* is a tale of people, not protons—the small community of gifted individuals whose accumulated work has, in the words of Nobel laureate Steven Weinberg, answered all the "easy questions, such as why is the sky blue and what is inside the nucleus of the atom?"

No single person ever wrote a book; for us, the process of writing *The Second Creation* has been a lesson on the object truth of this maxim. Our approach would have been impossible without considerable assistance by professional physicists. We have been fortunate to have had the help of many. Above all, Gerald Feinberg, Murray Gell-Mann, Sheldon Glashow, Abdus Salam, Robert Serber, and Steven Weinberg have tolerated repeated interviews, read many sections of the book, and given us extensive help and criticism. They must have come to dread our reappearance; the book exists thanks to their kindness—a kindness that cost them many hours away from their own work.

"Intellectual debts are odd, for they take hard work to accumulate and are a pleasure to acknowledge." We have many. In the course of research, we interviewed some 125 physicists, many repeatedly, and most at length. Their names are found in the notes, to all we give our thanks. Many scientists whom we interviewed read and commented, sometimes copiously, upon early drafts of chapters. These include Neil Baggett, Ulrich Becker, Hans Bethe, James Bjorken, Robert Brout, Min Chen, David Cline, Martin Deutsch, Samuel Devons, Richard Feynman, Howard Georgi, Gerson Goldhaber, Maurice Goldhaber, Jeffrey Goldstone, Gerard 't Hooft, John Iliopoulos, William Kirk, Willis Lamb, Richard Learner, Leon Lederman, Luciano Maiani, Robert Marshak, M. G. K. Menon, Robert Mills, the late Paul Musset, Yoichiro Nambu, V. S. Narasimham, Yuval Ne'eman, Abraham Pais, Robert Palmer, Francis Pipkin, David Politzer, Charles Prescott, I. I. Rabi, Burton Richter, George Rochester, Carlo Rubbia, Nicholas Samios, Mel Schwartz, John Schwarz, Julian Schwinger, Lawrence Sulak, Richard Taylor, Samuel Ting, Martinus Veltman, Victor Weisskopf, Frank Wilczek, Robley Williams, Frank Yang. Several physicists allowed us to dip into their private papers: Gary Feinberg, Murray Gell-Mann, Sheldon Glashow, Gerson Goldhaber, Richard Imlay, Paul Musset, Robert Palmer, Abdus Salam, Nicholas Samios, Mel Schwartz, Lawrence Sulak, Samuel Ting. In addition, several historians of science resisted the usual hostility of academe to interlopers by giving us suggestions, answering questions, and allowing us to pick their brains: Mara Beller, Stephen Brush, J. L. Heilbron, Andrew Pickering, Silvan Schweber. We have benefitted immeasurably from their writing. If despite the presence of so much eminent help we have introduced errors into this book, we claim full credit for them. We owe a particularly odd debt to Horace Freeland Judson, a man whom we have never met, for his book, *The Eighth Day of Creation*, a work that gave us a glimpse of what is possible in science writing for nonspecialists—and the quotation that begins this paragraph.

We are indebted to Carol Ascher, Mary Lee Grisanti, Dale McAdoo, Peter Menzel, Mark Plummer, Victory Pomeranz, Debbie Triant, Sophia Yancopoulos for a host of personal favors. *The Second Creation* bears the invisible scars from the skillful scissors of a brace of talented editors—Barry Lippman, Debbie McGill, and William Whitworth. Richard Balkin, *agente incredibile*, walked the contractual tightrope with aplomb.

We are also indebted to the American Institute of Physics, the Church-ill College Archives, the Oppenheimer Papers at the Manuscript Division of the Library of Congress, the National Archives, the Pauli Archives at CERN, the Archives for the History of Quantum Physics at the American Institute of Physics in New York City (where Spencer Weart helped us), the Science and Physics Libraries of Columbia University, the New York Public Library, the IMB collaboration (special thanks to Dan Sinclair and Larry Sulak), Brookhaven National Laboratory (especially Neil Baggett and Anne Baitinger), the Enrico Fermi National Accelerator Laboratory (where Dick Carrigan looked after us), the Stanford Linear Accelerator Facility (William Kirk and William Ash), the European Organization for Nuclear Research (Roger Anthoine and the incomparable Gwendoline Korda), the Bohr Institute (Eric Rüdinger), the Tata Institute in Bombay (B. V. Sreekantan), the International Centre for Theoretical Physics in Trieste (where Dr. Salam hosted us). One of us (R. P. C.) would like to thank the Department of Philosophy at Columbia University for tolerating his delayed dissertation.

Finally, each of us would like to thank the other for his wholehearted commitment to the sometimes frustrating, always fascinating task of writing this book.

I
Waves and Particles

1
Beginnings and Ends

A COLD WIND BLOWING IN FROM LONG ISLAND SOUND WHISPERED THROUGH the cranes and scaffolding over our heads. It was just after dawn. Snow had fallen a few days before, melted, and refrozen into glassy ovals now glittering with first light. Twenty feet from where we stood, a dozen men surrounded a metal object that resembled a giant wedding band, silver in color and fifty feet in diameter. The men wore hardhats and gloves and heavy boots; some had toolkits on their belts. Steam drifted from their lips and coffee mugs. Despite the cold, they were working carefully, even slowly. Inside the shiny band at their feet was a spool of high-precision cable lent by a laboratory in Japan and wound to a tolerance of thousandths of an inch; nobody wanted to ruin the alignment in the process of transporting it. Perhaps fifty feet away from the ring stood its destination—a metal shed. Inside its featureless gray walls, scores of physicists had spent almost ten years and millions of dollars in government funds to penetrate deeper into the heart of matter than humankind had ever been before.

A particle physicist named Gerry Bunce was meeting us at the site. Rangy and affable, he greeted us with casual, unprintable reflections on the state of the weather, the practice of rising before dawn, and the wisdom of locating a laboratory on a flat island with no nearby ski facilities. The bleakness of the day was appropriate, for the landscape that surrounded us was a trashed technological wonderland: great cylindrical gas tanks; pipes, ladders, and girders of tarnished aluminum and rusted steel; solid blocks of weather-stained cement bigger in every dimension than a human being; and twists of black-wrapped electrical cable hundreds of yards long. All these scraps were left over from experiments conducted by physicists like Bunce over the course of almost half a century; much of his current work in progress, under assembly in the shed, would undoubtedly join them.

The task of the morning was to transfer the fifty-foot ring from its present position on the ground into the building, where it would be inserted, Russian doll-style, into a still bigger ring that already circumnavigated most of the available floor space. The procedure was more difficult than it might seem. The entrance to the building was an ordinary door less than three feet across— and it was vital for the workers not to warp the delicate ring out of true while moving it. After much consultation, the physicists had decided to cut away the bottom ten feet from one wall to make an opening. They had assembled an impressive amount of gear: two big cranes (one outside and one inside the shed), a huge wagon wheel-like frame, an improvised cart that looked somewhat like a railroad push car, and a set of rails for the cart to ride on. The plan was to tie the ring to the wheel, hoist the rigid ring/wheel assembly with one crane onto the cart, and roll the cart down a set of rails through the cut-away gap in the wall. Once inside, the ring/wheel would be hoisted into place by the second, indoor crane.

Bunce wore a ski parka with a zipper encrusted by lift tickets, souvenirs of a recent vacation. A sling cradled his left arm—another souvenir. "Just as well," he said, indicating his useless arm. "The engineers get nervous when I try to help." A de facto supervisor, he drew us out of range as a crane approached, huge doughnut wheels slowly crunching through clumps of snow. The crane arm swung about, dangling a steel hook; hands locked it into place at the center of the yellow wagon wheel. Quickly the hard-hatted workers— engineers, technicians, and lab staff—lifted the wheel into position atop the ring, then clambered jungle-gym style over the wheel, attaching it to the ring with two-foot bolts. The bolts slid home at the end of the spokes, and the air filled with the clanging of wrenches.

"Up a foot!" the foreman shouted. Wheel and ring rose to shin height and remained there, swaying gently, like a giant levitating diadem. Transfixed by the unexpected elegance of the sight, everyone stopped moving for an instant. The moment quickly passed; the crane operator swung the ring over to the cart and set it down gently. A gruff command sent the crew back to the ring/wheel assembly. Wrenches clattered again as they bolted it to the cart.

"Good, good," Bunce said. Despite his imprecations about the early hour, he was enormously pleased to be at the scene. Bunce had worked at Brookhaven National Laboratory, in Upton, Long Island, for almost twenty years. But he had never lost the enormous, almost greedy enthusiasm for his subject that is the hallmark of a fine scientist; as he stomped his feet to keep them warm, it was hard not to think he was also dancing in glee.

When Bunce began his career, he performed "discovery" experiments— that is, experiments to discern whether an unknown phenomenon exists rather than to measure known quantities more precisely. Both discovery and measurement experiments are essential to science, but in the past the discoverers have tended to be brasher and attract more attention. These days, though, after a century of unprecedented scientific growth, the distinction between discovery and measurement has eroded, and scientists like Bunce find themselves seeking new truths by measuring well-known quantities with extraordinary precision. The hope is that under extraordinarily careful scrutiny these well-known quantities will turn out to harbor intimations of new worlds.

In this experiment, the measurement will be performed on a special breed of subatomic particles called muons. The constituents of ordinary matter are electrons, protons, and neutrons. Muons are something else. Created only in such special situations as exploding stars, they never form part of any atom; instead, they are solitary creatures that slip into nonexistence fractions of an instant after they are made. Far from being trivial, though, the behavior of these transitory entities provides hints to the deepest laws of nature.

During the last decades, physicists have carefully observed such fleeting bits of matter and have come up with rules of thumb for their behavior. Physicists have learned that subatomic particles are not tiny, marble-like dots but strange, beehive-like swarms of activity. Throughout the brief interval of its existence—about 2.2 millionths of a second—each muon is enveloped in a nimbus of even more short-lived particles, which it constantly emits and reabsorbs. These secondary particles in turn create still more evanescent tertiary particles, and so on, the whole forming a furiously active, almost corporeal cloud. Far from being extraneous to the muon, the cloud carries the particle's electric and magnetic force. Not only that, the juggle of interactions within the muon cloud is so complex and far-reaching that it contains traces of every possible elementary form of matter.

All particles everywhere are surrounded by such clouds, but Bunce and his fellow team members chose to study the muon because the theory used to calculate its behavior—quantum electrodynamics, or simply QED—has proven to be extraordinarily precise. According to QED, every particle of matter in the Universe behaves as if it were spinning like a top. Now imagine a spinning muon pushed into a perfectly circular orbit with the axis of its spin aligned with its direction of motion. In terms of the Earth, this would be as if the planet orbited the Sun with the North Pole leading the way. QED predicts that the hugger-mugger in its particle cloud will make the muon's axis of spin (the line between its north and south poles, so to speak) drift out of alignment. It will no longer point precisely in the direction of travel.

QED also predicts the exact amount of the deviation. It is tiny. After one orbit, the theory says, the axis of spin will point inward just a little bit more, as if the north pole were to tilt ever so slightly toward (and the south pole away from) the Sun. As the number of orbits rises, the poles will keep sliding around; by the fifteenth orbit, north and south will have switched places, and the axis of spin will be aiming backwards, away from where the planet is moving. By the thirtieth orbit, according to QED, the axis of spin will be back where it began, pointing in the direction of travel.

The calculations needed to arrive at this prediction for such small, strange bodies in this exact configuration are exceedingly complex. To make them precise requires years on state-of-the-art computers. It is harder than trying to predict the amount of momentum in a sharply-hit baseball due to a single stitch, harder than trying to judge the additional speed an eighteen-wheeler will gain when its driver throws an apple core out the window, harder than trying to figure out the change in the Earth's orbit caused by the takeoff of an airplane. Nonetheless, QED lays it out.

The precision of the theory goes still further, in fact. It accounts for the effects of the whole cloud by summing the individual effects of each type of particle in it. Which means, physicists had realized, that measuring the deviation caused by the cloud around each tiny muon could advance one of the principal goals of particle physics: drawing up a census of the fundamental constituents of the Universe. Because the predicted value for the advance of the spin represents the itemized effects of every type of known matter, any difference between its predicted and measured value—any deviation from the deviation, so to speak—would indicate the existence of some hitherto unknown type of matter. Because the cloud wasn't doing exactly what QED said, that is, something new must be inside it. Something yet unknown.

Performing this measurement was an eye-poppingly difficult task. The Brookhaven team not only had to compass the muons—objects millions of times smaller than the smallest pinhead—with enough precision to observe the even smaller effects of the cloud but also had to measure them precisely enough to observe deviations millions of times smaller. It would be, to continue an earlier analogy, as if they had to measure the change in the Earth's orbit caused by the takeoff of an airplane with sufficient accuracy to determine how many passengers are inside. Such exactness would be possible only in specialized conditions—conditions, Bunce hoped, that could be found in the fifty-foot magnetic ring that he and his collaborators were building in Long Island.

"It'll take them a minute," Bunce said, referring to the crew bolting the railroad cart to the wagon wheel and ring. "Let me show you where the particles come in." Apparently untroubled by his recently separated shoulder, Bunce loped across the asphalt staging ground to the large, snowy earthen berm that loomed over the area. The berm was created by heaping dirt over a particle accelerator, a machine that pushes particles to nearly the speed of light. In its shadow lay a stack of eight-foot corrugated metal tubes that resembled giant versions of storm drainage pipes; around them rose stacks of ten-ton concrete blocks like toys discarded by giant children. Bunce carefully searched the pool of asphalt. "Here," he said proudly. We gave him blank looks: he was standing at a place that seemed utterly unremarkable. "The particles come out of the accelerator and hit a target here," he said, explaining. He looked about, orientating himself. "No, wait," he said. He stepped about six inches closer to the berm. "Here."

When the particles hit the target, fragments of its atoms break off; with powerful magnets, physicists separate the fragments by type and guide them into paths. In this case, the tiny, violent impacts would eventually create muons. "They go along here," Bunce said, walking briskly, assuming the role of a muon tour guide. He beckoned us to follow him into a giant storm-drainage pipe leading into a tunnel perhaps sixty feet long. As Bunce disappeared inside, the harsh flash of an arc welder momentarily silhouetted him in the entrance. We rushed to follow.

Bunce slowed down to pick his way ("Excuse me, pardon me, excuse me") past half a dozen surveyors who were aligning the tube that would carry the muons by peering through the eyepieces of a number of surveyors' transits. Nearby, sparks flew as welders worked on the coffin-like metal boxes that contained complicated electromagnets; the acrid smell of ozone drifted through the air. At the far end of the corrugated tunnel was a metal wall from which protruded a small pipe. Anonymous as a plumbing connection, it was one of the more important items in the area. It would carry the muons into the gray shed that housed the experiment proper.

Once outside again, we discovered that the workmen had unhooked the ring and frame assembly from the crane and bolted it to the cart. The cart was sitting at one end of a fifteen-foot track; its job was to ferry the assembly to the other side. "Push, everyone!" a foreman called. Bunce grabbed an edge with his one good arm; hugging the sling to his chest, he gave the assembly a good shove. So, obediently, did we. We had expected that moving the fifty-foot assembly would be a strain. To our surprise, it was so carefully designed that all seven tons of metal glided forward with little more resistance than it would take to push a case of soda across a kitchen table. Fifteen feet later, the entire mass came gently to a halt against a metal bumper at the end of the track. Instantly the workmen were atop the wheel again, unbolting the cart. Meanwhile, amid much shouting of orders, the indoor crane rolled over.

Like many experiments in particle physics, this one had a triple objective. The first and simplest was to measure a number—in this case, the deviation in the muon's spin caused by the ghostly cloud around it. A second, more speculative goal was to use the number to establish the existence of what scientists call "new physics"—in this case, new forms of matter. The third goal lay behind the other two. Scarcely discussed by the physicists working on the experiment, it was nonetheless present that cold morning.

That goal is to create a theory that draws into one picture all matter and energy, from the hottest supernovas to the whirring fragments of the atom. This goal is known as "unification," and in a sense it defines the discipline of physics as a whole. Physics is a generous term that comprises the study of phenomena as disparate as friction, X-rays, the formation of ice crystals, and the evolution of the Universe; it is a canon of faith that these aspects of nature, no matter how dissimilar in appearance and scale, arise from a few unifying principles—perhaps just one—which branch out, interweave, and extend outward like the veins in a leaf, becoming finer and finer until they mesh indissolubly with the cosmos, becoming the stuff of the cosmos itself. Finding these unifying principles is equivalent to writing the equations of the world. Whether they knew it or not, all of the physicists who ever lived spent their days laying down bricks in the road to this goal. But the goal has proven seemingly endless, and history is littered with discarded notions of how to put it all in a single package; Einstein looked for such a theory for years and failed completely.

In view of this dismal record, the progress achieved in physics since the Second World War is remarkable. Within that relatively short period scientists have taken greater strides toward unification than all their predecessors put together. They created an experimentally-tested theory of matter, a framework that ties together all the bits and pieces of the atom and the forces that play among them. This group of ideas is known as the "standard model of elementary particle physics," and it is the piecing together of the standard model that makes physicists today most proud of their work.

Even as they put the finishing touches on the standard model in the 1970s, some physicists went further, proposing true unification theories. These moved beyond the standard model, which placed all the known constituents of the Universe into a single theoretical framework, by attempting to demonstrate that the disparate forms of matter and energy derived from a single source. They went by a variety of names: grand unification, supersymmetry, and the absurdly-titled string theory. At first dismissed as implausible, these unified theories had a compelling clarity which by the 1980s had caused many physicists to become converts.

By the 1990s, however, these physicists had been disappointed, for the pace towards unification had slackened. Although the standard model seemed to provide the theoretical language necessary to construct a true unification theory—and even, some felt, demanded that construction—no one had been able to go beyond it. Early tests of predictions made by unification theories had uniformly failed to support them; indeed, every attempt to reach beyond the standard model had only ended up confirming its predictions. In this way, physicists were victims of their own astounding success. Worse, attempts to continue toward unification unavoidably seemed to require larger, costlier, and more time-consuming experiments. Caught in budget crises, governments became wary of the hefty price tags. In 1993, to cite a particularly dramatic example, the U.S. Congress terminated support for a huge, partly-built particle accelerator, the multibillion-dollar Superconducting Supercollider. Thus physicists had to devise less costly experiments that might provide clues to physics beyond the standard model toward further unification. Hence the experiment Bunce was working on—an experiment with a long history.

In the nineteenth century, the great Scottish physicist James Clerk Maxwell took one of the first steps to unification when he showed that two apparently disparate phenomena, electricity and magnetism, were in reality aspects of the same phenomenon: electromagnetism. The scientific theory of electromagnetism is called electrodynamics. Maxwell's insight led to a host of devices now used all over the world, from the little electric motors in toy trains to the hulking electromagnets in Bunce's experiment.

Then came quantum mechanics and relativity, which seemed to change everything, electrodynamics no exception. Modifying electrodynamics to take quantum mechanics and relativity into account created a new theory: quantum electrodynamics, or QED. The old, pre-quantum version of electrodynamics had already predicted that the spins of tiny bodies with electric charges would advance with the orbit when traveling in the reach of a magnet. For particles like the muon, the axis of spin should shift exactly in step with the orbit. Quantum electrodynamics, however, predicted that the spin should advance slightly faster than the orbit, with the difference due to the minute, frenzied actions of the cloud of particles around the muon. In QED, the spin advance is referred to as "g." In the units used by physicists, g equals 2 when the spin moves in step with the orbit. As a consequence, the difference between the new electrodynamics and the old is "(g-2)," pronounced "gee-minus-two." Put simply, Bunce's experiment was trying to measure (g-2) more precisely than it had ever been measured before.

Despite its awkward name, (g-2) is an important number to physicists. It encapsulates the difference between old and new, between the nineteenth century and the twentieth, between a world in which one can ignore quantum mechanics and relativity and a world in which they are essential. If (g-2) were equal to zero—which is to say, if g were exactly equal to 2—much of twentieth-century physics would not exist. It would mean, roughly speaking, that there would be no strange cloud of ghostly particles around each particle. But a measurement of (g-2) almost fifty years ago was enough to demonstrate that QED, including the swarm of particles it depicted, was a brilliantly accurate description of nature. Physicists kept measuring (g-2) as tools improved, constantly using the number to test their understanding of the world. Today, (g-2) has been calculated to twelve decimal places and is arguably the most precisely known quantity in theoretical physics. Indeed, Bunce and his colleagues were hoping to use its very solidity as a means to launch themselves into the unknown.

Inside the building, welding tools and wiring lay scattered about. The floor was concrete and many feet thick, shielding the delicate equipment against disturbance. Temporary scaffolding arched over the partly-assembled experiment. The yellow bridge crane rode overhead on parallel tracks, and its hook began to lower slowly, like a *deus ex machina* in a medieval play. A hand reached up and snagged the hook. Bunce grinned. "It's great," he said, "that there's so much *physics* in this stuff." Bunce meant that the intellectual value of the enterprise legitimized how much fun he was having, but his exact language is worth noting. The word *physics* carries a certain weight in Bunce's lexicon; it means the clarity, depth, and beauty with which one can illuminate the workings of nature.

He is by no means alone. In the last few years, we have sought out a number of other practicing physicists and have been struck repeatedly by the passion with which they approach their subject and the eagerness with which they seek the thrill of the sudden, excruciatingly sharp insight into nature that is the ambition of all scientists. Over and over, we were surprised—naïvely, perhaps—by the similarities between these participants in the hardest of hard sciences and artists, and how often the growth and flowering of twentieth-century physics resembled the growth and flowering of an artistic movement. Like the best artists, the finest scientists work to bring forth a vision of the world; like artists, their search for that vision is guided by the times they live in and by their teachers, their individual tastes, and their personal proclivities. Over time, our conversations with physicists allowed us to see the compelling sweep of the work of the last several decades, the standard model of elementary particle interactions—an artistic landmark that physicists themselves still marvel at even as they attempt to supersede it.

In the big laboratories and small university offices where high-energy physicists do their work, researchers are shaking the surprise out of their heads, bemusedly realizing that they have been through an extraordinary time—a time that they fear may never happen again. Because they knew all along that they were in unfamiliar territory, there were no old answers to overturn, and hence this second wave of physicists—Einstein's children, if you will—has never called its accomplishments revolutionary. Rather, these men and women worked for decades in a frightful tangle of confusion, which suddenly resolved itself in a collective moment of insight that bound together the long work of many separate hands. When the standard model fell into place, it did so all at once, after much disorder, and physicists who had expected to spend their careers straightening up a small corner of the immense arena of their incomprehension abruptly found themselves closer to their goal than they had ever dreamed. The atom revealed its secrets only gradually, like a cleverly written play, allowing sparks of knowledge briefly to illuminate corners of the story but jumbling scenes and inferences so that illumination came only at the end, after many members of the audience had given up hope of understanding. Centuries hence, historians may still describe the past fifty years as the most exciting ever in the development of particle physics.

Despite the recent setbacks on the road to unification, the dogged enthusiasm of Bunce and his collaborators is undiminished, as is the pride they take in their collective enterprise. There is every motive for them to feel that way.

The history of physics in this century is astonishingly distinguished—who in this age of popularization has not heard of quantum mechanics and relativity? It is a tale of a steady drive toward knowledge that today pushes at the beginning of time and the end of matter. But there is another reason for the joy good scientists take from their work: They are members of a tradition, a collegial association that continually celebrates its past even as it seeks ever greater novelty.

Under the unflattering illumination of the lights in the shed, the giant silver diadem was floating downward. As the crane cable slowly ratcheted to the earth, the great ring settled into position with the ease of a flying saucer in a science-fiction movie. Finally touching down, the ring made a small, solid click. It was a fraction of an inch away from its final location—a distance that would take weeks of fine adjustment to bridge. Even as the team decoupled the crane hook, workers were rushing to measure the next step. The engineer filled out a form detailing exactly what had been accomplished. Bunce shook his head in amazement. "This," he said of the (g-2) experiment, "is the most thought-out project I have ever been involved with."[1]

□ □ □ □ □

Once before, physicists—or some of them, at any rate—thought that the next truths of physics lay in measuring numbers with ever-greater precision. But their motivation was totally different. They believed that all of physics was already known and that future generations of scientists would only buff and polish the insights of the past. Indeed, the certainty with which many physical scientists of the 1880s thought they had the fundamental puzzles nailed down is today a source of puzzlement to scholars. At Harvard University, for instance, the then-head of the physics department, John Trowbridge, felt compelled to warn bright graduate students away from physics. The essential business of the science is finished, he told them. All that remains is to dot a few i's and cross a few t's, a task best left to the second-rate.[2] In 1894, Albert Michelson of the University of Chicago, one of the most prominent experimenters of the day and the future recipient of a Nobel Prize, told an audience that "it seems probable that most of the grand underlying principles have been firmly established and that further advances are to be sought chiefly in the rigorous application of these principles to all phenomena which come under our notice. . . .[T]he future truths of physics are to be looked for in the sixth place of decimals."[3]

Michelson's timing was comically bad, as it happened. Before the conference proceedings were printed, the first evidence of the previously unknown phenomenon of radioactivity was discovered by one Antoine-Henri Becquerel, the third Becquerel in a row to occupy the chair of physics at the Musée d'Histoire Naturelle in Paris. A balding, irascible man with a fierce little Vandyke beard, Becquerel had spent his twenties and thirties performing undistinguished experiments on phosphorescent crystals. He got his doctorate at the age of thirty-five and almost immediately gave up research, settling into the comfortable respectability of his professorship. Becquerel was, to say the least, an unlikely candidate for celebrity; everything about him suggested that he was destined to be a footnote to future histories of science.

There are few scientific discoveries whose circumstances are known as minutely as those around the almost accidental finding of radioactivity.[4] On January 7, 1896, the great French mathematician Henri Poincaré received a letter containing several astonishing photographs of the bones in someone's hand. The bones belonged to Wilhelm Conrad Röntgen, a scientist Poincaré had never visited. The letter explained that the pictures had been taken with the aid of a new discovery, X rays, that Röntgen had turned up the previous month, and that he was publicizing his findings by mailing off prints all over Europe. Publicized they were: The photographs created a sensation across the globe. Within three weeks, little Eddie McCarthy of Dartmouth, New Hampshire, became a local cause célèbre when his broken arm was set by physicians armed with X-ray images of the fracture.[5] It is easy to imagine Poincaré's amazement—photographs of the inside of a human being!—and he quickly asked two local doctors if they could duplicate Röntgen's work. On January 20, they showed their own X-ray photographs to the assembled members of the French Académie des Sciences.[6] The reaction was immediate and extreme. In the next fortnight, five members of the Académie presented papers on the new phenomenon.

Becquerel, too, was sitting in the audience when the X-ray photographs were shown. He was fascinated by the strange ghostly images and the mysterious emanations that produced them. Both he and his father had studied the phenomenon of phosphorescence—the museum laboratory was filled with lumps of stone and wood that shone in the dark. The glow of X-ray emission put Becquerel in mind of the light in his study; although he had not done much active research in the last few years, he thought immediately of putting some phosphorescent rock on photographic paper to see if it would darken it in the same way as one of Röntgen's X-ray sources. It would not be all that much work.

What happened next has been recounted many times: how over the next month Becquerel tried a variety of phosphorescent stones, and found nothing; how one day he happened to pick up a chunk of potassium uranyl sulfate, a messy crystalline mix of uranium, potassium, sulfur, and other elements, which he knew from experience glowed under ultraviolet light; how he set the rock out on his balcony to be charged up by the ultraviolet rays in the winter sunlight; how he took a photographic plate, wrapped it up in thick black paper to shield it from the sun, and put it beneath the uranyl he called it "penetrating rays"—from the rock had glided through the paper and produced gray smudges on the plate.

Becquerel was certain that he had shown that X rays were somehow linked to phosphorescence. But he wanted to prove it scientifically—nail it down. Over the next few days he put coins and irregular pieces of metal between the uranyl sulfate crystals and the plates. Sure enough, they blocked the penetrating rays, showing up as coin-shaped spots of white in the darker gray. On February 24, Becquerel told the Académie of his results: Phosphorescence caused X rays.[7]

Becquerel's study was a model of the scientific method. It has come down to us, however, as a textbook of the practical difficulties in applying that method—which Becquerel was the first to find out. By February 26, the weather became dreary, as often happens in Parisian winters. While waiting for the sun, the professor put the plates, paper, and crystals into a file drawer. They lay in the dark for nearly a week; nothing could happen there, Becquerel knew, because the uranyl sulfate was not exposed to light, and hence could not phosphoresce. Nonetheless, on March 1, when the sun came back, he had one of those happy, once-in-a-lifetime thoughts: Why not develop the plates anyway? He had time. In the darkroom, he saw the darkest exposed blotches yet. Becquerel realized to his dismay that the photographic plates were *not* exposed by phosphorescence. There was something in the rock that did it. The uranium, it seemed, was spitting out X rays all by itself.[8]

This, too, was not entirely correct. In fact, the lump of potassium uranyl sulfate was emitting a whole spectrum of radiation, of which only a small portion was X rays. Nonetheless, the discovery caused a sensation, in part because it was so easy to duplicate. Almost every laboratory in the world had construction paper, photographic plates, and chunks of uranium ore. Within weeks, scientists across the Continent were looking in astonishment at the blurred black patches on their photographs. Becquerel became famous; judging by his contemptuous dismissal of rival claims to the finding, he seems to have enjoyed his sudden notoriety.[9] In 1903, Becquerel was given one of the new science prizes established by a posthumous bequest of the late Swedish industrialist Alfred Nobel.

Within weeks, news of Becquerel's findings had spread to Germany, Great Britain, Italy, and the United States, further exciting researchers already stirred by the discovery of X rays. Tests of the two phenomena were often conducted on the same workbench. The consequences of each discovery, however, were far different. X rays were found to be simply pulses of light—light of an intensity and power never before seen, but light nonetheless. Radioactivity, on the other hand, was something entirely new, something that did not fit anywhere. The existence of radioactivity—metal that somehow shot out energy!—was a direct attack on the most ardent beliefs of Becquerel and his colleagues. When the strange behavior of uranium was first noted, Becquerel wrote in his memoirs, "There was no reason to presume that the phenomenon was [anything but] a new example of a known type of energy transformation. Contrary to every expectation, the first experiments demonstrated the existence of an apparently *spontaneous* production of energy. . . ."[10] They had spent many years, those nineteenth-century scientists, establishing the law of conservation of energy: Energy was neither created nor destroyed. But every single piece of uranium seemed of its own accord to produce radiation that fogged photographic plates, electrified gases, and sometimes even burned physicists—and the energy needed to do these things evidently came from no place at all. The metal just sat there, its atoms quietly working away, continuously beaming out penetrating rays in seeming disregard for the conservation of energy.

As it happened, the first clue to the nature of radioactivity, although it was not recognized as such immediately, was found just a year after Becquerel's work, when Joseph John Thomson, an Englishman, deduced the existence of small objects later called *electrons*. The director of the Cavendish Laboratory in Cambridge, England, "J. J." had a gift for designing experiments, although his clumsiness prevented him from actually building the equipment. He was notoriously inattentive in matters of dress—his tie is askew in his official Cavendish portrait. To Thomson's own surprise, he had been chosen to head the Laboratory in 1884, when he was only twenty-eight.[11] By 1897, when Thomson made his discovery, the Cavendish had become the most prominent physics laboratory in the country. It had twenty full-time staff members, and had recently acknowledged the growing women's rights movement by allowing women to enter the Laboratory. Research was steadily turning away from practical matters such as telegraphy to such useless topics as the nature of electricity.[12]

At the time, a chief means of examining electricity was to pump out the air from a long glass tube, insert wires in both ends, and connect the wires to a battery. If one of the ends of the tube was painted with zinc sulfide or some other fluorescent material, a tiny, glowing spot would appear on the paint as soon as the battery was switched on. Obviously, *something*—or maybe a stream of somethings—was passing out of one wire, shooting across the tube, and smacking into the zinc sulfide. (The dot of light was like the one seen on old television sets the instant after they are turned off.) Because once the air was sucked out and there was nothing in the tube, scientists reasoned that whatever was going across and making the paint fluoresce must be a flow of electricity in its elemental form. If one could learn of what that flow was made, it might reveal the nature of electricity. Thomson and his subordinates had spent more than a decade, off and on, worrying about the problem.

After a series of experiments, Thomson gave a talk on Friday, April 29, 1897, in which he announced that he had the answer.[13] (A long paper was published six months later.) The glow, Thomson claimed, was caused by a stream of small particles—*corpuscles*, in the jargon of the day—each bearing a set amount of negative charge. They were, in a way, atoms of electricity: electrons, as they are now called. The corpuscles, Thomson said, were sailing off the wire into the zinc sulfide, and somehow the energy of their collision was causing the chemical to emit light. They were very small—too small to weigh by any known means. And the amount of electric charge on each corpuscle was also tiny—too tiny to measure. But Thomson found one property of these electric corpuscles he *could* evaluate: the ratio of their electric charge to their mass. Only the ratio of these two quantities, however, not the actual value of either. Indeed, for several years, the charge-to-mass ratio of these particles was their only precisely measurable property.[14]

Nevertheless, it was clear that Thomson's corpuscles were smaller than any known object, including the atom. The reaction to this notion was not positive: It is difficult to grasp how startling the notion of a subatomic particle was to nineteenth-century physicists, many of whom did not believe that atoms existed, let alone that they had constituent parts.[15] An influential and primarily German school of thought argued that physicists ought not to truck with creatures they could not see; because atoms could not be observed directly, they therefore should not be made the object of speculation. If this were true for atoms, it was all the more true for pieces of atoms. Years afterward, Thomson recalled his colleagues' lack of enthusiasm for his discovery: "At first there were very few who believed in the existence of these bodies smaller than atoms. I was even told long afterwards by a distinguished physicist who had been present at my lecture at the Royal Institution that he thought I had been 'pulling their legs.' I was not surprised at this, as I had myself come to this explanation of my experiments with great reluctance, and it was only after I was convinced that the experiment left no escape from it that I published my belief in the existence of bodies smaller than atoms."[16]

Atoms are building blocks of nature. Pile up enough of them and you get a lump of stuff that you can see or break or hit someone over the head with. With electrons it is an entirely different affair. No matter how many you gather together, it is impossible to make a lump of something that can be tasted or smelled or held in the hand. Little wonder that Thomson's colleagues did not think he was speaking seriously.

They were forced to accept Thomson's work by the increasing number of scientists who argued that the results of their experiments could only be understood by believing in the existence of subatomic particles. Becquerel, for example, made his last important contribution to science when he demonstrated, in 1900, that the emanations from the uranium in his laboratory included a considerable number of electrons.[17] But this discovery raised more questions; if there were negatively charged bits of matter in the atom, and if ordinary atoms were electrically neutral, there had to be some positively charged matter to balance out the negative. In other words, there had to be more pieces to the atom. Where were they, and how were they arranged? What are the ultimate constituents of matter and what forces play among them? Answering these questions required decades of work, which culminated in the 1970s with the standard model of elementary particle interactions and the prospect of unification.

The first hints were provided almost a decade later by Ernest Rutherford, a Thomson protégé then working at the University of Manchester. A large, confident man with a big red face and a walrus moustache, Rutherford was born and raised in colonial New Zealand. He was good with his hands, clever, ambitious, and hardworking; he did not want to spend his life grubbing about a subtropical farm. In 1894, he won a scholarship to Cambridge, which had just changed its rules to admit graduates of other schools, thereby permitting Rutherford to become the first foreign research student in the Cavendish laboratory. He arrived in England in September of the following year. He was just twenty-four years old.

As soon as Rutherford appeared in Cambridge, he went to Thomson and, according to the letters he dutifully sent his fiancée in New Zealand, "had a good long talk with him. He's very pleasant in conversation, and he's not fossilized at all. As regards appearance, he's a medium-sized man, dark and quite youthful still—shaves very badly and wears his hair rather long." Rutherford reported that Thomson "seemed pleased with what I was going to do."[18]

The Cavendish then was crammed into a Victorian neo-Gothic building, gray and bilgewater yellow, in the center of Cambridge. The experimenters worked cheek by jowl in grubby, crowded laboratories where Rutherford was apparently allowed to hear the mocking comments of Cambridge instructors about scholarship winners from the Antipodes. Nonetheless, he quickly impressed Thomson with his brashness, skill, and extraordinary drive, the same qualities that prevented him from making friends immediately. "One can hardly speak of being friendly with a force of nature," Rutherford's Cavendish colleague Paul Langevin is said to have remarked.[19]

Rutherford was a perfect man for the time, a hardworking, hardheaded scientist who was impatient with mathematical abstraction and complicated equipment. Consciously presenting himself as the epitome of an experimenter, he declaimed that he liked to discover facts the reliable way, through experiments, without a lot of theoretical pettifogging. The experiments themselves should be quick, simple, and performed on equipment scavenged from the basement of the laboratory. Each test should build on the one before. The object was to find good, solid facts—did X happen or not?—to find them first, and to let other people clean up decimal places. He said, "There is always someone, somewhere, without ideas of his own, who will measure that accurately."[20]

Rutherford possessed one of the talents that a good physicist must have, a sense for the right direction to pursue. At any given moment, there are literally thousands of experiments that could be done; a great experimentalist has a sixth sense for which one will lead to something profound, rather than merely informative. Rutherford had early on been intrigued by electromagnetic waves, and actually built a practical radio transmitter a little ahead of Guglielmo Marconi. At the Cavendish, however, Rutherford quickly decided to drop the radio, which was only practical, and chase after radioactivity, which might involve some real *physics*. With characteristic directness, he began to take the phenomenon apart. He learned that radioactive materials boiled with activity; they constantly spewed out huge numbers of particles, and the particles traveled at terrific speed—thousands of miles a second. Moreover, Rutherford found radioactivity had two distinct forms, which "for convenience" he named alpha and beta rays, after the first two letters in the Greek alphabet.[21] The alpha rays could be blocked by a piece of paper, but the beta rays had a hundred times the ability to punch through a shield. (Both had been emitted by the uranium in Becquerel's laboratory. Beta rays were later shown by Becquerel to be composed of electrons.)

In 1898, Rutherford was offered a professorship at McGill University, in Montreal, where he was "expected to do a lot of original work and knock the shine out of the Yankees!"[22] At McGill, Rutherford began what was to be his life's work, the study of alpha rays. For the next few years, Rutherford, Becquerel, and Becquerel's collaborators and friends, the young Marie and Pierre Curie, kept up an intense but friendly rivalry to be the first to comprehend the nature and behavior of alpha particles. The contest was sharpened by the Curies' discovery of radium, an element a million times more radioactive than simple uranium. Radium was fantastically rare; the Curies processed tons of uranium to get microscopic amounts of radium, and by 1916 the total world supply was less than half an ounce, parceled out in minute doses among the score of laboratories investigating its properties.[23] Using a few hot milligrams of the stuff sent to him by the French, Rutherford measured the charge-to-mass ratio of alpha particles, just as his mentor Thomson had done for electrons; after four years of work, Rutherford was certain they were positively charged helium atoms.[24] During this time, most scientists

vaguely thought the atom was a buzzing hive of thousands of electrons somehow held together by a positively charged glue. Rutherford decided that his
alpha particles were made by knocking out a few electrons from the glue in
helium atoms; the absence of the negatively charged electrons gave the
helium a net positive electric charge.

At about the same time Rutherford was establishing the identity of
alpha particles, Becquerel had performed an experiment that seemed to
indicate that alpha particles had the spooky property of increasing their
momentum as they pushed through the air.[25] This was bizarre: Alpha particles were spat out of atoms like so many minute bullets, but instead of
slowing down as they traveled, they seemed to speed up.[26] Rutherford, on
the other hand, found that alpha particles slowed down gradually.[27] The two
men challenged each other's findings, and both repeated their own experiments. Rutherford was correct. The argument would today be forgotten
except that it sparked Rutherford's curiosity. He derived no particular satisfaction from being the victor in a minor scientific dispute, but he *was* intrigued with the question of why his French colleague had gone astray.
When Rutherford turned his attention to the details of Becquerel's experiment, he was impressed with how difficult it was to measure precisely the
paths of the alpha rays. The lack of definition was "evidence of an undoubted
scattering of the rays in their passage through air."[28] In other words, as the
alpha particles sailed on their merry way, at least some of them bounced off
the molecules of air in their path. This finding was a critical step towards
Rutherford's discovery of the structure of the atom.

He didn't realize it at first. For several years, Rutherford thought that
the deflection was only another stumbling block that nature maliciously had
put in the way of experimenters who wanted to find facts without any bother.
In 1907, he accepted a post at Manchester and moved back to England,
where he continued working with alpha rays, using a much stronger source
of radioactivity he had obtained after much wrangling with colleagues. A
year later, he won the Nobel Prize. To his astonishment, it was for chemistry,
not physics.[29] Despite the move and the prize, he continued to work at the
same pace. But whenever he was forced to make a precise measurement—
despite his dislike, he would do it in a pinch—the scattering kept making the
job harder. For example, when he wanted to find out the precise charge of
alpha particles, he thought he would fire them one by one into a device
capable of assessing them individually. But the various measurements never
agreed with each other, and Rutherford and his team, annoyed, realized that
"the scattering is the devil" that was plaguing their work.[30] To the scientists'
dismay, alpha particles seemed to be ricocheting all over their equipment.

Exasperated, Rutherford told his assistant, Hans Geiger, that to avoid
further problems they would have to measure how much the alpha particles

were being jostled. Geiger was joined by an undergraduate, a New Zealander named Ernest Marsden. They beamed alpha particles through a thin metal foil into a thin metal screen that gave off tiny flashes whenever it was struck by the particles. Beforehand, the experimenters had to sit in the dark for a quarter of an hour to let their eyes adjust enough to see the flashes. The problem was, of course, the scattering: So many alpha particles were deflected by the air and the walls of the tube that it was difficult to discern what particles were bouncing where. One day in the early spring of 1909, Rutherford told Marsden to see if any particles would actually bounce back from the foil.[31] They quickly found that about one in eight thousand alpha particles would slam into a sheet of gold leaf and rebound.[32]

At first, Rutherford assumed that the alpha particles had, like billiard balls in a complicated shot at snooker, simply ricocheted off several atoms of gold. But over the next year the scattering apparently nagged at him; he simply did not think it likely that a fast little alpha particle could graze a few atoms and end up turning 180 degrees.

On the other hand, Rutherford could imagine an alpha particle rebounding off *one* atom—the guiding metaphor here might be shooting a bullet at an anvil. The problem was that atoms were not supposed to be like anvils. The most prominent physicist in England, J. J. Thomson, argued forcefully that the atom must consist of a ball of positive charge studded with electrons. The whole ensemble was often described, rather vaguely, as a spongy, doughy blob—a "plum pudding," as it was later termed, with electrons standing in for plums. Rutherford liked to have a clear pictorial image in his head of what was going on when he performed an experiment. He knew the alpha particles were drilling through the air at great speed; he could not imagine how such bullets could bounce back from a lump of pudding.[33]

By late November or early December of 1911, Rutherford had the first inkling of the answer. A quick seat-of-the-pants mathematician, Rutherford figured that if almost all of the mass of the atom were concentrated into a little charged node in the center, that would be enough to deflect an alpha particle. He wasn't sure whether the charged center was positive or negative—that is, whether it kicked away the alpha particle or whipped it around like a comet—but the opposite charge had to be in a sort of thin, gaseous sphere surrounding the middle of the atom. Although he was pleased with this image, Rutherford hesitated. He was not really certain of his ideas and, despite a well-deserved reputation for speaking his mind, he was leery of placing himself in the classic Oedipal situation of publicly disputing his mentor.

At this point, Rutherford had a stroke of luck: At about the same time he was chewing the matter over, another Thomson protégé, J. A. Crowther, announced that he had confirmed the plum pudding model by an experiment similar to the one done by Geiger and Marsden, except that Crowther had fired beta rather than alpha particles at metal foils.[34] The experiment gave Rutherford something to react against; it was psychologically much easier to go after Crowther than Thomson, although the end result was the same.[35]

On March 7, 1912, Rutherford first presented his theory at a session of the Manchester Literary and Philosophical Society.[36] He discussed the Thomson model and Crowther's experiment and then bluntly attacked both. The results of Geiger and Marsden, he contended, could not be explained by a plum pudding. Although it was a small effect, the deflection of one in eight thousand alpha particles occurred and must be caused by something; the alpha particles had to have slammed into something very small and very hard in the atoms of the target. Rutherford called that small, hard thing "a central electric charge concentrated at a point." We now call it the nucleus of the atom. Around the nucleus, Rutherford said, is a "uniform spherical distribution of opposite electricity," which today we know to be the orbiting electrons.[37]

□   □   □   □   □

The solar system–like picture of the atom has become familiar to us from the covers of numerous high school physics books and from the logo of the old Atomic Energy Commission. But this symbol is a domesticated, smoothed-over version of the real thing. Rutherford found that the atom, and therefore matter as a whole, consists overwhelmingly of empty space. If an atom were blown up to the size of a domed football stadium, the nucleus would be the size of a fly in the center; scattered throughout the enclosure is a sprinkling of even tinier electrons. More strangely still, the nucleus is incredibly heavy: It accounts for almost all the weight of the stadium, while all the grandstands and roof panels are as light as mist. The apparent solidity of everyday objects is due to the play of electrical forces among atoms and molecules, not the substance of the material itself; in truth, substance is one of humanity's most persistent illusions. With the discovery of the emptiness in matter, nuclear physics—indeed, the whole nuclear age—was born.[38] It was Rutherford's greatest accomplishment; we are still reaping the consequences.

At first, nobody paid attention. That spring, Rutherford had written to many of the physicists he knew, telling them about his model with characteristic ebullience. Reactions ranged from polite acknowledgment to indifference; after all, other than one rare type of alpha scattering, there was little evidence that the atom had a nucleus and that matter was mostly void. Rutherford seems to have been a little daunted by the lack of enthusiasm. At any rate, for a while he gave up proselytizing for his idea in favor of writing a big book summarizing the state of knowledge about radioactivity.[39]

But even if physicists had taken the nucleus seriously, Rutherford's model literally could not work. It was like a model airplane whose designer had included elements that would not fit together. In nearly all arrangements, the orbiting electrons would either be sucked into the positively charged nucleus or be ejected by the negative charge of fellow electrons. With considerable mathematical finagling, everything could be balanced, but even then the slightest nudge or disturbance would cause the whole system to go awry. If an atom like Rutherford's had ever existed, it would have torn itself apart in a fraction of an instant.[40] Any reasonable physicist, therefore, would have dismissed the idea after thinking about it for five minutes.

2
The Man Who Talked

ON A GRAY RAINY STREET IN THE CENTER OF THE GRAY RAINY CITY OF Copenhagen is a small cluster of buildings that protrudes into the side of a city park. Neatly tended and vaguely inhospitable in the Continental manner, Faelled Park is a stretch of wet greensward laced by gravel paths that run beneath stands of trees. The park is old, square, pristine, proudly aloof from city life—except where the complex on its edge has gobbled up a meadow and nibbled at the edges of a gathering of oak. As if to disguise their intrusion, the buildings cut into the flank of Faelled Park have inconspicuous slate-colored walls and red tile roofs and curtained windows like their neighbors in Copenhagen proper. They are, however, one of Europe's greatest centers of theoretical physics and a living monument to the torchbearer of the quantum revolution, Niels Hendrik David Bohr.

Bohr's working habits have become legendary among his successors, part of the lore of science along with Einstein's flyaway hair and Rutherford's remark that relativity was not meant to be understood by Anglo-Saxons. Bohr *talked*. He discovered his ideas in the act of enunciating them, shaping thoughts as they came out of his mouth. Friends, colleagues, graduate students, all had Bohr gently entice them into long walks in the countryside around Copenhagen, the heavy clouds scudding overhead as Bohr thrust his hands into his overcoat pockets and settled into an endless, hesitant, recondite, barely audible monologue. While he spoke, he watched his listeners' reactions, eager to establish a bond in a shared effort to articulate. Whispered phrases would be pronounced, only to be adjusted as Bohr struggled to express *exactly* what he meant; words were puzzled over, repeated, then tossed aside, and he was always ready to add a qualification, to modify a remark, to go back to the beginning, to start the explanation over again. Then, flatteringly, he would abruptly thrust the subject on his listener—surely this cannot be all? what else is there?—his big, ponderous, heavy-lidded eyes intent on the response. Before it could come, however, Bohr would have started talking again, wrestling with the answer himself. He inspected the language with which an idea was expressed in the way a jeweler inspects an unfamiliar stone, slowly judging each facet by holding it before an intense light.

His continual struggle with language extended to the most ordinary

20

acts. Bohr was one of the few people on earth to write drafts before sending postcards. His articles were composed with such care and precision that they sometimes verged on incomprehensibility, and were always late. He asked friends to read preliminary versions, and weighed their comments so thoughtfully that he would often begin over again; a frustrated collaborator once snarled to a colleague who had given Bohr a minor suggestion on a draft, prompting a seventh rewrite, that when the new version was produced, if "you don't tell him it is excellent, I'll wring your neck."[1] Bohr studied problems with the slow gravity of an earnest child; he was willing to appear foolish if it meant he might learn. He was utterly unable to tease. He was entirely without malice.[2]

Shy and pensive, he had a long, oval, big-cheeked Danish face and thick hair which he combed straight back from his forehead. Although he had been a brilliant student, Niels was consistently overshadowed by his younger brother, Harald, who was considered the Bohr child with real promise. Niels was an excellent soccer player, but it was Harald who in 1908 played halfback on the Danish Olympic team and brought home a silver medal. Harald was two years younger than Niels, but he entered the university just a year behind him and completed his doctoral dissertation a year earlier. Niels wrote strange, difficult, brilliant physics articles, but at first it took Harald's urging to make scientists look at them. Harald became a distinguished mathematician; even late in life, Niels claimed that his brother "was in all respects more clever than I."[3]

As a schoolboy, Bohr's worst subject had been Danish composition, and for the rest of his life he passed up no opportunity to avoid putting pen to paper. He dictated his entire doctoral dissertation to his mother, causing family rows when his father insisted that the budding Ph.D. should be forced to learn to write for himself; Bohr's mother remained firm in her belief that the task was hopeless. It apparently was—most of Bohr's later work and correspondence were dictated to his wife and a succession of secretaries and collaborators. Even with this assistance, it took him months to put together articles. Reading of his struggles, it is hard not to wonder if he was dyslexic.

Early in 1911, at the age of twenty-five, Bohr defended his dissertation and received a fellowship from the Carlsberg Brewery, which many Danes believe makes the finest beer in the world, to study in England for a year. He was excited about the prospect of working with the famous J. J. Thomson in the Cavendish. Bohr had thought a great deal about Thomson's plum pudding atom, and was sure that it could not possibly be correct; he could hardly wait to discuss his criticisms of it with the master. He came to Cambridge in September of 1911, sixteen years after Rutherford, and like Rutherford was at first elated to be there. A few days after his arrival he wrote to Margrethe

Nørlund, his fiancée, "I found myself rejoicing this morning, when I stood outside a shop and by chance happened to read the address 'Cambridge' over the door."[4]

Like many shy young people, Bohr often had to work up the nerve to speak with strangers, and the accompanying anxiety would cause him to blurt out whatever was on his mind. This failing, coupled with his then-unsteady grasp of English, made his first encounter with Thomson something of a disaster, and his stay at Cambridge discouraging. Unlike Rutherford, he could not learn to get along with the English. Bohr showed up in the lab, bumptiously anxious to talk of plum puddings, and was promptly and rudely dismissed. "I had no great knowledge of English, and therefore I did not know how to express myself," he said later. "I could only say [to Thomson], 'This is incorrect!' He was not interested in the accusation that it was not correct."[5] Thomson seems not to have known what to do about the young, anxious, inarticulate Dane; Bohr soon noted that whenever he managed to catch Thomson's attention "for a moment, he [Thomson] gets to think[ing] about one of his own things, and then he leaves you in the midst of a sentence (they say that he would walk away from the King, and that means more in England than in Denmark), and then you have the impression that he forgets all about you until the next time you dare to disturb him."[6] Fending off the young foreigner, Thomson promised to read his thesis, but never did. Bohr felt stranded in the coldly civil laboratory.

Unhappy, he attended the annual December Cavendish dinner, a boisterous, collegiate affair featuring music hall numbers the scientists wrote parodying themselves and their work. Wine flowed freely, and researchers bellowed physics jokes, like the toast, "To the electron! May it never be of any use to anybody!" Rutherford, who always enjoyed a good party, often came down from Manchester for the occasion; the Cavendish dinner of 1911 was no exception.

Bohr's feelings upon meeting Rutherford for the first time are easy to imagine. Standing up at the dais was a rugged man with the ruddy complexion and thick moustache of a country butcher; laying aside a pipe that spat out smoke and ash at a volcanic rate, he launched into a direct, humorous, even bawdy account of developments at Manchester. His manner was enormously heartening to a young man distressed by English formality. Later Bohr discovered that almost everyone had a favorite Rutherford anecdote; one Cavendish man told him that of all the physicists with whom he had worked, Rutherford was the one who could swear at the experiments most effectively.[7] In March 1912, Bohr moved to Manchester, staying in a little room at Hume Hall. In a short time, his attention drifted to Rutherford's model of the atom. Rutherford told his new assistant not to spend too much time wondering what went on in the nucleus; it was just an idea, and an idea was not worth as much as a fact. Bohr ignored the advice.

In their different ways, both men had the great gift of physical intuition, of being able to picture the doings of the unseen entities they were studying. Bohr was drawn immediately to the nuclear atom—"I just believed it," he said later—partly *because* it would not work; some extra new thing would be necessary to make the model fly. Bohr had the notion that a young man unimpeded by the received wisdom of his elders could provide the answer. Moreover, he seems to have been one of the first to guess that the structure of the atom and the behavior of its parts would be the question that would drive the progress of physics in this century, something not obvious at the time. By the end of spring, he was mulling over the suspicion that the nuclear model might be coherent if it were joined to something from an apparently unrelated idea of physics—the quantum.

A child of the century, the quantum was born on December 14, 1900, when a conservative German academic, Max Planck, reluctantly announced that certain experimental results could best be understood if it were assumed that substances emit light only of certain energies and not others.[8] Born in 1858, Planck came from a family of ministers and lawyers—upright, dutiful, honest people. He attended universities in Munich and Berlin, doing well but not strikingly well. His thesis adviser told him to look into another field, because physics in the 1880s was just about finished. Planck took a teaching job in Kiel and then, to his surprise and pleasure, was offered a prestigious post at the University of Berlin in 1889. He was not certain whether he believed in the reality of atoms.[9]

In the late 1890s, Planck spent six years studying the way substances emit light; an example is the blue glow of the gas in a neon sign. As can be imagined, the connection between light and matter was hard to understand at a time when many reputable scientists did not think atoms existed. Planck tried to avoid the whole question of what was making the light by treating light as if it were emitted by "oscillators" whose oscillation produced light waves in somewhat the way a plucked guitar string makes sound waves. One way of performing such calculations is to divide up the total energy of each oscillator into little pieces of approximately equal size, let them become infinitely small, and then use the techniques of calculus to add them all back up; the sum—or, more properly, the integral—would then be the original energy. Unluckily, in Planck's case it didn't work that way. If he were to make his result fit with the experimental data, the little pieces of energy *could not* become vanishingly small. They had to have some finite size—meaning that their sum, the total energy of the oscillator, could only have particular values.

The more Planck thought about this, the less he liked it. If his oscillators could not vibrate with any energy they pleased, then something made them choose certain values and prevented them from selecting others. This was apparently as senseless as claiming that a guitar string could be tuned to produce $C^\sharp$ and $B^\flat$ but no sounds in between. Planck was absolutely unable to justify his statement; he had to fudge the math to make the equations even *look* right.[10] On the other hand, Planck, like most physicists, was a practical sort, and he realized that his wild idea produced formulas that matched the charts and graphs of the experimenters.

After weeks of uncertainty, he finally performed what he later called "an act of desperation" and asserted that his oscillators could only have certain discrete energies, and therefore that they could produce light only in certain specified frequencies.[11] In modern language, Planck's claim is reducible to the expression

$$E = nh\nu$$

where E is the energy of the light source, n is a positive integer (that is, a number like 0, 1, 2, 3, and so on), ν is the Greek letter *nu*, which physicists use to mean frequency, and h is a small, unchanging number now known as Planck's constant.[12]

Simple in appearance, Planck's formula had great resonance. Quite literally, its strangeness threw the world into turmoil—although slowly, for truly great discoveries sometimes acquire their stature only in retrospect. If n must be an integer, then it cannot have a value between 0 and 1. Elementary multiplication thus shows that E, the energy, cannot have a value between νh and 0. Values like $\frac{1}{2}\nu h$, $\frac{1}{4}\nu h$, and $\frac{1}{10}\nu h$ are forbidden. They cannot exist. If light is emitted only in certain selected energies, Planck found himself saying, it must be packaged in little νh-sized units, which he called *quanta*, from the Latin for "how much."

Deeply uncomfortable with his own formula—why couldn't lightwaves have any energy under the sun?—Planck suggested that maybe light actually *can* have any energy value under the sun, but that somehow it is *emitted* only in quanta, just as milk is packaged only in pint, quart, and half-gallon containers, but, once bought, can be poured or spilled in any amount.

Although Planck's formula was bizarre, it *worked*. When physicists calculated with it, they got the right answers, and much more beside. Many scientists strove to explain the results in some other way. Planck, in particular, seems to have felt like Epimetheus, the mythical Titan who opened Pandora's box; although he was richly rewarded for his findings, the discoverer of the quantum spent years in a useless battle to remove it from the world. Years after Planck's death, his student and colleague, James Franck, recalled watching his fruitless struggle "to avoid quantum theory, [to see] whether he could not at least make the influence of quantum theory as little as it could possibly be—whether he could not, for instance, say it might be only the emission but not the absorption. I mean, a lot of things he tried out. He was really trained in classical physics, and if ever there was a classicist in character, it was he. He was a revolutionary against his own will. And I remember that he always came with attempts to see whether one could not avoid—with some resignation, but also with looking ahead. He finally came to the conclusion, 'It doesn't help. We have to live with quantum theory. And believe me, it will expand. It will not be only in optics. It will go in all fields. We have to live with it.' "[13]

Physicists did have to live with it. In 1905, Einstein made one of his first and greatest contributions to the field when he took Planck's idea more seriously than its creator. Light not only comes in quanta, Einstein argued, it *is* quanta.[14] This step was even crazier; as late as 1913, when Planck and three other physicists recommended Einstein for membership in the Prussian Academy, they stressed Einstein's great contributions to physics, even though "he may sometimes have missed the target in his speculations, as, for example, in his hypothesis of light-quanta." Even when Einstein's equation incorporating the light-quanta was proven by the American experimenter Robert A. Millikan in 1915, Millikan described the theory behind the equation as "wholly untenable."[15] But Einstein's reasoning was clear and precise, his assertions confirmed, and when the Swedish Academy finally awarded him the Nobel Prize, in 1922, this, not relativity, was the work they cited.

Today's physicists readily accept light-quanta, and call them "photons."[16] But back in 1911, the concept seemed far-fetched indeed. In Rutherford's opinion, the Continental theorists didn't want to explain how their fancy, highfalutin talk of the quantum translated into something real that he could find in an experiment. Rutherford was not satisfied by hypothetical "oscillators." He thought that such theoretical chatter shirked the essential task of physics, and the whole overly mathematical Germanic school of physics was suspect for it. At the same time Rutherford was putting together his model of the atom, he complained to a friend that "continental people do not seem to be in the least interested to form a physical idea of the basis of Planck's theory. They are quite content to explain everything on a certain assumption, and do not worry their heads about the real cause of the thing. I must, I think, say that the English point of view is much more physical and much to be preferred."[17] He would have been flabbergasted if someone had told him then that a Continental physicist would establish the physical basis of the quantum in Rutherford's own lab, and that the basis was, in fact, Rutherford's own model of the atom.

Nobody knows who first told Bohr about the nucleus, or how he learned that atoms with nuclei would be unstable. It is evident, however, that this instability, which made most non-Manchester theorists unwilling to consider the nucleus seriously, was what interested Bohr most. Convinced, perhaps irrationally, that the nucleus must exist, he soon realized that demonstrating its reality would require wholly new ideas. Indeed, he was certain that any explanation of Rutherford's model was impossible *without* these ideas. Sometime in the late spring of 1912, it occurred to him that quite possibly the explanation for the stability of this atom lay in the still growing domain of quantum theory. "It was clear," he said later, "and that was *the* point in the Rutherford atom, that we had something from which we could not proceed at all in any other way than by radical change."[18] *Radical change* was something that interested the young Bohr.

Bohr knew that Einstein had said that light consisted of tiny bundles called quanta; what if quantization was a fundamental property of all energy? Could this somehow be linked to the atom's stability? By June he thought he had it. Excited, he wrote Harald that "perhaps I have found out a little about the structure of atoms. Don't talk about it to anybody, for otherwise I couldn't write to you about it so soon. If I should be right it wouldn't be a suggestion of the nature of a possibility (i.e., an impossibility, like J. J. Thomson's theory) but perhaps a little bit of reality."[19] More confidently, he wrote his fiancée two weeks later, "It doesn't perhaps look so hopeless with those little atoms, even though the outcome of the calculations has its ups and downs."[20]

By this time he was working under some pressure, for he was planning to leave on July 24 for Copenhagen, where he was to be married a week later. Like every busy bridegroom who has tried to plan a traditional ceremony in the faraway hometown of his bride, Bohr had trouble juggling his personal and professional life. Nevertheless, he had time to prepare a little summary of his ideas for Rutherford.[21] In it he stated as a hypothesis the idea that the electrons around an atom neither pushed each other away from the nucleus nor fell into it because they simply could not do so unless something external—a photon, say—intervened. Bohr suspected that the fixed quantities of energy comprising light somehow corresponded to fixed electron orbits in atoms. He could give no reason why, other than that it seemed to ensure that Rutherford-style nuclear atoms did not fall apart.[22]

Rutherford was surprised that Bohr had taken the model so seriously, and drawn such far-reaching conclusions about the nature of the atom from his little picture of how it was organized. "He thought that this meagre evidence about the nuclear atom was not certain enough to draw such consequences," Bohr remarked later. "I said to him that I was sure that it would be the final proof of his atom."[23] Excited, Bohr convinced his wife that they should not, as they had planned, go to Norway on their honeymoon, but instead should spend the time in Cambridge, where he could get some work done.[24]

Bohr expected to write a paper fairly quickly from the memorandum he had given Rutherford. But seven months later, in February, he was still slowly working out his ideas, quietly juggling everything he knew about atoms and quantum theory—all without real satisfaction. His problems were resolved, he often said later, at a stroke, when a friend suggested that he look into a formula giving the frequency of light emitted by atoms. (Recall that the frequency of a lightwave is the number of wavelengths—crests and troughs, if you will—per second.) It had been discovered that heated materials gave off only certain specific frequencies of light, and not any others; gas in a neon sign, say, glows with particular shades of blue that are characteristic of its composition. These colors can be broken up into their individual components in the same way that a prism divides a ray of white light into a rainbow of red, yellow, blue, and green. If an element—for example, hydrogen—is examined in this way, the result is not a continuous spectrum, but a series of colored bands, called spectral lines, that represent the tones and frequencies hydrogen atoms are capable of producing. Spectroscopists stack these lines atop of each other in sequences that look quite like those ugly black marks printed on tin cans that are "read" by electronic cash registers; the pattern of horizontal lines is different for every substance and as individual as a fingerprint. In 1885, a Swiss high school teacher and amateur numerologist, Johann Balmer, had noticed that the frequencies of the light emitted by hydrogen atoms were mathematically related.[25] (Hydrogen is the lightest and simplest element, with just one electron.) The regularity interested scientists, but few were convinced that it was important. It is unlikely that Bohr had never seen Balmer's work before, but he had probably forgotten it. This time, a glance was enough to electrify him. "As soon as I saw Balmer's formula the whole thing was immediately clear to me," Bohr said.[26]

What he experienced at that moment was the single, intensely pleasurable instant of illumination when a great deal of hard thought abruptly coalesces into a vision, a process similar to the abrupt emergence of a painter's style from a morass of false starts and derivative juvenilia. In a sense, Bohr realized that he should take Planck's constant very seriously indeed. Planck's constant h is 6.62×10^{-27} erg-seconds, an incredibly small number with somewhat peculiar dimensions. The dimensions of a number are the units it is written in; for instance, the dimensions of velocity are distance divided by time, miles per hour. Erg-seconds, the dimensions of Planck's constant, are energy multiplied by time, which is identical to those of a quantity scientists call *action*. For this reason, h is often termed a quantum of action. The idea of action was elaborated by eighteenth-century astronomers, who found that they could simplify complicated problems of planetary orbits by introducing a new variable related to energy: action. For example, the action of the earth going about the sun is calculated by dividing the orbit into a series of points, multiplying the earth's momentum at each point by the change in radius from the point before, and adding up the result; engineers have to do this kind of arithmetic today to tell astronauts in the space shuttle when to deploy weather satellites. Later, astronomers learned that the easiest way to work with action variables was by incorporating them into functions called Hamiltonians, after their inventor, the nineteenth-century physicist William Rowan Hamilton.[27] Hamiltonians are a method for finding the minimum value of a given equation. Physicists use them to calculate orbits, trajectories, and the like because objects naturally follow the path of least action. Bohr won the Nobel Prize for the insight that when calculating the orbits of electrons he should stick in h, the quantum of action, every time he saw an action variable in the Hamiltonian.

Because h is a number with a fixed value, this was the same as saying that the electrons in the atom could only have orbits with specified values for the action, and that these values were multiples of Planck's constant. If this were true, Bohr realized, the electrons around the nucleus must exist in fixed arrangements—as if satellites could circle the globe only in certain orbits—and going from one arrangement to another must take or release a certain predetermined amount of energy. These amounts of energy, absorbed or released, come in the form of electromagnetic radiation—that is, light. An electron can absorb light (or, rather, a photon) if and only if the light photon has exactly enough energy to kick that electron from one state to another; no more, no less. Because $E = h\nu$, this is the same as saying that a given orbiting electron can only absorb light of certain frequencies. Similarly, when an electron falls from a high-energy state to one of low energy, it does so by squirting out a photon, again of a definite frequency and energy— the same frequency and energy required to get the electron there to begin with. Quanta of light are the tolls collected or paid by electrons as they jump about their permitted places in the atom. The spectral lines described by Balmer's formula are a set of subatomic hops, skips, and jumps, the jitterbug moves of hydrogen's lonely electron.

Like an artist who dithers over a canvas for years but quickly executes it once inspiration arrives, Bohr wrote up his insight rapidly indeed. He saw the Balmer formula in mid-February; on March 6, he put a draft of a long paper in the mail for Rutherford. In an attached note, Bohr said it was the first of several related articles.[28] Fifteen days later, Bohr sent a second, amplified draft. In the meantime, Rutherford had read the first and found it interesting, if not entirely plausible. He also thought it was much too long. He wrote to Bohr, explaining that "long papers have a way of frightening readers. It is the custom in England to put things very shortly and tersely in contrast with the Germanic method, where it appears to be a virtue to be as long-winded as possible." He kindly offered "to cut out any matter I consider unnecessary in your paper. Please reply."[29]

Bohr replied by taking the first available boat to England, going directly to Rutherford's office, and arguing doggedly against cutting a single phrase. He had spent weeks agonizing over the thing; each stilted sentence had a precise meaning. Rutherford experienced the astonishment, common to editors, caused by an ordinarily quiet writer's apparent willingness to kill over a comma. He relented, and Bohr's three long, historic papers were published in virtually unedited form in the *Philosophical Magazine* issues of July, September, and November of 1913.[30]

The Rutherford-Bohr vision of the atom—a tiny central nucleus surrounded by a pattern of electrons in different states—was confirmed by another Rutherford protégé, Henry Moseley. Bohr had predicted that when the innermost electrons moved from a very high to a very low energy state, they would give off high-energy light—X rays. The precise energy of the X rays would depend on the electric charge of the nucleus, for the stronger the positive charge attracting the electron, the more energy it would use up when it moved outward, or give off when it moved inward, and the higher the frequency of the corresponding X rays. (Because $E = h\nu$, energy and frequency are directly proportional.) Using Balmer's formula and his own ideas, Bohr realized that one ought to be able to work backward and, from the frequencies of the X rays given off, determine the charge on the nucleus. Moseley set out to test this by shooting a beam of electrons at different elements. In an experiment of classic simplicity and elegance, he sealed in a vacuum chamber a sort of toy train, which hauled samples of different elements back and forth in the line of an electron gun; it was as if he had put the samples in the middle of a television tube, blocking off the stream of electrons that generate the image. He planned to try every element, going one by one, step by step from lightest to heaviest. Some of the electrons shot from the gun and crashed into the inner electrons of the target atoms, knocking them free; when others rushed in to fill their place, they gave off light quanta—X rays—in the process. Moseley measured the results and discovered that as the elements got heavier the frequencies increased; as each element was replaced by the next, the frequency rose to match.

The result not only established the Rutherford-Bohr model of the atom, but also illustrated a growing phenomenon in the science of physics, the widening division between theory and experiment. Despite his real brilliance, Rutherford lacked the talent for purposive daydreaming that is the hallmark of the theoretician; if he had possessed it, he would not have been the experimenter he was. And Bohr, despite his enormous powers of concentration, did not possess the skill to overcome the brute intractability of matter necessary for performing experiments. His forte was talk; he liked to stop and think things over, and frequently took advantage of the theorist's freedom to discard or rework a troublesome idea. Experimenters like Rutherford, on the other hand, must invest themselves in a course of action and see it all the way through.

"Science walks forward on two feet, namely theory and experiment," said the American scientist Robert Millikan on the occasion of receiving the Nobel Prize for physics in 1924. He continued, "Sometimes it is one foot which is put forward first, sometimes the other, but continuous progress is only made by the use of both—by theorizing and then testing, or by finding new relations in the process of experimenting and then bringing the theoretical foot up and pushing it on beyond, and so on in unending alternations."[31]

□   □   □   □   □

The Cavendish laboratory, like the whole of English physics—for that matter, like England itself—was shaken by the war. The universities were emptied by the fight: research assistants given artillery commissions, students inducted, professors sent off to do something useful. Just before the hostilities, Geiger left for Berlin; one of Rutherford's best students, James Chadwick, went there to work with him and spent the whole of the war in an internment camp. Marsden served in New Zealand; Moseley refused Rutherford's offer of wartime scientific work and patriotically went to the front. In 1915, he died in the senseless battle of Gallipoli. Rutherford was drawn into working on antisubmarine warfare for the War Research Department, but managed to steal time now and then to perform his own experiments.

Once again, he was thinking about what happened when atoms were bombarded with alpha particles. After Geiger had left for Berlin, Marsden continued experimenting with, again, a target, a screen, and a source of radiation. Alpha particles from the source hit the target and were absorbed or reflected. By this time, people knew quite a lot about alpha particles; they knew, for example, that they had quite a short range—after shooting through a few inches of air, they petered out. What Marsden noticed was that the screen kept flashing even if he moved it farther away than alpha particles were supposed to reach. Whenever he brought a magnet near, the flashes moved in response. Physicists had long known that a magnet bends the paths of charged objects speeding by it—positively charged objects in one direction, negatively charged objects in the other. Just before returning to the Antipodes, Marsden ascertained that the sparks were caused by something with a positive charge, something as light as hydrogen, the lightest element. Rutherford and he quickly had the idea that they were seeing the nuclei of hydrogen atoms. It was logical to suppose that hydrogen nuclei, the lightest known entity of positive charge, might be the positive version of the electron, that is, the particle that had to exist to cancel out the negative charge of the electron. On the other hand, Rutherford realized, the flashes might be due to a previously unknown gas, one lighter than hydrogen. People were still discovering elements; Rutherford himself had found one.[32] He carried on alone, one of the few active physicists in the nation. By the end of 1917, he was fairly sure of the answer: The alpha particles were slamming into atoms of nitrogen in the air, and breaking off chips—hydrogen nuclei. If hydrogen nuclei were *inside* nitrogen nuclei, this was a strong indication that the hydrogen nucleus might be a fundamental building block of matter. It also meant that Rutherford, alone in the Manchester lab except for an assistant, was splitting the atom. Over the next months, he sat in a darkened room, counting minute flashes of light as he varied the experimental apparatus slightly to eliminate the possibility of error. He knew he was onto something important. Nonetheless, he was sufficiently engaged by war research that he did not submit a paper on the splitting of the atom until April 1919.[33]

A year later, he had progressed to the point of calling the ejected hydrogen nucleus a *proton* and arguing that nuclei in general must be made largely of assemblages of protons packed closely together.[34] Protons weigh almost two thousand times more than electrons, which is why the mass of the atom is concentrated in the nucleus. By this time, he had succeeded J. J. Thomson as director of the Cavendish. He ran the laboratory according to his own lights, which fit in well with Cambridge traditions: simple experiments, scavenged equipment, readily understandable ideas, strong facts, sharp inferences. Greedy as ever for quick, cheap, dramatic discoveries, Rutherford tried hundreds of experiments, a scattershot approach that reaped great immediate benefits even as it laid the seeds for the laboratory's long-term decline. The most important Cavendish work stemmed from Rutherford's conviction that even if the proton and electron were the only components of matter, with no more to be found, some close combination of the two might form a neutral particle, and he and his former student Chadwick made sporadic attempts to discover it. In 1931, they had a stroke of luck when the discovery was virtually handed to them on a platter. Irène Curie and Frédéric Joliot, the daughter and son-in-law of Mme. Curie, studied a new type of radiation made by bombarding beryllium with alpha particles.[35] This radiation—high-energy photons, they thought—was uncharged, but so powerful that it could knock protons flying at tens of thousands of miles a second. When Rutherford heard of such photons, he snapped, "I don't believe it!" Chadwick immediately suspected that the protons were reeling from collisions with massive particles, rather than with massless photons. By doing the experiment the Cavendish way, Chadwick was able to show that the radiation indeed consisted of neutral particles about as massive as protons. Neutrons were soon shown to be present in the nucleus along with protons.[36]

By 1932, all three components of ordinary matter—the electron, proton, and neutron—were known. But identifying the parts of the atom told as little about the structure of the nucleus and the forces holding it together as the knowledge that a building is made from brick would say about its architecture. The answers, Rutherford thought, would not be found "in this generation or the next and probably not completely for many years, if at all, or for many hundreds of years, because the constitution of the atom is, of course, the great problem that lies at the base of all physics and chemistry. And if we knew the constitution of atoms we ought to be able to predict everything that is happening in the universe."[37]

The bit of radium in the Cavendish was the heart of the quest, and the institution as a whole. It was kept inside a sort of oven made from lead bricks that rested in a skinny tower on the top floor of the laboratory. Low and heavy, the radium box was treated with the respect of an altar; only the trusted few were allowed near it. There was less than a gram of radium bromide sealed away inside, but that amount produced enough radiation to give the room the sharp, fresh odor of ionization. When an atom of radium splits off an alpha particle, what is left behind is an atom of radon, a heavy, inert, highly radioactive gas that seeps out of the source like fog from a marsh. (The radium was kept in an upper floor so that if the radon somehow got loose it could be fanned out of the window before it collected in the cellar.) Day in, day out, radon gradually accumulated in a sealed glass bottle whose walls were stained purple by radiation; experimenters routinely siphoned off doses thousands of times stronger than the largest permissible measures today. These early experimenters with radioactivity did not have the handicap of knowing how dangerous the phenomenon was; if their precautions had been better, it is unlikely they could have had sources of sufficient strength to get proper results. One of the last people to work with the Cavendish radium, Samuel Devons, told us once, "I would ask for, say, three hundred millicuries"—a curie is the amount of radioactivity put out by one gram of pure radium—"which today would be considered an absolutely absurd lethal—I mean, you wouldn't go within miles of it! The hullaballoo about it! Christ, I had that in my bare hands. Well, I had rubber gloves on."

We spoke to Devons at the Barnard College History of Science Laboratory, an historical re-creation of early experimentation, which he ran. He had an English manner—tweeds, a white beard outlining his mouth and chin, fiercely hooked nose. His fingers were smudged from working with the equipment in the laboratory. "There was a little flight of stairs going up the tower," he said. "Twenty steps. At the bottom you'd change your jacket for a jacket you kept on a peg and at the top you'd change that jacket and put a coat on and rub your hands with chalk for a little bit and put rubber gloves on. When you came out, there was a place at the top where you were supposed to wash your hands and another at the bottom—progressively getting cleaner. I don't know how much of this was important, but if I forgot to do any of the ritual at all, I could tell when I passed by Maurice Gold-haber's counters. He was doing neutron work with incredibly primitive sources. And as I walked past, just enough stuff would be on me to swamp his counter. If I picked up a Geiger counter after being in the tower and blew on it—wfff! the thing would go *brrrr.*" The result, of course, was that much of the laboratory was poisoned by radioactivity. "There were rooms where you could take a Geiger counter and run it down the baseboards and at a certain bench you'd hear that *brrrr!* They were painted over and scraped and painted, but they were never thrown out. It was too expensive. It wasn't all that bad, but oh yes, rooms got contaminated. Certain experiments that required a very clean background you couldn't do in a certain room. And the tower itself was so filthy, it was absolutely shocking."

Rain came down as Devons spoke, making fat wet asterisks on the office windows. His lab was closed for the day, but nobody was going anywhere until the shower was over. On his desk, his copy of the *Collected Papers* of Lord Rutherford lay open to a drawing of the equipment used in a scattering experiment, a simple collection of wire, wood, and brass that could have been, and was, constructed from scrap by a graduate student in a few hours. We asked Devons about the difference between working at the Cavendish and working in one of the enormous national laboratories today. Devons shrugged. The scale of the Cavendish, he said, was so small that people nowadays find it hard to believe. "There are more conferences going on today than there were scientists a hundred years ago. And each conference may have a few hundred, a thousand people in it, say. Physics has just *grown.*" He looked up for a moment at portraits of Rutherford and Thomson on the wall. A lamp made from a glass cider jug illuminated their faces. "You might say the underlying basis of it is still doing the same thing. It's like seeing the first Mr. Henry Ford humming away, being his own banker, his own bookkeeper, his own employer, and making cars, right? And the Ford Motor Company of today is still producing cars. But look at the organization today. The people at top, they've got manicured fingernails, and—you see? They've never smelled a furnace."[38]

3

A Children's Crusade

"PEOPLE," HOWARD GEORGI SAID, "STILL HAVE AN EINSTEIN COMPLEX." WE were sitting in his office in Lyman Hall, a lumpy pile of brick in an unlovely corner of the Harvard University campus that over the last thirty years has hosted many of the brightest luminaries in theoretical physics, Georgi among them. His window was open, and the sound of construction drifted across the small, paper-strewn room. Georgi is a tall, limber man with a wide face and an even wider russet beard curled around the edges of his smile. That morning he was wearing tennis shorts and sneakers; his white socks were pulled up nearly to his knees. He was talking about his generation of physicists, the successors to Werner Heisenberg, Bohr, and Planck, and explaining a dissatisfaction, common among his colleagues, with certain aspects of the public legacy of Albert Einstein. As he spoke, he gestured toward a small picture of the savant thumbtacked to a bulletin board. Rumpled, sorrowful, and ethereal, Einstein looked as if he were ready to sink beneath the weight of his own wisdom, whereas Georgi was the picture of American health—he had just come off a tennis court. A racket in its press lay across the papers on his desk. Yellow-green tennis balls dotted the floor, a menace to visitors. "Revolution!" Georgi said, dismissing the subject. "You're always getting asked if physics is having another revolution, if it's like Einstein all over again. But that's not the point at all. The things we've learned are so *interesting*—" here a quick burst of his infectious, high-pitched laughter "—that it's just *foolish* to ask about Einstein. Because that's not what it's about, really—revolution."[1]

In the fall of 1973, Georgi became one of the first physicists to look beyond the standard model when he and another Harvard theorist, Sheldon Glashow, wrote down the first completely unified theory of elementary particle interactions—the "grand unified theory," as they called it.[2] Although their particular theory did not pan out, their ideas changed physics forever. Among other things, they showed how the last mysteries of particle physics might provide an outline history of the Universe, from the first fraction of an instant after the Big Bang to the long cold slumber at the end of all things, as well as a complete inventory of its constituent parts. Indeed, the physicists moving equipment in the freezing dawn at Brookhaven were working in the light shed by Georgi and Glashow.

Despite the enormous scope and recent impact of his work, Georgi would be the last to call it revolutionary, or indeed to so regard most of the current unification ideas. The reason is that they are all couched in the language known as "quantum field theory," the same theoretical grammar that produced the standard model. Georgi saw the unification movement he started with Glashow more as a logical culmination of past ideas than as a striking novelty. Although it represented a significant departure, grand unification was implicit in earlier work.

A couple of years later, we took up the subjects of quantum field theory, the standard model, and unification with Georgi when he came into New York City to give a seminar at Columbia University. He amiably agreed to meet for breakfast at a local landmark, the Hungarian Pastry Shop, near the half-complete majesty of Saint John's Cathedral. The morning was cold, and the cafe's cappuccino machine was busy. Georgi had to lean over the table every now and then to make himself heard over the steam. We asked him about the way theorists approach the art of building unified theories.

Putting such theories together, he said, physicists are strongly guided by the necessity of expressing their thoughts in terms of quantum field theory, the theoretical language that was invented in the late 1920s by melding quantum mechanics and relativity. "What's happened since [that time] is that the lexicon was expanded. We discovered that we could build more theories of this kind. We seem to have found them all now—again, without changing the rules. One can, as I say, almost prove that the lexicon is exhausted, where what I mean by exhausted is that this is all you can do without a dramatic change in the rules."

You need constraints, he said. You need to operate within a framework. "The primary constraint comes from the language which we use. Quantum field theory is, as far as we know, the only way of combining relativity and quantum mechanics, at least without changing the rules in some more complicated way, like assuming there are ten dimensions. You can more or less prove that the only way of building a reasonable quantum mechanical theory that has special relativity in it and that doesn't violate causality and the various other simple principles is to build it in the form of a quantum field theory. So we're constrained by relativity and quantum mechanics to speak this funny language." The waitress called out our order, interrupting the discussion with hot apple turnovers and hotter coffee. "At some level—" Georgi took a healthy bite of his turnover "—it's description. Field theory is a language which you use to describe the interactions of the particles and fields that you see. You fit them into the grammar and syntax of quantum field theory, and then you're sure that you've succeeded in describing the interactions in a way that's consistent with all of the constraints of quantum mechanics and relativity. You can then ask—and sometimes answer—questions such as, why do the interactions have a particular form?"

Using quantum field theory must carry with it some implications, we said. He nodded in agreement. "The language is fundamentally absurd. What I mean by that is this: If you insist that relativity and quantum mechanics work down to arbitrarily short distances, then you're stuck with local quantum field theory down to arbitrarily short distances. What 'local' means is that the interactions which cause everything take place at single points." The cappuccino machine suddenly blasted, and Georgi had to shout. "A point is not a physical thing! A point is an infinitely small mathematical absurdity! So this assumption is crazy. It's clearly an act of hubris to assume that you know what's going on down to arbitrarily short distances." The point business, in fact, caused vast difficulty all the way along; it was only when a definitive end run was made around the problem that quantum field theory became solid enough to base ideas of unification on it.

We mentioned some models and theories we had heard about that had recently been developed in the search for unification—including one with the improbable name of "ten-dimensional string theory"—and asked whether these portended a complete description of nature.

"It depends on what you mean by 'complete,' " Georgi said equably. "I don't think we have the right to talk about a complete description of nature. The point is, one of the things we've learned is that our description of nature is organized by distance scale. Newtonian mechanics, for instance, is fine down to something like a billionth of a centimeter. After that, quantum effects come in. We certainly have the right to say that we have a complete description of nature at distances greater than 10^{-16} centimeters—that's a thousand times smaller than a proton—because we have some information about what's going on at distances greater than 10^{-16} centimeters. We might even guess that we have a complete description of nature down to some shorter scale, and that may not be unreasonable as long as (a) we're really just extrapolating from what we know, and (b) this description has experimental consequences that can be checked some day. That I regard as a sensible sort of physics. Unfortunately, you can also go the other way around and say, 'Look, damn it, the geometry of strings in ten dimensions is so elegant that this must be the unique theory of the world!'—and then try to work your way back to long distances. That seems to me to be rather silly." He laughed, took a bite. "You have to remember that theoretical physicists are parasites. The people that do the real work are experimenters. Theorists, of course, tend to forget that." More laughter; another bite. "A lot of the present speculations I'm really sort of unhappy with. I feel about the present status of grand unification a little bit the way I imagine that Richard Nixon's parents might have felt if they'd been around in the latter days of the Nixon administration. I'm very proud that my creation is so important and that people talk about it all the time, but I'm not really very happy with the things that it's doing at the moment." He finished the turnover. "It's gotten a bit too *theological*."[3]

At Cambridge, Georgi had spoken about the same ideas; they had been much on his mind in recent years. "We've really come an amazingly long way," he said. "I'm sometimes not sure if my colleagues realize that. Everything anyone has ever seen can be described in terms of the standard model. And all of it in some sense stems from relativity and quantum mechanics, plus a few experimental discoveries." He sat cross-legged in his swivel chair, tossing a tennis ball up and down as he considered the question. Above the construction noise was the sound of undergraduate laughter somewhere outside Lyman Hall. "I guess I'm saying that I think we're on the right track," he said finally. "You start with relativity and quantum mechanics, you make quantum field theory, and you proceed from there. If there is going to be unification, I suspect it will stem from that approach. We're not in for revolution. Progress will come from the experimenters, and from things that are already in our hands."[4]

□ □ □ □ □

Physicists today mention them in a single breath, but to their creators, relativity and quantum mechanics could not have been more different.[5] Whereas quantum conditions govern the properties of extremely small bits of matter, relativistic conditions govern the properties of matter traveling at extremely high speeds. Their difficult marriage produced what is called quantum field theory, a sickly child that eventually grew into the robust standard model of elementary particle interactions.

Relativity was developed almost entirely by one man, Albert Einstein, who, like Rutherford, labored alone during the First World War.[6] Einstein had developed some aspects of relativity earlier, in 1905, in a paper that introduced the famous equation $E = mc^2$, although in a slightly different form.[7] In subsequent articles, he struggled to make his ideas more general; he succeeded only at the end of 1915, when the war was in full swing.

Einstein developed relativity in the course of pondering a striking inconsistency between Newtonian mechanics, which describes the gravitational force that makes apples fall from trees and keeps the planets in their orbits, and Maxwell's electrodynamics, which show how electricity, magnetism, and light are different aspects of the single phenomenon of electromagnetism. One of the most impressive aspects of Newtonian mechanics is that its laws do not depend on the location or velocity of the system to which they are applied. Objects fall in the same way whether they are in New York or Tokyo, whether they are in the middle of a stadium or traveling thousands of miles an hour in a jet plane. If someone in the plane were to throw a ball out the window, it would be moving much, much faster relative to the ground than to the aircraft. Nonetheless, the trajectory of the ball could be predicted from Newton's laws, regardless of whether the point of view was the ground or the plane.

The situation is radically different in the four equations that sum up Maxwell's electrodynamics. Electricity, magnetism, and optics can only be tied together if light travels at specific velocities in specific media—186,000 miles per second in outer space, somewhat slower in air. For example, if instead of throwing a ball the airplane passenger shines a flashlight, the light, unlike the ball, would not be moving faster relative to the ground than to the aircraft—unlike anything in Newtonian mechanics.

Both theories were enormously successful, but both could not be completely correct. Starting with Maxwell, a number of prominent physicists tried to reconcile them, usually by postulating that the speed of light was not really an absolute quantity, but relative to some sort of invisible medium that carried it. This medium, which was called the *ether*, was everywhere the same, and Maxwell's laws held strictly for light's passage through it. Newton's laws also held for light, but they predicted variations in speed that were too small to be noticed.

This explanation was set back considerably when it was established that there was no ether pervading space. In 1895, the great Dutch theorist Hendrik Lorentz constructed a series of equations on an *ad hoc* basis that attempted to resolve the conflict between Maxwell and Newton by introducing changes in the definition of space and time.[8] This set of rules is today known as the "Lorentz transformations," and quantities that fit into them are said to be "Lorentz invariant," that is, they are not thrown off kilter by the redefinitions. Lorentz introduced his transformations simply to show that a mathematical schema could be built in which electromagnetic phenomena could be described independent of the point of view, as was the case for Newtonian mechanics, although one had to pay the price of doing odd things to space and time. Over a decade later, Einstein decided the Lorentz's redefinitions had much more meaning than their creator knew: Space and time truly were different than had previously been supposed.[9]

Einstein's fundamental reconception of both is the foundation of the theory of relativity, an achievement so great that his successors still marvel that one man could have accomplished it. It was accomplished in two stages, the first, special relativity, in 1905, and the second, general relativity, completed in 1915 after several false starts. With special relativity came a host of strange effects—time slowing down at high speed, the shortening of objects in the direction of motion, and so forth—all of which are mathematically derivable from the necessity of changing Newton to fit Maxwell. Small wonder that when Einstein first put forward his even stranger ideas about general relativity, which included the concept that space is curved, many physicists had not the faintest notion of what he was talking about. In practical terms, Einstein's legacy is the physicist's duty to work only with quantities and theories that take account of relativity, special and general, by being Lorentz invariant.

General relativity had few predictions that could then be tested by experiment. One of its deepest consequences involved the curvature of space. Although it was not widely known at the time, Newton's laws could be interpreted as predicting that strong gravitational fields would bend light in much the same way that strong magnetic fields cause charged particles to swerve off course. General relativity made a similar prediction, but added that the curvature of space around massive bodies would cause light to bend even more. Many scientists withheld their acceptance of the theory until this prediction was examined. The best way Einstein could think of to test this was to see how much the light from stars bent as it passed near the sun. Performing this experiment required observing the sun during a total eclipse; the stars within a few degrees of the sun, ordinarily invisible, would then be both viewable and out of position by tiny fractions of a degree. The fate of relativity depended on how much.

In 1912, Argentine scientists traveled to Brazil to measure starlight during an eclipse; the experiment was rained out. A German expedition to observe an eclipse at the Crimea in 1914 was foiled by the outbreak of war. As it happened, these failures were lucky, for in 1915 Einstein finally understood how the curvature of space bent starlight, and revised his predictions. The war also scotched a 1916 expedition to study an eclipse in Venezuela; an American team was unable to establish anything conclusive from an eclipse two years later. Two British teams finally ran a successful experiment that confirmed Einstein's prediction. They announced their results on November 6, 1919, at a joint meeting of the Royal Society of Sciences and the Royal Astronomical Society. J. J. Thomson himself was the chair of the gathering; although older and increasingly set in his ways, he proclaimed, "This is the most important result obtained in connection with the theory of gravitation since Newton's day," describing the result as "one of the highest achievements of human thought."[10] The following day, the London *Times* headlined[11]

REVOLUTION IN SCIENCE.

NEW THEORY OF THE UNIVERSE.

NEWTONIAN IDEAS OVERTHROWN.

Further in the article was the alarming subhead

SPACE "WARPED"

(In fact, Einstein had not demonstrated that Newton was wrong; he had not

d he did. Rather, relativity showed that factors had to be added to
alculations, and that these factors could be easily ignored except
... certain circumstances.) Three days later, the *New York Times* printed its
first article on relativity. Crossing the Atlantic, Thomson's praise inflated; he
was now quoted as saying the result was "one of the greatest—perhaps the
greatest—of achievements in the history of human thought."[12] A second *New
York Times* article, printed the next day, had a marvelously grandiose set of
headlines:[13]

LIGHTS ALL ASKEW
IN THE HEAVENS

Men of Science More or Less
Agog Over Results of Eclipse
Observations.

EINSTEIN THEORY TRIUMPHS

Stars Not Where They Seemed
or Were Calculated to Be,
but Nobody Need Worry.

A BOOK FOR 12 WISE MEN

No More in All the World Could
Comprehend It, Said Einstein When
His Daring Publishers Accepted It

During the next few years Einstein toured the world, speaking to packed
audiences. In the popular imagination, the man typified physics, even science itself. For the rest of his life he would remain "Mr. Physicist," in the
mainstream of the popular idea of physics. But he did not remain in the
mainstream of the discipline, for he rejected the quantum theory he had
done so much to establish. He spent much of the rest of his life on a fruitless
quest to unify gravity and electromagnetism, hoping to show that they are
two aspects of the same thing, within the framework of what he called a
"unified field theory." Einstein's attempt to unify was, to say the least, premature, and he failed utterly. The year 1919 represents the culmination of his
career; thereafter, as physicists became increasingly preoccupied with quantum theory, Einstein's views became for his peers a source of puzzlement,
sorrow, and finally indifference.

"Already in 1905," Einstein once said, "I realized what a *Schwein-
erei*—" what a stinking mess "—the quantum theory was."[14] Relativity, like
Athena, sprang full-grown from Einstein's head. Quantum mechanics, on the

other hand, had a difficult delivery that required dozens of midwives. The notion that many aspects of the atomic domain were quantized proved singularly difficult for physicists to grasp, although the best among them were absorbed by the challenge of interpreting it. They learned and relearned that atoms and their constituents have properties that can be described in terms of simple numbers— +1, 0, and −1, say, or perhaps −½ or +⅔—multiplied by constants like h (Planck's constant), c (the speed of light), or pi (the ratio of the diameter of a circle to its circumference), which implied that the relations among the bits and pieces of the atom were vastly different than the interactions among ordinary, nonquantized objects. Physicists found the puzzles electrifying, but the answers were dissatisfying, abstract, even nonsensical; as hard to understand as they were impossible to visualize.

Early in the 1920s, theorists realized that what a particle is and how it behaves can be completely identified by a small set of numbers. There just isn't anything more to describe; particles are too simple. In this respect, the pieces of the atom are vastly different from objects like tables and cans of chicken noodle soup. The Campbell's soup cans elevated by pop art into symbols of uniformity are in truth not uniform at all; each can has hundreds of physical characteristics that can be measured and described, such as its precise shades of red and gold or the exact dimensions of its label, and the soup cans stacked in a supermarket differ, however slightly, in all of them. Subatomic particles, on the other hand, have less than a dozen attributes. They have no color, taste, nor odor; they are neither hard nor soft, shiny nor dull. They have only a handful of numbers for a few simple properties. All electrons have exactly the same electric charge, −1; protons all have a charge of +1.[15] An electron at rest has a mass of about 10^{-30} kilograms; a proton's mass is about 1,836 times greater. As far as scientists know, there is no reason why electrons couldn't have a charge of, say, −1⅜, and protons a mass, say, nine hundred times larger than electrons. But they don't.

Oddest of all, however, some of the numbers describing how particles behave are not only identical, but also quantized; that is, they can have *only* certain particular values. Niels Bohr showed that orbiting electrons have specified energies. Moreover, electrons whirl about the nucleus with just a few preferred values of angular momentum ("angular momentum" can be loosely thought of as the momentum with which something, in this case an electron, goes around a curved path). Because the electron can be curving any which way, angular momentum is associated with direction. The direction—or, anyway, the part of it that can be measured—is also quantized. It, too, can only have particular values that physicists can write as 0, 1, 2, or 3.[16] The complete set of such numbers for a particle is called that particle's *quantum numbers*. Particles with the same quantum numbers are absolutely identical—they *cannot* be distinguished.

Freshly encountering this situation, it was of obvious interest to physicists in the 1920s to specify a complete list of the quantum numbers associated with the electrons in atoms. A decisive contribution was made by a twenty-five-year-old Viennese named Wolfgang Pauli. Brilliant, acerbic, and fat, Pauli was already well on his way to becoming one of the great physicists of the century; even as a young man, his intellect and passionate style of argumentation intimidated many of his colleagues. He had a kind heart, an irascible temperament, and a streak of melancholia that he overcame by driving at physical questions as if his life depended on it, which perhaps it did. Like many theoreticians of the day, Pauli was concerned with understanding the spectral lines emitted by atoms. Bohr's original model worked for the relatively simple patterns emitted by hydrogen, but heavier, more complex elements were much harder to understand. For example, cesium, strontium, and barium—several of what chemists call the alkaline-earth metals—produce spectral lines that upon close examination are seen actually to be split in two; these lines, called doublets, are made of two almost identical frequencies. In December of 1924, Pauli suggested that a complete list of the quantum numbers of an orbiting electron would include its energy, angular momentum, and orientation in space; in addition, to explain the alkali doublets, he suggested that there had to be a fourth quantum number, which he called, rather unhelpfully, *Zweideutigkeit*—two-valuedness.[17]

In the summer of 1925, Samuel Goudsmit, a young Dutch physicist, was trying to explain Pauli's ideas to a young countryman, George Uhlenbeck, who had been out of Holland for a while and was trying to get back into physics. Uhlenbeck had been told by his teacher that he should be briefed by Goudsmit. During afternoon talks, Goudsmit explained Pauli's four quantum numbers to Uhlenbeck. "I was impressed," Uhlenbeck recalled later, "but since the whole argument was purely formal, it seemed like abracadabra to me. There was no picture that at least qualitatively connected Pauli's formula with the old Bohr atomic model."[18]

It occurred to Uhlenbeck that Pauli's *Zweideutigkeit* was not really another quantum number, but simply another property of an electron. He suggested that perhaps an electron spins on its axis like a toy top; unlike a top, however, the spin of the electron would be quantized, and it could only turn at certain speeds. Looking at Pauli's formulas, Uhlenbeck and Goudsmit realized that if electrons had a second angular momentum associated with a spin, this would perfectly account for both "two-valuedness" and the double spectrum lines from the alkaline earths. The amount of spin was one-half $\hbar$, where $\hbar$ is physics shorthand for $h/2\pi$, that is, Planck's constant divided by twice pi. Although both men were struggling in the sea of quantization, they "appreciated right away that if the [spin] angular momentum of the electron was $\hbar/2$, one had a picture of the alkali doublets as the two ways the electron could rotate with respect to its orbital motion."[19] The two ways of rotation would give the electrons two slightly different energy levels, and electrons with two slightly different energy levels would create two slightly different frequencies of light.

They took their idea to Uhlenbeck's teacher, Paul Ehrenfest, who headed the physics department in Leiden. Ehrenfest made some suggestions, had them write up a short paper about spin, and then told them to take it to Hendrik Lorentz, the grand old man of Dutch physics. In addition to inventing the Lorentz invariance on which special relativity is based, Lorentz was the first man to construct a theory of the electron. In 1925, Lorentz was seventy-two and ostensibly retired, but he still taught a class at Leiden every Monday morning from eleven to noon. After one class, Uhlenbeck and Goudsmit showed Lorentz their paper, which was only a few paragraphs long.

"Lorentz was not discouraging," Uhlenbeck once said. "He was a little bit reticent, [but] said that it was interesting and that he would think about it." *Thinking*, for Lorentz, was apparently an active occupation. "It was so typical of Lorentz that he immediately made very extensive calculations on the classical theory of rotating electrons. I think the next week, but maybe two weeks later, he gave me such a *stack* of papers with long calculations. Large white paper, I still remember. He tried to explain it to me, but it was so learned that I . . ."[20] Uhlenbeck's voice trailed off. Lorentz had explained several problems, one of the most grievous being that if the electron really had a spin angular momentum of $\hbar/2$, this implied that it rotated with a particular velocity. The old man had figured out the speed: ten times faster than the speed of light. Because nothing can go faster than the speed of light, this was, Uhlenbeck and Goudsmit decided, a devastating critique. They were most unhappy.

We visited Uhlenbeck at Rockefeller University, a set of isolated buildings of funereal modernity on the Upper East Side of Manhattan. The university is one of the more carefully guarded in the nation, and we were detained by an armed guard for some twenty minutes. It was one of those clear mornings when the gray, boxy magnificence of the cityscape is just slightly fuzzed over by smog. He was eighty-four, a tall man with thinning hair scattered across his head like so much straw. Impatient with his growing deafness, he asked us to sit close. "Talk in my left ear," he said cheerfully. "The other one is primarily for decoration." He kept a pair of glasses in his hand, twisting the frames around his fingers like worry beads. We asked him what he had done with Lorentz's calculations.

"I thought, well, therefore it is all wrong, what we have done. And I went back to Ehrenfest and said, 'You better not publish that paper, because Lorentz has shown that it is not correct.' He said, 'I sent it out right away. It will come out next week.' "

Ehrenfest had mailed it off without waiting to hear from the master? Uhlenbeck roared with laughter.

"He *knew!*" he said. Ehrenfest had immediately seen the difficulties with spin. "But he said to us—he said it in German—*'Sie beiden sind jung genug sich eine Dummheit leisten zu können!'* Both of you are young enough to afford a stupidity!"[21]

Ehrenfest was not being entirely cavalier. It is important for theoreticians not to pay attention at all times to what is wrong with their ideas. They must not be too afraid to follow their intuition; every now and then, the best way to hit a target is to shoot from the hip. Spin, as an example, had been thought of earlier, by one Ralph de Laer Kronig, an American of Hungarian descent. Unfortunately, Kronig asked caustic Wolfgang Pauli for his reaction to the notion that *Zweideutigkeit* was due to the effects of a spinning electron. Pauli tore apart the idea and, to the later regret of both men, Kronig never published it.[22] Bohr, on the other hand, dismissed Lorentz's objections immediately, telling Uhlenbeck that the faster-than-the-speed-of-light problem would "disappear when the real quantum theory is found."[23]

As it turned out, Bohr was correct. When the real quantum theory was found, the spin problem disappeared. But the complete quantum theory, quantum mechanics, could not be established until the various pieces of quantum theory—spin, angular momentum, orbitals, and the rest—were expressed in a common mathematical language. Metaphorically speaking, every physical theory is woven on a frame supplied by mathematics. The frame provides the backing, the warp and woof on which physicists tie and knot their tapestries of ideas. Newton, for instance, could not have put together the laws of mechanics if he had not first invented a language in which to write them: calculus. Similarly, Einstein could not have described general relativity if a quiet German named Georg Friedrich Bernhard Riemann had not developed a strange kind of geometry in which parallel lines could meet: Riemannian geometry. As the realm of quantization grew apace, the need became evident for a mathematical formalism that could handle the eccentric requirements of the new physics.

One of those who felt this need particularly strongly was a young German named Werner Heisenberg. The son of a professor of Greek at the University of Munich, Heisenberg had the sort of good looks usually associated with the word *dashing*. His character was a museum of turn-of-the-century Teutonic virtues: romantic temperament, ironic patriotism, love of intellection and the natural world. He was thoroughly steeped in the Greeks;

he was an excellent pianist; he could recite Goethe from memory. His late autobiographical essays are full of evocations of the special quality of well-being he felt walking across a winter beach, the cold spray across his face, head full of the ticking mechanisms of the world.[24]

Just before Heisenberg's seventeenth birthday, Germany was swept by a leftist rebellion. Munich was a center of the fighting; when the revolution collapsed, the city spent the spring of 1919 in a state of starvation and violent anarchy. The re-formed national government assembled troops to seize control of the city, and Heisenberg, caught up in an adolescent sense of the moment, volunteered to act as a guide. After a few weeks of shooting, Munich was captured, and Heisenberg was assigned to guard the telephone exchange. The city began to calm down. Heisenberg realized that classes would soon start again.[25]

"I had duty during the night," he explained years later, "and it was a nice summer in 1919. During the morning at 4:00, nothing was happening at the [telephone] office, of course, and somehow I couldn't sleep, so I went up to the roof of the house into the sunshine. It was nice and warm. I had Plato's *Timaeus* with me. I studied the *Timaeus* partly to keep up with the Greek, because I had to know Greek for my examination, but partly also because I was really fascinated by atomic theory. You know all of Plato's atomic theory was in the *Timaeus*."[26]

The *Timaeus* baffled Heisenberg. The basic text of Greek cosmology, it asserts that the Universe will never be understood until the smallest components of matter are known, and that these tiniest pieces consist of tiny right triangles, jointed together in various arrangements to form all the regular bodies of solid geometry. Heisenberg thought Plato's ideas were ridiculous. Nonetheless, Heisenberg was left, as he wrote later, with the feeling that "in order to interpret the material world, we need to know something about its smallest parts"; furthermore, he continued, "I was enthralled by the idea that the smallest particles of matter must reduce to some mathematical form."[27] He decided to study atomic theory.

During the 1920s, there were three outstanding centers of theoretical physics: two older German schools, Munich and Göttingen, and an institute at Copenhagen founded by Bohr in 1921, which soon rivaled the others in importance. Each had its own character: Munich was known to be more physically oriented, Göttingen was one of the world centers of mathematics, and Bohr's institute was characterized predominantly by the philosophical attitude of Bohr himself. Heisenberg eventually spent time at all three.

In 1921, he attended Munich, where his father still taught, as an undergraduate, then went to Göttingen a year later. In the last months of 1923, he went back to Munich to obtain his doctorate—and almost flunked. One of his examiners, Wilhelm Wien, was annoyed by Heisenberg's indifference to experimentation, and amused himself during the orals by asking the Ph.D. candidate questions about experimental technique. How clear were the images from telescopes? Wien asked. Heisenberg had not the faintest idea. How does a battery work? Heisenberg didn't know. Wien expressed his incredulity. A lengthy argument ensued, during which Wien announced that Heisenberg had failed, and the other two examiners insisted that he was the best physics student at the school in many years. Compromise was finally achieved in the form of letting Heisenberg pass with a *rite*, the equivalent of the "gentleman's C" given to wealthy but indifferent scholars in elite American prep schools.[28] The humiliated Heisenberg then became a *Privatdozent*, a very junior teaching fellow, in Göttingen. At about the same time, he got a fellowship and split his time between Copenhagen and Göttingen.[29]

At Munich, he was taught what was being called *quantum mechanics*. But as a graduate he found that the new laws of Bohr, Planck, and the rest were often nothing but hunches and badly formed approximations—it was a great shock, he said later, learning that theoreticians *guess* at answers—and, in the winter of 1925, twenty-three-year-old Werner Heisenberg set out to create for himself a more rational quantum theory.[30] He was part of a *nouvelle vague* of young theorists, arrogant, self-assured, and convinced that the time had come for children to sort out the confusion bequeathed them by their elders. The quest, he thought, was best pursued by the young—an attitude certain to ruffle feathers in the authoritarian German university system.

Having struggled to make a picture in his head of the orbit of an electron—the fixed states Bohr had imagined, and that he and others had elaborated on—Heisenberg began to wonder if the game was worth the candle. Clearly, the electrons were not simply going around the nucleus. If that were all that was going on, the electrons would (as Bohr had known ten years before) either fall into the nucleus or push each other away. In addition, electrons seemed to be able to jump almost instantly from one orbit to another and back again, absorbing and releasing the necessary energy for the move in the form of a photon. This was like claiming that Venus could suddenly hop close to the Earth, stay a while, and then, without missing a beat, leap back to its original orbit. Heisenberg thought that calling that kind of behavior *orbiting* was stretching the term out of shape.[31]

Moreover, because you couldn't ever *see* the orbit, Heisenberg wondered if one shouldn't junk the idea of electron orbits altogether—temporarily, anyway—and think about what could be seen, namely, the spectral lines.[32] These lines were created by lightwaves of specified frequencies. If the frequency of a lightwave is known, it is an easy task to calculate the wavelength, the distance between two successive crests or troughs. With a bit more mathematical finagling, one could figure out the amplitude of the waves, that is, how high the crests are. Unlike orbits, the frequencies and amplitudes could be directly measured in the lab. Physics being an empirical science, Heisenberg liked basing his ideas on tangible quantities rather than intangible entities like "orbits." If a particular frequency was supposed to be emitted when an electron jumped from one "state"—Heisenberg's replacement for "orbit"—to another, one could make a little table representing all possible states and frequencies:

	S_1	S_2	S_3 $\cdots$
S_1	ν_{1-1}	ν_{2-1}	$\nu_{3-1} \cdots$
S_2	ν_{1-2}	ν_{2-2}	$\nu_{3-2} \cdots$
S_3	ν_{1-3}	ν_{2-3}	$\nu_{3-3} \cdots$
$\vdots$	$\vdots$	$\vdots$	$\vdots$

S_1 stands for "state #1," S_2 for "state #2," and so on, and ν, again, is physics shorthand for "frequency." When an electron goes from S_1 to S_2, it produces light of frequency ν_{2-1}.

At this point, Heisenberg wrote later, "My work along these lines was advanced rather than retarded by an unfortunate personal setback."[33] He was smitten by such a severe bout of hay fever that he had to ask his adviser, Max Born, for a two-week leave of absence. A dripping, sneezing, coughing wreck, Heisenberg fled for Helgoland, a small island in the North Sea, where he hoped the bracing, pollen-free sea air would help him recover. The landlady of his *Gasthaus* took one look at his swollen face and concluded that he had been in a fight. But she installed him in a quiet, second-floor room with a view of the nearby houses, the beach, and the dark expanses of the North Sea. There Heisenberg made the fundamental discovery that you can accomplish more when nobody bothers you and you don't have a phone.

Working in a solitary fury of excitement, he realized that he could construct equations describing the tables of frequencies, amplitudes, positions, and momenta—"quantum-mechanical series," as he called them— that related the quantities that experimenter *actually observed*. Adding in an extra assumption or two, he came up with an awkward but definitely workable scheme that did not rest on ineffable ideas like nonorbiting orbits. Then he realized that he wasn't sure if his scheme was consistent with the law of conservation of energy.

I concentrated on demonstrating that the conservation law held, and one evening I reached the point where I was ready to determine the individual terms in the energy table, or, as we put it today, in the energy matrix, by what would now be considered an extremely clumsy series of calculations. When the first terms seemed to accord with the energy principle, I became rather excited, and I began to make countless mathematical errors. As a result, it was almost three o'clock in the morning before the final result of my computations lay before me.

The energy principle still held. Sitting in his rented room that June morning, Heisenberg felt that oceanic sense of clarity, of luminous and special insight into nature, that is the greatest joy a scientist can experience. "At first," he recalled,

I was deeply alarmed. I had the feeling that, through the surface of atomic phenomena, I was looking at a strangely beautiful interior, and felt almost giddy at the thought that I now had to probe this wealth of mathematical structures nature had so generously spread out before me. I was far too excited to sleep, and so, as a new day dawned, I made for the southern tip of the island, where I had been longing to climb a rock jutting out into the sea. I now did so without too much trouble, and waited for the sun to rise.[34]

But later, when Heisenberg then tried to calculate something with his quantum-mechanical series, he discovered a distressing asymmetry: No matter how hard he worked, his new quantum theory seemed to violate one of the first mathematical principles learned by every schoolchild, the commutative law.[35] According to the commutative law, the order in which you multiply two numbers does not affect the result: $A \times B$ always equals $B \times A$. But quantum-mechanical series A multiplied by quantum-mechanical series B did not give you the same answer as quantum-mechanical series B times quantum-mechanical series A. Heisenberg did what any theorist would do when a horrible problem marred a wonderful idea: He swept the difficulty under the rug. In the paper he submitted that July, the noncommutativity is only mentioned, rather sheepishly, in a single sentence, which is quickly followed by an example in which, oddly enough, the problem does not show its ugly face.[36]

That summer, Heisenberg traveled to Berlin, Leiden, and Cambridge to talk about his ideas. At the Cavendish, where Heisenberg knew almost nobody, he stayed with Ralph Fowler, Rutherford's son-in-law and the house theoretician. After two months of constant travel, work, and hay fever, Heisenberg was dazed with exhaustion. When Fowler left for a day of meetings in London, Heisenberg went to breakfast and slept through the entire day at the table, terrifying the maid.[37] The next day, speaking to the Cavendish experimenters, Heisenberg merely stressed the need for a new theory of spectral lines; but when he talked privately to Fowler, he pushed his own recent work. Fowler asked him to send over a copy of the proofs of the article as soon as they were available.[38]

Back in Göttingen, Heisenberg's supervisor, Max Born, had in the meantime been poring over the quantum-mechanical series. Fascinated, he felt that there was something important in the paper, but that he wasn't sure what it was. In the middle of July, he suddenly realized that he had seen these quantum-mechanical series before. They were, in fact, what mathematicians call matrices.[39]

Although they are today taught in any linear algebra course, matrices were virtually unknown to physics then. A classic use of matrices is given in the "payoff" matrix depicting the possible outcomes of matching pennies, the children's betting game in which the two participants flip pennies simultaneously, Child #1 winning if the coins land with the same face showing, Child #2 if one penny comes up heads, the other tails. The matrix below shows the outcome for ten throws. Child #1 threw four heads, three of which came when Child #2 also threw heads, and six tails, three of which came when Child #2 also threw tails. Thus Child #1 won six times, which mathematicians derive by adding up the numbers on the diagonal.

		CHILD #1	
		Heads	Tails
	Heads	3	3
CHILD #2	Tails	1	3

Ordinarily, mathematicians write matrices with no marginal entries—

$$\begin{pmatrix} 3 & 3 \\ 1 & 3 \end{pmatrix}$$

—and parentheses to indicate that they are matrices. Other examples are

$$\begin{pmatrix} 1 & 2 \\ 3 & 4 \\ 5 & 6 \\ 7 & 8 \end{pmatrix} \quad \begin{pmatrix} a & b \\ b & a \end{pmatrix} \quad \begin{pmatrix} 1 & 0 & 0 \\ 0 & 1 & 0 \\ 0 & 0 & 1 \end{pmatrix} \quad \begin{pmatrix} 1 & 1 & 1 & \dots \\ 1 & 1 & 1 & \dots \\ 1 & 1 & 1 & \dots \\ \vdots & \vdots & \vdots & \ddots \end{pmatrix}$$

The last matrix is infinite, the entries running endlessly in all directions.

A French mathematician, Augustin Cauchy, was the first to write matrices in the rectangular form used today. His successors, especially Arthur Cayley, a Cambridge mathematics professor, realized that these tables had mathematical properties. They could be added together, subtracted from each other, and multiplied; they even had inverses. To find the product of two matrices

$$\begin{pmatrix} a & b \\ c & d \end{pmatrix} \times \begin{pmatrix} A & B \\ C & D \end{pmatrix}$$

Cayley multiplied and added the members of the individual rows and columns

$$\begin{pmatrix} aA + bC & aB + bD \\ cA + dC & cB + dD \end{pmatrix}$$

This had a strange implication. If he multiplied the two matrices

$$\begin{pmatrix} 1 & 1 \\ 2 & 2 \end{pmatrix} \qquad \begin{pmatrix} 2 & 2 \\ 1 & 1 \end{pmatrix}$$

the answer depended on the order in which he put the matrices. Despite the similarity of the two matrices,

$$\begin{pmatrix} 1 & 1 \\ 2 & 2 \end{pmatrix} \times \begin{pmatrix} 2 & 2 \\ 1 & 1 \end{pmatrix} = \begin{pmatrix} 2+1 & 2+1 \\ 4+2 & 4+2 \end{pmatrix} = \begin{pmatrix} 3 & 3 \\ 6 & 6 \end{pmatrix}$$

but reversing the order of the two produced

$$\begin{pmatrix} 2 & 2 \\ 1 & 1 \end{pmatrix} \times \begin{pmatrix} 1 & 1 \\ 2 & 2 \end{pmatrix} = \begin{pmatrix} 2+4 & 2+4 \\ 1+2 & 1+2 \end{pmatrix} = \begin{pmatrix} 6 & 6 \\ 3 & 3 \end{pmatrix}$$

Because $\begin{pmatrix} 3 & 3 \\ 6 & 6 \end{pmatrix}$ is not the same as $\begin{pmatrix} 6 & 6 \\ 3 & 3 \end{pmatrix}$, matrix multiplication is not commutative.

Born was one of the few physicists in Europe—perhaps the only one—with a good knowledge of matrix mathematics.[40] He realized that Heisenberg's quantum-theoretical series were nothing more, nothing less, than awkward manipulations of frequency matrices

$$\begin{pmatrix} \nu_{1-1} & \nu_{2-1} & \nu_{3-1} & \cdots \\ \nu_{1-2} & \nu_{2-2} & \nu_{3-2} & \cdots \\ \nu_{1-3} & \nu_{2-3} & \nu_{3-3} & \cdots \\ \vdots & \vdots & \vdots & \ddots \end{pmatrix}$$

Born was delighted. Rewriting Heisenberg's equations as matrices led to a whole new world of applications he could explore. The first thing he figured out was that the matrix q for position and the matrix p for momentum are noncommutative in a very special way: That is, pq is not only different from qp, but the difference between pq and qp is always the same amount, no matter what p or q you chose. Mathematically, he wrote this

$$pq - qp = \hbar/i$$

where $\hbar$, as usual, is Planck's constant divided by twice pi, and i is the special symbol mathematicians use for the square root of minus one.

Born was pleased with this result, but he was tired, and wanted someone else to help him develop the insight. He approached Pauli, a former student; the latter bluntly informed Born that he would end up ruining Heisenberg's lovely ideas with "tedious and complicated formalisms." Born turned to another pupil, Pascual Jordan; by the end of July, the two men had laid out the basic principles of what was latter called matrix mechanics. A skittish, egotistical man, Born was driven by the excitement of the chase into a nervous collapse that lasted through all of August 1926.[41] Jordan had finished most of the work by the time Born returned; their paper was submitted to the *Zeitschrift für Physik* a few weeks later.[42]

Principally, what the two men had discovered was a deep connection between classical physics and quantum mechanics. Just as relativity is an extension of nineteenth-century physics, the new mechanics of Born, Jordan, and Heisenberg stemmed from Newton. "The concepts for quantum mechanics can only be explained by already knowing the Newtonian concepts," Heisenberg remarked in an interview many years ago. "That is, quantum theory is based upon the existence of classical physics. This is the point which Bohr emphasized so strongly—that we cannot talk about quantum physics without already having classical physics."[43] Without Newton, quantum mechanics would have nothing to diverge from; without classical mechanics, the new physicists would not even have words and concepts to redefine. As physicists will, Born and Jordan made the connection in formal terms: After developing the matrix methods they needed for quantum theory, they argued that their matrices could be linked to classical theory by using them in Hamiltonians. (Bohr had plugged Planck's constant into the Hamiltonian; now matrices also went in.) Intrigued, Heisenberg wrote to Jordan, and the three men began a series of discussions that culminated at the end of October, when Heisenberg joined Born and Jordan in Göttingen for a week of intense work. Born had promised to lecture in the United States, and was anxious to publish; Heisenberg thought they had an appalling amount of work to do, and no time to do it. He stole a moment to dash off a letter to Pauli in Zurich: "I am almost entirely occupied just now with quantum mechanics, and I further doubt very seriously whether this problem—writing this three-man paper—really can be solved in a finite time."[44] Nonetheless, after a week the three men had a draft of a long article; the complicated paper was sent out in mid-November.[45]

But by that time, to their chagrin, they had company.

4
Uncertainty's Triumph

ONE OF THE MOST TRYING ASPECTS OF PRACTICING THE ART OF PHYSICS IS that the shape of the answer is not known from the outset. Although they can draw upon advanced experimental technology and the wealth of data collected by past scientists, physicists must work in the dark whenever they proceed close to the frontier of knowledge; they are aided only by a set of aesthetic prejudices, a few mathematical tools, and the knowledge that whatever they come across is unlikely to contradict directly the conclusions of the past, although it may modify them. The ideas of theoreticians must be at least somewhat amenable to being tested by others in the community; experimenters need to make it seem plausible that others could reproduce their work, and achieve the same results. But these guidelines leave more than enough room for error, and the scientists who make the most remarkable advances—perhaps especially the most remarkable advances—are almost inevitably haunted by doubt, anxiety, and the fear of being forgotten.

Consider Werner Heisenberg, residing temporarily at Göttingen, writing to his close friend, Wolfgang Pauli, on the very day that the three-man paper on matrix mechanics is received by the office of the *Zeitschrift für Physik*. Heisenberg encloses a copy, now apparently lost, of the typescript, which, like most products of Göttingen at the time, is laced with complex mathematics. (Heisenberg prefers the clear air of Copenhagen, where Bohr's approach is more conceptual.) The math, which comes mainly from Jordan and Born, makes Heisenberg a little uneasy. He awaits, a bit nervously, the famously critical Pauli eye. Here is how he feels just after completing the work for which he was later awarded the Nobel Prize:

I've taken a lot of trouble to make the work physical, and I'm relatively content with it. But I'm still pretty unhappy with the theory as a whole and I was delighted that you were completely on my side about [the relative roles of] mathematics and physics. Here I'm in an environment that thinks and feels exactly the opposite way, and I don't know whether I'm just too stupid to understand the mathematics. Göttingen is divided into two camps: one, which speaks, like [the prominent mathematician David] Hilbert (and [another mathematical physicist, Hermann] Weyl, in a letter to Jordan), of the great success that will follow the development of matrix calculations in physics; the other, which, like [physicist James] Franck, maintains that the matrices will never be understood. I'm always annoyed when I hear the theory going by the name of matrix physics. For a while,

I intended to strike the word "matrix" completely out of the paper and replace it with another [term]—"quantum-theoretical quantity," for example.

Fortunately, Heisenberg did not follow up this last suggestion. He continues, grumbling in an aside: "Moreover, 'matrix' is one of the dumbest mathematical words that exists." He wonders if the whole mathematical base of the theory had not been constructed too hurriedly. In the paper, Born, Jordan, and he have been satisfied with rough, approximate calculations; Heisenberg now tells Pauli that "when one *really* integrates," chunks of the theory reveal themselves to be nothing but "formal garbage."[1]

His apprehensions were more than justified. Matrix mechanics impressed his colleagues, but many, if not most, of them found the equations impossible to solve. In addition, they resisted the notion, propounded by Heisenberg, that the impossibility of visualizing matrices was actually one of the theory's more favorable qualities; this seemed as opaque as the Kierkegaardian dictum that the fact that one cannot comprehend how God could be is evidence for His existence. Other theorists pointed out that if quantum matrices were all based on the light emitted in transitions from one atomic state to another, the theory therefore admitted the existence of steady states—so much for Heisenberg's hope to get rid of the detested orbits!—and therefore should encompass some picture of exactly what these states are. (It did not.) Finally, there was the transition problem. In the everyday world—what physicists tend to refer to as the "macroscopic scale"—there is no need for matrices to describe the position and momentum of, say, the trees on a front lawn or the motorcycles whizzing by them. It was therefore necessary to specify at what point matrix mechanics came into the picture, and how one made the passage from ordinary variables to "quantum-theoretical quantities."[2]

Nonetheless, physicists had few alternatives; with varying degrees of enthusiasm, they gamely tackled matrices. Pauli, an early advocate of the approach, helped the matrix advocates along when he used the new methods to arrive at the Balmer spectrum lines for hydrogen.[3] Inelegantly, perhaps, as Pauli was the first to admit, but he did it; Heisenberg was elated.[4] Two other physicists independently did the same thing shortly afterward.[5] Unfortunately, the matrix methods were so hard that none of these very smart people could then go on to calculate the spectrum for the next simplest atom, helium, which has two electrons.[6]

The impasse was broken by Prince Louis Victor de Broglie, a young nobleman from one of the more illustrious families in France, an amateur in science unknown to the research community, who wrote a doctoral dissertation sufficiently farfetched that the Sorbonne faculty was unable to evaluate its correctness.[7] Whereas Einstein had insisted that light comes in particlelike packets governed by the sovereign relation $E = h\nu$, de Broglie turned the question on its head; insisting with equal vehemence that energy is energy, no matter the form in which it is encountered, he argued that particles have energy, and therefore they, too, are ruled by $E = h\nu$. This meant that, say, an electron is characterized by some frequency and wavelength; indeed, all material objects have both. Refrigerators, baseballs, subatomic particles—all have wave characteristics.*

Such a thesis would have seemed simply ludicrous had not de Broglie quickly made use of it to explain features of the Bohr atom that had puzzled theorists since its creation. Bohr had shown that electrons surround the nucleus in orbitals of particular sizes, but neither he nor anyone else had been able to figure out why nature had chosen those orbitals and not others. De Broglie claimed that the orbits in an atom are of exactly the right size to permit electrons to complete each "lap" around the nucleus in a whole number of wavelengths. This limits the number of possible orbits in a way that is easy for anyone with a handful of quarters to picture. Imagine squatting on the floor, a pile of coins nearby, placing one quarter next to another to make circles of quarters on the carpet. A circle of one size can be formed with, for example, five quarters; one of another size can be made with six; but it is impossible to form circles of some in-between size without overlapping quarters. There is no problem if the coins on the carpet overlap, but if *waves* overlap, they reinforce and interfere and cancel each other out higgledy-piggledy, as everyone who has splashed water in a tub can attest. For an electron wave to avoid becoming entangled with itself, it must whirl 'round the nucleus in a whole number of wavelengths, and only a whole number of wavelengths. Thus de Broglie's simple idea ingeniously explained

---

*The chain of reasoning can be followed by anyone with a few minutes and some knowledge of simple algebra. (We ignore a few factors of pi and the like.) The length between the successive crests of a lightwave is known as its wavelength and called *lambda* (λ). It can be calculated by taking the velocity of the wave (c) and dividing it by the number of waves per second (ν, the frequency).

$$\lambda = c/\nu \tag{1}$$

In the quantum world, $E = h\nu$. To find the value of ν, divide both sides by h.

$$\nu = E/h \tag{2}$$

Sticking E/h into the first equation,

$$\lambda = ch/E \tag{3}$$

Another fact: Einstein demonstrated that $E = mc^2$. Substituting for E,

$$\lambda = ch/mc^2 = h/mc \tag{4}$$

Mass times velocity—mc—is equivalent to momentum. Particles like electrons and protons also have momenta equal to their mass times velocity. It is well known that a particle cannot travel as fast as the speed of light. Therefore, de Broglie said, to find the wavelength of a particle, one should simply insert its actual velocity v—not to be confused with the frequency, ν—into Equation (4). Young de Broglie argued that this was not an algebraic trick but that particles really do have wavelengths and frequencies given by the relation

$$\lambda = h/mv \tag{5}$$

what Bohr had only been able to postulate: why electrons have fixed, quantized orbits.

One of de Broglie's thesis examiners knew Einstein and passed on the thesis to the great man, who in turn recommended it to another colleague, Erwin Schrödinger. Few people had paid attention to the thesis; Schrödinger changed all that. He pushed de Broglie's ideas far enough to change the face of physics when, in March of 1926, he published a single equation purporting to explain almost all aspects of the behavior of electrons in terms of de Broglie waves, rather than of matrices.[8] Schrödinger proposed that matter could best be understood as a collection of waves adding up, interfering with each other, creating nodes, and so on; their behavior was described by a new branch of physics he later called *wave mechanics*.[9] A theorist then at Zürich, Schrödinger was a few years older than the matrix physicists, respected in the field as a serious craftsman. He was a dreamy, troubled man, a chain-smoker, competent in six languages and, offstage, an incessant womanizer. He hated Hitler and is said to have intervened when he saw Jews assaulted by the Gestapo; shortly thereafter Schrödinger fled Berlin, where he had been teaching.[10] Like Einstein, he spent many of his later years in a futile attempt to reach unification.

The Schrödinger wave equation was an enormous if somewhat disturbing achievement. In a series of papers that appeared with the speed of a drumroll, Schrödinger tied his wave equation to almost every aspect of the new physics; in his second paper, for example, which appeared just three weeks after the first, Schrödinger rewrote his equation in terms of the Hamiltonians used by Bohr, thereby converting most theorists to his approach in a single blow.[11] Moreover, the mathematics Schrödinger used was much easier for physicists to understand. In school, they all had *studied* waves. They had taken *classes* about waves. If it was hard to imagine how a solid object like an atom could really be made out of waves—what was making the waves?—many physicists had confidence that Schrödinger, a clever fellow, would figure out the answer. Meanwhile, theorists had formulas they knew how to manipulate. They could *solve* Schrödinger's equation; they could obtain exact answers. They could not do this with Heisenberg's version of quantum mechanics. Moreover, the trick of envisioning quantum phenomena as the result of crests and troughs, nodes and interference patterns meant that physicists could visualize the atom in terms of continuous processes—the ripple and flow of long-standing waves—whereas with matrices they had to deal with Heisenberg's assertion that the nature of the microworld was discontinuous and impossible to picture. Little wonder that many physicists threw away their matrices and started working with Schrödinger's methods![12] Even Max Born, who had helped develop matrix mechanics, wrote to Schrödinger that after reading the first wave paper he had become so excited "that I want to defect—or, better, return—with flying colors to the camp of continuum physics. After my whole course [through quantum mechanics], I feel myself drawn to the place from where I set out, namely, the crisp, clear conceptual formulations of classical physics."[13] Solving an individual atom, he said, was simply a matter of solving its particular wave equation. No need for all this stuff about quantum leaps! As Schrödinger wrote (using the now-obsolete term "vibrational mode" instead of "standing wave"), "It is hardly necessary to point out how much more gratifying it would be to conceive of a quantum transition as an energy change from one vibrational mode to another than to regard it as a jumping of electrons."[14]

Unfortunately, the very success of Schrödinger's approach pushed forward the question of what was supposed to be waving in the wave equation. For those who wanted to know what these waves were supposed to be and how wave theory proposed to explain the evidence, accumulated since J. J. Thomson's discovery of the electron, that the atom was made from particles, Schrödinger had an answer: A particle was in reality nothing but "a group of waves of relatively small dimensions in every direction," that is, a sort of tiny clump of waves, its behavior governed by wave interactions. Ordinarily, the bundle of waves was small enough that one could think of it as a dot, a point, a particle in the old sense. But in the microworld, Schrödinger argued, this approximation broke down. There it became useless to talk about particles. At very small distances, "we *must* proceed strictly according to the wave theory, that is, we must proceed from the *wave equation*, and not from the fundamental equation of mechanics, in order to include all possible processes."[15] This was an implicit attack on the matrix methods—and Heisenberg recognized it as such. Nonetheless, Schrödinger kindly mentioned his hope that wave mechanics and matrix mechanics "will not fight against one another, but on the contrary . . . will supplement one another, and that the one will make progress where the other fails. The strength in Heisenberg's program lies in the fact that it promises to give the *intensities of the spectral lines*, a question that we have not approached as yet."[16]

It is not difficult to imagine Heisenberg's reaction to this last statement, and to wave mechanics in general. He was terribly aware of the failings of the matrix methods he had initiated; although Schrödinger in Zürich may not have known it, the physicists at Göttingen and Copenhagen had already proven unable to calculate the intensities of the spectral lines.

In the first months of 1926, Werner Heisenberg was twenty-four years old. He was already one of the most prominent physicists of the day, a position that, of course, he enjoyed. His reputation was due almost entirely to his work on quantum mechanics, a subject that had occupied the whole of his short professional career. He had spent a year and a half on the train of reasoning that culminated in the three-man paper, and eighteen months is a long time to a young man in a hurry. When Schrödinger's wave equation appeared, Heisenberg reacted in a fury; he must have envisioned all his work being consigned to oblivion. He berated Born for deserting matrices and told Pauli that the wave business was so much "crap."[17] It was, he said later, "too good to be true." He hoped it was wrong.[18]

It was not. On April 12, Pauli sent a lengthy letter to Jordan in which he proved that the two approaches were identical.[19] Schrödinger himself proved the same thing, a little less completely, a month later.[20] (It was the third wave paper in less than two months.) The attitude of the Göttingen-Copenhagen group seems to have surprised and annoyed him; thirty-eight-year-old Schrödinger didn't like their barely suppressed scorn for their elders and their conviction that exploring quantum theory was a quest for young men— a game Schrödinger was already too old to play.[21] In the equivalence paper, Schrödinger mentioned, *pro forma*, that it was really impossible to decide between the two theories—and then went on to argue fiercely the merits of wave mechanics.

Far from being a good point, he said, the impossibility of picturing matrices was "disgusting, even repugnant." It meant that it was terribly difficult to imagine how one would go from the macroworld (where objects could be seen and measured) to the microworld (where, according to the matrix methods, objects per se might not even exist). On the other hand, Schrödinger said, waves existed in both domains; the wave equation was ideally suited for the basic question of future research—that of the interactions between the atom and the electromagnetic field.[22]

Privately, Schrödinger was even more deprecatory: He suspected— correctly, as it turned out—that Heisenberg, Pauli, and the rest of the matrix people had papered over difficulties in their mathematics.[23] When they calculated the spectral lines for hydrogen, they had fitted their work to the data (unconsciously, to be sure). Later, when the dust had long since settled, it was shown that if the matrix technique then promoted by Heisenberg, Jordan, and their associates was used strictly and carefully, it produced the wrong answer.[24]

To Heisenberg's dismay, yet a *fourth* Schrödinger paper—this one submitted in May—showed how to calculate the Balmer lines for hydrogen *and* their intensities.[25] (The success with the intensities should have particularly galled Heisenberg, for the Göttingen-Copenhagen physicists had been unable to come up with them.) That summer, Heisenberg recalled in his memoirs, there was a large conference in Munich, where he planned to confront Schrödinger directly. After the latter's lecture, Heisenberg leaped to attack his rival. "I . . . raised a number of objections, and, in particular, pointed out that Schrödinger's conception would not even help explain Planck's radiation law [that is, $E = h\nu$]." Heisenberg's reasoning was, apparently, that continuous waves could not produce discrete packets of energy. "For this I was taken to task by Wilhelm Wien," the experimental physicist who had attempted to flunk Heisenberg on his dissertation,

who told me rather sharply that while he understood my regrets that quantum mechanics was finished, and with it all such nonsense as quantum jumps, etc., the difficulties I had mentioned would undoubtedly be solved by Schrödinger in the very near future. Schrödinger himself was not quite so certain in his own reply, but he, too, remained convinced that it was only a question of time before my objections would be removed. My arguments had clearly failed to impress anyone—even [Arnold] Sommerfeld, who felt most kindly toward me, succumbed to the persuasive force of Schrödinger's mathematics.[26]

The measure of Heisenberg's consternation perhaps is most fully demonstrated by his neglect, in this and other autobiographical writings, of the fact that Wien, Schrödinger, and Sommerfeld were absolutely right to wave off his objections. Even half a century later, Heisenberg apparently did not enjoy recalling that after the Munich conference Schrödinger promptly published a paper showing that Planck's radiation law could indeed be derived by his methods.[27] Heisenberg was furiously determined to eject waves from the microworld; light consisted of photons, not waves; electrons were particles, not waves; continuous equations were wrong, discrete and unvisualizable matrices were right.

Dismayed, Heisenberg wrote Bohr that there would soon be nothing left of the quantum theory that had been gradually built up since Bohr's model of the atom thirteen years before. Although tending to agree with Heisenberg, Bohr was troubled by the entire wave-particle question. How could an electron be a particle that bounced and ricocheted like a tiny bullet yet be described by an equation for a wave? Bohr asked Schrödinger to visit Copenhagen in October of 1926. A man of pronounced intellectual honesty, Schrödinger was happy to debate; in addition, he seemed to be on the winning side, and could afford to dally in his rival's headquarters. He was not prepared for the welcome he received. Schrödinger met Bohr at the railroad station and immediately was involved in an argument—a discussion that went on for days, early in the morning until late at night. Bohr had arranged for Schrödinger to stay at his house, so that nothing could interrupt the confrontation. Heisenberg once recalled,

Bohr was an unusually considerate and obliging person, but in this kind of discussion, which concerned epistemological problems which he thought were of vital importance, he was capable of insisting—with a fanatic, terrifying relentlessness—on complete clarity in all argument. Despite hours of struggle, he refused to give up until Schrödinger had admitted his interpretation was not

enough, and could not even explain Planck's law. Perhaps from the strain,
Schrödinger got sick after a few days and had to stay in bed in Bohr's home. Even
here it was hard to push Bohr away from Schrödinger's bedside; again and again,
he would say, "But Schrödinger, you've got to at least admit that . . ." Once
Schrödinger exploded in a kind of desperation, "If you have to have these damn
quantum jumps then I wish I'd never started working on atomic theory!"[28]

With Heisenberg at his side, Bohr was able to browbeat Schrödinger into a
temporary retraction. But it didn't last, and by the end of October, Heisenberg had worked himself into that desperate, creative frenzy that precedes
much of the finest scientific work. Like Rutherford fueling himself with an
attack on the Thomson "plum pudding" atom, Heisenberg had railed against
wave mechanics to the point where he was now ready to entertain crazy
ideas.

The intensity of the debate at the birth of quantum mechanics was
more than a matter of careers, temperaments, and prejudices; the participants were gripped by the conviction, endemic to the science, that they
were arguing about the shape of the Universe itself, and that the picture they
were forming had profound philosophical resonances. In some respects, the
belief that discerning the laws of the quantum world is equivalent to deciphering the most primary code of nature is naïve and reductionist; but in
other ways it is exactly what they were doing, and the physicists of the
time—as well as their successors today—have rightly been caught up in the
breathtaking implications of their quest.

Many physicists have experienced a moment, usually in their childhood, in which they perceive the Universe as a house of law, a domain whose
architecture can be learned by human beings. The American physicist Isidor
Isaac (I. I.) Rabi, for example, spoke of such an experience vividly, although it
occurred when he was just twelve years old. When we spoke with him, Rabi was
among the last survivors of the first generation of quantum physicists. (He died in
1988.) He was raised by almost violently orthodox Jewish parents—"God was
in almost every sentence," he once told us—in pre–World War I Brooklyn, a
time and place in which it was still possible for a boy never to learn that the
earth went around the sun, and to believe that the sunrise was a daily
miracle. One day in 1910, Rabi stumbled across a book on Copernicus in the
library. There, in the stacks, he discovered the grand movements of the solar
system and was filled with a perception of the cosmos as a composition of
musical beauty and sculptured precision. He passed through a private version of the Copernican revolution, went home, and bluntly informed his
parents that there was no need for God. "I was too young to appreciate what a
terrible blow it was for them," he said to us.[29]

Rabi was a short, pugnacious man, whose sharp opinions and caustic laughter were never dimmed by age. We spoke with him in 1984 in New York City, near Columbia University, in his apartment across the street from Riverside Park. We sat around a silvery coffee table that had been presented to him upon his retirement from Columbia some seventeen years before; its top had once formed part of a particle detection chamber that operated in Columbia's physics laboratory for many years. Sunlight streamed in through the window over the treetops. Unlike most New York City apartments, this one was almost noiseless save for the wind blowing in through the window and the occasional noise of a bus rumbling by on Riverside Drive. Rabi was uninterested in questions about the chronology of physics developments in the 1920s. "What the hell do you want?" he snapped, waving his hand scornfully. "It's been fifty years!" More important, he thought, was why he or anyone could care so much and for so long about such an abstract subject as physics; he wanted to talk about the role physics had played—still played—in his life.

"Quantum mechanics and relativity affected me deeply—personally. It affected my attitude toward the world. I've always thought of physics as a sort of ivory tower, from which you venture forth into all other human affairs, of all kinds. That's why I became a physicist. I could've earned more money as a lawyer." He laughed a little, hunching forward over the silver table.

We pointed out that the vantage he was speaking about is one more traditionally associated with religion, philosophy, or art.

Rabi nodded in agreement. "I believe in the importance of literature," he said. "The basic things like the Russian novel, for instance. I'm very impressed by philosophy. On the other hand, this is one world, as far as our experience goes. And I never know how much of philosophy as such is a play on words and the moving power of language. Like in German, *Volksgefühl*." *Volksgefühl* means "folk wisdom." Some philosophers only articulate the common wisdom, the *Volksgefühl* of their people or their country, whereas others seek something unchanging and deeper. Rabi had been unsure he could distinguish between them; he turned to physics, where he was confident of his own ability to work on a fundamental level. "With science I felt I could grab on to actual things and try to understand them. And then they turn out to be so extraordinarily *mysterious!* Newton's laws of motion, the laws of the electromagnetic field, relativity—they're so far removed from experience, but yet there it is. It's a measure of all the other things that I look at. It gives you an approach to the human race, apart from these inherited things of nationality and whatnot, which you can't take very seriously. That's what science was for me—a citadel. I know some place where I can find out things which are *so*, and not trivial. Far from trivial." He was speaking slowly, hunting for the right phrases. Behind him, wind shook the treetops of the park. "So relativity can have an enormous effect on how I regard myself in the world. It's hard to communicate to other people who haven't that experience. Since it is, as far as we know, a universal human possibility to investigate nature, and the nature of discoveries is so remarkable, so wonderful—if you want to think of the goal of the human race, there it is. To learn more about the Universe and ourselves. In physics, the newest discoveries, like relativity and the uncertainty relation, uncover new modes of thought. They really open new perspectives." A sudden, sad look passed over his face. "And I thought that, say, fifty years ago, that this would happen, that these revolutions and advances in science would have an effect on mankind—on morals, on sociology, whatever. It hasn't happened. We're still up to the same things, or, well, I think, regressed in values. There's this terrible thing— between the United States and the Soviet Union, we might destroy the world. What differences are there between them, except some ideas—not too well-founded—about the distribution and manufacture of the world's goods? And one side has Reagan, who solemnly believes in the God of his fathers, and the other side, Marx and Engels. And they're such unimportant things in comparison with what actually *exists*—the wonder and mystery of the world."

There may have been no revolution in morals, but the physicists of Rabi's time found wonders and mysteries aplenty. Perhaps the greatest and most widely known of these is the indeterminacy relations—popularly called the "uncertainty principle"—which were discovered by Werner Heisenberg in the months after Bohr's confrontation with Schrödinger. Pushed by the apparent failure of matrix mechanics, his own ambitions, and the drive to find the beautiful and profound secrets in the domain of the quantum, Heisenberg spent the autumn of 1926 in a furious attempt to make sense of the two approaches.

He was not alone. Max Born, too, spent the last half of the year working on the problem. Born had been chagrined when Schrödinger's equations appeared, for he was convinced that he could have found them first if Heisenberg and Jordan had not led him astray with their matrices.[30] But he modified his initial enthusiasm after he began working with wave mechanics; by the fall, he was convinced that the waves in Schrödinger's papers were not real, three-dimensional waves, as their creator thought, but something else. Born suggested they might be probability waves, rising and falling with the likelihood that a particle might be in a given place.[31] His basic idea was right, but it was Wolfgang Pauli, in a letter to Heisenberg from the middle of October, who put it into its final form.[32]

Pauli, too, was preoccupied with the puzzle of resolving the wave and matrix approaches. He, too, had been struck by the noncommutative relationship $pq - qp = \hbar/i$—an equation that hung like an icon over quantum mechanics, full of unfathomable meaning. (Here, as before, p stands for momentum, q for position, $\hbar$ is Planck's constant divided by twice pi, and i is a symbol meaning the square root of negative one.) It was difficult to know what this noncommutativity *meant*. For instance, if physicists followed their natural inclinations and put p and q into formulas, the result depended on the order in which the terms were arranged. This implied that one equation could produce several answers; conversely, if you knew, say, p and the answer, but wanted to find q, you ended up with an array of possible values. In the quantum domain the position and momentum of a particle (or bundle of waves) were inextricably related, but what, Pauli asked, was the meaning of it? What was hidden in $pq - qp = \hbar/i$? Pauli felt the problem like a missing filling; he could not stop tonguing the hole. In October, he wrote to Heisenberg, a lengthy missive in the famous Pauli style, a mixture of the abstruse, the elegant, and the gossipy. "But now comes the obscure point. The p's must be assumed to be *controlled*, the q's *uncontrolled*. That means one can always calculate only the probabilities of particular changes of p's with fixed initial values, [and only when they are] *averaged over all possible values of q*." The more you learned about one, the less you could say about the other. Pauli spent a page and a half demonstrating this rigorously.

"So much for mathematics," he continued. "The physics of this is unclear to me from top to bottom. The first question is why can only the p's (and not simultaneously both the p's *and* the q's) be described with any degree of precision? This is the old question of what happens when you know (with reasonable accuracy) the direction of the velocity and the greatest distance from the nucleus of the orbiting electrons. About this I know nothing I haven't known for a long time. Always the same question . . ."

In a burst of annoyance, Pauli summed up: "One cannot simultaneously hook together both the p-numbers and the q-numbers with ordinary c-numbers [that is, regular classical variables]. You can look at the world with p-eyes or with q-eyes, but open both eyes together and you go wrong."[33] And he still had no idea what this meant.

Nine days later, Heisenberg gratefully responded. He had shown Pauli's letter to Bohr and several other physicists at Copenhagen, where it had stimulated heated discussions. Heisenberg, too, had been intrigued by Pauli's "obscure point." It seemed to him that what his friend was implying was that the $pq - qp = \hbar/i$ relation meant that the individual p's and q's couldn't be precisely defined in the old way. Reaching for an analogy, he wondered if talking about a particle's momentum p at a point q might not be like trying to determine the length of a wave if you only had measured one point. (Less exactly, this is like trying to guess the height of the Empire State Building when you know only that the front door is about ten feet wide.) Perhaps momentum and position were large-scale concepts, qualities that required a little breadth of scale before they took on meaning. Then he confided, "Above all, I hope there will eventually be a solution of the following type (but don't spread this around): That time and space are really only statistical concepts, something like, for instance, temperature, pressure, and so on, in a gas. It's my opinion that spatial and temporal concepts are meaningless when speaking of a *single* particle, and that the more particles there are, the more meaning these concepts acquire. I often try to push this further, but so far with no success."[34]

In the meantime, Heisenberg's friend and collaborator, Pascual Jordan, had been thinking about what it meant when physicists said that they wanted to restrict their theories only to quantities they could observe. Shy, amiable, and apparently bereft of personal ambition, the twenty-four-year-old Jordan was easily sidetracked into helping others on lengthy projects of no particular profit to himself.[35] In school, he had flunked out of elementary laboratory physics; his first paper, which sought to modify Einstein's approach to the light quantum, was quickly demolished by the old master himself.[36] In 1926, he was the youngest of the important physicists in Born's circle; he had studied physics seriously for only four years, and regarded the twenty-five-year-old Heisenberg as his senior colleague. He liked fundamental questions; this inclination meant that he soon drifted from quantum physics into cosmology. (On the other hand, he also wrote a humorous paper about the jaw motions of cows chewing grass.)[37] At the time, a number of physicists were wondering about the nature of measurement.[38] Several had claimed that even if a microscope were developed that was precise enough to observe individual atoms, the endless random tumbling and rattling of the microscope's own atoms would render it impossible to measure exactly what other atoms were doing. Jordan argued that if one could somehow freeze the microscope to absolute zero, the temperature at which there is no molecular movement at all, then one could, at least in theory, use the instrument to make a certain measurement of both the position and the momentum of an atom or its constituent electrons.[39] Therefore, Jordan argued, these variables are not, in principle, unobservable; they are just hard to see.

Heisenberg hadn't thought of this. Jordan's paper excited him because it was a direct challenge on his home ground. In the beginning of February, Heisenberg wrote jokingly to Pauli that he "was still occupied with the logical foundations of the whole phony $pq - qp$ business." Jordan, Heisenberg said, talked about the "probability of finding an electron in a particular space," but nobody had a good definition for the idea of "the place of a particle." Heisenberg thought that this definition was something to think about. A bit unkindly, he signed his letter, "With complete youthful unconcern for continuum tricks, W. Heisenberg."[40]

In Copenhagen Heisenberg stayed in the garret apartment of Niels

Bohr's brother, Harald, a small place with walls that sloped toward the ceiling. Niels liked to come in after supper, pipe in hand, and the two men would argue about physics until the early hours of the morning. Their routine was ordinarily quite fruitful, but the difficulties with the wave and matrix approaches to quantum mechanics proved so intractable that by February of 1927 Bohr and Heisenberg were tired and snappish in each other's company. Bohr wanted to understand the wave-particle confusion; Heisenberg wanted to get rid of the waves in any form. Feeling their friendship under strain, Bohr left to go skiing in the mountains north of Oslo. Late one February night, Heisenberg took a solitary walk in Faelled Park, behind the Institute, before going to sleep. As he walked, Heisenberg was thinking that if one multiplied p and q, you simply could not be sure what the result would be. He was thinking that he did believe in the matrix theory. He was thinking about how one would discover where a particle was and how fast it was moving. He was thinking about Jordan's frigid microscope. And then, in the park, it occurred to him that if his scheme were right—that is, if the laws of quantum mechanics were truly correct—he ought to see how nature fulfilled them.[41]

For example, suppose you did want to look at a subatomic particle cooled to absolute zero. Looking at an object in this way means bouncing light quanta—photons—from it and then catching the reflected photons in the lenses of the instrument. If you are looking at something small, such as an electron, the energy of the photons will shove the electron away from its original position in somewhat the way that a rolling billiard ball will knock aside another, stationary ball. You inevitably change the particle's momentum in the act of measuring its position.

To prevent changing the momentum, Heisenberg realized, would require using a photon with a tiny amount of energy. The energy of a photon is given by Planck's formula $E = h\nu$, where ν is the number of waves per second. If E is small, then ν must also be small. (Planck's constant h cannot be changed.) To have a small number of waves going by every second, the waves must be long, just as fewer limousines packed bumper to bumper pass by in a given time than tiny sports cars. But the longer the wavelength of the photons, the less precisely they pinpoint the location of the electron; this leads to the absurd situation where you are trying to measure something by bouncing off it, not a billiard ball, but an enormous whiffleball.

Enormously excited, Heisenberg spent the next little while writing a fourteen-page letter to Pauli about his idea. He described his imaginary microscope and the "thought experiment" of trying to look for particles with it.[42] "One will always find that all thought experiments have this property: When a quantity p is pinned down to within an accuracy characterized by the average error Δp, then . . . q can only be given at the same time to within an accuracy characterized by the average error $\Delta q \geqslant \hbar/\Delta p$."

Today, this now-famous uncertainty principle is usually phrased

$$\Delta p \Delta q \geqslant \hbar$$

where the triangular symbol Δ is the Greek letter delta, and Δp is read "delta-pee." In other words, the position and momentum of an electron cannot be determined exactly—a small but irreducible margin of uncertainty is unavoidable. If the error in the position (that is, Δq) is very small, then the error in the momentum (that is, Δp) must increase to keep their product, $\Delta q \times \Delta p$, larger than $\hbar$. If the position of an electron is measured with such precision that the error is almost zero, then the corresponding error in the momentum becomes almost infinite—the momentum is completely indeterminate. Not only that, Heisenberg showed Pauli how this was a direct consequence of $pq - qp = \hbar/i$—at last giving the equation some physical meaning. Heisenberg found that now that he could visualize some aspects of the matrices, he didn't mind at all.

Heisenberg later recalled that Pauli's reply from Hamburg was enthusiastic. "He said something like, '*Morgenröte einer Neuzeit*,' "—the dawn of a new era.[43]

With such encouragement, it must have been a shock to Heisenberg when Bohr returned from vacation and showed Heisenberg that his paper contradicted Heisenberg's own interpretation of quantum affairs. Bohr reminded Heisenberg that even in the subatomic domain energy and momentum are conserved. If you bounce a photon from a motionless electron and set it moving in the process, you can still know the precise momentum of the electron by looking at the recoil of the photon. As long as you measure the momentum of the photons bouncing off the electrons, all uncertainty vanishes.

However, Bohr said, Heisenberg's idea was still correct. It works because you *can't* determine the momentum of the recoiling photons precisely. You can't because they spread out and diffuse in a wavelike manner—which is precisely why you need a microscope lens to collect and focus them. Even if an electron was used instead of a photon, Schrödinger's equation would ensure that it would diffuse in exactly the same way. But to acknowledge this meant acknowledging that Schrödinger-style waves played an essential role in the theory—a step Heisenberg was so little prepared to take that he burst into tears with frustration.[44] He could establish the physical basis of quantum mechanics only by firmly entrenching his rival's place in it.

Heisenberg and Bohr had a terrible quarrel. "I did not know exactly what to say to Bohr's argument," Heisenberg said years later. "And so the discussion ended with the general impression that now Bohr has again shown that my interpretation is not correct. Inside I was a bit furious about this discussion, and Bohr also went away rather angry. . . ."[45] The two men met again a few days later; Bohr flatly told Heisenberg not to publish the article.

Ignoring his friend's advice, Heisenberg wrote his paper with the old, incorrect arguments. At the end, Heisenberg tacked on a curious little postscript advising the reader that Bohr had found problems, but not disclosing what they were. On March 23, 1927, the article, "On the Visualizable Content of Quantum Theoretical Kinematics and Mechanics," arrived in the offices of the *Zeitschrift für Physik*. It was published at the end of May.[46]

On July 4, Schrödinger wrote to Max Planck, who had recently retired from his position at the University of Berlin. Schrödinger had taken his place, and in his conscientious way he wanted to keep the old man well informed about the discipline to which he had given his life. It had not escaped Schrödinger that the indeterminacy relations restored visualizability to physics, and that wave functions were an integral part of the theory. He and Planck had talked often, it seems, about waves and matrices; now he wrote, "In a brand new article by Heisenberg, my-much-smiled-at wave packets are said to have at last found their correct interpretation as 'probability packets.' . . . Well, as God wills. I keep quiet."[47]

A kind and honorable man, Schrödinger kept true to his word. And in December of 1933, he and Heisenberg found themselves on the same podium, smiling fraternally as the Swedish Academy awarded them both the Nobel Prize for physics—Heisenberg the delayed awardee for 1932, Schrödinger the co-winner for 1933. Although there is no record of any love lost between the two men, their conjoined work became part of the common lore of physics. Schrödinger's wave methods are used by all the workers in the field, while his premises have been rejected; despite the celebrity of Heisenberg's premises, the matrix methods he pioneered are now rarely used for the purposes he had hoped.[48]

□ □ □ □ □

To this day quantum mechanics has not lost its ability to provoke. Many, if not most, students have trouble swallowing the ideas when they learn them in class; it has been said that physicists never understand quantum mechanics, they just learn to use it. At times, thinking about the subject exhausted our faculties, and we consulted a physicist named Robert Serber. We were impressed by Serber's sure grasp of the heart of a question. Every so often we met him for a few hours in his office at Pupin Hall, in Columbia University. Serber is a patient man, and, not unnaturally, we liked him because he was patient. We would talk for a couple of hours, and then he would look at his watch and say he had to pick up his son from the sitter. A short, wiry man with thinning hair and strong blunt-fingered hands just beginning to bend with age, Serber is one of the physicists who keep the field going, making solid, unflashy contributions that rarely come into the public eye. Typically, he would wear a loose turtleneck shirt, faded jeans, and a Western belt buckle studded with turquoise and wide as any made; his well-used glasses have thick black frames. The kind of man whose presence in a room is not immediately noticed, he speaks softly, in narrow circles, continually curling back to the subject, reformulating, clarifying, breaking off a sentence—maddeningly, in mid-phrase—to erase an earlier equation, to modify a parameter, to restate a definition, to think it over again from the beginning; a kind of deep stammer that involves whole phrases and yards of discourse and is impossible to reproduce on the page. It took a little while to realize that what we heard was the audible record of a man thinking through a physical question, pushing hard as he could at nature. If we said something he did not think was so, he would make a barely perceptible grimace, and then say, "Ah, yeah . . ." Then, about ten seconds later, he would say politely, "Well, I'm not sure." Then he would say, "That's just all wrong," a little apologetically, and get up and draw on the blackboard whatever the right idea was.

On one occasion, we came to talk about a few books that we had read. These books, which are quite popular, assert that there is a connection between quantum mechanics and some forms of Eastern mysticism. Behind the connection, more or less, is the view of quantum mechanics hammered out by Niels Bohr and subsequently known as the Copenhagen interpretation. Because position and momentum cannot be measured simultaneously, the Copenhagen interpretation claims that experimenters choose which one shall have a definite value by the act of measuring it. Physicists are therefore part of the reality they are probing; the element of conscious choice, some theorists believe, means that any picture of an observed phenomenon must include the mind of the observer. The Nobel Prize–winning physicist Eugene Wigner described the logical conclusion: "[I]t was not possible to formulate the laws of quantum mechanics in a fully consistent way without reference to the consciousness [of the observer]. . . . [Remarkably,] the very study of the external world led to the conclusion that the content of the consciousness is the ultimate reality."[49]

The books we had read took this a bit further, insisting reductively that the connection between physicists and the minute particles with which they work demonstrates that we are all floating in a sea of mind.[50] The philosophical resonance of quantum mechanics, it has been said, is, if anything, "psychedelic."[51]

Quantum mechanics is complicated enough without being psychedelic, and we thought we might talk to Serber to shore up our shaky understanding of the subject. Serber learned about the uncertainty principle three years after its formulation, in 1930, as a first-year graduate student at the University of Wisconsin, in Madison, under the able tutelage of John H. Van Vleck, one of the first and foremost expositors of the new physics in the United States. For some years now, Serber has been retired, though he still works on various physics projects.

Clean, orderly, seldom used, Serber's office has the blue, melancholy lighting of retired scholarship. A bookcase stands to the side, its glass doors framing the reflections of the Chagall and Klee prints on the opposite wall. Over an empty blackboard hangs the urgent message: *DO NOT WASH.* Tucked into a corner is a frightening souvenir of the atomic era, a piece of wallboard from a school in Hiroshima with the shadow of a windowframe burned into its surface. Although faded by time, the burn is a clear reminder of the practical consequences of abstruse physical theories. We talked with him a while, and then, on another day, Serber had thought of a better way to say what he meant and we returned to the subject. (Some of the remarks that follow are taken from his reformulation on the second day.)[52]

We asked right away if it was true that the uncertainty principle brings the observer into physics. He shook his head; no; shook it again, even more emphatically. "I mean, the whole idea of observers—that's a pedagogical thing. It's very convenient when you're trying to understand something to imagine doing an experiment or think about an observer doing something. But that doesn't have anything to do with physical law, that has to do with the understanding of the law. A physical law is a description of nature, not a description of observers. Using the word *observer* in any place in physics at all—it's irrelevant. It's *never* part of a physical theorem." He took his glasses off and rubbed his eyes, which were slightly reddened from lack of sleep. He had been up the night before with sick children. "Look," he said, "they apply quantum mechanical laws to the Big Bang! There were no observers there!"

What made the uncertainty principle so important, he said, was the way it helped physicists reconcile the two apparently incompatible concepts of particles and waves. Einstein had shown that light waves can behave like particles, de Broglie had shown that particles can behave like waves, and physicists had become terribly confused. "Heisenberg realized that there wasn't necessarily a contradiction between the two views," Serber said. "In any real physical situation, you're never dealing with a single wave—that's a mathematical abstraction—but with a wave packet." He went to the board, found a stub of white chalk, and drew a series of waves whose crests coincided at one point, which he labeled *A*.

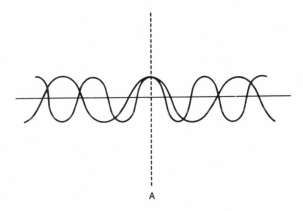

A

Ordinarily, waves in such a jumble, overlapping each other every which way, would tend to cancel each other out—go out of phase, as scientists say. If circumstances are right, however, they may all pile up for a brief moment, as they do at A. "If you add the way these waves interfere," Serber said, "you wind up with something like this." He superimposed the sum on the first sketch.

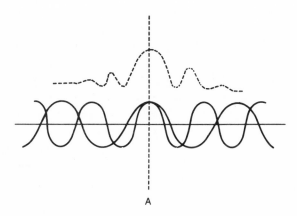

A

The monadnock in the center of the drawing is very narrow because the various waves rapidly cancel each other out as they spread from Point A. The sum of the waves, that is, the wave packet, is therefore only detectable close to Point A, and the wave function represented by the picture describes something quite like the tiny dot conjured by the word *particle*. If the waves did not interfere with each other right away—if the central bump smeared out into a mesa or a series of ripples—the Schrödinger function would describe something more and more like a conventional *wave*. The speed with which the component waves of the packet cancel each other out depends on the difference in their wavelengths. A large difference, and one has a particle; a small difference produces a wave. The underlying reality is the same for both.

Now comes the uncertainty principle, Serber said. "The spread in wavelength is proportional to the spread in momentum, right? De Broglie showed that. So a small difference in the wavelength of the components corresponds to pinning down the momentum precisely. But that means the wave is not localized in space very well—you don't know where it is, really. If you want to make the wave packet narrow in space, on the other hand, you have to make the momentum spread very big, so that the waves interfere right away.

"You can't have both at once. Heisenberg said that either you don't have a definite momentum or you don't have a definite position. And the fact that you can't have both at once is what removes the contradiction between wave behavior and particle behavior." The famous formula $\Delta p \Delta q \geq \hbar$, Serber said, is simply a mathematical consequence of Schrödinger's equation; all Heisenberg did was to figure out its physical meaning.

What was the role of quantum mechanics and the uncertainty principle in the present drive toward unification? "In the 1920s, there wasn't any push toward unification to begin with, right? People were just learning a lot of new physics. They had no idea that it would lead to any kind of unified theory. Einstein was looking for a unified theory, but, if you really get down to it, it didn't make any sense. All he was talking about was electromagnetic theory and gravity. But that isn't the whole story. It was obvious that you couldn't expect to get a complete solution from an incomplete picture in the first place. You had to understand a lot more about the nature of the world before you could proceed. You had to know about strong interactions, weak interactions, what the symmetry properties were, what the fundamental particles were before you could start. Just to unify the weak and electromagnetic interactions, you had to know an awful lot about electrodynamics that wasn't known until the late 1940s. Then you had to know an awful lot about the weak interactions that wasn't learned until the 1960s. So all this—" the equations for the uncertainty principle he had put on the blackboard "—is the preliminary stages, when you're learning the pieces that you have to put together, that you have to unify. Quantum mechanics is the groundwork.

"It's astonishing, really. Look, quantum mechanics was created to deal with the atom, which is a scale of 10^{-8} centimeters. That was in the twenties, right? Then in the thirties, forties, fifties, they applied it to nuclear physics. Instead of 10^{-8} centimeters, it's now 10^{-13}, and quantum mechanics still worked—that's another leap. Then after the war, they began building the big machines, the particle accelerators, and they got down to 10^{-14}, 10^{-15}, 10^{-16} centimeters, and it *still* worked. That's an *amazing* extrapolation—a factor of a hundred million!"

Serber once showed us the notebooks he used in graduate school in 1930. They were artist's notebooks with neat red spines and inner covers papered with Florentine designs. His handwriting is an impeccable tribute to the rigidity of prewar grammar schools. Van Vleck, his teacher, had an approach to the subject that would seem ludicrously complicated today. He explained Heisenberg's microscope and gave another example, deriving the uncertainty principle from a prism. On one page, nestled in a tangle of integrals, is a reference to a popular essay by the Harvard physicist Percy Bridgman, which appeared in the magazine *Harper's Bazaar* two years after the discovery of the uncertainty principle. Entitled "The New Vision of Science," it is a exposition of quantum mechanics for the nonphysicist as good as any that has been published in the intervening six decades. (The lack of success of scientific popularization may be adjudged by the fact that if Bridgman's article were published today it would be equally illuminating to the general public.) One of the dreams which humanity refuses to let die is that some new breakthrough—usually from the realm of philosophy, art, or political revolution—will one day permanently and unalterably change humanity for the better. For Bridgman, the startling discoveries of quantum mechanics, and especially the uncertainty principle, were proof that this saving power would come from physics. The collapse of the commonly understood law of cause and effect, Bridgman correctly predicted, would at first "let loose a veritable intellectual spree of licentious and debauched thinking"; to fix this, new methods of education would have to be introduced to teach the young the right way of understanding concepts foreign to everyday experience. But once this is accomplished, Bridgman thought "understanding and conquest of the world will proceed at an accelerated pace"; indeed, humanity will improve, acquiring a certain "courageous nobility" in the face of this new wisdom. "And in the end, when man has fully partaken of the fruit of the tree of knowledge, there will be this difference between the first Eden and the last, that man will not become as a god, but remain forever humble."[53]

II

Particles and Fields

5
The Man Who Listened

THE STANDARD MODEL OF ELEMENTARY PARTICLE INTERACTIONS WAS pieced together by three distinct intellectual generations of twentieth-century physicists. Each arrived suddenly, of a piece, in the course of two or three years, a group of young men—women as well, in the case of the third—which emerged with their style of play fully developed, in confident command of the tools of the trade. Like the abstract expressionists, whose intemperate urgency and immediate prominence in postwar New York City stunned their elders, the new physicists startled their contemporaries with the sweep and precision of their attack and the fierceness with which they demanded to be heard. Although, like the expressionist Willem de Kooning, the new workers may actually have labored unrecognized for years, it seems to the community at large that a movement has abruptly formed, fast and bright as a stroke of lightning, and that everything has changed. The construction of quantum mechanics, in the mid-1920s, was the work of the first *nouvelle vague*; names like Heisenberg, Pauli, and Oppenheimer, Rabi, Schrödinger, and Jordan, suddenly appeared on papers that had to be read by every serious practitioner in the field. Meanwhile, the old hands, the Rutherfords and Bohrs, were pushed to new accomplishments. (Some never adjusted; J. J. Thomson was one, and—sadly, grandly—Einstein was another.) The years immediately after the Second World War marked the entrance of another group of scientists—young, hungry, predominantly American; the beginning of the 1970s has witnessed the rise of a third.

In each of these brilliant, collective entrances, there has been one player whose superior insight into the structure of field theory has led his confreres to adjudge him, rightly or wrongly, the most brilliant of the lot. These men are not well-known outside the field—in a related discipline, how many members of the public know of David Hilbert, arguably the finest mathematician this century has produced?—and they may not produce as much physics as others of their generation, but nonetheless the acuity of their vision and the formal tools under their control have awed their contemporaries. Physicists have as much difficulty with complex mathematics as anyone else, and it is little wonder that the legends should accrue around mathematical physicists. In the movement to assemble the standard model that gave impetus to unification, the name of Gerard 't Hooft stood out; in

1948, the honor went to Julian Schwinger. During the short, happy heyday of quantum mechanics, the reputation went to Paul Adrien Maurice Dirac.

At a time when young physicists created stirs with their first papers and every dissertation opened a new field, it was Dirac who most shaped the resulting science. If Einstein, with his disdain for the quantum theory, was truly the last classical physicist, then P. A. M. Dirac, as he always signed his work, was the first completely modern one. In an article written soon before Dirac's death in 1984, the physicist Silvan Schweber remarked, "Dirac is not only one of the chief authors of quantum mechanics, but he is also the creator of quantum electrodynamics and one of the principal architects of quantum field theory. All the major developments in quantum field theory in the thirties and forties have as their point of departure some work of Dirac's."[1]

Circumstances seem to have conspired to make Dirac painfully shy and taciturn. He was born on August 8, 1902, in Bristol, England, then, as now, a commercial town on the confluence of the Avon and Frome rivers. His father, a Swiss émigré, was retiring to the point of being antisocial; the family never had guests, and the Diracs never went out. Dirac ate with his father in the dining room, while the rest of the family had dinner in the kitchen; the boy would have preferred to be with his mother, sister, and brother, but there were not enough chairs in the kitchen. At home, Dirac once recalled, "My father made the rule that I should only talk to him in French. He thought it would be good for me to learn French in that way. Since I found that I couldn't express myself in French, it was better for me to stay silent than to talk in English. So I became very silent at that time."[2] Quiet and introverted, Dirac spent much of his time outdoors, taking solitary walks through the English countryside. He liked order and symmetry, the meticulous compression of mathematical relationships. "A great deal of my work is just playing with equations and seeing what they give," Dirac said later. "I don't suppose that applies so much to other physicists; I think it's a peculiarity of myself that I like to play about with equations, just looking for beautiful mathematical relations which maybe don't have any physical meaning at all. Sometimes they do."[3]

Although Dirac's father had no appreciation for the importance of social contact, he did realize the use of a good education, and encouraged his son's mathematical bent. Moreover, this proclivity was fostered by an accident of history: As a teenager, Dirac was pushed into a higher level than was normal for his age, because the more advanced classes had emptied out when older students left to go to war. He liked the Merchant Venturer's School, which shared quarters with the engineering college of Bristol University, partly because the curriculum deemphasized philosophy and the arts, subjects that he found nearly incomprehensible for most of his life.[4] Afraid that there would be no jobs in mathematics, Dirac chose to specialize in engineering when he attended the university. He was a good student, but only the theoretical aspects of the field really interested him; his early on-the-job training ended disastrously when his employer found that Dirac "lacked keenness, and was slovenly."[5] Although his only brother was working in the same factory, the two boys never exchanged a word.

In the fall of 1921, having completed an engineering degree, Dirac found himself unable to obtain employment. The Bristol mathematics professors, who had made known their disappointment that a brilliant mathematician was taking engineering courses, offered Dirac free tuition; because he had nothing else to do, he accepted. The only other student in the honors program of mathematics was a woman firmly intent on studying applied mathematics, especially that which could be used in physics. Because he did not have any firm convictions, Dirac went along with her wishes; in this way the career of one of the great physicists of the century began.

From the haphazard beginning of his career to the end of his days, Dirac was convinced that mathematics is the key to progress in physics. In one of his last addresses, Dirac explained his credo: "[O]ne should allow oneself to be led in the direction which the mathematics suggests . . . [o]ne must follow up [a] mathematical idea and see what its consequences are, even though one gets led to a domain which is completely foreign to what one started with. . . . Mathematics can lead us in a direction we would not take if we only followed up physical ideas by themselves."[6]

At Bristol, Dirac became acquainted with relativity, which electrified him. He won a B.Sc. and, buttressed by two grants, entered Saint John's College, Cambridge, in 1925. By 1927, at the age of twenty-five, his contributions to quantum mechanics had ensured that he was one of the most important physicists in the world.

It should be said that he was not changed unduly by the growth of his reputation; he continued to be so laconic that people who met him often thought him rude. Although an honored member of the Cambridge physics group, he had few students, established no school, and talked rarely to experimenters. Samuel Devons, who spent the late 1930s at the laboratory, once told us, "There was a Cavendish Physical Society meeting, a sort of semiformal gathering once every two weeks. Some lecturer would come in, and Dirac would sit in the front row and listen. Very rarely would he open his mouth. Sometimes he would be prodded by Rutherford, who'd say, 'Now what do you theoretical people think?' Rutherford had this notion that theory was some sort of speculation—the real facts were in the experiments. And Dirac would sit and say nothing."[7]

Dirac spoke so precisely and carefully that he approached the Delphic; when he taught quantum mechanics, he stood behind the lectern and read to the class from the book he had written on the subject, believing that he had set down his point of view there as well as he could. In 1928, he gave a series of talks at Leiden, where Paul Ehrenfest, the Dutch theoretician who had quickly submitted the idea of spin before its creators could get scared, was frustrated by the Olympian manner of Dirac's presentation.[8] H. B. G. Casimir was in the audience. Each lecture, he recollected later, "was presented in perfect form. You know Dirac's habit—if you didn't understand things, he would not offer any explanations but would very patiently repeat exactly the same thing. Usually it worked, but it wasn't quite Ehrenfest's way of doing things." Ehrenfest wanted always to see the human being behind the work. "I—" Casimir again "—remember that once Ehrenfest put a question to Dirac to which Dirac had no immediate answer. And so Dirac began to work it out on the blackboard. He covered the entire blackboard with very small [writing]. And Ehrenfest was [standing] right behind him, trying to see what he did, and exclaiming, '*Kinder, Kinder! Schaut jetzt zu! Jetzt kann man sehen, wie er es macht!*' " Kids, kids—look at this! *Now* we can see what he's up to![9]

In August of 1925, R. H. Fowler of the Cavendish received the proofs of an as-yet unpublished paper by Werner Heisenberg that, in the opinion of its author, created "some new quantum mechanical relations."[10] It was Heisenberg's first paper on quantum mechanics, the paper that inspired Born to think of matrices. Fowler gave the proofs to his young research assistant, Dirac. Dirac's interest vanished as soon as he saw that Heisenberg had prefaced his work with several paragraphs of philosophical musings about the importance of considering observable quantities. Noticing that the principal example of the new methods Heisenberg had provided was of no real import, Dirac decided that there was no reason to pay much attention. He came back to it a week or so later, and this time realized that Heisenberg's quantum-theoretical series, the variables with which the equations were formulated, were noncommutative, that is, $A \times B$ did not equal $B \times A$. Dirac saw that this was the key to Heisenberg's entire scheme, and that the reason he had not appreciated the article's significance was that Heisenberg in his anxiety had chosen to demonstrate his idea with a sample calculation in which the noncommutativity didn't show; because noncommutativity was central to the whole conception, he had been forced to provide a trivial example. Dirac remarked later that Heisenberg "was afraid this [noncommutativity] was a fundamental blemish in his theory and that probably the whole beautiful idea would have to be given up. . . . At this stage, you see, I had an advantage over Heisenberg because I did not have his fears. I was not afraid of *Heisenberg's* theory collapsing. It would not have affected me as it would have affected Heisenberg. It would not have meant that I would have had to start again from the beginning."[11] In Dirac's opinion, the originators of new ideas are too protective of their brainchildren to develop them properly; creativity implies a corresponding lack of perspective.

Like Born, Dirac quickly realized that $pq - qp = \hbar/i$. But unlike Born, Dirac did not see at first that p and q were associated with matrices. He regarded them simply as strange versions of position and momentum, and looked for some means to connect Heisenberg's new quantum mechanics with the classical mechanics Dirac felt he understood.

At this time [September of 1925] I used to take long walks on Sundays alone, thinking about these problems and it was during one such walk that the idea occurred to me that the commutator A times B minus B times A was very similar to the Poisson bracket which one had in classical mechanics when one formulates the equations in the Hamiltonian form. That was an idea that I just jumped at as soon as it occurred to me. But then I was held back by the fact that I did not know very well what was a Poisson bracket. It was something which I had read about in advanced books of dynamics, but there was not really very much use for it, and after reading about it, it had slipped out of my mind and I did not very well remember what the situation was.

The French mathematician Siméon-Denis Poisson had introduced this idea in 1809, as a curiosity, to help him make some calculations about the motion of a planet in its orbit. Although Poisson brackets had drawn occasional interest from physicists, until Dirac's insight they were rarely used.

Well, I hurried home and looked through all my books and papers and could not find any reference in them to Poisson brackets. The books that I had were all too elementary. It was a Sunday. I could not go to a library then; I just had to wait impatiently through the night and then the next morning early, when the libraries opened, I went and checked what a Poisson bracket really is and found that it was as I had thought. . . . This provided a very close connection between the ordinary classical mechanics which people were used to and the new mechanics involving the noncommutating quantities which had been introduced by Heisenberg.[12]

A lengthy letter in Dirac's minute, fussy handwriting arrived in Göttingen on November 20, 1925. Written in English, it explained to the astonished Heisenberg how in a few concise steps his version of quantum mechanics could be reformulated in classical terms using a mathematical device he had never heard of. Moreover, working alone, Dirac had come up with a version of quantum mechanics much more general and complete than that produced by Heisenberg, Born, and Jordan in their just-completed joint paper. Stunned at this authoritative communiqué from a complete stranger, Heisenberg replied immediately—in German, a language that Dirac was barely able to follow:

I have read your extraordinarily beautiful paper on quantum mechanics with the greatest interest, and there can be no doubt that all your results are correct as far as one believes at all in the newly proposed theory. . . . [Moreover, your paper is] really better written and more concentrated than our attempts here.[13]

Casually, Heisenberg mentioned that he and others have been working along some of the same lines.

I hope you are not disturbed by the fact that part of your results have already been found here some time ago and are being published independently in two papers, one by Born and Jordan, the other by Born, Jordan and me. [He didn't mention that "some time ago" meant about a month.] Your results, in particular . . . the connection of the quantum conditions with the Poisson bracket, go considerably further.[14]

He then peppered Dirac with questions about the applicability of Poisson brackets to quantum theory. In the next ten days, Heisenberg sent off a postcard and two more letters, all with further questions.

Unfortunately, Dirac's replies are lost. They were among the papers that American military authorities confiscated from Heisenberg at the end of World War II, and, despite all pleas, never returned. But Dirac recalled later that, in one way or another, he was able to answer all of Heisenberg's objections and show that Poisson brackets do, indeed, allow quantum mechanics to be cast neatly into Hamiltonian form by simpler and more classical methods than those Heisenberg used. "That," Dirac said, "was the beginning of quantum mechanics so far as I was concerned."[15]

Three months later, in March of 1926, Schrödinger's wave equation appeared.[16] Like Heisenberg, Dirac was annoyed, but for entirely different reasons. Dirac was hot in pursuit of the mathematical analogies he had unearthed between Heisenberg's quantum mechanics and classical mechanics, and he resented having to turn his attention to a formidable-looking new set of basic ideas whose value was not clear.[17] But he, too, was eventually unable to avoid the wave function.

By the time Dirac had completed his Ph.D. in the spring of 1926, he had completely reorganized quantum mechanics, and was ready to begin pushing its frontier forward. Following graduation, Dirac had the chance to travel to further his studies. His first thought was, unsurprisingly, Göttingen, the home of matrix mechanics. But Fowler, who was fond of Bohr, strongly encouraged him to go to Copenhagen. Torn, Dirac decided to spend several months at each place, going first to Copenhagen. He arrived in the second week of September 1926.

Despite Bohr's garrulity, his overwhelming concern with the philosophical implications of physics, and his notorious difficulties with precise expression, he hit it off immediately with Dirac. "I think mostly Bohr was talking and I was listening," Dirac recalled. "That rather suits me because I'm not very fond of talking."[18] At late-night talkfests in Copenhagen taverns, Dirac met Heisenberg, Pauli, and the rest of the "quantum mechanics" for the first time—a loquacious, snobbish, simpatico bunch that appears to have gotten him to open up a little. Like many of the shy, smart people in that time and place, Dirac had violently colored political views; with passionately lofty detachment, he told his Continental colleagues that there was no reason for the poor to suffer, that he saw little purpose in rewarding the greedy with wealth, and that organized religion was a ludicrous sham.[19] After one such disquisition, Wolfgang Pauli, who had mystical leanings, is supposed to have remarked, "Dirac has a new religion—there is no God, and Dirac is His prophet!"

In Copenhagen, Dirac began to work on some ideas that he hoped would reconcile quantum mechanics and relativity. Quantum mechanics had much to say about the principles of action in the subatomic domain, but it did not take into account the special effects that occur at speeds near that of light. Ever since Einstein put together relativity, physicists had known that theories which did not take it into account were doomed to be incomplete or even wrong. A proper theory of light and matter thus had to describe quantized, relativistic behavior. Dirac set out to create such a theory during the last months of 1926 and the first weeks of 1927, the same period in which Heisenberg wrestled with the uncertainty principle. In the optimism of the time, Dirac had little notion of what a tangle he was to create, and how many years—decades, in fact—it would take to unravel.

Dirac's principal tool was an old idea, that of a *field*. A field is a region of space in which particular quantities are precisely defined at every point. For instance, the pattern of arcs that iron filings form around a magnet illustrates the magnet's field. Each point along each arc is subject to a force of particular strength and direction. The alignment of each filing indicates the direction of the field at that point, and the density of the filings indicates the field's strength. Fields can be defined for such diverse domains as temperature, sound, and matter. The idea of a field was slowly developed by nineteenth-century physicists, and culminated in James Clerk Maxwell's demonstration, in 1861, that light of all varieties could be described as a pattern of electric and magnetic fields.[20] (For this reason, visible light, radar waves, X rays, and infrared beams are all called electromagnetic radiation.) In 1905, Einstein showed that electromagnetic fields are associated with quanta—photons—that act as the agents of the field.[21] If the field is like the domain of a feudal overlord, then photons can be thought of as the soldiers, tax collectors, magistrates, and factota who make the wishes of the ruler felt in that area. By the time Dirac considered the problem of the interaction of matter and electromagnetic radiation, physicists fully realized that to describe the comings and goings of electrons and photons it would be necessary to treat the field associated with the photons. And this is exactly what he set out to do.

Like Planck, Heisenberg, and Schrödinger before, Dirac had recourse to systems of oscillators. But unlike his predecessors, he thought of the atom *and* the field in these terms. Using Heisenberg's quantum mechanics, Dirac was able to come up with a Hamiltonian for the field that was fully compatible with the Hamiltonian for the atom from quantum mechanics.[22] Dirac was thus able to say that the Hamiltonian for the entire process could be found by adding up the separate Hamiltonians for the atom, the field, and the interaction. Moreover, Dirac showed that by juggling the Hamiltonians through an appropriate mathematical procedure, he could prove a law that Einstein had discovered, which gave the probability that a given atom in a given state that sat in a particular field in a particular configuration would absorb or release a photon.[23]

The result was the first real quantum field theory. Because it linked quantum theory with the dynamics of electromagnetic fields, Dirac called it *quantum electrodynamics.* Another way of describing Dirac's accomplishment might be to say that in the beginning of 1927, human beings understood fairly accurately for the first time what happens on the atomic level when someone shines a light in a mirror and the glow is reflected back.

Highly satisfied, Dirac submitted his paper to the *Proceedings of the Royal Society* in the last days of January of 1927, just three weeks before Heisenberg wrote his long letter to Pauli about the uncertainty principle.[24] One of the most influential papers in the history of twentieth-century physics, "The Quantum Theory of the Emission and Absorption of Radiation" begins with the prescient assertion that there is little further work to do in quantum mechanics.

The new quantum theory, based on the assumption that the dynamical variables do not obey the commutative law of multiplication, has by now been developed sufficiently to form a fairly complete theory of dynamics. . . . On the other hand, hardly anything has been done up to the present on quantum electrodynamics.[25]

"Hardly anything" is a classic bit of Oxbridgian understatement. *Nothing* had been done before on quantum electrodynamics.[26]

The questions of the correct treatment of a system in which the forces are propagated with the velocity of light instead of instantaneously, of the production of an electromagnetic field by a moving electron, and the reaction of this field on the electron have not yet been touched.

Dirac said often that he was beset by fears when he came up with his new theory. Perhaps. The tone of the article doesn't betray it.

[I]t will be impossible to answer any one question completely without at the same time answering them all.[27]

In any case, he soon discovered that his fear was justified. Quantum electrodynamics was indeed a great step forward, but it came at a great price. Dirac had set down the beginnings of the modern theory of electromagnetism—the first solid piece of the standard model—but he had also unwittingly let loose an onslaught of conceptual demons that would change our views of space and matter. As a step to quantizing the electromagnetic field, Dirac hypothesized that his oscillators did not disappear when there was no field. Rather, they went into a "zero state," in which they existed but could not be detected. Associated with the zero state oscillators were zero state photons; these, too, could not be detected. It didn't trouble Dirac that his mathematical scheme implied that empty space should contain billions of invisible photons. As long as they didn't show up, their presence made no difference.[28]

But when Dirac's theory was interpreted in light of the uncertainty principle, it turned out that in some sense the zero state photons *do* show up. According to the uncertainty principle, the energy of a field over any given time cannot be determined exactly. (The indeterminacy relations apply equally to particles and fields.) Paradoxically, the less time one spends measuring, the greater this margin of error becomes. It is therefore conceivable that within a time interval on the order of trillionths of a trillionth of a second the zero state oscillators of a field actually might not be at the zero state. In fact, they might have a vast amount of energy that escapes detection. As Einstein's equation $E = mc^2$ demonstrated, mass and energy are two forms of the same thing. Thus the uncertainty principle dictates that if any small area can contain undetectable energy, then, according to Einstein's equation, it can contain undetectable matter. This basic uncertainty is not just a lacuna in our knowledge. Mathematically, there is no difference between this uncertainty and actual random fluctuations in the energy (or matter) measured. Therefore, at least in theory, because any space *might* harbor particles for a short time, it *must* do so.

As strained through the uncertainty principle, quantum field theory exposed a frightful chaos on the lowest order of matter. The spaces around and within atoms, previously thought to be empty, were now supposed to be filled with a boiling soup of ghostly particles.[29] From the perspective of quantum field theory, the vacuum contains random eddies in space-time: tidal whirlpools that occasionally hurl up bits of matter, only to suck them down again. Like the strange virtual images produced by lenses, these particles are present, but out of sight; they have been named *virtual particles*. Far from being an anomaly, virtual particles are a central feature of quantum field theory, as Dirac himself was soon to demonstrate.

At first, however, he does not seem to have realized the full implications of quantum electrodynamics. Instead, he worried about how to make his work jibe with that of his colleagues.[30] There were several major difficulties. First, quantum electrodynamics was based on Dirac's formulation of quantum mechanics, which was itself not entirely in accord with the dictates of relativity. (Dirac's electromagnetic field was relativistic, but his matter field was not.) Deeply sympathetic to the urge for consistency and order that was one of the wellsprings of Einstein's thought, Dirac thought the discrepancy was terribly bothersome. Moreover, in recent months two German physicists, Walter Gordon and Oskar Klein, had produced a relativistic version of Schrödinger's equation for the electron.[31] A little while before his death, Dirac said that the Klein-Gordon equation "could not be interpreted in terms of my general [version] of quantum mechanics and was therefore unacceptable to me. Other physicists with whom I talked at the time were not so obsessed with the need for having quantum theory agreeing [sic] with the general . . . theory, and they were rather inclined to let it go as it was. But I just stuck to this problem."[32] He stuck to his last until the end of 1928, when he returned to England.[33] By then, he had developed yet another equation for the electron's motion—the one used today. To reconcile his own version of quantum mechanics, relativity, Klein-Gordon, and the experimental data, Dirac produced a single equation for an electron traveling through space that had four components, which, taken together, describe the spin and energy of the particle.[34]

The Dirac equation, as it is called, suddenly explained a host of puzzles about the electron. To give one example, it confirmed Bohr's old prediction that the problem of the velocity of electron spin would vanish when the *real* quantum theory was found. According to Dirac's theory, an electron is not localized in a particular place, but has a set of probable locations that are scattered around a point of maximum probability like the cluster of holes around the bull's-eye of a sharpshooter's target. Moreover, these locations circulate around the center as if the target were spinning; according to Dirac's equation, this motion is the spin of the electron. Lorentz had envisioned a little ball rotating when he calculated the electron spin velocity, and found that it turned faster than the speed of light—a flat impossibility. But Dirac said that you should use the average radius of the cluster instead, which is more than a hundred times bigger than the electron envisioned by Lorentz. This made the spin velocity much slower, and removed the contradiction with relativity. Not only that, Dirac's theory predicted that this circulation would create a tiny magnetic field around each electron, and accurately predicted its strength, something previous descriptions of spin had been unable to do.

It was difficult for physicists not to be impressed, even awed, by the Dirac equation.[35] Like children shaking fruit from a tree, they extracted equations describing the collision of two electrons, the interaction of photons and free electrons, and the correct formulas for the spectral lines of the hydrogen atom.[36] The results came so quickly that some members of the field were sure that it was only a matter of time—months, perhaps—before Dirac or someone else would come up with an equation for the last remaining piece of the puzzle, the proton. With photons, electrons, and protons disposed of, scientists would soon knock off the nucleus, and then, except for filling in a few loose ends, all of matter and energy would be explained. Unification would occur, and physics would be over.

"There was one of the regular conferences in Copenhagen," the physicist Rudolf Peierls has recalled. "It may have been the conference of thirty-two or thirty-one; I don't know. The interesting point is that there was a general feeling among some people there, not everybody, that physics was almost finished. This looks ridiculous looking back, but if you look at it from the point of view of the time, practically all of the mysteries had resolved themselves—nearly all. Everything that had bothered one about the atom and molecules and solids and so on, had suddenly fallen into place. . . . Now, I'm not saying this was the common view. I don't think I ever shared it, really. I don't think Niels Bohr, for example, would ever have had any such illusions. But there were sort of over-lunch—or sometimes quite serious—discussions about what we would do when physics was finished. By 'finished' was meant the basic structure; of course, there are all the applications. The majority of people said that that would be the time to turn to biology. Only one person really took that seriously and did turn to biology, and that was Max Delbrück, who certainly was present at these discussions."[37] Delbrück did well in molecular biology, winning the Nobel Prize for medicine in 1969 for beginning the chain of investigations that led to the discovery of the doubled helix of DNA. On the other hand, physics was anything but over.

In the meantime, a small physics community had begun to form in the United States, but it was as yet a paltry thing compared to its equivalent on the other side of the Atlantic.[38] America had the resources to do experimental work, but except for isolated savants like Josiah Gibbs of Yale, the theory came entirely from Europe.[39] More important, the context of physics was set by Continental researchers. Classes at Caltech and Columbia University were assigned textbooks written in Göttingen and Leiden, and graduate students in Cambridge, Massachusetts, angled for the opportunity to travel to Cambridge, England. Aided by grants from the newly formed National Research Council, a pool of European-trained physicists slowly accumulated in this country; from them would later come the leaders of the next generation, one dominated by Americans. I. I. Rabi was among them.

Rabi got his dissertation in 1927, and promptly left for Europe with his wife, Helen. "I found I was actually better prepared than most Europeans of the same level," he told us. "What we lacked—and my generation was to supply it in this country—was a kind of understanding and feeling for the subject, which is hard to get if you're not in contact somehow with a verbal tradition, with people who are making the subject. [You have] to see the living tradition." After a month in Munich and two in Copenhagen, Rabi settled down in Hamburg to work with Wolfgang Pauli. There he met all the creators of quantum mechanics and quantum field theory. He took Dirac to Hamburg's well-known Hagenbeck Zoo, and was startled to discover that the Englishman's love for order extended to an insistence that they see the exhibits *in seriatum.* For Rabi's benefit, Max Born listed all the accomplishments of the past few years. "It was heady stuff," Rabi admitted, a half-smile playing on his lips. "Born said, 'We *have* all that. I think in six months we'll have the proton, and physics as we know it will be over.' There would be a lot to do, but the heroic part was finished. I was just a fresh postdoc [a postdoctoral fellow], and he was foreclosing on the field!"[40]

Not everyone was as sanguine as Born. Heisenberg and Pauli, for example, were troubled by some of Dirac's assumptions, and spent 1928 and 1929 working on their own version of quantum electrodynamics.[41] (To the relief of other physicists, the two versions were subsequently shown to be identical, much as Schrödinger wave mechanics and Heisenberg matrix mechanics were discovered to be different ways of describing the same thing.)[42] Dirac was perhaps the least satisfied of all. It pleased him that two of the four components of his electron equation corresponded to the spin of the particle. But the two other components, those corresponding to the energy of the electron, were more difficult to interpret. When the Dirac equation is solved for the energy of an individual electron, there are two answers, one positive, one negative, in a manner analogous to the way that the square root of a number can be either positive or negative. (The roots of 49, for instance, are 7 and -7, because both 7×7 and -7×-7 equal 49.) The negative answer was trouble: negative energy, itself a puzzling idea, implied, by $E = mc^2$, negative mass—an absurd impossibility. Ordinarily, physicists would assume that the world had started off with all electrons in positive energy states, in which they would remain, and no harm would come from the theoretical existence of negative energy.[43] The problem was that a positive energy electron should be able to emit a photon with enough energy to drop into a negative energy state. In fact, *most* electrons should end up with negative energy.[44]

Alas, "negative energy" is not an easily understood concept. Nobody knew what negative energy electrons could be. One theorist called them "donkey electrons," because they always do precisely the opposite of what you want.[45] Since energy and mass are equivalent, negative energy implies negative mass. What does it mean to say that a subatomic particle can have less than zero mass? As Dirac admitted, the whole subject "bothered me very much."[46] He wrestled with negative energies through all of 1929, trying to see if there was some way to keep the Dirac equation but get rid of the negative energy states.

And then [Dirac recounted later] I got the idea that because the negative energy states cannot be avoided, one must accommodate them in the theory. One can do that by setting up a new picture of the vacuum. Suppose that in the vacuum all the negative energy states are filled up. . . . We then have a sea of negative energy electrons, one electron in each of these states. It is a bottomless sea, but we do not have to worry about that. The picture of a bottomless sea is not so disturbing, really. We just have to think about the situation near the surface, and there we have some electrons lying above the sea that cannot fall into it because there is no room for them.

In other words, we don't notice the negative energy electrons because they are omnipresent—as undetectable to us as water is to a fish. But wait:

There is, then, the possibility that holes may appear in the sea. Such holes would be places where there is an extra energy, because one would need a negative energy to make such a hole disappear.[47]

Because of quantum randomness, in other words, photons should hit some of the electrons in the sea with enough energy to make them jump out of it— that is, become positive energy electrons—leaving a hole in their former location. This hole would appear as a sort of "opposite" electron: positively charged, because it corresponds to an absence of negative charge. Therefore, Dirac realized, even if he accommodated the negative energy electrons, he was forced to predict the existence of particles just like the electron, except with a positive electric charge.

Here Dirac's courage failed him for one of the few times in his career. He refused to follow the naked mathematics. Although logically the hole particles must have the same mass as electrons, because they are created by them, Dirac's paper on the subject said they must somehow be protons— which are almost two thousand times heavier.[48] He was also attracted by the hope of thus explaining both known elementary particles with the same theory. (The neutron would not be discovered for another two years.) Dirac's suggestion was promptly flattened by, among others, the mathematician Hermann Weyl, who reported from Göttingen that whatever the holes might be, they could not be protons.[49] Unafraid of *Dirac's* theory collapsing, Weyl said that if Dirac wanted to salvage his quantum electrodynamics, he was going to have to predict a previously unheard-of type of matter with the same mass but the opposite charge as an electron.[50] Moreover, these new particles should be all over the place. Not one had ever been noticed.

No matter how good a theory might look on paper, Dirac thought in dismay, if it predicts particles that don't exist there is something terribly wrong. Discouraged, he even wondered if all of quantum field theory might have to be scrapped. "I thought it was rather sick," he said afterward. "I didn't see any chance of making further progress."[51]

It is pleasing to report that the answer came quite literally from the clouds. Decades before, Charles Thomson Rees Wilson, a young and impoverished Scotsman with a scholarship at Cambridge, had gone to the highest mountain in the Highlands, Ben Nevis, where there is a meteorological observatory. He had been entranced with the beauty of the clouds surrounding the hilltop, the shimmering rings that suddenly burst into colored existence when the sun burst through the cumuli, or the great shadows cast by the peaks on the foggy masses below.[52] When Wilson returned to the Cavendish with the announced intention of studying the process of cloud formation, J. J. Thomson was willing to let him try to make clouds himself in the Cavendish. The fruit of these studies, made between 1896 and 1912, was the cloud chamber, which for decades was the primary means of studying subatomic particles.

Clouds, Wilson knew, had something to do with water vapor in the air. The amount of water vapor that a given volume of air can hold depends on its temperature; it can contain more when hot than when cold. Wilson also knew that if a volume of air suddenly expands, its temperature drops. When the compressed gas in an aerosol can is released, for instance, it expands out the nozzle and chills immediately—the reason that spray deodorant feels cool on the hottest of summer days. While Thomson and others at the Cavendish puzzled over radioactivity and X rays, Wilson constructed glass tanks with ingenious pistons tightly fitted into the inside walls. Wilson discovered that if he sprayed the air inside his tank with mist until it could absorb no more water, then suddenly pulled out the piston, expanding and cooling the air, the surplus water vapor "fell out" and formed a miniature cloud.[53]

The condensation was started by droplets of water collecting around dust particles in the air. Wilson tried cleaning the particles out of the chamber to see if that prevented clouds from forming. It did not, although the clouds emerged only if the piston were withdrawn a lot more.[54] Now what was the water coagulating around? In February of 1896, less than two months after the discovery of X rays, Wilson shot a beam of them into his glass tank. It was much easier to make clouds. Wilson was sure that the X rays smacked into air molecules, creating the charged objects known as ions, and the ions caused the clouds.

A year later, when J. J. Thomson discovered the electron, it would be clear that the X rays knock loose electrons, which are easily swept onto tiny droplets of water in the air. When several electrons collect on one droplet, the net charge begins to exert an effect on nearby molecules of water vapor. Water molecules (H_2O) are shaped like a broad V, with the oxygen nucleus at the vertex and the two hydrogen nuclei—protons—sticking out the ends. The electrons of the hydrogen atoms are yanked toward the oxygen atom by its much bigger, positively charged nucleus, which means that the tips of the arms of the V tend to end up with a net positive charge. The positive charge is easily drawn toward the negative electrons on the droplet, with the result that in rapid order a lot of water molecules are pulled to the charged droplets, gas condenses to form liquid, and the number and size of the droplets grow. A tiny cloud is born.

With the excitement in the Cavendish caused by radioactivity, it was soon discovered that water vapor would also condense around the electrons left behind by alpha particles speeding through the chamber. For a few instants—long enough for a photograph to be taken—the paths of these particles are visible as slender white lines that arc across the chamber like the contrails of a jet squadron. Alpha particles leave thick tracks, for these slow and doubly charged entities dislodge many electrons; the electrons from beta radiation have a smaller charge, are much faster, and leave light, wobbly tracks reminiscent of scratches from a cat's claw. If a magnet is wrapped around the chamber, particle identification is even easier: The magnetic field forces negatively charged electrons to curve in one direction, whereas positive alpha particles and protons are sent off in the other. The degree of curvature earmarks the particle's momentum. Thus Wilson could put a lump of uranium by his cloud chamber, pull out the piston with a thump, quickly click the camera shutter, and produce a photograph of the tracks left by the spray of radiation. Using a ruler and compass, he could discern that such-and-such a thick line that curved across the photographic plate in such-and-such a direction was left by an alpha particle with such-and-such momentum, whereas a light little line was a beta electron with a different momentum. Cloud chambers soon became standard equipment in laboratories where radioactivity was studied, and Wilson was awarded the Nobel for his invention in 1927.

In 1930, Robert Millikan, the head of the laboratory at the California Institute of Technology, asked Carl D. Anderson, a fresh young Caltech Ph.D, to build a new cloud chamber for use in studying cosmic rays, the recently discovered high-energy radiation that bombards the earth from space.[55] With the cloud chamber, Anderson built an electromagnet so powerful that the laboratory lights dimmed when the cloud chamber was in use.[56] For the next two years, Anderson expanded the chamber at random times, photographed the tracks of any cosmic rays that happened to pass through, and studied the results.

The great majority of the photographs were blank—no cosmic rays came through. But from the very beginning Anderson also started to find that a small number of the photographs had something strange, tracks from light particles that could either be negatively charged particles traveling upward or positively charged particles traveling downward. Anderson wrote later:

In the spirit of scientific conservatism we tended at first toward the former interpretation, i.e., that these particles were upward-moving, negative electrons. This led to frequent and at times somewhat heated discussions between Professor Millikan and myself, in which he repeatedly pointed out that everyone knows that cosmic-ray particles travel downward, and not upward, except in extremely rare instances, and that therefore, these particles must be downward-moving protons. This point of view was very difficult to accept, however, since in nearly all cases the [thickness of the track] was too low for particles of proton mass.[57]

To settle the argument with Millikan, Anderson inserted a metal plate in the center of the chamber. Any particle passing through the plate would lose momentum, slow down, and thus be bent more sharply afterward, thereby indicating its direction. Furthermore, Anderson could calculate the particle's mass from the momentum lost passing through the plate.

On August 2, 1932, Anderson obtained a stunningly clear photograph that shocked both men.[58] Despite Millikan's protestations, a particle had indeed shot up like a Roman candle from the floor of the chamber, slipped through the plate, and fallen off to the left. From the size of the track, the degree of the curvature, and the amount of momentum lost, the particle's mass was obviously near to that of an electron. But the track curved the wrong way. The particle was *positive*. Neither electron, proton, or neutron, the track came from something that had never been discovered before.[59] It was, in fact, a "hole," although Anderson did not realize it for a while.[60]

The identification was made by two chagrined English experimenters—chagrined because they had actually seen the particle in their apparatus before Anderson and had been told by Dirac what it might be, but had waited much too long to be sure.[61] Anderson called the new particle a "positive electron"; *positron* was the name that stuck.

Positrons were the new type of matter—antimatter—Dirac had been forced to predict by his theory. (The equation, he said later, had been smarter than he was.) Physicists soon realized that electrons and positrons would annihilate each other when they met, producing two photons—two flashes of light. Similarly, a photon going through matter could split into a virtual electron and a positron.[62] From an embarrassment the negative energy states were transformed into a triumph for a quantum electrodynamics, the first time in history that the existence of a new state of matter had been predicted on purely theoretical grounds. Dirac won the Nobel Prize in 1933; Anderson went to Sweden three years later.

The canonization of the new developments occurred at the Solvay Conference of 1933, in which the evidence for both the positron and Chadwick's just-discovered neutron was discussed eagerly by Bohr, Curie, Dirac, Heisenberg, Pauli, and the rest of the luminaries of subatomic physics. (The Solvay Conferences were periodic gatherings of prominent physicists that were paid for by a rich Belgian industrial chemist named Ernest Solvay.) With a negative, positive, and neutral particle of real matter and the discovery of antiparticles, the mood was of extraordinary excitement. A form of hydrogen, called deuterium, had been discovered which had a proton *and* a neutron in its nucleus, and scientists were beginning to build machines, called particle accelerators, that promised to unlock even more discoveries. Some physicists thought (again) that the subject was rushing to a close.[63] One of the few sour notes came from Ernest Rutherford, who remarked of the positron that he "would find it more to [his] liking if the theory had appeared *after* the experimental facts were established."[64] Nonetheless, despite the feverish pace of development and the triumph of the positron, there was one idea shared unanimously among the conference attendees: Something was very wrong with quantum electrodynamics.

6
Infinity

WOLFGANG PAULI READ THE FIRST OF DIRAC'S TWO PAPERS ON THE ELEC-
tron with characteristic care as soon as the *Proceedings of the Royal Society*
came in the mail. Pauli understood immediately that an equation for a single
electron floating through space could be nothing more than a starting point,
for most situations in the real world involve many electrons interacting with
each other; he seems also to have realized that the constant presence of
virtual particles in the subatomic domain made the very idea of talking about
one particle by itself unrealistic—unphysical, in the scientist's phrase. Pauli
promptly dispatched a letter to Dirac informing him of the necessity to
formulate quantum electrodynamics without such a dubious assumption. In
the middle of a discourse on the version of quantum electrodynamics that he
and Heisenberg were working on, Pauli broke off to ask, "I would like to ask
your opinion about what is essentially a physical difficulty that Heisenberg
and I have run into and can't get around." It seemed that their calculations
were being thrown off whenever "the problem of 'a particle interacting with
itself' rears its ugly head."

In certain conditions, an electron is affected by the electromagnetic
field it gives off, in somewhat the way that a boat can be rocked by its own
wake, or an airplane shaken by the sonic boom it has created. Working ever
deeper into the thickets of quantum electrodynamics, Pauli and Heisenberg
had become increasingly concerned at the apparent inability of the theory to
account for "a particle interacting with itself." A satisfactory theory should be
able to deal with this phenomenon. As far as Pauli could see, quantum
electrodynamics instead produced nonsense: "If a single electron is present,
and its energy is taken into account, . . . you get an ever-expanding (even
infinitely increasing) equation."

In other words, quantum electrodynamics apparently predicted that
the interaction of an electron with its own field would be infinitely strong—
as if a plane were smashed to bits in its own sonic boom. The same odd
prediction was tucked into Dirac's version of the theory. Pauli believed that if
an equation that is supposed to produce a definite answer instead gives an
infinite result, then something is wrong. Worried, he asked Dirac: *"What do
you think about this?* I don't have a satisfactory way out yet. And I even think
we will have to make fundamental changes in our perspective if we are going
to avoid these difficulties."[1]

In a sense, Pauli's letter was old news. The interaction of a particle with its own electromagnetic field is known as its "self-energy," and the self-energy of the electron had puzzled physicists ever since the particle's discovery in 1896. Initially, the difficulties were due to the problems inherent in applying physics rules derived from ordinary objects with definite shapes and sizes to the electron, an entity which apparently has neither. As far as is known, electrons are points in space, without breadth, width, or depth. In high school physics, students learn that the way to figure out the energy of the field produced by an electrically charged object—the metal ball of a Van de Graaf generator, say, which shoots out the sparks in Frankenstein movies—is to add up the energy of the field at every point in space. Carrying out this measurement is next to impossible, but nineteenth-century physicists came up with an easy way to get the answer nonetheless.[2] The energy of an electric field at any individual point is equal to the electric charge creating the field (which is often indicated by the letter e) divided by the fourth power of the distance (written r^4) between the point being measured and the center of the field. The whole thing is multiplied by $\frac{1}{8\pi}$. In textbooks, the equation looks like this:

$$E_{\text{point}} = \frac{1}{8\pi} \frac{e^2}{r^4}$$

Using the techniques of calculus, physicists can add up all the little individual points to get the total energy, which is

$$E_{\text{total}} = \frac{e^2}{a}$$

where a is the radius of the spherical object that is producing the field. With this equation, it requires only thirty seconds and a hand calculator to find the total energy of the field—provided you know what a and e are. Simple! Except that the formula does not work for electrons. Electrons are points, and thus a is zero. It is simply not possible to divide by zero; the answer becomes infinitely large. Thus, physicists discovered that their equations, which work very well for Van de Graaf generators and other big things, predicted that infinitely close to tiny charged objects are infinitely large electric fields.

The puzzle of infinite self-energy was augmented by another oddity of physical theory, electromagnetic mass, which surfaced when scientists still believed that electromagnetic waves travel through the ether, a mysterious fluid that was once supposed to permeate all of space. In 1881, J. J. Thomson noted that the ether would resist the passage of an electrically charged object in much the way water resists when you swish your hand rapidly around in a bathtub.[3] Just as your hand feels heavier from the drag of the liquid, the charged body acts as if its mass had increased due to the interaction of its electric field with the ether. This extra mass is today called "electromagnetic mass."

The same year that Thomson did his calculations, Albert A. Michelson began the first of a series of experiments that would establish definitively the nonexistence of the ether. Theorists continued to work on the concept of electromagnetic mass, and around the turn of the century several realized that even without the ether *the same effect still happens*; in a vacuum, a charged particle effectively acquires mass because of its interaction with its own field, that is, its self-energy.[4] Clearly, if the field close by a point electron is endlessly large, as physicists' equations suggested, then its electromagnetic mass must also be infinite. Both conclusions are patently false.

In an ingenious but inelegant argument published in 1904, the great Dutch theorist Hendrik Lorentz tried to dispose of the two infinities by taking the measured mass of the electron, assuming it to be entirely of electromagnetic origin, and feeding it into formulas for electromagnetic mass.[5] Cranking the calculational handle, he came up with what the energy of the field would have to be to generate the electron mass experimenters measured in their equipment. He then plugged this number into the equation $E = e^2/a$, and came up with a radius for the electron of two hundred and eighty quadrillionths of a centimeter, or, in scientific notation, 2.8×10^{-13} centimeters. Lorentz knew enough about relativity to know that objects become pancake-shaped and do other odd things when they travel at high speed, and he fashioned his model accordingly. But accounting for relativistic phenomena made it hard to apply the electromagnetic formulas consistently, and the answers he found depended on his method of calculation. Such mathematical inconsistency is the hallmark of a diseased theory, and for the next several decades the infinite self-energy remained inexplicable.

After the birth of quantum mechanics, physicists hoped that if they worked on other parts of quantum theory, the self-energy infinities might somehow disappear. Such hopes were not as foolish as they may seem. The self-energy complications occurred because of the point nature of the electron, and the idea of a point—a position, in other words—was much changed by quantum mechanics.[6]

Pauli, too, at first nourished this hope. Perhaps, he thought, by juggling around the new equations of quantum electrodynamics, the self-energy could be eliminated.[7] When he discovered that it could not, an alarm sounded: An infinite answer is an infinite answer, and something was terribly wrong. When he wrote Dirac that physicists needed to make "fundamental changes in their perspective," he was intimating that quantum field theory might have to be replaced by something else.

The impact of Pauli's argument had as much to do with the real diffi-culties with self-energy—difficulties which were to grow ever more intracta-ble the more physicists examined them—as with his remarkably forceful personality. Pauli read everything, knew everyone, and wrote thousands of elegantly phrased, passionate, acid-tongued letters to everybody he knew. His criticism was often brutal, a chill wind that occasionally had the effect of freezing out good ideas but killed off many more bad ones; if an idea was useful, Pauli felt, it would survive despite him. His fierceness was legendary, and no one was spared; as a student, he is reported to have astounded his classmates at a seminar by beginning a question with the words, "What Professor Einstein just said is not so stupid." Pauli told one assistant, "Your duties will be very light. The only duty you will have is to contradict me when I propose anything, with all the facts at your disposal." He terrified Born and badgered Bohr; he sometimes signed his letters, "The Wrath of God."

His intuition was more critical than creative. Although he had many good thoughts of his own, he preferred to place the work of others in perspec-tive, to set the research agenda, to root out confusion and lack of clarity where he found it. For three decades, he was the arbiter of the science, a position that suited a man with a thirst for judgment. In 1958, when Pauli died after a short illness, one of his colleagues called him "the conscience of physics."[8]

While still a twenty-year-old student, Pauli established a reputation by writing a definitive presentation of relativity for the *Encyklopädie der Math-ematischen Wissenschaften*, an important encyclopedia of science.[9] Four years later, in 1925, he formulated the "exclusion principle," which says that no two nearby electrons can be in exactly the same state.[10] One of the primary laws of quantum mechanics, the exclusion principle helps dictate the complicated patterns formed by the electrons in the atom. Because the chemical properties of an element derive from the behavior of its orbiting electrons, the exclusion principle is one of the pillars of chemistry. For its discovery Pauli won the Nobel Prize in 1945.

After 1925, Pauli worked almost constantly with Heisenberg, a collab-oration that reached one of its peaks while they pieced together their version of quantum electrodynamics. As they wrestled with the theory, the two men kept trying to rid the equations of infinities—"divergences," as mathemat-icians call them. Heisenberg and Pauli thought it possible that even if quan-tum mechanics alone had not done the trick, a properly formulated quantum field theory might avoid the divergences of the past. As the two theorists worked on their long and painful proof of the relativistic invariance of quan-tum electrodynamics they began to suspect the existence of new, more intractable infinities. The prospect was so dark that Pauli threatened to quit physics and live in the countryside, writing utopian novels.[11] Heisenberg left for a lecture tour of the United States. Gloomy and depressed, Pauli explained the problem to the assistant who helped him complete the paper.

The assistant was J. Robert Oppenheimer. Born into a wealthy family, he was a sickly, precocious, over-indulged child with a stutter.[12] He was given the best education money could buy, and he dipped intermittently but deeply into religion, languages, poetry, and music. Eventually he became known in the community for his fascination with Hinduism and the *Bhagavad-Gita*.[13] He was extraordinarily quick—fatally so, for his physical intuition was all too frequently spoiled by arithmetical errors. As an adult, he grew to have a magnetic, rumpled charm; when Oppenheimer became head of the Manhattan Project, he ran the Los Alamos laboratory with such commitment and enthusiasm that many of the physicists there remember constructing the atomic bomb as one of the high points in their lives. He raced through his thoughts, speaking in a rapid tumble of erudite phrases and ironic slang that annoyed his teachers but enchanted his fellow students. When Oppenheimer graduated *summa cum laude* from Harvard in 1925, he went immediately to Europe—the dream of every U.S. physicist—where he studied with Rutherford for more than a year, toured the Continent, and received his doctorate from Göttingen. He went back to Europe in 1928, working under Paul Ehrenfest in Leiden. Ehrenfest soon discovered he could not abide Oppenheimer.[14]

Oppenheimer was an impatient, neurotic young man with brilliant prospects and occasional fits of generosity. Ehrenfest begged Pauli to do something with Oppenheimer. To prevent the situation in Leiden from exploding, Pauli agreed to have the young American come to Zürich to work with him.[15] Pauli thought that Oppenheimer was full of ideas, but all too prone to shoot off on wild, poorly thought out tangents. "Oppenheimer's physics is always interesting," Pauli told his friends. "But Oppenheimer's calculations are always wrong."[16] Still, something certainly could be done with him. He needed a steady problem—like working with infinities.

The exact problem Pauli gave Oppenheimer concerned that favorite subject of physicists, the spectrum of light emitted by a hydrogen atom. When Heisenberg and Pauli produced their version of quantum field theory, they, too, ended up with the Hamiltonian:

$$H_{total} = H_{field} + H_{particle} + H_{interaction}$$

Simply adding the respective Hamiltonians for the field and the particle and ignoring the interaction was enough to account reasonably well for the spectrum of hydrogen. Indeed, this was a triumph for the theory. But there remained the last term, the interaction Hamiltonian, which includes the self-energy. In quantum field theory, the self-energy of a charged particle can be thought of as what occurs when the particle emits a virtual photon and quickly reabsorbs it—a process that happens fast enough to stay within the bounds of the uncertainty principle. Because there are an infinite number of ways in which this emission and reabsorption can occur, the self-energy is not easy to evaluate. Nonetheless, it should have some effect on the behavior of the electron in the atom. Because the spectrum of hydrogen is produced by its electron jumping from one state to another, the energy level changes induced by emitting and absorbing virtual photons should displace the spectral lines ever so slightly. It was this displacement that Oppenheimer set out to calculate.[17]

When Oppenheimer came back to the United States at the end of 1929, he accepted a joint appointment at the California Institute of Technology, in Pasadena, California, and the University of California at Berkeley, outside San Francisco. His first published paper upon returning was the self-energy calculation.[18] Modestly entitled "A Note on the Theory of the Interaction of Field and Matter," the article claimed to show that if quantum electrodynamics was true, the endless number of possible virtual photon interactions should lead to an infinite displacement of the spectral lines of hydrogen—that is, the theory had the ludicrous consequence of predicting that there should be no lines at all because the self-energy effects would knock the electron to kingdom come. Appearing a month after the final formulation of quantum electrodynamics by Heisenberg and Pauli, Oppenheimer's paper seemed both to confirm their worst fears and to expose a terrible flaw in the whole enterprise. A theory may be wrong and still be useful. It cannot be ridiculous.

Few things are more difficult to understand than a brilliant man without a voice. Oppenheimer was strikingly perceptive, but not careful; thoughtful, but not original; decisive on the small scale, but paralyzed by doubt over the long run. Although he accomplished much, this great man never quite made the fundamental contributions to knowledge that his gifts promised. His career was brilliant, but Oppenheimer's talent and energy were such that merely brilliant seemed disappointing, and his very real accomplishments were always shadowed by his even larger potential. When a challenge came from the outside, he met it nobly and well: He built the bomb; he created the American school of theoretical physics; he fought cancer until, in 1967, it killed him. But scientists—good ones, anyway— must have their own visions of nature, and how to tease out her secrets. Oppenheimer did not have his own agenda. It is not easy to say why. The physicist Abraham Pais, a friend for more than fifteen years, told us once that Oppenheimer was in Pauli's company at a time when Pauli was particularly pessimistic and Oppenheimer was particularly impressionable. Clasping

Pauli's received discontent to his bosom, Oppenheimer simply never became his own man.[19]

In any case, Oppenheimer was convinced throughout the 1930s that quantum field theory was not good enough, not deep enough, to penetrate the secrets of matter and light. His concern infected his students, and some of them, later, felt that it slowed the progress of science. Certainly he was despondent in 1930. From his self-energy calculation he could have concluded merely that the infinities were an unexpected blemish on the surface of field theory. Instead, he took them to mean that physics was wholly off track.

Unfortunately, Oppenheimer missed a wonderful opportunity, as did all the other physicists who read his brilliant and influential paper. The dictates of quantum mechanics, and in particular the Pauli exclusion principle, mean that electrons can only reside in particular "neighborhoods" around the nucleus. Hopping from one neighborhood to another is accomplished by emitting or absorbing a photon. Over the years, these neighborhoods, which scientists call "orbitals," have been exhaustively catalogued; tables of orbitals fill fat appendices in atomic physics textbooks. As it happens, some orbitals have different quantum numbers but the same energy. The $^2S_{1/2}$ ("two-ess-one-half") and $^2P_{1/2}$ orbitals of hydrogen, two neighborhoods occupied fairly frequently by the atom's single electron, are examples of this phenomenon.[20] If Oppenheimer had thought about the $^2S_{1/2}$ and $^2P_{1/2}$, he could have subtracted one identical energy from the other, thus getting rid of all the infinite terms in the equation and leaving a finite residue that can be calculated exactly and matched against experiment as a test of the theory's validity. Had Oppenheimer thought of this, he might not have been so distrustful of quantum electrodynamics, and the recent history of physics might have been considerably different.

Three years later, Oppenheimer and a postdoctoral fellow, Wendell Furry, put together a lengthy, overcomplicated proof demonstrating that a simple change in notation would allow one to forget about the infinite sea of negative energy electrons and simply talk about electrons and positrons.[21] But even here, while restoring some degree of visualizability to the subatomic world, the authors are almost gleefully pessimistic about quantum electrodynamics. Getting rid of Dirac's sea, they note, does not get rid of the self-energy difficulties, "which rest ultimately on *an illegitimate application of the methods of quantum mechanics to the electromagnetic field.*"[22] Having disposed of all quantum field theory in a phrase, they claim the divergences "are not to be overcome merely by modifying the [equations for the] electromagnetic field of the electron within these small distances, but require here a more profound change in our notions of space and time. . . ."[23]

Oppenheimer's gloom was more than habitual fatalism; in that paper,

Furry and he had come across perhaps the strangest consequence yet of quantum field theory, the interaction of electrons with the vacuum—that is, with *nothing*.[24] Nine thousand miles away, Dirac had independently explored the subject a month or two earlier, but with even more distressing results. He stunned his colleagues at the seventh Solvay Conference, held in Paris, with a demonstration that quantum field theory seems to predict that the interaction between the electron and the vacuum should have the peculiar result of reducing the total charge of the particle to zero.[25]

Delivered in the perfect French that was a legacy of his father, Dirac's address began by recapitulating the hole theory, which still confused many if not most of the physicists who encountered it. Taking his audience through the whole chain of logic, he demonstrated that his "holes" really did correspond to positrons. He then pointed out that an electron could "jump into a hole, which it would then fill up. We then have an electron and positron annihilating one another," producing two photons with the combined energy of both particle and antiparticle.[26] Similarly, a photon passing through matter could split into an electron and a positron, each with half the energy, momentum, and spin of the original photon. Therefore, the vacuum, which can, according to the uncertainty principle, randomly produce energy in the form of virtual photons so long as it doesn't stay around long enough to be observed, may also create pairs of virtual electrons and virtual photons.

And even the virtual particles, the evanescent electron-positron pairs, can have real effects. Dirac asked his audience to consider an electron floating through empty space—or, rather, not-so-empty space. At any given moment, the particle is surrounded by a swarm of virtual photons, electrons, and positrons, buzzing like ghostly bees around the particle in its progress. In the brief moment of their existence, the virtual positrons, which are positively charged, are pulled toward the real electron; the virtual electrons are in the meantime repelled. The result is that the electron has a cloak of positrons; it is surrounded by a shimmer of ghostly antimatter. As the distance becomes shorter, the number of virtual particles grows, in accord with the uncertainty principle. The result is that extremely close to the electron, its charge is blanketed by an ever-increasing, indeed endless snarl of, virtual positrons. Dirac called the process of attracting virtual positrons and repelling virtual electrons the "polarization" of the vacuum; it was yet *another* infinity in quantum electrodynamics.[27] (This infinity has a completely different source than the self-energy infinity, but physicists often include the various divergences in quantum electrodynamics under the loose rubric of "the self-energy problem.")

Dirac was far from giving up. The presence of vacuum polarization implied that whenever experimenters measured the charge of an electron, they never measured the "real" charge, which self-energy rendered infinite, but the "real" charge *minus* the offsetting infinite vacuum polarization. Nature, in other words, had subtracted one infinite quantity from another, and obtained a finite result. Although subtracting one infinity from another is fraught with mathematical ambiguity, Dirac proposed that theorists do likewise, and balance the divergences in their equations against each other. It wasn't a solution—infinite quantities were still present, and no one knew how to get rid of them—but theorists were ready to embrace any technique that would even sweep them under the rug.

Although Pauli ridiculed Dirac's "subtraction physics" and the arabesque of infinities in the hole theory, it was the only game in town.[28] By this time, he had a new assistant, Victor Weisskopf, whom Pauli had selected when he could not get another young man named Hans Bethe. After baldly explaining that Bethe had been the first choice, Pauli gave Weisskopf a little problem. About ten days later, he asked for the results. Weisskopf showed them to him. Pauli said, "I should have taken Bethe!" It was the beginning of a friendship.[29] Viennese like Pauli, Weisskopf was an intuitive thinker, a scientist who sought physical reasons rather than correct formal expressions. Pauli must have liked his new assistant's way of working, for he gave Weisskopf a fundamental problem: Calculate the self-energy of the electron, but now take into account the hole theory and the offsetting vacuum polarization discovered by Dirac. Weisskopf finished the figuring early in 1934, and the result was bad news. The self-energy was just as divergent as before.[30]

Shortly after the paper appeared, Weisskopf had the dreadful experience of receiving a letter from Wendell Furry, Oppenheimer's collaborator at Berkeley, that gently informed him of a foolish mathematical error in his work. Once the mistake was removed, the answer changed dramatically; the self-energy was still divergent, but only logarithmically. This was vastly different. The logarithm of a number—say, 100—is the power to which a second number (often 10, because we use the decimal system) must be raised to equal the first. Thus, the logarithm of 100 is 2, because $10^2 = 100$. The logarithm of 1,000 is 3, because $10^3 = 1,000$, and the logarithm of a trillion (10^{12}) is 12. Clearly, a very big number can have quite a small logarithm. Furry pointed out that the self-energy did not depend on x^N, but on $\log_x N$—and that this crawled up to infinity at the pace of the proverbial snail.

Weisskopf today is an emeritus professor at the Massachusetts Institute of Technology, in Cambridge, Massachusetts, a former director-general of CERN, the giant international particle physics laboratory in Switzerland, and one of the founding fathers of nuclear physics. He was, to his sorrow, intimately involved in the effort to build the first atomic bomb, and for years has been a prominent advocate of arms control. A distinguished career did not seem likely to him in 1934, when he received Furry's letter. He was humiliated by his error. "Pauli gives me a fundamental problem to solve and I do it wrong," Weisskopf told us in his small office at CERN. "I went to Pauli and said, 'Pauli, I have to give up physics! It's terrible!' And he said, 'Oh, no, don't worry. There are many people who made mistakes in their papers—I, never.'"

We laughed and asked if Pauli really had said that.

"'I, never,'" Weisskopf insisted, smiling at the memory. He paused for a moment, thinking; many of his sentences were graced with small, reflective rests. Then he said, "I find it wonderful, you know. It shows the courtesy of physicists at that time, that Furry didn't publish. He merely wrote to me. I published a correction, of course, thanking him.[31] But that wouldn't happen today. Today, the corresponding Furry would publish a paper and say that I was wrong. So this is why on every possible occasion I say how grateful I am to Furry. Because some of my fame comes from this—it's always connected with my name—and actually it was Furry."[32]

Heisenberg, for his part, took little comfort in the shrinking of the divergences.[33] Infinity is still infinity, he thought, and quantum electrodynamics will remain a cheat until the divergences are gone. Heisenberg took his turn at exorcising them. He still hoped that if he set down the full and complete theory of quantum electrodynamics without using any of the approximations dear to theorists, the infinities would cure themselves. The resultant paper, "A Remark on Dirac's Positron Theory," is one of his greatest, but one of his least conclusive.[34] Unlike anyone before, he explicitly wrote down the quantized field equations for the electron and the electromagnetic radiation—both at once—and the interaction. With this article, modern quantum field theory was born, although Heisenberg derived little satisfaction from it.

Long, knotty, awkward, and intense, the article contained in a crude form the rules by which physicists can subtract infinities one from another. There are several such rules, the most important being that the calculations cannot be just a little relativistically invariant; they have to take relativity into account every step of the way.[35] ("Relativistic invariance," again, means that the quantities in an equation are completely faithful to the dictates of general relativity.) Conscientiously subtracting, Heisenberg tried to deal with all of the divergences that grew like weeds in the crevices of the theory: the infinite field produced by the sea of negative electrons, the infinite mass from self-energy, the infinite vacuum polarization, and yet another paradox, the inability of a particle surrounded by an infinite cloud of virtual particles to interact with anything else.[36] But after a lengthy journey through the brambles of the theory, he was still in divergent terrain. When he treated the matter field and the electromagnetic field on an equal basis, he found the same infinities in both—the photon, like the electron, had self-energy problems. The reward for writing down the full theory, treating matter and energy on an equal footing, was discovering the same set of divergences in both. Dismayed by Heisenberg's results, Pauli complained that he was "drowning in the Heisenberg-Dirac hole equations, and would grasp at any straw that was offered."[37]

As striking as the noble mess of Heisenberg's paper is the fact that it took seven years from when Dirac first quantized the electromagnetic field to the first time anyone tried to write down the theory in a complete way. Why the delay?

"It shows how *crazy* the whole theory was considered," Weisskopf said in Geneva. His fingers tapped softly on a copy of a speech he had given recently in opposition to the militarization of outer space. "I mean, the filling of the vacuum and Dirac's theory of the positron—nobody really believed it at the time. And the theory was so ugly! All these absurd ideas and this terribly complicated mathematics, it all looked very ugly. In a sense, I suppose the problem [of the infinities] seemed academic because for a long time nobody knew where the physics was in it. And a lot of people just said, 'To hell with it, I'm going to do some nuclear physics.' "[38]

On a summer afternoon not long ago, we walked through the crowded city campus of Columbia University toward the malachite dome of Pupin Hall, the seat of the physics and astronomy departments. Pupin Hall has several floors of dusty laboratories, a small observatory on the roof that is nearly blinded by the lights of Manhattan, and a grim brick facade that stands in crabbed contrast to the Gothic reaches of nearby Riverside Church. Baffled by the plethora of infinities, we turned again to Robert Serber, one of the handful of physicists during the 1930s who came very close to solving the divergences of quantum field theory.

Serber met Oppenheimer for the first time in 1934, at a summer physics seminar at the University of Michigan, in Ann Arbor. Like many of his contemporaries, Serber was immediately enthralled by Oppenheimer's intensity and articulateness. "No one thought quicker than he did," Serber told us. "People used to say that you never finished phrasing a question to Oppenheimer, because he'd answer before you got to the end of your sentence." Hadn't Pauli said that his calculations were always wrong? "That's because he would whip through something in five minutes on the back of an envelope, doing just the essential calculations, and leaving out all the numerical terms. Somebody like Dirac would take a few hours and get all the twos right, and the pis in the right place." Oppenheimer was marvelously, poetically impatient; rushing along in an enthusiastic wave, he didn't like the painful details such as whether there was a three in the numerator or denominator.[39]

For the first few days at Ann Arbor, Serber kept an awed distance.

Then the participants of the summer school were all invited out for drinks to the home of an Ann Arbor physics professor named David M. Dennison. Upon their arrival, the company discovered that Dennison had taken Prohibition seriously, and that "drinks" meant lemonade. In that instant of shock, Serber happened to catch Oppenheimer's glance, and both simultaneously rolled their eyes heavenward. Serber had been on his way to Princeton as a National Research Council Fellow; he decided to turn right around and follow Oppenheimer to Berkeley instead, where he arrived in September of 1934.[40] A decade later, he went with "Oppie" to Los Alamos to build a bomb; a few years after the mushroom clouds over Hiroshima and Nagasaki, he was hired by Columbia, where he has been ever since.

We asked him about Oppenheimer's first self-energy paper. How many of the divergences had he known about at that time?

"*Well,*" Serber said. He pulled a cigarette out of his breast pocket. It had been many years. Oppenheimer's paper lay on the desk; Serber looked at it for a minute, smoking, while the traffic rumbled below on 120th Street. "There are three basic divergences in electromagnetic theory. There's the proper energy, and there's the vertex term, and there's the vacuum polarization." He went to the blackboard, hunted around for a piece of chalk. "You have an electron coming along––I'll use a diagram." He drew a line across the board to represent the path of the electron, then a second, wavy line arcing out of the first and then falling back into it.

"This is an electron emitting a virtual photon and then absorbing it. That gives rise to the proper energy of the electron. The proper energy is a big part of what is usually lumped together as 'self-energy.' That's one infinity. Another one is if the electron interacts with a photon—say it gets scattered by a photon." He drew another electron line, but this one bent at an angle when the electron bounced off a photon. "Then you could put in a virtual photon like this." The chalk squeaked as he made a curved, wiggly line joining the two halves of the electron line, pre-and post-collision.

"That's a higher-order term. The electron emits a photon, the scattering takes place, and then the photon is reabsorbed—you see? Now, this has nothing to do with proper energy, although we didn't realize that for an awfully long time. This is the vertex correction. If the electron absorbed the photon before being scattered, it would be—" he tapped the first diagram "—part of the proper energy.

"The third type is vacuum polarization, which is . . ." Serber paused; vacuum polarization is not as easy to diagram. "A photon comes along and creates a virtual pair." Another wiggly photon appeared on the board, this one divided by a circle. "The polarization is that if you have a charge—" he quickly put a fat dot on the board to symbolize an electron "—its electromagnetic field produces not only direct photons, but photons that make pairs." For good measure, another few photons and pairs were added to the picture.

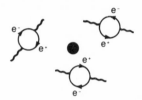

"So if this central point is a negative charge, the net effect is that you have a bigger—it will bring a positive charge closer and push a negative charge further way. So it's just like the vacuum were behaving like a dielectric." A dielectric is the material that keeps the positive and negative plates of a capacitor from discharging into each other. "That is, the negative charge here induces a polarization in the vacuum, the positive charges being attracted and the negative charges being pushed away." Properly speaking, only the first two are "self-energy," but in what follows we shall follow the sloppy habit of physicists and lump the vacuum polarization under the same name.

"In the complete theory, they are all mixed up with one another, so it took some time for everyone to understand that there are just these three primary processes, and many variations of them." To calculate the full self-energy in quantum electrodynamics, then, requires adding up the likelihood of these three interactions and each of their infinitely many variations and combinations, because all have a slight chance of occurring at any given moment. As there are an endless number of ever more complex self-energy diagrams, with ever-branching trees and loops of electrons and photons, evaluating their total contribution to the theory is a horrendous mathematical quagmire.

"That was the problem," Serber said. "Nobody knew how to evaluate these things directly—we still don't." Above his three diagrams, he wrote

$$H_\text{total} = H_\text{field} + H_\text{particle} + H_\text{interaction}$$

He put his hand across the final term, blocking it from view. "This is the one that caused the trouble," he said. "At the beginning, they just ignored it;

they really didn't know what else to do. You could actually calculate fairly precisely if you pretended the interaction term wasn't there, so you knew that even if it *looked* infinite the interaction term had to be fairly small. But the whole business was a mess. We knew that. And when we finally started to work with the whole theory, we said, 'Okay, we can assume that the interaction term, the term with the infinities, is really only a small part of the total. Then you can calculate the whole thing [H_{total}, in other words] as a slight change in the sum of this and this' "—the Hamiltonians of the particle and its electromagnetic field.

The calculations were and are extremely cumbersome. They can be done by using what is called a perturbation expansion, which works by viewing each possible self-energy interaction as a small change, a perturbation, of some known state. By adding up the infinite number of such small changes, a final answer can be obtained. However, this answer is finite only if the successive terms become smaller, or, more exactly, if the series of terms *converges* toward a particular number. If the successive terms become bigger, the series *diverges*, and you end up with infinity. In the early 1930s, the series seemed to diverge off the map.

When Serber came to Berkeley, Oppenheimer gave him his self-energy paper, the article he had written with Furry that got rid of the sea of negative electrons, and Heisenberg's just-published "remarks" on Dirac's subtraction theory—the papers everyone in the group was talking about. Much as the presence of a single energetic play director is sometimes enough to create an entire theatrical community in a city where there had been little before, Oppenheimer created a theoretical community in Berkeley, and lent his weight in addition to the Caltech and Stanford departments. In his office was always a coterie of awed graduate students and National Research Fellows, many of whom imitated the light cough, mumble, and penchant for incredibly hot food that were Oppenheimer trademarks. Oppenheimer had no desk; he worked on a table in the center of the room covered with paper and cigarette ash. A blackboard covered one wall, densely packed with the equations whose constants Oppenheimer never could remember. There were another six or seven people at nearby Stanford with Felix Bloch, a Nobel Prize-winner. Everyone congregated in 219 Le Conte Hall, Oppie's office, intent and informal young men with the dark narrow formal ties and thin-collared shirts of the time. A hundred miles outside of town was *Grapes of Wrath* country and the Depression at its fullest; the air in Berkeley bore the whiff of radical politics and Oppenheimer's blackboard chalk.[41]

Once a week, the Journal Club discussed what had come out during the past week in journals; at these meetings, the theorists and the experimentalists who were working on the first, beginning particle accelerators could make close contact. In late 1934 and early 1935, the Berkeley theorists heard reports at the Journal Club that some spectroscopists at Caltech were finding slight deviations from the predictions of quantum electrodynamics for the Balmer series of the hydrogen atom; it looked as if some of the lines were not where they were supposed to be. Oppenheimer suggested that Ed Uehling, another National Research Fellow, should use Heisenberg's subtraction techniques to see if vacuum polarization could account for the new data. Serber joined the discussions, and eventually found himself writing a companion paper. The two articles were published back-to-back by the *Physical Review* in July 1935.[42]

Uehling took Heisenberg's theoretical framework as a given, but he wanted to provide some of the numbers he thought Heisenberg had been too busy to get around to. Because of the polarization of the vacuum, as Uehling noted, "the energy levels for the [hydrogen] electron are slightly displaced. This displacement may be calculated . . . as a small perturbation," and Uehling then did the mathematics. The answer was anything but satisfying. Both he and Serber found that there would be a change in the spectral lines, but it was much too small and in the wrong direction from the experimental results.[43]

A few months later, the Berkeley group was delighted to host P. A. M. Dirac on one leg of a round-the-world trip. Knowing that Serber and another Oppenheimer protégé, Arnold Nordsieck, were working on a variant of the subtraction technique originally suggested by Dirac, Oppenheimer arranged for them to meet the great man. Elated at the opportunity, the two young Americans explained their work for over an hour in a small Berkeley conference room. Dirac uttered not a word. At the end of the talk, a lengthy silence ensued. Finally, Dirac said, "Where is the post office?"

Exasperated, Serber asked if he and Nordsieck could accompany Dirac to the post office. Along the way, perhaps, Dirac could tell them his reaction. "I can't do two things at once," Dirac said.[44]

Weisskopf, on the other hand, was extremely interested in what Serber and Uehling had done. In 1936 he left Pauli to work with Niels Bohr in Copenhagen, where he tried to further generalize the calculations for vacuum polarization.[45] He argued that if quantum electrodynamics is right, the infinite polarization obviously has to be there; equally obviously, it is not seen directly. In most situations, Weisskopf said, the "electron" seen in experiments actually consists of an electron with an infinite cloak of virtual positrons that shields most of its charge. But if another particle bores in very close to the electron, shouldering through the mass of positrons, it discovers that the real charge of the "bare" electron is much higher than it usually seems. In fact, the "bare" charge is infinite. The infinite charge is almost always hidden beneath an infinite shield of vacuum polarization, for, like the slight tremors that constantly ripple across the junction of the two straining sides of a geological fault, the small observed charge of the electron is just the visible result of two opposite and almost equal infinities. In quantum field theory, the world is full of infinite quantities that nearly negate each other, and nature is a balance of divergences.

Although published in a relatively obscure Danish journal, Weisskopf's paper attracted much attention. The arguments were in many respects similar to those of Heisenberg, Serber, and Uehling, but Weisskopf's clear style and simple presentation made them accessible to colleagues still uncomfortable with the complicated formalism of field theory. The spectre of vacuum polarization that Dirac had conjured up three years before at the Solvay Conference in Paris had been reduced to the trivial role of eliminating another, otherwise unobservable infinity.

One of the three infinities on Serber's blackboard had been killed off, but the other two were as alive as ever. In an article that sharply criticized Heisenberg's subtraction program, Serber set down the rules for calculating the proper energy.[46] He carried the calculations through at the lowest level, but did not publish them because he could see so many other divergences coming down the pike that he concluded the task was hopeless. Something else would have to be tried.[47]

We asked Serber one day why he had given up canceling infinities when from today's vantage he was so close to the answer. There were many reasons why, he said, shrugging, and he had not been the only one. People had been distracted. The formalism was difficult. And he, at least, had not clearly understood the difference between the proper energy and the vertex correction, between his first and second diagram. If the problem isn't identified correctly, it's insoluble. Did anyone make the distinction?

"Well, yes and no," Serber said. We were at a restaurant high above the wintry slopes of Morningside Heights. Glasses tinkled, dishes scraped, and Serber's meal lay almost untouched before him. "Heisenberg's paper at the time didn't describe the problem of the vertex infinities, although he may have been aware that they existed. Then the next thing was that [Felix] Bloch and [Arnold] Nordsieck solved the [low-energy vertex] difficulties, which really went a long way to show how you could do things.[48] And then Bloch and Oppenheimer tackled the vertex corrections on the high end. They didn't do the actual calculation themselves, they gave it to Sid Dancoff to do. Sidney had a fellowship with Bloch. We had some of the right ideas, distinguishing between proper energy and vertex infinities, but Dancoff made a mistake."[49] Serber pointed out the error, but Dancoff insisted it made no difference. It *did* make a difference, and Oppenheimer always felt afterward that he might have nailed all the divergences and maybe won a Nobel Prize if Dancoff had corrected his paper. "We were getting discouraged too easily.

Essentially two-thirds of it was solved by '37, but we didn't really understand that. Then there was still the vertex, and Sid Dancoff almost got that right. A lot of it was *there*. It just wasn't put together properly."

The process of shaking the divergences out of a theory is now known as *renormalization*—getting a theory back to making ordinary, finite predictions. The word may have been coined by the Berkeley-Stanford group; its first appearance in print is apparently in Serber's papers on proper energy.[50] "The interesting thing about renormalization," Serber told us over lunch, "is that the thing never happened all at once. Sometimes science works like that, sort of staggering along. There were a lot of steps and developments, but the *real*" —he laughed, interrupting himself—"the earlier steps didn't make much difference. The people who did it right later on didn't even know about them. Or I don't think they did; it's very hard to tell about things like that. Some things you hear in the summer and haven't understood, and then all of a sudden it'll come back to you three years later and you finally get the point. It may even be that the people who did it later heard our ideas—in the background, in a speech—and it didn't sink in until they used them later. I don't think I consciously had much influence on someone like Schwinger, when he was working it out."[51]

By 1937, European physicists had other things to think about than self-energy. Jewish scientists had to protect their lives and their families; many non-Jews were preoccupied with saving their colleagues. Bohr ran an "underground railroad" for young Jewish theorists, piping them from Copenhagen to England or the United States; Weisskopf, who went to the University of Rochester that year, was one of "Bohr's refugees."[52]

But physicists like to do physics, and a few tried to keep working on renormalization despite the war clouds scudding overhead. In retrospect, they were handicapped by their belief in the radically new. With a kind of hopeful despair, Heisenberg and the other founders of quantum physics hoped the infinities indicated the need for another revolution, another set of apparently crazy ideas to save the theory. They did not recognize that the science had shifted, and needed a different sort of originality, a different sort of perseverance. Heisenberg, for example, tried to resolve the divergences by arguing that the electron was not a point because space itself was quantized. There was a "universal length," and the continuum was grainy, like the dots in a half-toned newspaper photograph.[53] Pauli spent years trying to figure out another way to solve the equations of quantum field theory, but got nowhere.[54] Dirac convinced himself that trying to rid quantum field theory of its divergences was a case of working on too many questions at once, and came up with an infinity-free classical theory. To do it, he had to introduce "negative probabilities," events which had less than no chance of occurring, a concept that even Dirac could not make sense of.[55] Other theorists speculated that the self-energy effects might be canceled out by a second, previously unknown field invented by nature to do just that.[56]

By and large, physicists washed their hands of the problem. Something was wrong, but the theory seemed to produce more or less correct answers if the infinities were simply ignored. The difficulty was oddly abstract; the mathematics was convoluted; and many scientists were seduced away from the problem-by the emergence of nuclear physics, and the exhilarating, appalling discovery that the atom could be split.

The last man to deal directly with self-energy before the war was Victor Weisskopf. In a speech before a New York meeting of the American Physical Society and a subsequent paper, Weisskopf demonstrated that the "peculiar interaction between the electron and the vacuum" reduced the self-energy problems to logarithmic divergences not only in simple situations, but in the most complex interactions.[57] The divergence was incredibly weak, but it was still there. Like Lorentz four decades before, he tried to make an end run around the problem by using the infinity to tell him how big the electron radius would have to be. It was 10^{-58} times smaller than the classical electron radius—but it was still not a point. Four months after the article appeared, the Germans invaded Poland, and physicists, like everyone else, had other things to worry about than the size of the electron.

We asked Weisskopf one time what he had been missing in 1939. He was so close to renormalizing the theory. What had he lacked?

"Persistence." He laughed. "Persistence! I was much too much interested in other things. I always say, you know, because I really believe it made me miss the Nobel Prize, that I think I'd gladly pay the Nobel Prize for having this general overview of physics which I have from those years. I was interested in too many things. I prefer it this way. I prefer to know nothing about everything rather than everything about nothing. At that time life was not so easy, and there was a lot of time I had to spend helping refugees and things like that. If I had sat down and said, 'This is my purpose in life [renormalizing the theory],' and forgot everything else—all the nuclear physics and the other things—I probably could have done it. But I haven't got that persistence.

"I don't regret it at all. I always tell my students when they say they cannot work on physics in these terrible days that physics is what kept our equilibrium then. Physics was a wonderful thing! If it were not for physics, my character would have come *apart* at that time. I was a Jewish refugee, my family lived in terror—and physics was the great thing that kept us *human* at that time." He thought a moment, his face turned to the ceiling with a curious mixture of emotion. "But I would say, look, I'm not the type who can work on one thing. I'm just not persistent enough."[58]

7
The Shift

EVERY NOW AND THEN, A FACT, A SINGLE ASPECT OF NATURE, ASSUMES AN importance to physicists far outside of its intrinsic significance, and it makes or breaks theories, careers, reputations. After the contretemps is over, the fact becomes a footnote in future papers, and the irregular means by which it was brought to light is forgotten. The next generation of physicists learns the fact in graduate school as an accepted part of the world; it seems to have appeared in experimental equipment when required, its importance never in doubt, and to have been there always—had past scientists only possessed the wit to seek it.

Such a fact is the exact energy level of the $2S_{1/2}$ ("two-ess-one-half") electron orbital in atomic hydrogen.* One of the lowest, simplest excited states of the lightest and simplest atom, its location is a corollary of quantum theory. The study of this orbital was the subject of a dozen experiments during the 1930s, all of which bore directly on the worth of quantum field theory in general and the renormalizability of quantum electrodynamics in particular. But, because science is a human enterprise of fallible people, ideas were not put together, connections short-circuited, and the result was a slow tragicomedy on the experimental side that accompanied the contortions of the theoreticians. While theorists spent the 1930s alternating between pretending the infinities weren't there and fruitlessly trying to grapple with them, experimenters passed the same decade painfully trying to decide if the predictions of quantum electrodynamics could be confirmed. Self-energy was the root of both problems. At the end, quantum field theory seemed to be vindicated, and yet again physicists had the hope that this time they were at last on the road to a complete picture of nature. They were, but in a different way than they had imagined.

Many of the students of the hydrogen spectrum and the $2S_{1/2}$ wanted to measure a magic number known as alpha (α), which had nothing to do with

*The terminology for energy levels is somewhat antiquated and specialized to the point of incomprehensibility. Orbitals are referred to as S, P, D, or F, after the old spectroscopist's language for the type of spectral lines they create—"sharp," "principal," "diffuse," and "fundamental." The left-side number 2 in $2S_{1/2}$, for instance, denotes the principal quantum number; the right-hand subscript denotes the angular momentum. (Actually, the $2S_{1/2}$ is technically called the $2^2S_{1/2}$, with the superscript 2 on the left related to the total spin s by the formula $2s + 1$, but we have followed common practice and dropped the superscript for convenience.) We apologize to our non-spectroscopist readers for the necessity of inflicting them with this language.

alpha particles but everything to do with quantum mechanics and the atom. Like parsley in cooking, alpha appears everywhere in quantum electrodynamics, for the number pops up in the perturbation expansions necessary to calculate the self-energy terms in quantum electrodynamics. These equations express the probability that the electron will emit a virtual photon, which in turn depends on the strength of electromagnetism as a force. The relative strength or weakness of any force is measured by experimenters and expressed by a number called a *coupling constant*. For no special reason, the coupling constant for electromagnetism is called alpha, and thus every calculation of the interaction of photons and electrons has one or more αs in front of it. By 1930, many experimenters had measured alpha. It was almost exactly 1/137.

The constant 1/137 was, as it happens, first described by Arnold Sommerfeld, who supervised Heisenberg's dissertation at Munich. He gave the coupling constant alpha yet another name, "the fine structure constant," because its size plays a role in determining the separation of the fine, bunched-together lines in the hydrogen spectrum. The Balmer lines of the hydrogen spectrum on which Niels Bohr constructed his model of the atom had subsequently turned out to be composed of many different lines of slightly different wavelength—a fine structure. Their exact placement was described by equations couched in terms of alpha. More important, however, alpha put together many of the mysteries of twentieth-century physics. Alpha is the ratio $e^2/\hbar c$—where e is the charge of the electron, $\hbar$ is Planck's constant divided by twice pi, and c is the speed of light. It was hard not to think that the magnitude of alpha must somehow be *necessary*, must be a consequence of deep and hidden connections among e, $\hbar$, and c.

"People were *fascinated* by that number," remembered Markus Fierz, Pauli's assistant and co-worker in the late 1930s. "Now, e is electrodynamics, $\hbar$ is the quantum theory, and c is relativity. So in this one constant all the fundamental theories are related. The hope was that if one could figure out why this number had its particular value—1/137—the whole thing would be solved. It was a magic number!

"It's hard to realize how important this all was just from the publications. They made many attempts to get it that were wrong, and they didn't publish those. Great physicists don't publish their abortions."[1] Heisenberg and Pauli spent years trying to understand why the fine structure constant was 1/137 and not, say, 1/136.[2] Convinced that unexplained factors of 137 have no role in a proper theory, they believed that quantum electrodynamics could be renormalized only if they understood alpha; at the same time that they struggled in the slough of hole theory, they chased up blind alleys in search of the fine structure constant. They were not alone. Sir Arthur Eddington, a well-known astronomer whose exposition and testing of general relativity were partly responsible for its rapid rise to fame, viewed alpha as the key to the way nature hangs together. At first, the fine structure constant was thought to be 1/136, and Eddington said the Universe was an "E-matrix" with 136 parameters. Later, when subsequent measurements moved the number to 1/137, he added another parameter, restoring agreement with alpha itself. *Punch* published a cartoon identifying him as "Sir Arthur Adding-One."[3] When Eddington went public with his numerology, he embarrassed his colleagues and exiled himself from the main currents of science.[4]

Nevertheless, Eddington did help stimulate a vogue of interest in the fundamental constants, and a number of measurements transpired.[5] The first experiment after the onset of quantum electrodynamics was announced by Frank Spedding, C. D. Shane, and Norman Grace, all of Caltech, in the middle of 1933.[6] They hoped to use the fine structure of the Balmer lines of hydrogen, which the Dirac equation seemed to predict exactly, as a standard for determining the fine structure constant. To their evident surprise, they found alpha to be 1/138—"distinctly greater," as they noted, than the results from other experiments. "This discrepancy," they noted, "does not necessarily imply erroneous values for e and h but may arise from an incompleteness of the theory of the fine structure." By "theory of the fine structure," of course, they meant quantum electrodynamics.

A second Caltech team, William Valentine Houston and a visitor from China, Y. M. Hsieh, also examined the fine structure of hydrogen. They, too, found that some of the fine structure lines were about 3 percent off the predictions of the theory.[7] The discrepancy was "large," they said. It was caused by "a deficiency in the theory." Houston had become an assistant professor of physics at Caltech in 1928, and was widely regarded as a man to watch. When he said the experiments might not show what the theory predicted, the theorists listened.[8]

At the same time, R. C. Gibbs, an experimenter at Cornell University, in Ithaca, New York, and one of his graduate students, Robley Williams, were measuring the same thing. Unlike the Californians, Gibbs was not initially interested in determining the fine structure constant. He had merely noticed that an earlier measurement of the fine structure by three experimenters from Boston University was riddled with technical errors.[9] The two men decided to redo the experiment and see if the spectrum was where theory said it should be.[10] At first, the readings seemed to agree with theory, but as the experiment went on the Cornell group, too, saw discrepancies. They gave an initial report on their work at the same time as Spedding, Shane, and Grace, but weren't certain of their results for another six months.[11]

Theory said that when you plugged in all the numbers and solved all the equations—or, rather, all the equations except the interaction terms—the primary line in the Balmer series was supposed to be split up into two major and three minor components formed by different transitions among the 2*S*, 2*P*, 3*S*, 3*P*, and 3*D* orbitals.[12] The two most intense components (#1

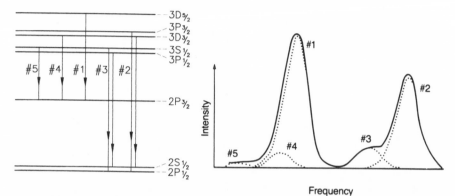

A portion of the hydrogen spectrum (*left*) gives rise to experimental values for intensity and frequency (*right*). Complex "selection rules" govern the permissibility of transitions: *D* levels can rise or fall to *P* levels, but not *S* levels, for instance.

and #2 in the drawing) were the ones that spectroscopists could see with relative ease. Gibbs and Williams knew the theoretical value; they got numbers about 6 percent lower. On Epiphany Day, 1934, they sent off a little note to the *Physical Review*; it was published on February 1.[13] More was coming, they told their readers.

They were beaten by Houston and Hsieh, whose long and excellent paper on the fine structure of the Balmer lines appeared in the following issue of the *Physical Review*.[14] Claiming that their measurements were sufficiently accurate that they had "attained to that degree of precision in which the theory is no longer satisfactory," they make what is in hindsight an astonishing suggestion: "One possible explanation of this [discrepancy] is that the effect of the interaction between the radiation field and the atom [that is, the self-energy] has been neglected in computing the frequencies."[15] This explanation, which is now thought entirely correct, had come from two theorists: another member of the Caltech physics department, J. Robert Oppenheimer, and a distinguished guest, Niels Bohr, who was visiting the United States at the time. Both men emphasized to Houston and Hsieh that all the theoretical predictions they used ignored the self-energy, which nobody yet understood. Few experimenters were aware of such theoretical niceties, but, once informed, Houston and Hsieh argued that "it seems to us very probable that this [self-energy] is the cause of the discrepancy we have observed."[16]

Everyone interested in field theory in those days was talking about the newly discovered polarization of the vacuum. Oppenheimer, one of its co-discoverers, was particularly excited; he allowed his new National Research Fellow, Ed Uehling, to see if the separation found by Houston and Hsieh could be explained by vacuum polarization.[17] Unfortunately, as mentioned, Uehling found that the vacuum polarization gave an effect, but it was of the wrong sign and more than ten times smaller than the observed discrepancy.[18] Having already given up on proper energy infinities, Oppenheimer gave up on vacuum polarization as well, and began to doubt whether the experimental results were real.

In Cornell, Gibbs and Williams published *their* long paper two weeks after Houston and Hsieh.[19] They picked out the cause of the discrepancy, a slight shift upward of the $2S_{1/2}$ level, but not the cause of the shift, self-energy. Gibbs was not theoretically minded; if he had even heard of the divergences of quantum electrodynamics, he would have been unlikely to connect them with something as physical as spectrometer readings. Nonetheless, by mid-1935 there was a rapidly developing consensus that something was wrong with the predictions of quantum electrodynamics. Then, at the end of the year, Spedding, Shane, and Grace recanted.[20]

Having, to some extent, started the fuss, the three men now said the theory was all right. Their experimental data had not changed very much, but their interpretation was radically altered. This time, they found the "right" result for the fine structure constant, 1/137; when corrected by a "more detailed method" of analysis, their readings fit theory exactly. Perhaps wishing to avoid antagonizing their colleagues at Caltech, Spedding, Shane, and Grace said little about Houston and Hsieh, but attacked Gibbs and Williams for insufficient statistical expertise. The three men also suggested that the Cornell experimenters might not have carefully controlled the pressure of the hydrogen that produced the spectrum.

Gibbs and Williams were by now in contact with the California groups.[21] They responded immediately and indignantly, telling the New Year's meeting of the American Physical Society that the separation of the line components was "remarkably uniform" no matter what the experimental conditions.[22] Later, Houston, too, stuck to his guns.[23]

The three groups used the same equipment in similar conditions to test for the same phenomenon, but came to opposite conclusions—Houston, Gibbs, and Williams in favor of the line displacements, Spedding, Shane, and Grace against. Looking back, one sees that the disagreement was partly due to the minute size of the shift and, more important, the nature of the hydrogen atom itself. Like everything else, hydrogen does not emit light in ordinary circumstances. It does so only when stimulated by energy in the form of something like an electric current, which kicks the electrons into higher orbitals. When the electrons fall back to the ground state, they emit light. The discharge is a kissing cousin of the familiar glow of a neon sign.

Unfortunately, there is a conflict between brightness and clarity. To make the light bright and easy to measure, you pass a large current through the gas—but this heats up the little hydrogen atoms so much that their furious jiggling smears out the spectral lines you are trying to measure.[24] To get sharp lines, the hydrogen must be cooled to the temperature of liquid air, about −380°F (−190°C), which means using a minimal current and therefore getting a dim, hard-to-photograph line. If you try to circumvent the trouble with a long exposure, you run into other problems—temperature drifts, vibration, and so on begin to swamp the tiny effects you are looking for. In sum, making cold hydrogen emit bright light was no easy task.

Physicists had to make compromises if they were going to work with hydrogen. They would get results, but the numbers would necessarily be affected by the conditions in which they were obtained. The argument among the spectroscopists was over how the data could be corrected for these perturbations—how close to reality the raw measurements were.

The spectrometer used by the three experiments was called a Fabry-Pérot interferometer, after Charles Fabry and Alfred Pérot, the turn-of-the-century French physicists who invented it.[25] The device takes light emitted by chilled hydrogen and bounces that light through a collection of mirrors and lenses in such a way that the crests and troughs of the waves reinforce each other; once amplified, the crests and troughs create a series of glowing concentric rings that fall on photographic paper. The intensity and position of the rings are determined by an electric eye. After processing through a Fabry-Pérot interferometer, the hydrogen alpha line comes out to be a slightly asymmetric double curve that looks like the back of a camel. During the 1930s, spectroscopists like Houston, Gibbs, and Williams spent a lot of time staring at these double curves. The curves are net readings of all five parts of the primary Balmer line; by carefully examining the slight ripples and deformations of the curves, physicists could infer where the two major and three minor components fell. Experimenters would make plausible guesses, raising and lowering this curve or that one, comparing one series of curves to a second and a third, until they came up with something that felt right. How one team fussed with the data could radically alter its conclusions, as Spedding, Shane, and Grace showed when they withdrew their earlier findings.

By this time, Gibbs had become chairman of the Cornell physics department, a post he took seriously enough that it nearly ended his active career as an experimenter. Working alone, Williams redid the experiment, obtaining essentially the same measurements. At this point, however, he was waylaid by an unfortunate accident—the arrival of the distinguished physicists James Franck and Hans Bethe at Cornell. Franck was the first to directly measure the quantized nature of energy, an achievement for which he won the Nobel Prize in 1926; Bethe, a Nobel laureate in 1967, was the author of a definitive and widely quoted theoretical article discussing the fine structure of hydrogen.[26] Interested by the new spectroscopic findings, Bethe and Franck quickly suggested several reasons that Williams's shift might be explainable by theory after all.[27] "They really had discovered a shift in the energy level," Bethe told us. "But I was too stupid to recognize it. Their initial explanation was even the correct explanation, but there was no theory at the time which predicted a shift of the $2S_{1/2}$ level, so I was looking for other reasons."[28] Williams went back and rechecked his data on the basis of Bethe's ideas. The shift was still there. But by this time he had finished his Ph.D., and was ready to go back to his first love, astronomy. He went to the University of Michigan observatory, and his supervisor, Gibbs, was left with the paper. Gibbs was loath to publish a measurement which he could not explain. He was not a theoretician, and did not know if the discrepancy was really important.[29]

Over two years later, in mid-1938, Williams's paper finally saw the light of day.[30] The tone was more cautious than that of the first paper, but he had found that a third component of the line, in addition, was not where it was supposed to be. The interval between line #1 and line #2 was about 2.7 percent smaller than the value predicted by quantum electrodynamics. All the discrepancies, although Williams did not say this, were associated with transitions from the $2S_{1/2}$ orbital.

By then, the theorists had become tired of banging their heads against the wall of quantum electrodynamics. The spectroscopists seemed to be disagreeing with each other, and few people outside the experimental world could tell whether Spedding, Shane, and Grace or Houston and Williams were right. "I don't know whether my paper was actively disbelieved or whether it was mostly ignored," Williams says now. He shifted out of astronomy shortly after the war, and became a distinguished molecular biologist. Now a professor emeritus at the University of California at Berkeley, he has little but fond memories from his prewar spectroscopy. "Don't forget the water had been real muddy up till then. Spedding, Shane, and Grace came up with something that was just disgraceful. They didn't *begin* to show the kind of detail they should have seen with this material—it was just lousy spectroscopy. Why? I guess their mirror surfaces weren't very good. And then this [Boston University experiment] was *terrible*. Houston and Hsieh, that was good work. So the situation had been kind of muddy. I suppose that helped our work not to be believed. I don't think anyone insisted we had made a mistake, it was more that it was thought to be kind of marginal.

"And also don't forget, we and the others had a formidable bunch of competition, namely fellows like Heisenberg and Schrödinger and Dirac and a few other types, who predicted what that interval should be—and we damn well better get that interval."

What do you mean, we asked him, that you had to get that interval?

"Those guys had it," Williams said ruefully. "They were very capable, and who was going to doubt the theory?"

In the first paper, though, he *had* doubted the theory. Did he drop the idea because of Bethe's prestige?

Williams was still chuckling. "No, I don't think so," he said. "In a way, our experiments were just marginally ahead of their time. Had you been able to do them on a calculating machine with no fuss or muss and no curve fitting to do, maybe they would have been accepted or believed. It's like any research, I guess, and we were somewhat on the edge of what could be done—I hope so. And we got a result that didn't agree with theory and the result then must be wrong. Well, gee, that's not unprecedented, not at all."

We asked Dr. Williams how he felt a decade later when his measurements were repeated and the Nobel Prizes were being given out.

"Then I was mostly measuring tobacco virus and nucleic acids and stuff. I had dropped all this entirely. So I was pleased that they found something agreed with me."

That's all? Simply pleased?

"Oh, have I ever said that a good idea for the Nobel Prize would have been to share it between the two people who found what the real answer was? Yes, I've thought that at times. But not very hard."[31]

One of the few people interested in Williams's result was Simon Pasternack, a theorist who had recently arrived at Caltech. Pasternack talked to Houston, read the papers, and arrived at the same conclusion Gibbs and Williams had come up with in 1934: the $2S_{1/2}$ level must not be where Dirac said it was supposed to be.[32] Pasternack came across the discrepancies in 1938, when theorists had largely given up trying to understand the infinities. Instead of invoking self-energy, he ascribed the shift to "some perturbing interaction between the electron and the nucleus." His note caused a little stir, but neither he nor anyone else knew how seriously to take it or the experiments it was based on. Williams, who first met him after the war, remembered him as a curious man who had never "worked with his fingers." They sat for hours in the University of Michigan observatory, talking about the fine structure of hydrogen. The experimenter explained to the theorist why the measurements were good; the theorist explained to the experi-

menter that the results were very, very important, and that the validity of
quantum electrodynamics hung on them. Williams was astonished. It was
the first time that he learned of the implications of his own work.[33]

In London, a big name weighed in. Sir Owen Richardson, a pupil of
J. J. Thomson, a colleague of Rutherford and Dirac, and the recipient of the
1928 Nobel Prize in physics, was an internationally recognized expert on the
spectrum of hydrogen. By 1939, he had become aware of the new controversy
over the hydrogen fine structure; he decided to redo his measurements.

Richardson was joined by a colleague, William E. Williams (no relation), who had designed a new type of interferometer that worked in a
vacuum, unlike the Fabry-Pérot machine. Williams built the equipment, and
Richardson assigned one of his graduate students, John Drinkwater, the task
of performing the experiment. Drinkwater was not particularly interested in
arcane theoretical problems; he wanted to go into industry, where his work
would have immediate, practical results. A man with an Olympian view of
collaboration, Richardson was rarely in the lab with Drinkwater, and told
him little of the issues involved. Richardson said nothing, and wrote the
paper entirely alone.[34]

The result is a most curious document.[35] For sixteen pages, Richardson
carefully, even laboriously, explains the motivation and means of the experiment and lists the spectroscopy readings. At first, the paper seemed to
promise excitement. Pasternack has suggested a shift in the $2S_{1/2}$ level;
Drinkwater's observations "apparently support this hypothesis." However,
Richardson writes, "there are certain other factors that appear to rule out the
possibility of this explanation."[36] And, incredibly, he spends the last four
pages of the paper on a farrago of absurd suppositions that wish away the
discrepancy. Although he spent little time in the laboratory, Richardson
calmly assumes that Drinkwater must have used hydrogen that was an im-
proper mix of its atomic (i.e., H atoms) and molecular forms (H_2). The taint of
H_2 generated "secondary" spectral lines that gave rise to the apparent discrepancy. "We conclude," Richardson writes, "that no real evidence has yet
been obtained to show that the fine structures depart substantially from the
values calculated from Dirac's equations."[37]

"That killed it off," Dick Learner, one of Richardson's successors, told
us in the University of London. We had gone there in the hopes of finding
out what had happened in the experiment; Richardson had been dead for
more than thirty years, and his surviving collaborators remembered little of
the experiment. At Imperial College, we met Learner, who uses the paper in
a class on errors in measurement as an example of what *not* to do. A heavy,
genial, dark-haired man, Learner had a clear picture of what had happened
to Drinkwater, W. E. Williams, and Richardson. "It's the clearest example I

know of somebody refusing to believe his own data," Learner said during our long conversation. "I lecture on this in my course, and the odds of a bad result, according to his own figures, are something like ten thousand to one. He had to work *hard* to explain the effect away."

He was asked why Richardson would have worked so hard.

"Reading between the lines, I would say that he thought Dirac had achieved an outstanding synthesis of atomic physics and relativity—a theory of seductive elegance. Richardson knew Dirac personally, remember." Learner shrugged. Easy to see what was happening there. A few minutes later, he added, "In those days, the professor and the head of the department was barely distinguishable from God. England was still in the Germanic tradition, where the professor said do it and you bloody well did it or else. Richardson was really one of those old despots. Now Drinkwater's in the position of a grad student whose supervisor talked him out of a Nobel Prize."

Why did Richardson's experiment succeed in killing off the others?

"The general feeling after that experiment was that Dirac was right, oh, good, thank God. After that experiment, if you'd applied for funds to re-measure the fine structure, they'd have said, 'Forget it!' It wouldn't have got past the referees on the first chucking-out level. Particularly when it's a Nobel Prize winner whose conclusion is that there is nothing to find. The NSF would blow you a big, gray raspberry, wouldn't it?"

Is that all there was to it?

"People always forget an absolutely fundamental point, and that is that physics has got to be *paid* for. It's always assumed that critical experiments just get done, but they only get done if they get funded. After Richardson, this was obviously an unfundable experiment. Nobody in their right mind would have pushed for it.

"I teach this experiment to remind the people that the theorists are often wrong. If you're an experimenter, you get the illusion that the theorists are all such *smart* bastards. But many theorists have no idea what's going on in an experiment. If you stand a theorist next to an apparatus, it breaks. Some members of the group here, for example, are good in all directions, but among the rest are a few really banana-fingered gentry." Learner abruptly swung to face us. "The point to all this," he said, "is that you must remember that you know more about what you are seeing and how you are seeing it than they do. Experimental physics, alas, has an inferiority complex."[38]

Not every good physicist went to Los Alamos or the other centers of military research during the war. One of those who didn't was Willis Lamb, an Oppenheimer protégé who taught at Columbia. Lamb had a rocky relationship with Oppenheimer; he was fascinated by Oppie, but the two men did not really get along. While at Berkeley, Lamb read his mentor's 1930 paper on self-energy and the displacement of spectral lines; he also heard discussion of Houston's measurements of the fine structure of hydrogen. But his dissertation subject lay in another area of physics, and he never really studied the question. After finishing his thesis in 1938, Lamb met I. I. Rabi, who arranged an instructorship at Columbia for the then-princely salary of $2,400 a year. There he taught beginning courses in engineering physics, worked on theoretical nuclear physics, and came across Pasternack's explanation of Houston's spectroscopic data. He'd also read the paper by Drinkwater, Richardson, and W. E. Williams, and concluded from it that the level shift envisioned by Pasternack was an intriguing possibility, but one unlikely to exist in the real world.[39]

Rabi was then working on experiments involving beams of ions and molecules. Lamb did various calculations for the molecular beam. As the war work began, the molecular beam shut down because most of the experimenters, including Rabi, went off to build weapons or design radar stations. Because Lamb's wife, Ursula, had fled Europe in 1935—"for all the usual reasons," according to Lamb—she was classified as an enemy alien.[40] This prevented Lamb from obtaining the requisite security clearance, and he continued to teach at Columbia.

In November 1943, Lamb received a phone call from Oppenheimer. Security problems were over, Oppenheimer said. Did Lamb have any idea what they were working on in New Mexico? Lamb said he had a fairly good idea. Would Lamb like to work on a very important and very secret project? Lamb said no. He did, however, join the Columbia· Radiation Laboratory, which was housed in Pupin Hall, blocked from the rest of the campus by an armed guard. In the Radiation Laboratory were magnetrons, the devices that produce the microwaves used in radar tracking. "We had a rule there," Rabi recalled, "that everyone who worked there had to make a magnetron. So that's how Lamb got introduced to experimental work."

Lamb became interested in magnetrons, but quickly discovered that if he wanted to have his own ideas about them tested, he would have to do the experiments himself. He knew none of the necessary sophisticated metal fabrication and vacuum techniques; he learned. He also taught atomic physics, which made him think about what are called the selection rules for transitions between orbitals in the hydrogen atom. Because the spin, momentum, and other aspects of the electron are quantized, jumps between one orbital and another can only take place if the orbitals involved possess the right quantum numbers. Lamb noted that for selection reasons it is difficult for the $2S_{1/2}$ level to fall into the $1S_{1/2}$, which is the lowest energy state of the hydrogen atom. The $2P_{3/2}$, however, can do so readily. Because the $2S_{1/2}$ takes a long time to drop down to a lower energy level, it is called a "metastable" state—almost stable.

In the summer of 1945, three thoughts came together in Lamb's mind. First, he realized that if he could get a large number of atoms in the $2S_{1/2}$ state and popped them into the $2P_{3/2}$ state, they would fall so quickly into the $1S_{1/2}$ level that there would be a sudden reduction in the percentage of $2S_{1/2}$ atoms. Second, he noted that going from the $2S_{1/2}$ to the higher $2P_{3/2}$ involved a photon of almost the same microwave frequency as that emitted by the magnetrons in the radiation laboratory. Third, he could delicately fiddle with the energy of the hydrogen atoms by running them through a tunable magnetic field. Combining these insights, he was sure, would allow him to measure the difference between the $2S_{1/2}$ and $2P_{3/2}$ energy levels with unparalleled precision. (The size of the gap was predicted by quantum electrodynamics, but it is important to recall that the calculations neglected self-energy.)

Lamb was kept busy by the radiation laboratory and his students at Columbia, but every now and then went back to the idea of making a microwave measurement of the fine structure. By July 1946, he had a vague idea of what to do, and put an order into the machine shop for the parts. His experience was with magnetrons so his design looked like a magnetron. Even before he finished putting it together he realized it would not work.

That September, Lamb met a graduate student named Robert Retherford, who had decided to return to academics after a spell of working for Westinghouse. Retherford knew a lot about the techniques necessary to build the apparatus; delighted, Lamb and he quickly put together a proposal for a somewhat different method. [41]

The two men planned to bombard a jet of hydrogen with electrons from an electron gun similar in principle to the cathode ray tube used by Thomson to discover the electron half a century before. Gaseous hydrogen usually consists of H_2, two hydrogen atoms linked into a molecule. The idea was that the incoming electrons could have just enough energy to break the molecules into atoms and kick the atoms into the metastable $2S_{1/2}$ state.

The detector took advantage of the special properties of metastable atoms. The energy they cannot release can be thought of as being like water in a child's water balloon: Under most circumstances, the water can't escape, but if the balloon smacks into something, the water splashes out. In the 1930s, it had been discovered that a metastable atom will fall to the ground state if it comes close enough to a metal atom to knock out one of its electrons. When the ejected electron is close to a positively charged plate, the electron, which has a negative charge, will be attracted to the plate. If there was a large number of collisions producing a large number of electrons, the result would be a small electric current. Thus Lamb knew that if he beamed microwaves at the stream of $2S_{1/2}$ hydrogen atoms, he would be able to tell when he had popped them to a $2P_{3/2}$ state because the electric current from the detector would suddenly drop.

The two experimenters spent the next few months building the experiment on a metal stand in Pupin Hall. When they finally switched it on for the first time, the detector registered no current at all. It was not clear the hydrogen was even hitting the plate. They painted the plate with a sooty yellow mixture of molybdenum oxide, which combines with atomic hydrogen and turns blue. They switched the oven on again. The plate turned blue. They were getting hydrogen, but their method of exciting them into the $^2S_{1/2}$ state was failing.

They decided to rebuild the apparatus. This time, they proposed blowing a thin stream of H_2 into a small oven made from tungsten and heated to a temperature of over 2,000°C.[42] The heat would rip the hydrogen molecules apart, and the individual atoms would be spat out of a tiny slit in the side of the oven. Most but not all of these atoms would be in the ground $1S_{1/2}$ state. The electron beam would now only have the task of pushing the atoms into the $2S_{1/2}$ state. Still, the experiment had to be fiddled with until it worked. They had to keep adjusting the electron gun until they started getting a halfway decent reading.

On Saturday, April 26, 1947, the experiment finally succeeded. Lamb and Retherford set the magnet and microwaves to the energy which conventional quantum theory predicted would make the hydrogen atoms jump into the $2P_{3/2}$ state. The instrument registering the detector current—a common galvanometer with a beam of light that fell onto a long paper scale—stayed motionless. Surprised, they fiddled with the microwave beam. At a frequency level about 10 percent less than they started, the galvanometer beam dropped. The $2S_{1/2}$ was going to $2P_{3/2}$—but at the wrong place. Which meant that the $2S_{1/2}$ was not where it was supposed to be to begin with. Excited, they realized that the discrepancies fitted Pasternack's analysis. Houston, Gibbs, and Williams had been right.

An experienced theorist, Lamb was fully aware of the implications of the result. Late that night, after Retherford had gone home, Lamb went over alone to the laboratory and tried to confirm what he had seen earlier. He discovered it was impossible for one person to operate the machinery alone and called his wife Ursula to help. They could see the galvanometer spot moving. He woke up the next morning feeling good indeed. Unless he was badly mistaken, the spot of light in Pupin Hall meant that something was going to have to be done about quantum electrodynamics.

A few months after the atom bomb was dropped on Hiroshima, Victor Weisskopf was offered a job by the Massachusetts Institute of Technology. His hiring was in itself a testament to the dramatically changed fortunes of physics in the United States. Before the war, American universities had an informal quota system for the hiring of Jews. The development of the bomb and the subsequent end of the war made heroes out of physicists; the Holocaust made anti-Semitism increasingly less respectable. Backed by the greater prestige of science, Jewish physicists were sought by the academic temples that had previously looked askance at their applications. Once in the Ivy League, the physicists' new stature was enhanced by their address.

Weisskopf immediately went to work on quantum electrodynamics. He decided to go back to the beginning, retracing Oppenheimer's footsteps, and see what the self-energy of the electron in the hydrogen atom would do to its atomic spectrum. About Halloween of 1946, he gave the problem to his graduate student, J. Bruce French, and suggested that he pay particular attention to the $2S_{1/2}$ and $2P_{1/2}$ levels, which were supposed to be coincident.[43] They worked at the problem very slowly, hampered by the extreme difficulty of the perturbation expansion, matching up one infinity with another and trying to see if they canceled each other out.

As the work went on, Weisskopf received an invitation to a small conference on the foundations of quantum mechanics hosted by Duncan MacInnes of the Rockefeller Institute for Medical Research, in Manhattan, and Karl K. Darrow of Bell Telephone Laboratories, in Murray Hill, New Jersey.[44] MacInnes, the president of the New York Academy of Sciences, had arranged for the National Academy of Sciences to provide five or six thousand dollars for two conferences, one of which was to be on theoretical physics. Darrow, who had been secretary of the American Physical Society for years, offered to help MacInnes arrange it. After thinking about it for a while, they decided to arrange a sort of physics retreat—twenty to twenty-five physicists, mostly young and promising, for three to four days in an isolated spot. The subject would be "the foundations of quantum mechanics." It would be held on Shelter Island, off the eastern end of Long Island, New York. There would be no formal papers, no fixed agenda, no published proceedings. To ensure the conference would have some prestige, Darrow and MacInnes first invited Oppenheimer. Oppenheimer thought the whole thing sounded peculiar, but was willing to go along.[45] This settled, Darrow and MacInnes invited Weisskopf, who found "the whole idea of a few quiet days in the country together with Heisenberg's 'uncertainty relations' . . . exceedingly attractive."[46] Einstein, too, was invited, but he was unable to attend, and in any case was out of the mainstream of physics.

Weisskopf, Oppenheimer, and a Bohr disciple, Hendrik (Hans) Kramers, were to be discussion leaders. All three men were asked to prepare outlines of what they thought should be talked about. Weisskopf's deserves to be quoted at some length, for it represents the viewpoint of a leading theoretician on the state of play in the spring of 1947.

The theory of elementary particles has reached an impasse. Certain well-known attempts have been made in the last fifteen years to overcome a series of fundamental problems. All these attempts seem to have failed at an early stage. An agenda for a conference on these matters contains, necessarily, a list of these attempts. After returning from war work, most of us went through just these attempts and tried to analyze the reason of failure. Therefore, the list [of problems] which follows will be well known to everyone and will probably invoke a feeling of knocking a sore head against the same old wall. The success of this conference can be measured by the extent it deviates from this agenda.

Weisskopf listed the problems of quantum electrodynamics:

1) *Self Energies. Attempts to remove infinite self energies. . . . Why do logarithmic divergences defy most of these methods?*
2) *How reliable are the "finite" results of quantum electrodynamics derived by means of a subtraction formalism? Polarization of the vacuum and related effects. Is there a high energy limit to quantum electrodynamics?*[47]

He could have written the same list in 1934, thirteen years before.

Kramers and Oppenheimer, too, wrote outlines for the discussion. Oppenheimer focused primarily on cosmic rays, which interested physicists at the time. Kramers, on the other hand, turned to electrodynamics. But, unlike Weisskopf, he was optimistic. The reason was that he thought he was on the track. Kramers had collaborated with Bohr since 1916, and had occupied the chair of theoretical physics in Leiden since 1934. Before the war, he had become interested in the divergences.[48] But he, like Dirac, was convinced that avoiding the infinities in quantum field theory required solving too many problems at once, and he put his efforts into a classical theory, with no quantum mechanics. Along the way, however, Kramers had come to the conclusion that the infinite proper energy of a particle was mostly invisible; just as the observed charge of an electron is the finite residue of the "bare" charge and the vacuum polarization, so the observed mass is really the combination of a "bare" mass and the mass due to self-energy. Therefore, Kramers thought, if one could write a theory wholly in terms of the experimental mass and charge, rather than the bare mass and charge, the infinities would be avoided from the outset, and the theory be renormalized.

When the Nazis invaded Holland, Kramers remained in Leiden, secretly aiding the Jewish physicists whose locations he knew. In broken health, he came to the United States in late 1946, where his ideas began to generate interest in theorists returning from the war. Weisskopf, who was discussing his self-energy work with Bethe, wanted to hear what Kramers was up to. There were rumors floating around that one of Rabi's people had measured the hydrogen fine spectrum and got results that strongly deviated from theory. Weisskopf was pleased to learn that both Rabi and the experimenter, Lamb, were coming to the Shelter Island conference on the foundations of quantum mechanics. At the end of May 1947, Weisskopf took the train to New York with another conference attendee, one of the young comers whom the meeting featured—Julian Schwinger, an astonishingly gifted theorist who had been made a full professor at Harvard at the age of twenty-nine. If the tales from the experimenters were right, Schwinger and Weisskopf thought, then the deviations very likely had their origins in the self-energy.[49]

They met Darrow, MacInnes, and Oppenheimer at the American Institute of Physics, then on East 55th Street in Manhattan. There most of the twenty-five physicists piled into a bus and headed for Long Island. Lamb drove his old friends from California, Serber and Nordsieck.[50] Bethe went to New York City, where he borrowed his brother-in-law's car for the very long trip to eastern Long Island.[51] He missed a spectacular ride: Oppenheimer's prestige ensured that the last part of the trip was made in a police motorcade, sirens a-whirl, whizzing through the village stop signs.

Low, dry, and gnarled, Long Island runs parallel to the Connecticut shore for more than a hundred miles. Its eastern tip branches into two long peninsulas, known respectively as the north and south forks. Between the tines is a small, sandy triangle known as Shelter Island. Unlike the rest of Long Island, Shelter Island is still much as it was fifty or even a hundred years past. In marked contrast to the parade of boutiques that now crowd the fashionable Hamptons on the south fork, Shelter Island still has the plain, unornamented quality of the Quakers who first settled there. There are weathered rural stores, small roads, few cars; the air seems guarded, as if the island were poised to ward off inundation by the tides or, more probable, affluent summer people. Shelter Island is edged by long promontories—"heads," in the local parlance—that spike outward like the curiously curved swords of the old Ottoman Empire.

One of the most remote of these spits is Ram's Head, accessible only by a causeway that winter flood tides leave awash in brine. The telegraph poles lining the road are covered by the untidy evidence of osprey nests. Debouching from the causeway, one quickly sees the Ram's Head Inn, an unpretentious clapboard country hostel. The roses were out when we visited, curling over a split rail fence in front of the establishment; oak and maple trees swept invitingly over the back porch. The entranceway, edged with white trim, leads to a small hall with the eponymous stuffed ram's head on prominent display. The lounge behind has a piano, a fireplace, and a bar; its walls are cluttered with a comfortable jumble of maritime foofaraw.

Beside the door we came across a large scroll of the sort invariably found in rural hostels, celebrating forgotten football championships and the long-ago strikes of a prodigious bowling team. Made curious by the yet-untarnished condition of the metal frame, we approached the placard and discovered it to be the only physical memento of a landmark in twentieth-century intellectual history.

FIRST SHELTER ISLAND CONFERENCE
ON THE FOUNDATION OF QUANTUM MECHANICS

2–4 June 1947

The first Shelter Island Conference on the Foundations of Quantum Mechanics, held at the Ram's Head Inn in 1947, is remembered as the starting point of a series of remarkable developments in physics that have changed our views of the basic structure of matter and given us a new cosmology. . . .

To present-day physicists, the name Shelter Island is synonymous with a sense of remembered excitement; the congress was one of those rare, jeweled occasions where the machinery of the scientific imagination ran faultlessly, and with speed and quiet ideas were put together whose connections had eluded so many minds in the decades before. Years later, Richard Feynman, one of the younger physicists present at Shelter Island, said, "There have been many conferences in the world since, but I've never felt any to be as important as this."[52] Oppenheimer thought it was the most successful scientific meeting he had ever attended. Its total cost was just $850.[53]

The conference opened early in the morning of Monday, June 2, 1947, with Lamb's presentation of his experimental work. By that time, he and Retherford had succeeded in determining that the $2S_{1/2}$ level was about 3 percent off where it was supposed to be—an even greater separation than found by Williams. Rabi then explained the work of his junior colleagues, John E. Nafe and Edward B. Nelson, who used a somewhat similar apparatus to examine the magnetic behavior of the electron in hydrogen and its isotopes.[54] Here, too, they had found discrepancies.[55]

Kramers, Oppenheimer, Schwinger, and Weisskopf argued that the discrepancies were not due to actual errors in the theory of Dirac, Heisenberg, and Pauli, but to an inadequacy in the way the theory was *used*. That is, the theoretical values had been arrived at by ignoring the last term of the total Hamiltonian. If one calculated the way one ought to, using the full equation, one should arrive at something like the values found by the Columbia experimenters. Unfortunately, as everyone present knew, this was more easily said than done.[56] Getting rid of the infinities—renormalizing the theory, in the jargon—would be a lot of work.

Until late in the night the men hashed out the question, gulping down meals in a fury of technical discussions, wandering through the corridors in groups of two or three. They were the only guests in the inn, which had opened early to accommodate the famous J. Robert Oppenheimer. For many of the participants, the meeting was the first chance since Pearl Harbor to dive into physical waters untainted by weaponry; little wonder they took to it with a giddy pleasure. As Schwinger said later, "It was the first time that people who had all this physics pent up in them for five years could talk to each other without somebody peering over their shoulders and saying, 'Is this cleared?' "[57] A mark of a good physicist is a sort of surprised delight at the working of nature. Every man there was experiencing again the deep pleasures of curiosity. Why didn't the galvanometer beam in the tenth floor of Pupin Hall slip down when it was supposed to?

On the second day, they talked of cosmic rays. Here, too, the discussion was frenetic, animated, sharp.[58] The third day swung back to renormalization, and the group splintered off, its members filled with a sense of mission. Oppenheimer left on a seaplane to take an honorary degree from Harvard, bringing with him Schwinger, Weisskopf, and an Italian cosmic ray expert, Bruno Rossi, on what proved to be an alarming voyage. The pilot was unhappy at Oppenheimer's casual instructions to land at the New London Coast Guard Station, which is not open to civilian aircraft—so unhappy, in fact, that he almost sank the plane. They were met by an armed and infuriated coastguardsman. In Schwinger's recollection, Oppenheimer hopped out of the cockpit and calmed the enraged officer with the magic words, "My name is Oppenheimer." "*The* Oppenheimer?" was the gasped response. They were escorted to the train by an honor guard.[59]

Bethe, meanwhile, drove back to New York. He stayed overnight with his mother and took the train to the General Electric laboratories in Schenectady, New York, where he was then working. The Lamb shift excited him enormously. For years he had believed that the complexity of quantum electrodynamics made it unlikely that he would ever contribute anything.[60] Now, he had an idea of how he might account for the level shift in a simple way.

We asked him once to tell us about the Lamb shift. "It was a wonderful new effect—completely unexpected!" he said. "It tied in with the old puzzle of the infinite self-energy, and especially with a talk by Kramers on the idea of renormalization of the electromagnetic interaction. Kramers hadn't done any calculations, but he had the idea of renormalization. He had the idea that the self-energy of a free electron traveling in space was not observable, but part of its mass. To calculate the shift in the spectral lines, one should only consider the difference between the self-energy of a free electron, and the self-energy of an electron in a field, like the electron in the hydrogen atom." The difference is all we can see, Bethe thought, and that may be finite. For a quick calculation, he simply ignored the relativistic effects. "A relativistic calculation was far beyond my ability to do in any short time," he told us. "I knew—and said so in the article—that the self-energy diverges log-arithmically in the relativistic case. That had been known since 1934. So I said, all right, I'll get the nonrelativistic part. I finished the calculation in the course of the train ride."[61] Actually, Bethe didn't quite finish: He had to go to the library the next morning to look up a factor and a logarithm. Forgetting about relativity made the calculation much simpler, and, to his joy, he was able to account for 95 percent of the effect. Five days after the close of the conference, on Wednesday, June 9, 1947, he was ready with a preliminary draft.[62]

The calculation chagrined his colleagues. "I instantly saw what Bethe had done," Lamb remembered, "and mentally kicked myself for not being clever enough to do it first."[63] Crestfallen, the physicists saw that (1) Bethe was obviously right; (2) Bethe had done something that any of them could have done fifteen years before. "Any of the Berkeley people—Serber, Uehl-ing, Oppenheimer, and the rest—could have done what Bethe did," Lamb said on another occasion. We asked him why they had not. "They weren't thinking of the right thing at the right time. It was also because of the sheer simplicity of the calculation. Why do something nonrelativistically when you knew how to do sophisticated things?" In the 1930s, physicists knew a correct theory had to obey the dictates of relativity. It hadn't occurred to them that a simple nonrelativistic calculation might be enough to point the way. "Then there was the experimental fact of the shift," Lamb said. "If you knew there is a substantial effect involved, then there is much more of an incentive to calculate it than if the effect is tiny. It's like the atom bomb—it was a much better secret before anybody made one than after."[64]

8
Killing the Hydra (Part I)

JULIAN SCHWINGER WAS BORN IN NEW YORK CITY ON FEBRUARY 12, 1918. HIS remarkable abilities became quickly apparent. At the age of fourteen, he attended a lecture by Dirac on the hole theory; two years later, he wrote but did not publish his first article on quantum electrodynamics. His high school, Townsend Harris, was a teaching adjunct to City College. The teachers there soon suggested that he should be going to City College itself.

By chance, Lloyd Motz, a Ph.D. candidate at Columbia, became friendly with Schwinger's brother, Harold, who was at Columbia Law School. Motz had already heard through the grapevine that there was a high-school-age prodigy at City College. When he found out the prodigy was a friend's younger brother, he made it his business to look up Julian Schwinger. Motz discovered that although Schwinger had only completed his freshman physics courses, he was completely familiar with relativity and quantum mechanics and nuclear physics and hole theory and the rest of it and was doing advanced computations for his professors. Impressed, Motz asked Schwinger if he would like to collaborate on a paper. They started talking together, and Motz discovered that despite Schwinger's unquestioned talent, he was not doing well in school. Part of the reason was that he was working too hard to pay much attention to his classes. Motz decided he'd better introduce Schwinger to his own thesis adviser, I. I. Rabi.[1]

"The circumstances were rather romantic," Rabi said to us. We were sitting in his apartment, watching Rabi use a portable phone to fend off demands for his time. "I'd read a paper by Einstein, Podolsky, and Rosen. My way of reading a paper would be to get a [doctoral] student to explain it to me. So Lloyd Motz walked by, and I called him in and talked about it. After a while, he said, 'There's somebody waiting for me outside.' And there was this kid, Schwinger, who was a student at City College at the time. Lloyd brought him in. I said, 'Sit here,' and he sat there, and we went on with our discussion. Then Schwinger *settled* the argument by an application of the completeness theorem. I said, 'Who's *this?*' This kid! Then I heard about how he had difficulties at City College, and I suggested he change to Columbia. I got him a scholarship—not without difficulty."[2]

Schwinger said he would transfer to Columbia if he did not get into an honors graduate program at City College. Motz, who was teaching part-time at City, met with the department chairman. A sophomore in an honors graduate program? Out of the question! In the spring of 1935, Schwinger entered Columbia to the accompaniment of a lecture by Rabi on the importance of going to his classes.

Ordinarily less than impressed by mathematical dexterity, Rabi knew that Schwinger was a special case. He quietly brought eminent visitors— Pauli, Bethe, Fermi, and Uhlenbeck among them—to Motz's office, where they listened in stupefied silence to the words of an undergraduate physics major. Bethe wrote Rabi afterward that Schwinger already knew ninety percent of all physics, and that the remaining ten percent would only take a few days. In the summer of 1937, just after graduating from college, Schwinger wrote his Ph.D. thesis; Rabi told him that he should stick around and fulfill his residence requirements. "One always receives this bad advice," Schwinger said later.[3]

Schwinger amazed the physics department but annoyed the mathematics department. Although it was understood that he knew the subject material thoroughly, Schwinger's math professors refused to pass him unless he showed up for his classes. Rabi asked George Uhlenbeck, then spending a semester at Columbia, if he would do something about it. Uhlenbeck taught statistical mechanics, one of the many courses Schwinger never attended. "He never came," Uhlenbeck recalled with amusement. "At the end of the semester I had to give the usual exams. I asked Rabi, 'What should I do with Julian? Because he never came! I'm willing to give him an A, because I know already that he's the best.' But Rabi said, 'No, no, no, you give him an E!' I was appalled. Then he thought and said, 'No, Julian will take an examination from you at the end of this week.' It was three or four days away. Julian got my lecture notes from other students and studied them. And after three days he came, and I gave him the regular exam. He of course knew everything, except that he could do several of the derivations in a simpler way than I had shown in the course."[4]

"He had this notebook that was just *filled* with results and calculations," Motz said to us. "Every time a new paper came out, he'd show me he'd already worked it out in his notebook. He used to give seminars that were absolutely perfect. There was no one I knew in physics like him, including Fermi—and I knew Fermi very well, we wrote a book together. Schwinger was to physics what Mozart was to music."[5]

Bearing his new Ph.D. and a National Research Council fellowship, Schwinger arrived in Berkeley in the fall of 1939, and was soon deep into nuclear physics and collaboration with Oppenheimer. Two years later, in 1941, his grant expired, and he went to Purdue University, in Lafayette, Indiana. The Japanese attacked Pearl Harbor on December 7. Soon thereafter, Hans Bethe swept through the Purdue campus on a physics recruiting mission for the war effort. Schwinger joined the M.I.T. Radiation Laboratory. In 1943, he was asked to follow Bethe and join Oppenheimer's team in Los Alamos. Schwinger had a good idea of what was transpiring there, as did most nuclear physicists in the United States. Hesitantly, he asked if he could first put his toes in the heavy water by joining the group at the University of Chicago, which was designing reactors. After a few months, he began to comprehend the proposed size and yield of the bomb. He backed out and returned to M.I.T. "I give myself high marks for gut reactions," he said to us.

We were talking in a restaurant a few minutes' drive from the campus of the University of California at Los Angeles. Schwinger was a small man with heavy, leonine features and an almost Middle European air of elegance. He drove us to the restaurant in an immaculately maintained Italian sports car; his suit was cut with an attention to style rare in a physicist. A perfectly knotted silk tie hung straight down before his shirt. Almost inevitably, he reacted to his status as a prodigy by becoming hard to reach in his maturity, a practice his colleagues attribute variously to shyness, aloofness, or the simple wish for quiet. While at Columbia, Schwinger began to work at night, a habit that Rabi, with affection and respect, surmised was due to the desire to avoid his mentors—chiefly Rabi himself. (Schwinger died in 1994, a solitary giant.)

"Shelter Island was the first gathering of the clan of research physicists after the war," he said. He spoke readily and quickly, in concise sentences. "It was not a gathering of theorists proclaiming their theories. It was centered around the experimental discoveries that came from Willis Lamb and Rabi and so on." Lamb's work had been circulating about the grapevine for several weeks; before Shelter Island, Weisskopf and Schwinger talked about the level shift. They agreed it was an electrodynamic effect, and that the correction should be done relativistically. "The theories should account for it," Schwinger said. "There was this effect. It wasn't infinite and it wasn't zero. The reaction to the divergences was to say that the theory was wrong, let's throw out these wrong things, so everything that gave an infinity was set equal to zero. It was just thrown away. So here was an experiment saying it's not infinite, it's not zero, it's small, it's finite—we've got to understand it."

He was asked if the Lamb effect forced people to take their own theories seriously.

No, Schwinger said. "The point is, you don't get rid of history as dead wood. Electrodynamics was conceived to be *wrong*. There was a whole generation dedicated to changing it." He listed a few of the ways theorists had tried to rectify quantum electrodynamics. "All this had run wild," he said. When Lamb's work came out, theorists immediately rushed to their cutoffs and classical models and fundamental lengths. "People began immediately, of course, to say, 'Oh, I have a finite theory, now let me calculate this effect.' Very few people—and I count myself first and foremost—said, 'Let's go back and look at the original theory.' I mean, yes, people said it, but it wasn't the first thought of the majority of them. Not at all." He was modest about his own special qualifications. "What I brought to the problem was a physical instinct and an ability to do calculations. That's all that was needed."

Unlike Bethe, Weisskopf, and most of the other people at the Shelter Island conference, Schwinger's imagination was captured not by the Lamb shift but by the discrepancy in the magnetic behavior of the electron. What Nafe and Nelson had measured was the effect of the continual emission and reabsorption of virtual photons on the spin of electrons moving in a magnetic field, producing a discrepancy with the value predicted by Dirac's theory. Known as the anomalous magnetic moment, this phenomenon affects other particles as well, and is exactly the effect, in connection with muons, being measured by Brookhaven's (g-2) experiment.

"That was much more shocking," Schwinger said. The Lamb effect, as Bethe showed, could be accounted for almost entirely without the use of relativity. "The magnetic moment of the electron, which came from Dirac's relativistic theory, was something that *no* nonrelativistic theory could describe correctly. It was a fundamentally relativistic phenomenon, and to be told (a) that the physical answer was not what Dirac's theory gave; and (b) that there was no simpleminded way of thinking about it, that was the real challenge. That's the one I jumped on."

Were the others as interested in this?

"Not as much. That was my particular hangup. You know—holy cow!" He laughed. "But it was the right one, because you had to have a relativistic theory to get that right. Whereas any basically nonrelativistic theory would give the Lamb shift to better than an order of magnitude."

We asked Schwinger how long it had taken him to renormalize the theory. He had, we knew, spent two months after Shelter Island touring the country with his new wife.

"Well, I presume the little gray cells were ticking even as I was driving through California and various other places," he said. "But I began to work on it, I would say, in September. The first thing I began with was the magnetic moment, as I said. That was a real challenge. I had the answer in three months. I then turned to the Lamb shift—I couldn't stop doing calculations!—and I got the wrong answer. I got the wrong answer because in my calculation the magnetic moment that you recognize when the electron is at rest and the magnetic moment that you recognize when it's moving in the atom under the action of its electric field turned out not to be the same. Which meant that the methods I was using violated relativistic invariance. An electron in motion and an electron at rest are the same thing, and they better be that way in the theory."

Working with his student J. Bruce French, Weisskopf had also calculated the Lamb shift. They arrived at an answer and compared notes with Schwinger and Richard Feynman, another young theorist looking at the question. Unfortunately, both Schwinger and Feynman had made the same mistake and got the same erroneous answer. A flurry of cross-checking ensued. Trusting the math abilities of the younger generation more than his own, Weisskopf postponed publication. The discrepancy between the answers remained, and Weisskopf wrote Oppenheimer just before Christmas, 1947, that he was beginning to wonder if they had at last reached the true failure of quantum field theory.[6] He needn't have worried: Schwinger soon came up with the right answer. Feynman, too, found his mistake and handsomely apologized by juggling the footnotes of his paper until the admission of error appeared in the "appropriately numbered" footnote 13. But in the meantime, Lamb and his student Norman Kroll had scooped everyone with the right calculation.[7]

"The technique I was using before was primitive," Schwinger said in Los Angeles. "I was driven to invent a new method that was explicitly relativistic so I would get the same numbers no matter when I looked at the electron. I had the first ideas of that by the very end of December." His thoughts came together all at once; Schwinger worked night and day. "I gave a talk at Columbia at the end of January 1948, in which these results were announced[8]—the agreement with experiment as it stood then for the magnetic moment, the Lamb shift, and the first suggestions of this relativistic theory. I was on Cloud Nine, or whatever the unrenormalized number was back then."[9]

On Tuesday, March 30, 1948, a second Shelter Island conference convened, this one at Pocono Manor Inn, in Pocono Manor, Pennsylvania. Most of the members of the previous conference were there, but among the newcomers was Niels Bohr. The high point of the occasion was an extraordinary five-hour talk by Julian Schwinger. Covering the portable blackboards in the hotel lounge with equations, he led the assembled physicists through a dazzling and complete reformulation of quantum electrodynamics in which every infinity was subsumed into the observed charge and mass of the electron, and which avoided entirely the concepts of the "bare" mass, the "bare" charge, and the electromagnetic additions to them. Having recently been burned by relativity, he began with an explicit demonstration that every expression in his equations completely fitted the dictates of relativity, in the process taking some of his audience into byways of geometry and differential calculus they would just as soon not have traveled—and then derived the theory from there. Some of the physicists present reacted ambivalently to the glittering array of mathematical skills that Schwinger marshalled into play. They described his talk as the epitome of virtuosity, more technical flash than music, a beautiful but cold piece for a solo voice.[10] Most of the stuff he had been scribbling on the blackboard was merely a record of how he came to the theory, they said, not really *physics*. (Schwinger didn't see it this way.) Nonetheless, every man there was conscious of being present at a historic occasion. A new generation of physicists had finally taken over the reins. Before Niels Bohr's eyes, a young man—Schwinger was then not quite thirty years old—had vindicated field theory and renormalized quantum electrodynamics.

Delighted as he was by Schwinger's speech, Oppenheimer was even more pleased to find on his return to Princeton a missive from Japan informing him that a young theorist named Sin-itiro Tomonaga had mostly renormalized the theory some years before but had been prevented by the war from making the work known in the West. Working in appalling conditions— "perfectly isolated from the progress of physics in the world," as Tomonaga wrote—the Japanese had essentially duplicated Schwinger's reasoning in 1943.[11] When a copy of *Newsweek* arrived with news of the Lamb shift, Tomonaga, too, had been electrified; like the Americans, the experiment galvanized him into a fury of calculation.[12] Oppenheimer sent a copy of the letter to every participant and fired off a cable to Tomonaga by return post.[13]

GRATEFUL FOR LETTER AND PAPERS STOP FOUND MOST INTERESTING AND VALUABLE CLOSELY PARALLELING MUCH WORK DONE HERE STOP STRONGLY SUGGEST YOU WRITE A SUMMARY ACCOUNT OF PRESENT STATE AND VIEWS FOR PROMPT PUBLICATION PHYSICAL REVIEW STOP GLAD TO ARRANGE STOP MOST CONSTRUCTIVE DEVELOPMENT HERE APPLICATION BY SCHWINGER OF YOUR RELATIVISTIC FORMALISM TO SELF-CONSISTENT SUBTRACTION TO OBTAIN SEVERAL DEFINITIVE QUANTITATIVE RESULTS STOP BEST GREETINGS ROBERT OPPENHEIMER

By mid-May, Tomonaga had the summary ready for Oppenheimer, who got it to the *Physical Review* by June 1.[14] Although it was published shortly thereafter, Tomonaga was still six months behind Schwinger.[15] Nonetheless, American scientists, including and especially Schwinger himself, hastened to give Tomonaga credit. Physicists were in a generous mood; twenty years after its initial formulation, quantum electrodynamics was at last on its feet. Victor Weisskopf wrote later that

The war against infinities was ended. There was no longer any reason to fear the higher approximations. The renormalization took care of all infinities and provided an unambiguous way to calculate with any desired accuracy any phenomenon resulting from the coupling of electrons with the electromagnetic field. It was not a complete victory, because infinite counter-terms had to be introduced to remove the infinities. . . . It is like Hercules' fight against Hydra, the many-headed sea monster that grows a new head for every one cut off. But Hercules won the fight, and so did the physicists.[16]

To be sure, the show was not entirely over. Schwinger's calculations, so elegant in their initial impact, proved resistant to practical usage—or, rather, many theorists found them that way. There seemed little purpose to renormalizing the theory if the formalism was too difficult for anyone but a Schwinger to use. Pauli eventually subjected the work to severe criticism; Dirac let it be known that he thought the whole business was a cheap trick designed to paper over fundamental problems.[17] (He held that opinion until the day he died.) And everyone, Schwinger included, knew that checking every possible infinity in every possible interaction would be a wearisome job.

Most particularly, the show was not over because an entirely unexpected thing occurred at the second Shelter Island Conference: a *second* renormalization theory was displayed, this one with such a wholly different approach that Bohr, among others, thought that its creator, Richard Feynman, had not understood elementary quantum mechanics. In fact, the superiority of Feynman's methods was so enormous that the generation of aging young iconoclasts who had created quantum mechanics at first found them incomprehensible.

Feynman, too, was something of a prodigy, although one of a diverse stripe. Born in Far Rockaway, Queens, famous throughout the city as the end of the A train, Feynman was brought up by his father to be a scientist. He was a puzzle nut, an inventor, a practical joker, the local fixer of radios and typewriters; he didn't like the symbols in his high school mathematics texts, so he made up his own.[18] Some of his difficulty at the second Shelter Island Conference was due to the fact that his theories were couched in his own, brand-new system of notation. He had given up his teenage efforts to create new formalisms because he realized nobody else would know what he was talking about; by 1948, he was determined to *show* them what he meant.

He was a man without an internal censor, a formidable, hawk-faced presence who said exactly what was on his mind the instant it occurred. In conversation, he was unnervingly present; a prankster, he seemed constantly on the point of making faces and indeed had composed an autobiographical set of anecdotes strangely disserving to this original and serious thinker—it was devoted to the curious notion that a thrice-married Nobel laureate in particle physics was a regular old hell-raiser, a sort of Lucy-like figure who got into madcap scrapes with amazing regularity. While a young man in the atomic bomb project, he met Niels Bohr, whose presence awed everyone else at Los Alamos. Bohr made some remarks about a technical problem. Feynman said flatly, "No, it's not going to work," or something to that effect. Bohr said afterwards that it was the only honest reaction he heard during the day, and asked Feynman to be his talking partner while they were together at Los Alamos.[19] Feynman acquired some notoriety on the project for picking locks and blithely informing people that top-secret materials could be found in easily opened safes. He was, at the time, a curiously tragic joker: His first wife, Arlene, was sick with tuberculosis in an Albuquerque hospital. (She died in 1946; Feynman died in 1988.)

Feynman ultimately became a symbol for plain speech and going his own way. "I have an impatience and great difficulty in reading other people's papers," he said to us. "For me it's much easier to work a thing out from the beginning than it is to read another paper. Especially if I understand the idea. If I read the paper to give me an idea of the thing, then I prefer to work it out than to follow the equations, in most cases."

He was sprawled in a creaking chair, cradling his head in one arm, his loose plaid shirt sagging, eyes as bright and wild as a bird's. From the years before Los Alamos, he had been interested in the divergences besetting quantum electrodynamics. Feynman was not the type of person to let a problem go. He had some mathematical tricks—in particular a technique called "path integrals"—and after the war, while occupying a professorship at Cornell, he set out to use them. Or tried to, at any rate. Perhaps he had not recovered from the death of his wife, perhaps the destruction of Hiroshima had put a stain on science, perhaps any number of things, but he couldn't get to work.

Depressed, he turned down an offer from the Institute for Advanced Study in Princeton, New Jersey, because he felt he had nothing to contribute. Then he turned it around, he said, realizing that *he* was not to blame for *their* misconception about his abilities. He was just going to do whatever he liked in physics. "And only that afternoon," he said once, "while I was eating lunch, some kid threw up a plate in the cafeteria which had a blue medallion on [it]—the Cornell sign. And as he threw up the plate and it came down, it wobbled and the blue thing went around like this." He made a fluttering motion with his hands. "And I wondered—it seemed to me that the blue thing went around *faster* than the wobble, and I wondered what the relation was between the two. You see, I was just playing. No importance at all. I played around with the equations of motion of rotating things and I found that if the wobble is small, the blue thing goes around twice as fast as the wobble goes round. And then I tried to figure out if I could see why that was directly from Newton's laws instead of through some complicated equations. I worked that out for the fun of it." Pleased, he approached Hans Bethe, who asked him what was the use of finding the rate of rotation of cafeteria plate wobbles. None, Feynman said. He was reminded of another, similar problem, related to the spin of an electron, which in turn took him back to quantum electrodynamics. "It was just like taking the cork out of a bottle," he said. "Everything poured out."[20]

After that, of course, Hans Bethe's calculation of the Lamb shift gave him, he said, "a kick in the pants." Instead of a Hamiltonian, Feynman used another type of equation known as a Lagrangian, which was familiar from classical physics.[21] The Lagrangian was what physicists had tended to use before quantum mechanics resuscitated the Hamiltonian. In Lagrangian form, the basic electron-photon interaction in quantum mechanics consists of a term for the electron, a term for the photon, and a term for the interaction, each of which is made up of several wave functions and matrices. For complicated interactions, the equations got very long, and Feynman fell into the habit of making little doodles to help him keep track of the terms. For each electron term in the expansion, for example, he drew a line:

To make photons, he put down a wiggly line:

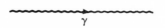

When they interacted, he had them meet at a point.

"I don't know when I started to make such pictures," Feynman said. He couldn't seem to keep still in the chair. "I do remember at one particular stage, when I was still developing these ideas, making such pictures to help myself write the various terms—and noticing how *funny* they looked." He was sitting on the floor of his room in Telluride House, a residential house at Cornell, late one night, a young man with a pencil and a head full of stick-up hair. (Early photographs of Feynman show him dressed in what today would be called "modified punk"—skinny little tie, white shirt, and thick hair erupting an inch and a half from the skull.) There was no room on the desk to work, almost no room on the floor: The room was awash with calculations and doodles. Pictures everywhere of complicated crossings among electrons and photons, positrons represented by reversing the arrow of the electrons. Thus, in Feynman's little diagram, an electron and positron collide to make a photon, which in turn briefly becomes a pair:

"I was sort of half-dreaming, like a kid would, that it might someday be interesting, that it would be funny if these funny pictures turned out to be useful, because *the damn Physical Review* would be—" he was laughing incredulously "—*full* of these odd-looking things. And that turned out to be true."

Feynman discovered that his little bookkeeping device had mathematical properties. By sorting through the diagrams he could quickly guess whether the associated expression, the sum of the lines and vertices, would be convergent or not. There were simple rules of thumb one could evolve. Highly excited by the ease and proficiency this brought to otherwise dreadfully complex calculations, he took his new methods to the conference in the Poconos.

Feynman spoke after Schwinger, and his exhausted listeners were anything but receptive to hearing more ideas they could not follow. Compounding the difficulty, Feynman has arrived at his conclusions through guesswork, intuition, unfamiliar mathematical tricks, *ad hoc* rules, and hours of cut-and-try calculation. He had invented a branch of mathematics called "ordered operators" to contain his ideas, but the only way he knew it was right was that he always got the right answers. Unable then to deduce his ordered operators from established theory, he tried to present his prescriptions without explanation and proceed by solving examples. This landed him in deeper trouble, because in each case Feynman had already done the problem and knew he didn't have to bother about this principle or that—but his audience was composed of the very physicists who had invented the principles, and who wanted to ensure that they were respected. These men were not satisfied by *ex cathedra* promises from a junior professor that things would turn out all right in the end.

Forced into a dreadful tangle of exposition, Feynman started to explain with diagrams. But as soon as he started drawing his pictures, Niels Bohr interrupted to say that the little line of the trajectory in Feynman's diagram was *exactly* the sort of thing that the uncertainty principle showed was impossible. Feynman replied that the diagram wasn't intended to show a real trajectory, but was, you see, in fact, a bookkeeping device for terms in the Lagrangian—

Bohr would not listen; he loathed Feynman's pragmatic, seat-of-the-pants approach to the sacred ritual of the Copenhagen interpretation. He sternly informed Feynman that there was a lot more to physics than what you get off the top of your head.[22] "They didn't understand it very well," Feynman said. "It was too much stuff to explain. So Bohr said that we already knew in nineteen-whenever that we can't talk about trajectories and the quantum mechanical laws don't permit that any more. I realized—sort of, I mean—I suddenly said, 'That's enough.' " He laughed, sharply and intensely. "They said that this idiot didn't understand quantum mechanics at all. My ideas were consistent. I knew that his objection was wrong. I realized that I'm going to have to write it for them. That's the way I remember it, okay?

"Now, Hans Bethe tells me that I was very depressed, and that he had to hold my hand and put me back together and all this kind of stuff. That psychological trauma I don't remember at all. I just decided—" his voice parodied bureaucratic resignation "—I'd just have to write it up, that's all, because they're not understanding." Then he hunched forward, leaning over his desk. "On the other side, I talked to Schwinger in the halls. And we had *no difficulty* with each other. We had *faith* in each other. If he would say something, I wouldn't argue that his equations might be wrong. If I'd say, 'I'd do it this way,' he didn't bother. He didn't even try to understand it. And I didn't try to understand his equations. But we could understand each other, if not the math. Because we had done the same problem. We talked about different kinds of terms that came in, compared notes, had a nice conversation. So I knew that I wasn't crazy."

It is there in the notes of the participants, this sense of two young men leaving the old guard in their wake. Physics was changing indeed when Niels Bohr, the man who personified the taste for the radical solution, found himself in the role of the ossified fogy. Feynman and Schwinger sat down with their notebooks—they each kept notebooks full of unpublished work—and compared notes. Each man's methods was opaque to the other; to their mutual pleasure, they still kept getting the same answers. "We talked to each other," Feynman said. "We compared notes, we exchanged ideas, I helped him, he helped me, I stole—not stole, I *used*—some of his clever tricks for doing integrals, but I referred to it, I liked it, I thought he was clever, I think he's a smart guy, and vice versa. I don't feel like I felt strongly competitive. In an amusing way, yes. In a mildish sort of way, it's hard to explain, an *amusing* competition like when two friends are racing."[23]

The competition may have become a bit more serious over time, as physicists read Feynman's papers, learned to draw pictures, and forgot many of Schwinger's methods. Freeman Dyson, a physicist now at the Institute for Advanced Study, gave many theorists much more confidence in the whole enterprise when he proved what had been only hoped, that the methods of Feynman, Schwinger, and Tomonaga were mathematically equivalent.[24] They all led to the same result, all proved the same thing, all demonstrated that quantum electrodynamics made sense. Quantum mechanics and relativity could become the cornerstones of a field theory; at long last, physicists had a place to stand on. For their contributions to quantum electrodynamics, Richard Feynman, Julian Schwinger, and Sin-itiro Tomonaga shared the Nobel Prize for physics in 1965.

□   □   □   □   □

And what about alpha, the fine structure constant, $e^2/\hbar c$, the famous 1/137? The latest tables of the physical constants give its magnitude as 1/137.03604, but we know no more than before *why* it has this value, and not another. In token of the mystery, Julian Schwinger's fine Italian sports car had vanity license plates with the number 137; Wolfgang Pauli died in a hospital room with that same number on the door.[25] Sir Arthur Eddington thought that the cosmos contained exactly $(137 - 1) \times 2^{256}$ protons and $(137 - 1) \times 2^{256}$ electrons; he may even be right, although it would be difficult to find a physicist nowadays who supports this notion.[26]

The fine structure constant remains, tantalizing and seductive to those theorists afflicted with the old physicist's dream of explaining the immensity and beauty of the cosmos on the basis of pure and simple whole numbers. A few years ago, Victor Weisskopf met Gershon Scholem, a prominent scholar of Jewish mysticism. Scholem told Weisskopf that one teaching of the Kabbalah is that every Hebrew word has a matching number with a meaning that can be decoded. Later, when Scholem asked about some of the questions that perplexed physicists, Weisskopf immediately thought of the fine structure constant. "His eyes," Weisskopf said later, "lit up in surprise and astonishment: 'Do you know that one hundred thirty-seven is the number associated with the Kabbalah?' " To this day, Weisskopf likes to remark, it is the best explanation that he knows.[27]

There is a hoary old particle physics joke about Wolfgang Pauli's first day in Heaven. At the pearly gate, Saint Peter says, "Pauli, God wants to meet you right away," and shows him the proper direction. At God's palace, the Deity says, "Pauli, you've been a good man, and I want to reward you in some way. Ask me any question you like." Without hesitation Pauli says, "Explain the fine structure constant." So God goes to the blackboard and starts writing, and Pauli listens in pleasure. But after two minutes Pauli stops smiling; after five minutes, he's shaking his head—and suddenly he's up on his feet, hissing, *"Das ist ganz falsch!"* That's *completely* wrong!

□   □   □   □   □

We could not count ourselves among the old baron's friends; we met him only once, at his apartment in the old town of Geneva, on a frigid gray day in February, 1984. A discreet brass plate outside the door bore the words: E. C. G. Stueckelberg, 20 rue Henri Mussard. Stueckelberg's wife ushered us into a close, dark, oppressively hot office, and left to tell her husband, whom she referred to as the "professor," of our arrival. Lighted only by a window facing the sleet-gray foothills of the Jura Mountains, the small room was jammed with the impedimenta of Victorian life: grand piano, rolltop desk, oversized sofa piled with wooden boxes. Portraits of Stueckelberg's titled ancestors marched across the walls, their frames done in faded gilt and black. Scattered everywhere were hundreds of books: philosophy, physics, biology, theology, and genealogy; Goethe, Swinburne, and Pauli; English, French, German, and Danish; a life's accumulation of learning, heaped in voluptuous and dusty confusion.

Stueckelberg appeared in the doorway, supporting himself on two canes. His sport coat hung on him slackly; plastic bags of tobacco and pipe cleaners were taped to his canes and on the armrest of his chair; his hair was mostly gone. The window light played across his long, angular face, the features thinned by his long fight with gravity. He fumbled in the bags and, after a few moments, extracted a fat, solid meerschaum pipe. "I'm living entirely on medicaments," he said suddenly. "I have terrible arthrosis, arthritis, whatever you call it." He had spoken English rarely since he taught at Princeton a half-century before. From his shirt he withdrew a thick pair of glasses. A match flared; he sucked the pipe into smoky life with evident satisfaction.

Ernst Carl Gerlach Stueckelberg, Baron Souverain of the Holy Roman Empire of the Teutonic Nations of Breidenbach at Breidenstein and Melsbach, professor of particle physics at the universities of Geneva and Lausanne, and the man who just may have first renormalized quantum electrodynamics, was born in Basel on February 1, 1905.[28] His father's family had been citizens of the canton since the fourteenth century; his mother was the baroness of a minute, forgotten fiefdom in central Germany. Stueckelberg acquired his Ph.D. in Munich at the age of twenty-two, under the aegis of Arnold Sommerfeld. Sommerfeld's name was sufficient to win him a post at Princeton University, in Princeton, New Jersey, where he taught until the Depression forced the school to let him go.

Once back in Switzerland, Stueckelberg had the first bit of what was to be a long run of misfortune. He discovered that even though he had been an associate professor at Princeton, he did not have the academic qualifications to teach in Switzerland, and thus was obliged to write another thesis. For many years after, he could find a job only as a *Privatdozent*, a poorly paid teaching assistant, at the University of Zürich. To make matters worse, he foolishly invested and lost the considerable wealth that belonged to his first wife, whom he married in 1931. Facing bankruptcy, he was forced to go into the military service to earn his keep, further delaying his academic career and making it difficult to leave Switzerland, something of a backwater, to meet his colleagues. Under considerable financial and personal pressure, Stueckelberg began to exhibit symptoms of what would today be called manic-depressive behavior. Most of the time he was rational, even brilliant, but occasionally he would feel a fit coming on and pack himself off to the asylum for a few weeks. Over the years, he had a score of different treatments, including electroshock therapy. Nothing helped.

Personal troubles notwithstanding, he gradually managed to acquire a small reputation for the originality and difficulty of his work. Unfortunately, the reputation had to be spread by word of mouth, because many of Stueck-elberg's most important thoughts were dismissed out of hand by his colleague in Zürich, Wolfgang Pauli. He predicted the first of the hundreds of subatomic particles discovered shortly before and after the war, but did not publish the idea after Pauli told him it was ridiculous. (Later, the Japanese physicist Hideki Yukawa received a Nobel Prize for this idea.)[29]

When Stueckelberg did publish, his papers were written in a convoluted style that not even Pauli could understand; they were further complicated by his habit, not uncommon among the most mathematically inclined theorists, of inventing a special notation to replace the helter-skelter of mathematical symbols that is the common language of physics. Stueckelberg switched the ordinarily used terms for variables with those for parameters, put the indices on the opposite side of the symbols, and filled his equations with an incomprehensible forest of curved arrows and colored letters. Moreover, his papers were usually in French and published in the venerable but not widely read Swiss journal *Helvetica Physica Acta.* "I practically always published there," he told us in the course of a long conversation. "It was easy—also, my secretary knew only French. This was one reason that my papers were never read. At that time, German and English were common languages for physics, but French was not. I also must admit that when I reread my papers later on, I saw that they were very complicated. I don't know why, but I had a very complicated style. By the way, my friend—he has since died—Professor [Jean] Wiegle always put on the introduction and the summary, and these are understandable."

Sometimes, too, he courted obscurity by his enthusiasm for questionable research programs. Convinced that all reality should be described by real numbers such as one, three-sevenths, or the square root of two, Stueckelberg devoted years to a quixotic attempt to eliminate imaginary numbers, such as the square root of minus one, from the equations of quantum theory. Unfortunately, imaginary numbers, whatever the difficulty one has in picturing them, are a central feature of contemporary physics, as firmly embedded in modern theory as pi in geometry. In the midst of such dubious schemes, he would sometimes toss out another, almost unrelated idea of fundamental import. As an aside to a paper on the atomic nucleus, for example, he postulated that the number of "heavy particles," by which he meant protons and neutrons, in the Universe never changes. If they could decay into other, lighter particles, matter itself would be unstable, ever so slightly radioactive, and the world as we know it would eventually disintegrate. Long ignored, the old man's postulate about the stability of matter suddenly resurfaced in the 1970s, when it became a key to the move toward unification.

In the mid-1930s, he began to consider the divergences in electrodynamics.[30] The infinities became a focal point of Stueckelberg's career; he lavished years upon their removal. Even decades later, when we met him, his face came alight when he discussed his theoretical strategems, the short-cuts he had devised, the cutoffs and approximations he had brewed. Reciting equations from memory, he traced their symbols in the air with small move-ments of his long, tobacco-stained fingers. Profligate with his ideas, he fol-lowed two separate tracks, a quantum and a classical approach.[31] Each was idiosyncratic; neither was understood; both were ignored. He explained his ideas to Pauli and Weisskopf; neither understood the presentation. They left Switzerland, and Stueckelberg, who was still in the army, was almost totally isolated from physics. Nonetheless, he apparently wrote up a lengthy pa-per—in English, for once—that outlined a complete and correct description of the renormalization procedure for quantum electrodynamics. Sometime in 1942 or 1943, he apparently mailed it to the *Physical Review*. It was rejected. "They said it was not a paper, it was a program, an outline, a proposal," Stueckelberg remembered. "Afterward, I was told that our friend and teacher, Gregor Wentzel—he was the expert [referee]—he got my paper." He rejected it? "Oh, it was done in an extremely obscure style," he said. Stueckelberg was not a bitter man. "Later, he took the manuscript and wanted to have it published to show that I got it before." We asked if he had the manuscript, which would help him establish priority. "I never cared much about that question," he replied. "I don't know what happened to the original copy. I lost it, it completely disappeared."

War swept over Europe. Even in neutral Switzerland, Stueckelberg was mobilized, although he obtained special dispensation from the army to teach his seminar every other week. It was his only contact with science. Nonetheless, he struggled to carry out the program rejected by the *Physical Review*. By the end of the war, in 1945, he seems to have done it.

The triumph, if there was one, was short-lived. His wife divorced him a year later. Long before, he had agreed to her family's demand for a marriage contract; now he courted ruin when he was forced to restore the fortune he had lost. When there was time between his need to scare up money and his sessions in the hospital, he wrote up bits and pieces of his ideas.[32] Eventually they were presented in a complete form in a chapter of the thesis of one of his students, Dominique Rivier.[33] But by then Schwinger had come out with his program, and Stueckelberg, who had the ideas first, published afterward.

He continued to do important work. In 1951, for example, he and his student, André Petermann, invented something called the renormalization group, which is now essential to the construction of grand unified theories.[34] Stueckelberg brought his dog, Carlo III, to seminars at CERN, the new particle accelerator laboratory outside Geneva. When Carlo barked, people would turn expectantly to Stueckelberg. He would survey the blackboard—

"There's always a mistake on them," he said—and point out the error. The rumor grew that somehow Carlo spotted the problems. As Stueckelberg grew older, he appeared less often at seminars. By the mid-1960s, years of experimental medication had slurred his speech, interfering with his thinking. Crippled by arthritis, he was carried to colloquia in the arms of his former students—a painful procedure that Stueckelberg described to us in detached, ironic detail, chuckling every now and then at his own frailty. He married again. He turned to the embrace of the Roman Catholic Church.

After we had talked for a couple of hours, he abruptly extended a hand as light and dry as a dead leaf. He was tired; the interview was over. The array of barons on the wall glowered in the deepening twilight. The old man gathered up his two canes and painfully lifted himself out of his chair. A clock ticked away. Stueckelberg nodded toward it and said, "I look forward every day to my eventual journey to Heaven." A heavy gold cross dangled from his thin neck. He was trembling slightly from the effort of standing. "We live too long," he said.[35]

Seven months later, on September 4, 1984, Ernst Stueckelberg was buried in Geneva at Plain Palais, the cemetery where Calvin had been laid to rest three centuries before.

Strange Interlude

9

From Deepest Space

ON MARCH 30, 1910, A COLD DAY IN PARIS, FATHER THEODOR WULF OPENED the elevator door at the top of the Eiffel Tower and pulled his equipment onto the platform.[1] A physics instructor in a Jesuit high school at the Dutch town of Valkenburg, Wulf was also an amateur scientist. A thousand feet above the Champ de Mars, Wulf spent the day measuring with glass and metal instruments how well the air conducted electricity. The results surprised him no end.

Like many other turn-of-the-century scientific dilettantes, Wulf was fascinated by radioactivity, which had remained novel and mysterious since its discovery fourteen years before by Antoine-Henri Becquerel. In the course of his plodding, methodical inventory of the properties of radioactive materials, Becquerel had found that the lumps of uranium in his laboratory made the air nearby conduct electricity—something that the atmosphere does not do under ordinary circumstances.[2] It happens because the alpha and beta ray emissions from radioactive materials knock out electrons from air molecules, creating *ions* that conduct electricity. The higher the level of radioactivity, the greater the air's conductivity.

The instrument then used to measure electric charge was the electroscope, a device consisting of two little strips of metal foil hanging like two stiff flags from a central rod. Strips and rod are sealed inside a glass jar, and the whole ensemble looks something like a TV aerial in a bottle. Electric charge spreads apart the two lower ends of the strips—a similar but opposite effect is "static cling"—with the distance apart indicating the strength of the charge. When the electroscope is charged up near some uranium, the conductivity of the air around the two strips causes them to discharge, and their ends fall limply back together. Scientists realized that this effect made it possible to press electroscopes into service to measure radioactivity, with the rapidity of discharge indicating the strength of the source.

Early radioactivity buffs had found, however, that their instruments "leaked," slowly discharging even when no uranium was around.[3] Called *residual discharge*, this slow dissipation interfered with measurements.[4] Annoyed, experimenters tried to get rid of it—and couldn't.[5] Even insulating the electroscope with five tons of pig lead could not stop it from losing its charge.[6] By 1905, it was clear that something was going on. The physicists

147

who had used electroscopes to measure radioactivity were now staring in confusion at the electroscopes themselves.

Father Wulf made his first contribution to the puzzle in 1909, when he invented an electroscope so sensitive that it was rapidly adopted by scientists everywhere. The precision of the Wulf electroscope of course highlighted the fact that it inexplicably and inevitably lost its charge, and the hunt for residual discharge began in earnest. Meteorologists and geologists as well as physicists took part, because for all anyone knew the cause might have to do with climate or the structure of the earth. Scientists took Wulf electroscopes all over the world to measure the discharge in different places, climates, and times. Wulf himself took one on a tour of Germany, Austria, and the heights of the Swiss Alps.[7] Commander Robert F. Scott's ill-fated expedition to the pole in 1911–12 included a meteorologist who took measurements over the open sea and on the Antarctic continent with a Wulf electroscope.[8] Residual discharge showed up everywhere, in different amounts, and scientists concluded that it must somehow be caused by low-level background of radioactivity in the crust of the earth.[9] Wulf wanted to make sure, by going to the top of the bizarre, skeletal steel structure that was then the highest building in the world: the Eiffel Tower.

Wulf knew that the thousand feet of air between him and the ground would absorb nearly all the earth's radioactive emissions, and that radioactivity from the Tower itself was almost nonexistent. But throughout his four days there, the electroscope still discharged. Something was causing the loss, but it was not in the Tower, the ground, or the electroscope itself. By August, Wulf concluded that there must be "either another source [of radioactive emissions] in the upper portions of the atmosphere, or their absorption by the air is substantially weaker than has hitherto been assumed."[10]

The remarks of the good Jesuit intrigued Victor Hess, a recent arrival at the newly founded Vienna Institute for Radium Research. Like many early students of radioactivity, Hess was careless about radium, eventually losing his thumb to radiation burns.[11] After checking to see that the air really should have absorbed the radiation from the ground before it arrived at the top of the Eiffel Tower, Hess began to "believe that a hitherto unknown source of ionization may have been in evidence in all these experiments. . . ."[12]

A tenacious and stubborn man, Hess decided that the only good way to test Wulf's results was to go even higher. In 1911, that meant taking a balloon, a dangerous procedure at best. The tiny and unpressurized baskets of fin-de-siècle balloons were easily buffeted by wind; worse, a single spark could blow up the hydrogen in the gas bag. Indeed, the two previous attempts to take electroscopes up in balloons had been plagued by instrument failure.[13]

Still, Hess could think of no other way to proceed. From 1911 to 1913, he went up ten times. Like Father Wulf, he found that the residual radiation decreased with altitude, but not as much as it ought to if it were caused by terrestrial gamma rays.[14] (Gamma rays, a third type of radioactive emission along with alpha and beta rays, are high-energy photons.) An hour or so after dawn on August 7, 1912, Hess began his ninth flight by taking off from Aussig, outside Prague. Six hours later, the orange-and-black balloon landed in a pasture about thirty miles east of Berlin. There is a photograph of Hess peering out through the ropes of the balloon, smiling against the sun, a wave of farmers lapping against the side of the basket. He was just twenty-nine, and had discovered something important.[15]

As in previous flights, the ionization of the electroscope had slowed down until the balloon reached a few hundred meters and then leveled off. At six thousand feet, however, the electroscopes began to discharge more rapidly than ever before. At fifteen thousand feet, the rate of discharge was more than twice ground level. Hess came to the unsettling conclusion *"that rays of very great penetrating power are entering our atmosphere from above."* At first Hess guessed that these penetrating rays came from the sun. But half the balloon flights had taken place at night, and there was no difference between the results of the day and night flights. Clearly, the sun was not the source of the rays.[16] They must come from outer space.

Hess's ideas were greeted with derision. The idea that interstellar rays were continually bombarding the earth with enough strength to shoot through several feet of lead was too wild for many scientists. It was intimated that the low pressure of the heights had confused Hess's instruments; that high-altitude electrical effects overwhelmed the electroscope; that Hess was simply incompetent.[17] While Wulf worked to improve his electroscopes, Werner Kolhörster, a German, made five balloon flights, culminating in a marathon ascent to thirty thousand feet, slightly higher than Everest, on June 28, 1914. He found the ionization level there twelve times that at sea level.[18] Hess was right.

Unfortunately, the same day that Kolhörster took his electroscopes to record heights, the heir to the Hapsburg throne was assassinated on the ground, igniting a chain of events that led with giddy speed to warfare throughout Europe. The balloon flights ceased; the mountain laboratories were abandoned; and all research into the strange radiation from above stopped as Western civilization turned its attention to destroying itself.

Thus began the tale of what would later be called *cosmic rays*, a powerful emanation which bathes our planet from deep space and that scientists are only now beginning to understand. The story of cosmic rays, in a sense, is a tale apart; beginning almost accidentally, even trivially, with the inability of laboratory equipment to stop leaking electricity, cosmic ray physics grew into a hybrid discipline of its own, populated by adventurous sorts who were unwilling and even unable to confine themselves to the laboratory.[19] Traveling across the globe and into the gelid heights of the stratosphere, they discovered that cosmic rays smash into ordinary atoms with enough power to change energy into matter and create dozens of new particles with strange new properties. Ultimately, cosmic rays provided the raw material for the standard model—the matter it classifies, the interactions it describes. Karl Darrow, the organizer of the Shelter Island conference, once wrote, "The subject is unique in modern physics for the minuteness of the phenomena, the delicacy of the observations, the adventurous excursions of the observers, the subtlety of the analysis, and the grandeur of the inferences."[20]

And, Darrow might have added, for the depth of the confusion. Right ideas were often held for wrong reasons, wrong ideas for right reasons; accidents, feuds, and ideologies rather than logical deduction shaped the scientific conclusions. After the war to end all wars, when cosmic ray studies resumed, the greatest share of woes befell a newcomer to the field, Robert A. Millikan, the most famous American scientist of the day.[21]

The son of a Congregationalist minister, Millikan won the Nobel Prize for physics in 1923—only the second bestowed on an American.[22] (He was the first to measure the charge of the electron accurately.) Much of his celebrity was due to his ardent attempts to mediate between the religion of his father's generation and the scientific accomplishment of which Millikan was an exemplar at a time when conflict between them seemed unresolvable. He was a deeply Christian man of science in the 1920s, a time when the nation was preoccupied by Prohibition, the Ku Klux Klan, and the Scopes trial. He believed that religion without science had in the past produced "dogmatism, bigotry, and persecution," but he also argued that science was less important than *"a belief in the reality of moral and spiritual values."*[23] To the pillars of the social order this was welcome news; the *New York Times*, for one, suggested that Millikan's moral stance was "even more significant" than his physics.[24] A rigid, uncompromising man, Millikan lacked the essential quality of doubt; he treated scientific ideas as if they were tenets of a creed. The lapses caused by his dogmatism were magnified by his genius for self-publicity, which ensured that he was always surrounded by a virtual cloud of reporters.

Millikan had read the work of Hess and Kolhörster in 1914, and decided to look for penetrating radiation himself. Believing that manned flights could not reach sufficient altitudes, he set out to develop extremely light, self-recording Wulf electroscopes for unpiloted balloons. His work was stopped by America's entry into the war, but when the conflict ended he and a graduate student built tiny, seven-ounce instrument packages, each with a Wulf electroscope, temperature and pressure gauges, and a camera to record the readings. In the spring of 1922, Millikan launched them from Kelly Air

Force Base at San Antonio, Texas; one flew to fifty thousand feet, almost twice as high as Kolhörster's highest flight. Millikan found that the rate of discharge was much less than the Europeans had reported. Although he had been prepared to confirm the existence of penetrating rays, he concluded that he had "definite proof that there exists no radiation of cosmic origin having such characteristics as we had assumed."[25] A second experiment, conducted during a snowy week in a hut atop Pike's Peak, 14,100 frigid feet above sea level, produced similar results.[26]

Millikan's data were right, but his conclusions were wrong. As scientists discovered later, cosmic rays are distributed unevenly over the earth. By chance, Millikan performed his experiments in regions of the American West with unusually low levels, and he incorrectly concluded that the Europeans were mistaken. Annoyed, Hess argued that Kolhörster's data were more trustworthy than Millikan's, while Kolhörster carried out further experiments on Alpine glaciers to confirm those of his balloon flights.[27]

Not one to tolerate even the suspicion of error, Millikan looked a third time in 1925. Cosmic rays, if they existed, were assumed to be gamma rays from the stars. Millikan knew that a gamma ray able to travel through, say, a foot of water before being absorbed would also be able to traverse 1,116 feet of air, because water is that many times denser. In August that year, he took his equipment to two deep, snow-fed lakes in the San Bernardino range of southern California: Muir Lake, 11,800 feet above sea level, underneath the brow of Mount Whitney; and Arrowhead Lake, three hundred miles south and 6,700 feet closer to sea level. In terms of gamma ray absorption, Millikan calculated that the big difference in altitude should correspond to six feet of water. If the penetrating radiation really comes from the stars, Millikan reasoned, he should detect the same amount of radiation six feet beneath the surface of Lake Muir as he did at the shore of Lake Arrowhead, for the cosmic rays would have traversed an equivalent amount of matter. The levels were the same. Millikan reversed himself, and asserted that he had never *really* come out against cosmic rays.[28] (In hindsight, he was right for the wrong reason, because cosmic rays are not gamma rays, do not behave like them, and are not equally absorbed by equal masses of air and water. Had Millikan's equipment been more sensitive, he would have discovered the different absorption levels, and once again drawn the wrong conclusion.)

Millikan's results excited the public, convinced the American scientific community, and pleased his European colleagues—at first. But Hess and Kolhörster were angered when Americans soon began to call Millikan the *discoverer* of cosmic rays. Although Millikan himself had coined the name "cosmic rays," the *New York Times* suggested editorially that they be called "Millikan rays" to honor "a man of such fine and modest personality"; American journals spoke of "M rays."[29] There ensued a ludicrous feud between

Millikan and the European cosmic ray workers as to who had produced the first definite evidence of their existence.[30] Although most American scientists came to realize that Millikan had been the latecomer, a history of cosmic rays published in the United States as late as 1936 described Hess and crew as a "nationalistically minded but misguided group of scientists."[31] But that year, almost a quarter century after his epochal balloon flight, the Nobel Prize in physics was awarded to Victor Hess, "for his discovery of cosmic radiation."

Throughout the 1920s, cosmic rays were assumed to be a more powerful form of gamma rays; Millikan in particular was a champion of this view. The first experiment to challenge this assumption was made by Kolhörster and another German physicist, Walther Bothe. Bothe had been initiated into physics by Hans Geiger, who taught him about a new instrument for studying radiation, the Geiger counter.[32] Named after its developer, the Geiger counter was a vast improvement over the old particle detectors—Wulf's electroscopes and Rutherford's scintillation screens. It consists basically of a gas-filled metal tube with a thin wire down the axis. The wire is positively charged; the tube walls have a negative charge. When a charged particle shoots through the tube, it strips away some of the electrons in the gas atoms. The negative electrons can be pulled toward the positive wire with such strength that they actually collide into more gas atoms on the way, knocking off additional electrons. Domino-style, the cascade spreads, until by the time it reaches the Geiger counter wire the flood of electrons causes a noticeable jolt in the current. In some models, the current surge is translated into the "click" now famous from movies and television shows. Geiger counters can detect charged particles with impressive efficiency but are not nearly as good at detecting uncharged photons, meaning that the counters are useful for studying only certain types of interactions.

In the spring of 1929, Bothe and Kolhörster used a brilliantly simple arrangement of Geiger counters to test the nature of cosmic rays.[33] They stacked one counter on top of a second, and counted the number of coincidences, or times the two registered at precisely the same instant. Although there would be an infinitesimal number of "sheer" coincidences caused by two simultaneous cosmic rays, nearly all the signals would be due to single particles, which Bothe and Kolhörster assumed would be electrons struck by gamma rays with enough force to be thrown through both counters. Bothe and Kolhörster knew that materials with lots of protons in their nuclei, such as gold, are particularly good at catching electrons. To eliminate penetrating electrons, the experimenters therefore inserted a two-inch slab of gold between the Geiger counters. If, as was generally believed, cosmic rays are photons, the coincidences should vanish; it seemed unreasonable to imagine that an incoming photon would hit an electron in one counter, tunnel through two inches of gold into the second counter, and then hit a second electron.

Unexpectedly, the slab barely put a dent in the number of coincidences: Fully three-quarters remained. To Bothe and Kolhörster, this strongly suggested that the Geiger counters were not picking up recoil electrons, but the high-energy cosmic rays themselves. This meant the conventional wisdom was wrong: Cosmic rays are not gamma rays. They are charged particles powerful enough to punch through gold without blinking an eye.

Many cosmic ray specialists thought the experiment somehow must be wrong—especially Millikan, who insisted cosmic rays were photons. When reporters who followed Millikan asked whether cosmic rays might be charged particles after all, he snapped, "You might as well sensibly compare an elephant and a radish."[34] What Bothe and Kolhörster observed, Millikan argued, was not *primary* cosmic ray particles, but *secondary* particles, particles that had been hit by cosmic rays.[35] Bothe and Kolhörster could not disprove this. But they did know that if primary cosmic rays consisted of charged particles, their paths would be affected by a magnetic field. The problem was to find a magnet that could reach out into space far enough so that physicists could study its effect on incoming cosmic rays. There was no magnet big enough on earth except one, the earth itself.

In the 1920s, the Norwegian Fredrik Størmer had discerned that charged particles spat out in solar flares are bent by the earth's magnetic field to the poles, where their shivering aerial dance creates the aurora borealis.[36] If cosmic rays were indeed charged particles, they, too, should ride the magnetic lines and come to ground in greatest numbers at the poles. A Dutch physicist, Jacob Clay, took electroscopes on boat trips between Holland and Java in 1927, and found that there was half again more radiation in northern Europe than near the equator.[37] Millikan had not found any latitude effect on his 1926 journey by boat to Bolivia—one reason for his conviction that cosmic rays were photons.[38] Bothe tried to resolve the disagreement himself with a sojourn in Spitzbergen, a large Norwegian island a few hundred miles away from the North Pole. He found no evidence of any latitude effect, but ascribed his failure to his inability to make headway with Størmer's calculations.[39]

The controversy inspired European scientists to measure cosmic ray intensities in the most diverse locations and the highest altitudes. Hess and some colleagues set up electroscopes in mountain observatories in the Alps.[40] Kolhörster took his equipment to the salt mines in Stassfurt, now part of East Germany.[41] The development of pressurized balloon cabins allowed physicists to bob up to unprecedented altitudes in search of cosmic rays. The first balloon flight to the stratosphere occurred on May 27, 1931, when Auguste Piccard, a former pupil of Einstein, and Charles Kipfer, his assistant, took an electroscope up from Augsburg, now part of West Germany. When the hot sun of the heights baked the seven-foot cabin, the crew survived by licking drops of water off the walls. Italian poet Gabriele d'Annunzio, moved by the thought of courting death in the name of knowledge, begged Piccard to be taken on a flight; never one to miss the opportunity for an extravagant gesture, D'Annunzio announced his readiness to be thrown overboard as ballast if necessary. Far better to die the noble death of being tossed from a balloon, the poet said, than to pass away "shamefully between two sheets."[42]

Such flights contributed to little except the romance of science. The practice of randomly measuring cosmic ray intensities in picturesque settings was an impossibly inefficient way of resolving the confusion over the latitude effect. Ultimately, these studies of the upper atmosphere became the absurd pretext for an early version of the space race in which European, American, and Soviet aeronauts battled feverishly to fly ever more advanced stratospheric balloons. Inevitably, disasters occurred. Three Russians soared to a record height of thirteen miles before their bag burst and they plummeted to the ground; according to some reports, their last words were, "We have studied the cosmic rays." They were buried in the Kremlin Wall.[43]

In the meantime, the venue for cosmic ray research had shifted definitively to the United States, where Robert Millikan was locked in yet another widely publicized fight, this time with a former student, Arthur Compton. The third American physicist to win a Nobel Prize, Compton earned his award in 1927 for describing what happens when a photon and an electron collide, a phenomenon now called the Compton effect. Bothe and Kolhörster's experiment had strongly impressed on him the possibility that cosmic rays were charged particles. Like Bothe, Compton believed that the issue would only be settled by discovering whether there was a latitude effect.

In 1930 Compton asked the Carnegie Institute, in Washington, D.C., for money to mobilize a worldwide study of cosmic rays. Once funded, Compton divided the world into nine regions and sent a different expedition to each. In all, over sixty physicists took part. Allen Carpé of the American Telephone and Telegraph Company laboratory in New York City was sent to Alaska, where he and another man climbed the slopes of Mount McKinley, fell into a crevasse on the Muldrow Glacier, and died. (Their equipment was found and the readings used in the paper.) Compton himself filled in all the gaps by embarking on a round-the-world trip with his wife and teenage son in March 1932.

Millikan, too, applied to the Carnegie Institute in late 1931 for money to perform his own series of experiments. Now working with Henry Victor Neher, a young Caltech Ph.D., Millikan developed a new and more sensitive type of electroscope, which they tested for steadiness by giving it a ride down some bumpy roads in a 1928 Chevrolet. During one test drive, Millikan peered into the eyepiece while Neher struggled with the unfamiliar clutch; a lurching shift into first smashed the electroscope into Millikan's nose, barely missing his eye and bequeathing a permanent scar. In September 1932, Neher set sail with his electroscopes on a ship going from Los Angeles to Peru.

Millikan missed the latitude effect on his 1926 trip to the Andes because just south of Los Angeles there is a cosmic ray "shelf" where the intensity abruptly dips and levels off; he happened to switch on his electroscope after the boat had passed it.[44] By sheer bad luck, Neher missed the "shelf" on his trip, too. Neher had two electroscopes with him, but only one was working properly at the start of his journey. That one malfunctioned within forty-eight hours of the launch, and became completely useless shortly thereafter. To his dismay, Neher discovered that the seas were too rough to allow him to repair the other until the boat anchored at Mazatlan, on the Mexican coast. By then, he had already passed the critical point. When Neher arrived in Panama, he sent a telegram to Millikan with the news that there was no latitude effect.

On September 14, 1932, Compton announced that cosmic rays varied considerably in intensity from the equator to the North Pole. Speaking before a crowd of reporters, Compton said, "Obviously if the north magnetic pole has any effect on the rays, they must be electrical in nature instead of a wave, as Dr. Millikan contends [that is, charged particles and not uncharged photons]. The difference shown by my experiments will be a severe blow to Dr. Millikan."[45] Millikan refused comment, saving his fire for the annual meeting of the American Association for the Advancement of Science (AAAS). Three days before his talk, he received a telegram from Neher: SEVEN PERCENT CHANGE RETURNING STOP CONCEALED BEFORE BY BROKEN SYSTEM AND DIFFERENT SHIPS STOP NEHER. Coming up to New York, Neher had again passed the cosmic ray "shelf"; this time, however, his electroscope was working.[46] Faced with overwhelming evidence, Millikan still blasted Compton in his speech. But hours afterward, reason returned. He conceded the point, and sent a long and savage telegram of denial to the *New York Times*, which had correctly reported his earlier declaration that cosmic rays are photons. In the telegram, Millikan insisted that Compton and he were in complete agreement. When published, the AAAS speech was completely rewritten, thus hurting Millikan's reputation further. Indeed, in his later autobiography, he made but passing reference to cosmic rays, the subject that had absorbed twenty years of his life.[47]

□   □   □   □   □

There were two main topics of conversation at the cosmic ray section of the International Conference on Physics at London in 1934, both with profound implications for the future of physics. The meaning of one passed almost unnoticed for forty years; that of the second was seized upon immediately. Five years before, a Franco-Soviet collaboration discovered that cosmic ray collisions sometimes occurred in connected bursts that sprayed across cloud chambers like shotgun blasts. By the London conference, people had counted showers of up to twenty particles and realized that the energy needed to produce the initial cosmic ray is literally astronomical, the kind of energy found in the sun and stars.[48] In fact, the figure was so high that most physicists considered the relation between subatomic particles and cosmic phenomena to be out of the purview of physics altogether.[49] Only in the 1970s, forty years later, would the connection between the smallest constituents of matter and the greatest reaches of the Universe at last be made.

Much more readily acted upon was the puzzling absorption of cosmic rays by various amounts of matter. Several conferees pointed out that cosmic rays seemed to come in two groups, with markedly different energies.[50] Cosmic rays of the first group, dubbed the "soft" component, were easily absorbed by a few inches of heavy metal, and were evidently a mixture of electrons and positrons. The other component was much "harder," and could pass happily through several meters of lead without slowing down. Nobody in London knew quite what to make of the "hard" particles. They were not heavy enough to be neutrons or protons, yet they punched through lead more readily than light little electrons. A little while later, Caltech experimenters began the tradition of irreverent nomenclature in subnuclear physics by announcing that the "solution" was to say that the soft component consisted of "red" electrons, and the hard of "green" electrons.[51]

Some physicists felt sure that both soft and hard cosmic rays were electrons, but that the latter were moving at such great speeds that their behavior fell outside of the purview of quantum electrodynamics. This was an eminently reasonable guess: Charged particles radiate energy when they accelerate or decelerate, affecting the motion of the particle, in a way reminiscent of self-energy.[52] Negligible at low speeds, the effect rises with the energy of the particle. Because physicists were then wholly unable to explain self-energy, many expected that as experimenters studied particles moving at higher and higher energies, at some point the theory would begin to give wrong predictions. Theorists even thought they knew where—when the electron energy reached $137mc^2$ (m is here the mass of the electron, and mc^2 is the energy of the electron when it is at rest).[53] Seeing the approach of alpha, some researchers were sure it signaled the end of the theoretical road.

After the London Conference, Carl D. Anderson, the discoverer of antimatter, and a graduate student, Seth Neddermeyer, set out to nail down the identity of the hard "green" electrons, using the same cloud chamber

and magnet that Anderson had employed to discover the positron. Once
back at Caltech, they inserted a thin metal plate in the chamber, with Geiger
counters above and below that set off the chamber and a camera whenever a
charged particle passed clean through the plate. The trick of using Geiger
counters to trigger cameras was invented in 1932 by P. M. S. Blackett and
Giuseppe Occhialini; by the time Anderson and Neddermeyer looked for
hard electrons, experimenters routinely took thousands of photographs. The
magnet's field would make the incoming particles curve to one side. By
studying the degree of curvature, the range, and the level of ionization in the
tracks, it was possible to determine precisely a particle's mass and charge. If
particles were pushed one way, they were positive; if they arched in the
other direction, they were negative; and the heavier the particle, the less it
was affected by the magnetic field.

The Caltech team spent two years photographing the chamber and
gradually convinced themselves that the particles seemed to be disobeying
the laws of quantum electrodynamics, not because the theory failed, but
because the experiments were finding something that had never been seen
before—"particles of a new type," as Anderson put it—heavier than the
electron but lighter than the proton.[54] They were dissuaded by J. Robert
Oppenheimer, who flatly informed them that they were just looking at fast
electrons; his superior mathematical acumen and confident lack of confidence intimidated them. In August of 1936, Anderson and Neddermeyer
published several photographs that might have been left by new particles,
but, to their regret, they refrained from claiming discovery.[55]

The article created a small sensation across the Pacific Ocean, in Japan,
where a theorist named Hideki Yukawa had actually predicted the existence
of such a particle. Moreover, Yukawa had predicted that this particle, which
weighed less than a tenth as much as the proton, should be the agent that
kept the nucleus together despite the mutual electrical repulsion of the
protons. Convinced immediately that Anderson and Neddermeyer had
found his particle, Yukawa told his Japanese colleagues that they had to find
some of the things themselves before Anderson and Neddermeyer tumbled
onto what they had.

Yukawa was born in 1907 as Hideki Ogawa. Shortly after his first birthday, his family moved from Tokyo to Kyoto, where he spent the rest of his
childhood. He was a conservative, painstaking, methodical child, insisting in
grammar school that he could not work unless his desk was aligned parallel to
the stripes of the tatami mats on the floor.[56] A short boy with a shy round
innocent face, he spent so much time in solitary silence that his father, a
professor of geology at Kyoto Imperial University, considered recommending
that Hideki, alone of his five sons, not go on to higher education.[57] In 1932,

Hideki's father arranged his marriage to Sumi Yukawa, the daughter of a local physician; following a Japanese custom, Hideki was adopted by his wife's father and assumed the family name.

Soon after entering the university, Yukawa disappointed his father by dropping the practical study of geology and embracing theoretical physics of the most abstract variety. One of his university classmates was Sin-itiro Tomonaga, who later played a major part in renormalizing quantum electrodynamics; the two young men helped each other through quantum mechanics, and stayed on afterward as unpaid assistants in the physics department. In the spring of 1931, Yukawa attended a lecture by Yoshio Nishina, a researcher at the elite Institute of Physical and Chemical Research of Tokyo (usually known as Riken, the abbreviation of its Japanese name). A towering figure in Japanese physics, Nishina worked at both the Cavendish and Göttingen, and spent six years at Copenhagen with Bohr; returning to Japan at the end of 1928, he was a one-man incarnation of the spirit of the new physics.[58] Much of Nishina's lecture concerned quantum electrodynamics, and the half-revealed intricacies of photons and electrons made Yukawa decide to pursue the subject himself. Immediately he ran into the problem of the infinities.

Sitting in an office overlooking the pastures of the agriculture school, Yukawa struggled futilely with the divergences in quantum mechanics; while he worked, mountain sheep bleated mockingly beneath his window. His account of that frustrating time is an echo of his European colleagues' struggle:

> As I fought daily with the devil called infinite energy, the call of those sheep sounded to me like the sneers of that devil. Each day I would destroy the ideas that I had created that day. By the time I crossed the Kamo River to my home in the evening, I [would be] in a state of desperation. Next morning, I would leave my house feeling renewed strength, but again would return at night thoroughly discouraged. Finally, I gave up that demon-hunting and began to think that I should search for an easier problem.[59]

Yukawa turned his attention to the atomic nucleus.

Few physicists of the time found the nucleus to be "an easier problem." Heisenberg had made the most recent attempt to tackle it, in a lengthy, three-part article appearing in the *Zeitschrift für Physik* in 1932 and 1933.[60] The chief theoretical difficulty with the nucleus was to describe why the mutual electrical repulsion of the protons did not make it fly apart. Physicists since Rutherford had argued that something else must hold it together—electrons, most likely. Nobody had been able, however, to figure out how the proton-electron attraction could actually cancel out the proton-proton and electron-electron repulsion. In the last few years, theorists had taken to speaking vaguely of a "nuclear force" that held the protons together. Having no influence on the everyday world, the nuclear force must have an extremely short range; beyond a distance of 10^{-13} centimeters, it has no effect at all—as if the gravity that holds people to the surface vanished the instant they climbed a footstool, allowing them to float into the clouds. In his three papers, Heisenberg treated the attraction among protons and neutrons as analogous to the relatively common situation in which two hydrogen nuclei form a molecule by passing an electron back and forth. Although he did not have a clear picture of the process, Heisenberg imagined the nucleus as held together by a game of intranuclear "catch" in which the protons and neutrons rapidly toss a rather strange sort of electron among themselves.

Yukawa knew that the two known fundamental forces, gravity and electromagnetism, were both described by field theories (although, to be strictly accurate, general relativity is a classical field theory and quantum electrodynamics is a quantum field theory). He applauded Heisenberg's decision that the nucleus was held together by another fundamental force and that it, too, needed a field theory; Yukawa even translated part of Heisenberg's work into Japanese, adding his own introduction.[61] He liked the scheme well enough to worry about the problems that cropped up if one took it literally. In the model, neutrons can emit electrons, turning into protons in the process. (The total electric charge, zero, is the same before and after.) The electron is attracted to a nearby proton, which absorbs the electron and turns into a neutron. With a constant flow of virtual electrons flickering between them, the two particles are bound to each other. Neutrons and protons have the same spin, however, so creating electrons involves making a quantum of spin appear out of thin air—violating the laws of conservation of energy and momentum. For this reason, Heisenberg knew the exchanged particle could not possibly be a real electron. He was willing to entertain the possibility of a new "spinless electron" to get around the problem, but Yukawa thought this preposterous. Some other tack was needed.

The obvious strategy was to think more carefully about the type of particle that would be required to carry the nuclear force. Stated this way, Yukawa wrote later, the answer almost jumps at you, "but my brain did not work so quickly. I had to take a wrong path first, before I could arrive at my destination. Those who explore an unknown world are travelers without a map; the map is the result of the exploration. The position of their destination is not known to them, and the direct path that leads to it is not yet made."[62]

Yukawa spent two years on blind alleys. Depressed and harried, he exacerbated his loneliness by his formal, withdrawn mien and the necessity of commuting between two lecture positions in Kyoto and Osaka. In Kyoto he had a new office overlooking the road to the railway station; the bleating of sheep had been replaced by the rumble of heavy trucks. All the while, he struggled to construct a theory of the nuclear force that would not compel him to make up the existence of some particle that nobody had ever seen.

His spirits were raised unexpectedly in the summer of 1934, when he chanced upon an Italian journal containing an article by Enrico Fermi on beta decay, the form of radioactivity responsible for beta rays.[63] Beta decay occurs when a neutron emits an electron and turns into a proton. Although beta decay had been known to exist since the discovery of radioactivity in 1896, all theoretical explanations of the process had been plagued by pre-cisely the same difficulties that troubled Heisenberg's theory of the nuclear forces. During beta decay, particles in the nucleus actually seemed to create electrons with apparent disregard for the conservation of energy and spin. Fermi championed a solution that Pauli had thought of three years earlier but had been reluctant to commit to paper.[64] Pauli had wondered if the spin of the created electron might be canceled out by the simultaneous emission of another particle, electrically neutral and previously unknown to science, with the opposite spin. He, too, had been unsettled by the thought of making up a fundamental constituent of matter solely to resolve a theoretical difficulty, and had never quite wanted to publish it. This particle was called, variously, the "neutron," "Pauli particle," and "neutrino." The last name, coined by Fermi as a joke, is the one that stuck.[65] Hoping that the new particle might also play a role in the nuclear force, two Soviet theorists rushed into a futile effort to produce a neutrino theory that would account for the stability of the nucleus. They ended up by demonstrating that the nu-clear force could not be linked to the neutrino, a conclusion that disheart-ened them—but not Yukawa.[66]

Learning about the neutrino, Yukawa wondered if he were not being unimaginative. He had been trying to arrive at the field by looking at the known particles; perhaps he should start from the field and see what particles were necessary to transmit it. But how did one go about this?[67] Sitting one sleepless October night in his small room, Yukawa realized that with some simple calculations he could figure out the mass of the particle that held together the protons and neutrons in the nucleus.

The calculation proceeded from the known range of the nuclear force. If the particle responsible for that force takes a while to decay, it can travel a long way in that time, and the force will be felt over a considerable distance. Because the range of the nuclear force is very short, its particle must quickly wink out of existence. Yukawa could thus come up with an approximate lifetime. He then recalled Heisenberg's uncertainty principle, which can also be phrased in terms of energy and time.

$$\Delta E \, \Delta T \geq \hbar$$

If the time (ΔT) approaches zero, then the energy (ΔE) must rise to giddy heights to keep the product near $\hbar$. Knowing the lifespan (in other words, ΔT) and the uncertainty relation, Yukawa could calculate the energy of the virtual particle (ΔE). If one considered the famous equation

$$E = mc^2$$

it is easy to find m, the mass of the particle at rest. Given that the range of the nuclear force is about 10^{-13} centimeters, Yukawa thought it should be transmitted by a particle that weighed about two hundred times as much as the electron, or, equivalently, about a tenth as much as a proton.

At the beginning of November, Yukawa began to put together a paper in English, which was published, many drafts later, early in 1935.[68] With its heavy sentences and slightly skewed syntax, the finished article bears the mark of Yukawa's imperfect mastery of the language. He begins with the assertion that it is natural to model the nuclear force on electromagnetism, and thus to describe it as a field, the "U" field. "In the quantum theory this field should be accompanied by a new sort of quantum, just as the electromagnetic field is accompanied by the photon."[69] This newly minted U quantum is the nuclear "glue" that shuttles electric charge among the protons and neutrons in the nucleus. Unlike the neutral photon, the U must come in two varieties, positive and negative. Thus, when a proton emits a positive U, it becomes a neutron; when a neutron emits a negative U, it becomes a proton. The scheme is tidy, but, as Yukawa admitted, "As such a quantum with large mass and positive or negative charge has never been found by the experiment, the above theory seems to be on a wrong line."[70] The answer to this objection, Yukawa said, is that this particle decays too fast to be spotted without great effort. Next, Yukawa brilliantly tried to kill two birds with one stone by arguing that beta decay is nothing other than the decay of a virtual U particle into an electron and neutrino. Almost in passing, Yukawa mentions one place the U quantum might show its face: the highly energetic interactions found in and around cosmic rays.

Yukawa gave a few lectures on the paper, but almost nobody but Nishina took it seriously.[71] Optimistically, Yukawa sent reprints to several European and American physicists, including Oppenheimer, and awaited their comments. Few came. When Niels Bohr came to Japan, Yukawa told him about the U quantum. Bohr's only answer was to ask if he enjoyed making up particles.[72] In the meantime, Yukawa wrote later, "I felt like a traveler who rests himself at a small tea shop at the top of a mountain slope. At that time I was not thinking about whether there were any more mountains ahead."[73]

He began to think again about the new particle when he came across the photographs by Anderson and Neddermeyer in the *Physical Review*.[74] Yukawa identified their photographs as his U quantum even before they were ready to claim they had discovered a new form of matter. Nishina and two other experimenters quickly tried to come up with some of the new particles themselves. During the 1930s, Nishina's Riken lab was a private organization largely supported by the industrial patents its engineers had won for inventing new methods of distilling sake.[75] It could not afford an electromagnet strong enough to bend cosmic rays. Discovering that he was competing with richer foreign groups, Nishina, an ardent nationalist, convinced the Imperial Japanese Navy to let him use a generator that charged submarine batteries.[76] Taking thousands of photographs at a frantic speed, the Riken team managed in the summer of 1937 to turn up a single cosmic ray track that stopped in the chamber, allowing them to estimate the mass of the particle; it was between 180 and 360 times that of an electron, exactly as Yukawa had said. But despite their efforts, they were beaten to the punch.

A few months before, Anderson and Neddermeyer had traveled to MIT, where they learned that two Harvard experimenters, Jabez Street and E. C. Stevenson, had used the same plate-in-the-cloud-chamber technique and were thinking of declaring that they had found the same new particle. His hand forced, Anderson dashed off a paper to the *Physical Review* claiming that "there exist particles of unit charge, but with a mass (which may not have a unique value) larger than that of a free electron and much smaller than that of a proton."[77] He was very annoyed with Oppenheimer.

The paper was published in May. Three months later, the disappointed Japanese sent their single picture to the *Physical Review*, which promptly sent it back. The editors refused to print the photo unless the accompanying text was cut drastically. A squabble ensued, delaying the appearance of the article and allowing Street and Stevenson to come in ahead of the Japanese.[78]

Until the discovery of the new particle, Yukawa's paper—a speculative article by an unknown Japanese theorist published in the obscure *Proceedings of the Physico-Mathematical Society of Japan*—received scant attention in the West. Anderson and Neddermeyer's announcement jogged a few memories. Oppenheimer pulled Yukawa's article off the shelf where he had put it two years before, and with Serber, sent a note about it to the *Physical Review*.[79] Reversing himself, Oppenheimer argued that the new particle did indeed exist, and, moreover, was Yukawa's U quantum, the particle that transmitted the nuclear force, although he and Serber did so with much hemming and hawing. (The hems and haws, as we shall see, were entirely justified.) The next issue of the journal included an article by Stueckelberg, who made the same identification in a more positive tone.[80] (Unlucky as ever, Stueckelberg had independently invented the U quantum, but was apparently talked out of it by Pauli. When he finally published, the paper was ignored.)[81] Other papers equating Yukawa's particle and the cosmic ray particle soon followed; Yukawa's shares soared on the physics bourse.

Yukawa's name for the new particle—the *U* quantum—neve̶ on, and a variety of bizarre terms began to circulate, including "penet̶ "dynatron," "heavy electron," "X-particle," "barytron," "Yukawa particle̶ and even "yukon." Taking advantage of their status as discoverers, Anderson and Neddermeyer attempted to end the confusion by sending a letter to *Nature* proposing the name "mesoton" (*meso* is derived from the Greek for "middle"). Anderson's mentor, Robert Millikan, soon objected. Years later, Anderson recalled,

At the time, Millikan was away, and after his return we showed him a copy of our note to Nature. *He immediately reacted unfavorably and said the name should be mesotron. He said to consider the terms* electron *and* neutron. *I said to consider the term* proton. *Neddermeyer and I sent off the* r *in a cable to* Nature. *Fortunately or not, the* r *arrived in time, and the article appeared containing the word* mesotron. *Neither Neddermeyer nor I liked the word, nor did anyone else that I know of.* [82]

Because of the general lack of enthusiasm for "mesotron," the matter was put to a vote the next year at an international meeting of cosmic ray researchers in Chicago. Millikan, a classicist manqué, stridently complained about the popular contraction *meson*, because it is an ordinary Greek word meaning "middle one." [83] No name received a plurality, although *mesotron* and *meson* led the pack. [84] Confusingly, scientists used both, although after the war *meson* eventually prevailed.

For a brief, shining instant—a few months at the end of 1937—cosmic rays made sense. The soft component was comprised of electrons that followed the laws of quantum electrodynamics, whereas the hard component consisted of mesotrons whose actions were described by an emerging branch of physics called "mesotron theory" or "meson theory." The state of the art of meson theory was presented in a lengthy paper by Heisenberg and Hans Euler in 1938. [85] They said that mesotrons are created at high altitudes by collisions between the still-mysterious incident cosmic radiation and the molecules of the atmosphere. A fraction of the mesotrons reach sea level, forming the hard component observed five years before. En route, some of the hard component interacts with nearby atoms, producing cascades of electrons—the soft component.

Most important, physicists believed that mesotron theory would account for the interactions among protons and neutrons, as quantum electrodynamics had taken care of interactions among electrons and photons. Just as the photon was the quantum of electromagnetism, the mesotron was the quantum of the force that held the nucleus together. If one chose to ignore the divergences in quantum electrodynamics—and many did in 1937—it was easy to believe once more that a grand synthesis accounting for all known particles and forces was not far off.

s for the remark that the Creator is subtle, but not
ists think this true, but their belief was sorely tested
\oped-for triumph of mesotron physics came apart in
\iic rays were not what they seemed. Far from being
\iesotron was an undreamed-of mystery now called
was not the key to the nucleus, not the solution to
____y, \iit was what Oppenheimer, in a bitter tribute to nature's un-
kindness, called a "ten-year joke."[86]

The joke was exposed under dramatic circumstances. In July 1938, Mussolini passed "racial laws" that prohibited Jews from holding such governmental positions as university professorships. Stripped of his job, the physicist Bruno Rossi left in October, and after some turbulence was offered a position at the University of Chicago by an old friend, Compton.[87] Bohr broke the traditional secrecy surrounding the Nobel Prize to tell Fermi that he was to be the next recipient, allowing Fermi, whose wife was Jewish, to plan his escape. After collecting his award in Stockholm, Fermi went directly to Columbia University.

The Italian physics community was devastated by the departure of Fermi, Rossi, and practically every other good physicist in the country.[88] As war approached, Edoardo Amaldi, the last remaining professor of physics at Fermi's Istituto Guglielmo Marconi, realized that if physics was to continue, he would have to band the remaining researchers into a single group. In that small, depressed group were Marcello Conversi and Oreste Piccioni, two young men who viewed themselves with the arrogance of their years as part of a "new generation" of physicists. Priding themselves on their skills with vacuum tubes and electronic components, they scorned the older generation, which venerated the arts of glassblowing and carpentry.[89]

They knew of the mesotron theory, and also knew that it had to be checked with precise measurements of the properties of the new particle. Believing that previous attempts to ascertain the mesotron's lifetime were flawed by inadequate electronics, Conversi and Piccioni threw themselves with almost religious zeal into the task of producing a circuit that could measure time differences to a ten-millionth of a second. Though one of the glassblowing older generation, Amaldi gave them whatever help he could.

Conversi avoided conscription due to amblyopia of his left eye; Piccioni was drafted but managed to be stationed in Rome; Amaldi gave lectures at six-thirty in the morning so that physicists in the service could attend before boot camp. At night, Conversi and Piccioni bought RCA tubes on the black market and started to work on their circuits, building the fastest electronic circuits on earth from scratch and stolen wire.[90] Their first workshop was on the university campus, where they could collaborate with their colleagues.

But the university was located alongside an important military target, the San Lorenzo freight station. After the Allied invasion of Sicily in July 1943, American warplanes appeared above the Palatine hill for the first time. Dozens of bombs rent the university campus. Conversi and Piccioni spent the next day carrying boxes of resistors and oscilloscopes to a deserted high school close to the Vatican, where the aura of holiness and political neutrality offered some measure of protection to those nearby. Working between air raids in a city on the point of starvation, they rebuilt their equipment in a basement used by the antifascist resistance as a weapons cache. They watched over underground transmitters and guarded rifles; grateful resistance leaders contributed scavenged tubes.

Early in September, with the Allies in Calabria, the Italian government fell and the Nazi occupation began. The two men pressed on—"Our work was the only pleasure we had," Piccioni explained—constantly afraid of being caught by the Germans.[91] (He was, once, but was ransomed by a friend's father for a pile of silk stockings.)[92] They stacked up counters and layers of metal in a score of arrangements to determine how much matter it would take to stop a mesotron. Just before the Allied liberation of Rome in June 1944 they showed that mesotrons lived slightly more than 2.2 microseconds before falling apart, a short lifetime but one that was still many times longer than Yukawa had predicted.[93] In their basement in the ruined city, Conversi and Piccioni realized that something was amiss.

At the cessation of hostilities, Conversi and Piccioni were joined by another young man, Ettore Pancini, whose nascent physics career had been interrupted by the necessity of getting out of the fascist army and into the partisan groups in the north. The three men planned a new experiment to study the puzzling mesotron. If the particle discovered by Anderson and Neddermeyer was indeed the particle in Yukawa's theory, the positive and negative mesotrons should behave differently when they were slammed into matter. The positive mesotrons should be repelled from the positive nuclei by the electromagnetic force, but attracted by the nuclear force; because electromagnetism has a long range, it would win out, and the particles pushed away and left to decay. On the other hand, negative mesotrons should be attracted by both forces, and would in almost all cases be absorbed by nuclei before having a chance to decay.[94] Using a set of magnetic "lenses," the three experimenters deflected cosmic ray particles into a carbon rod. In the carbon, all the positive mesons decayed with the usual rate—but so did many of the negatives. In other words, the negative mesotrons were not absorbed by the nuclei, but captured and put into orbit until they decayed.[95] The three men knew that Yukawa's nuclear force particle should plummet directly into the nucleus; the mesotron was therefore not the transmitter of the nuclear force.[96]

In the summer of 1939, as war loomed, the English government sent a group of physicists from Manchester to man a secret radar installation, one of several that had been built along the eastern and southern coasts.[97] Among the scientists was George Rochester, who had switched to cosmic ray work at the instigation of P. M. S. Blackett, a Rutherford student who did not share his master's predilection for the small and the cheap. Blackett had brought to Manchester a cloud chamber with an electromagnet that weighed eleven tons.

In 1984 we spoke with Rochester, now an emeritus professor at the University of Durham, in northern England, about that hectic time at the dawn of particle physics. He hadn't lasted long at the radar station, he said. They wanted physicists for other purposes. A slightly built man with a mild round face, Rochester was dressed in warm tweeds against the cold day. He showed us the Durham cathedral, which was visible from his office window. High, stately, immaculately preserved, the church had never been bombed; the entire north had suffered little from the air war. "At some point," Rochester said, "the Manpower Commission decided that scientists would be needed as the war effort went on. They'd made a frightful mistake in the First World War by pushing everyone out onto the front, and lots of the best young scientists were killed. So they sent me back to Manchester to run the honors school in physics.

"Now, each large establishment with a lot of buildings, like a factory or a university, was asked to form its own fire brigade, because the city fire brigade would have too much to do in the event of a real air raid. I led the university squad every sixth day and on many weekends. As there were few air raids on the city of Manchester, we had much spare time. Also on duty, but in Civil Defense, was a Hungarian physicist called Lajos Jánossy, whom Blackett had brought to Manchester. And whilst most of the other people on duty played bridge and billiards, Lajos and I talked about cosmic ray physics the whole time. Oh, there were training periods, and I would get my squad out and we'd run them up some ladders—that was quite a lot of fun—but then Jánossy and I would talk about the creation of mesotrons. Jánossy had already come to the conclusion that mesotrons must be created in large explosions—nuclear events, not electron showers—and had devised a method of selecting such events by a clever disposition of Geiger counters in a great mass of fifteen tons of lead. The lead cut out spurious events, like large electron showers." After 1940, they investigated what they called "penetrating showers," showing that they were produced by protons and neutrons, rather than photons. "All sorts of strange things appeared," Rochester said. "Penetrating charged particles, strange deflections in the gas—but we could not identify the particles because we had no magnetic field. We weren't allowed to use the Blackett magnet because it took too much power." Rochester and Jánossy stared at the cloud chamber as if they were watching a movie with the sound switched off, certain that something fascinating was happening but helpless to tell what it was. [98]

Japanese physicists, too, were conscripted for the war effort. Nishina led the Japanese drive to build an atomic bomb, which was severely hampered by material shortages and the conviction that the thing could not be built. [99] Although the Japanese had no uranium and few of the requisite technological skills, Nishina struggled to put together a bomb until the destruction of Hiroshima announced that the conflict was over. Tomonaga was employed by the Imperial Navy to develop microwave techniques for radar. Profiting by the esteem in which he was held, Yukawa avoided military duty and spent the war doing what he had done in peacetime, commuting between his teaching positions in Kyoto and Tokyo. The trains were often halted by bombing raids, and Yukawa would don a helmet and shoulder pads and quietly walk along the tracks to the next station. [100]

The last issue of the *Physical Review* entered Japan in July 1940; thereafter, the fledgling group of Japanese theorists was completely cut off from their Western confreres. [101] As the war exhausted the country, some scientists ran out of paper to write on, and were forced to calculate on the backs of previously used scraps. [102] Even before Pearl Harbor, most cosmic ray workers had realized that something was wrong with the mesotron: It wasn't absorbed easily enough and it lived too long. [103] The emerging discrepancies worried Yukawa, and a "Meson Club" was organized around Nishina to work on the problems. [104] In June of 1942, one club member, Shoichi Sakata, made the simple and ultimately correct proposal that there were *two* mesotrons, the one that experimenters saw in cloud chambers and the one that Yukawa had predicted, which Sakata assumed had not yet been seen. He wrote up the suggestion in Japanese that year, but it took a long time to reach the rest of the world. [105]

An atomic bomb fell on Hiroshima. Terrified by the devastation, the Army command asked Yoshio Nishina, Yukawa's mentor, to tour Hiroshima to decide if Truman's claim that there was an atom bomb was correct. Nishina arrived the day after the bomb into a scene that no human being has ever described successfully. With shocked calm, he went about the macabre business of estimating the size of the blast by inspecting the burns on corpses. In a smashed hospital he found some unused photographic plates for X rays. Like Becquerel a half-century before, Nishina developed them. They were black from radiation. He used a Geiger counter to test the bones of the dead men in the street. If this deeply patriotic man was humiliated that he had lost the race to build the bomb, he never talked about it. He told the American investigators that they should study the mutations in the area. He never admitted trying to make such a device himself.[106]

During the American occupation, Japanese physicists were closely supervised, access to journals was limited, and there was little chance for dialogue with foreign scientists. Yukawa managed to start publishing an English-language journal, *Progress in Theoretical Physics*, on the brownish spotted paper that was all that he could scavenge. The first issue contained both Tomonaga's demonstration of the renormalizability of quantum electrodynamics and a paper by Sakata and a colleague on the two-meson hypothesis.[107] The journal did not arrive in the United States until December 1947.[108] But by then the Americans had figured both ideas out for themselves.

In addition to the experiments on self-energy, the participants at the Shelter Island conference in June of 1947 spent a day debating the findings of Conversi, Pancini, and Piccioni in Italy. At the height of the discussion, Victor Weisskopf proposed the rather peculiar idea that an encounter with a cosmic ray might somehow make a proton or neutron "pregnant," capable of emitting mesotrons; the period of gestation, he suggested, might explain some of the discrepancies involving the lifetime of the particle.[109] The awkwardness of this solution prompted Robert Marshak, a theorist from Cornell, to make the simplest explanation he could think of, namely, that there are two kinds of mesotrons with different masses. One is produced in the upper atmosphere by cosmic ray collisions, while the other results from the decay of the first.

Marshak was pleased with the reception to his idea at Shelter Island, but was startled upon his return from the conference to find that the most recent issue of *Nature* contained a photograph of precisely the process he had described.[110] The picture had been taken by a team of four scientists in Britain with a specially developed photographic emulsion that was sensitive to most of the charged particles in cosmic rays. The resultant thin black lines could be measured in exactly the same way as the tracks in a cloud chamber, with the blackening and scattering of each track providing clues to the particle's mass and energy. The scientists had exposed many emulsions in this way, and on one of them, a mesotron had slowed to a halt in the middle of the film, whereupon it decayed into another, somewhat lighter particle. Marshak immediately set to work on a new paper, enlisting Hans Bethe's help because of the latter's extensive knowledge of cosmic rays.[111]

The first and heavier of the two particles is today called the pion or pi-meson. It corresponds to the Yukawa particle, and is exchanged by the protons and neutrons in the nucleus. It is about 273 times heavier than the electron and has a lifetime of about 10^{-8} seconds, roughly what Yukawa estimated it to be. Experimenters were shortly to discover three pions: positive, negative, and neutral.

The second and slightly lighter particle, created in the decay of the pion, is an entirely different matter. It is the particle Anderson and Neddermeyer discovered. It plays no role in the nuclear interaction, nor in any process essential to ordinary matter. It is about 207 times heavier than the electron; its accidental similarity in mass to the pion created the confusion between the two particles Oppenheimer described as a "ten-year joke." This second particle is nothing more, nothing less, than a plump, short-lived version of an electron—a second electron, that exists for no known reason and has no known function. Unpredicted by theory, unneeded by nature, the muon, as it is now known, turns into an electron a millionth of a second after it is made. It may not last long, physicists thought, but why is it there at all? I. I. Rabi summed up the general feeling of bafflement in the physics community with the snarl, "Who ordered *that?*"[112]

The muon was just the first surprise. After Rochester and C. C. Butler in Manchester finally were allowed to switch on the Blackett magnet, they began to photograph the so-called mesotron showers in a new cloud chamber. With this setup, they found, in October 1946, the traces of what was apparently a new particle. They happened on a second set of tracks seven months later. The first was interpreted as the decay of a neutral particle; the second was charged; both were about a thousand times the mass of the electron. Because both particles made forklike tracks in the cloud chambers when they decayed, they were christened "V particles."[113] Many scientists were initially inclined to dismiss the two photographs as the sort of glitch that occurs in experiments; before endorsing the existence of new states of matter, they waited for Rochester and Butler to produce some more of them.[114]

"Unfortunately," Rochester said later, "although the dear Lord was very kind to us in sending two of these, on the other hand He was extremely unkind in sending no more than two."[115] The two men spent another year unsuccessfully trying to snare more V particles, and growing ever more embarrassed when none showed up. In the summer of 1948, Rochester crossed the Atlantic to attend a conference in honor of Robert Millikan's eightieth birthday. He spoke about the V particles to a polite but unenthusiastic audience. "Then," he said to us, "at Pasadena I met Carl Anderson. And Carl was extremely interested, and after this conference Carl put his best chamber *straight on* to this problem on top of the White Mountain." At such an altitude, V particles, if they existed, should materialize much more often because there were many more cosmic rays to make them. To

Rochester's considerable relief, he and Blackett received a letter from Anderson saying that he had turned up "about 30" V particles, and that he agreed with their analysis.[116]

The discovery of the V particles, like most scientific discoveries, prompted those who missed it to reexamine their evidence. Sure enough, earlier photographs of high-energy cosmic ray showers were found to be sprinkled with V particle tracks. The discovery of yet another exotic type of matter—besides the pion and the troublesome muon, sent researchers in Europe, Japan, and the United States to the heights. By the beginning of the 1950s, cosmic ray laboratories dotted mountaintops across the globe. Conditions were rugged when they were not impossible, and experimenters had to stay for weeks at a time in tiny mountain huts. Nonetheless, some physicists found a certain Hemingwayesque thrill in their splendid isolation. Louis Leprince-Ringuet, a French physicist with the aristocratic mien of a nineteenth-century explorer, spent years eleven thousand feet up the frigid Aiguille du Midi, snowbound in tiny rooms stocked with scientific equipment and emergency brandy. There is a certain Gallic fluidity to his account of the virtues of high-altitude life:

The work goes on day and night, without stopping, without the slightest disturbance, without any of the quotidian annoyances of ordinary life: in such splendid isolation, one can do a year's work in a month. All the while the storm blows ceaselessly; for two solid weeks, we could not step over the threshold of the door, remaining shut up in the snow, unable to see out the window, for days on end. But after our seclusion was over, we had the rapturous pleasure of putting on our skis and darting, heedless of the crevasses, across the fine, crystal-sparkling powder snow of the heights. On beautiful days, the mornings . . . were magnificent: there was the mist rising up from the plains, and the perfectly calm and level sea of clouds, real seas whose billows were set and animated by a slow rocking motion, creamy bluish seas whose depths were tinted pink by evening. The great peaks were simply parts of our world; Cervin, Mont Rose, and Mont Blanc seemed so close—almost near enough to touch—that the rest of the earth seemed to collapse like Atlantis beneath the sea of clouds.[117]

Others, however, were less entranced, as we discovered when we talked to Antonino Zichichi, an Italian experimenter with a photogenic mane of white hair. In 1953, when he was twenty-four, Zichichi was sent from Amaldi's institute in Rome to the Jungfraujoch in the Swiss Alps.[118] On his arrival in the wasteland of wind and snow, Zichichi was greeted by the lab supervisor, who introduced himself as Hans Wiederkehr. A tyrannical sort, Wiederkehr kept the laboratory functioning at the price of total obedience from the intimidated physicists in his charge. When Zichichi first arrived, Wiederkehr pointed to an exit and explained at considerable length that it should never be used. Zichichi said that he could read and that there was a large *DO NOT EXIT* sign on the door. Grumpily, Wiederkehr indicated that the new arrival was to accompany him to the library, a splendid wood-paneled room with a large bay window that overlooks a glacier. Journals, charts, and maps rested in cozy disorder on the shelves, hauled up by an ancient funicular from the ski town below.[119] Wiederkehr indicated a gallery of neatly framed black-and-white portraits on a side wall. The subjects were physicists who had worked at the laboratory. The conversation, in Zichichi's recollection, went as follows:

WIEDERKEHR (snarling): *See these gentlemen? They're all dead.*
ZICHICHI: *What happened?*
WIEDERKEHR: *They didn't* obey the signs!

"And then," Zichichi said, "he explained to me that one physicist a year died on the average there. In fact, when I was there my group leader, Dr. [Anthony] Newth from Imperial College, went out one day—one nice day—on his skis. And he fell into a crevasse. He survived because at 5:30 Mr. Wiederkehr said, 'Where is Dr. Newth?' So he went out, followed the tracks, and pulled him out of the crevasse. Dr. Newth told me he thought he was dead. I was absolutely terrified! One physicist a year dying, on the average! And all their pictures were in the library. And the incredible thing is—" he leaned closer "—nobody understood a thing they were finding in all that time!"[120]

□ □ □ □ □

Between 1947 and 1953, cosmic ray researchers in a score of cramped mountain laboratories discovered a whole new world of physics. Opening and closing the pistons of their cloud chambers or the switches of their electronic counters, they photographed the tracks of muons, pions, and V particles; beyond that, they found half a dozen *more* particles, all born from the evanescent shimmer of high-energy cosmic rays. The most interesting photographs were made by standing up parallel sheets of photographic film and letting cosmic rays shoot through them, leaving little white trails in their passage. As more people worked with this method, they began to find— inevitably, perhaps—rare, spectacular events. (In the jargon, a particle interaction is referred to as an "event.") Scratched across the film, the lines showed *something* that broke up into three widely separated pions; piecing together another series of photographs, a second set of researchers might see that an unknown entity had disintegrated into a scatter of muons and electrons; yet a third might find that a nucleus in the film emulsion had apparently been hit by a cosmic ray particle, creating a starlike splash of light with subatomic tracers flying away in mysterious profusion. Baffled by the unexpected profusion of particles and decays, physicists might apply different names to one particle, misidentify a second, and deny the existence of a third—only to reverse themselves when they saw data from other laboratories. Within five years of the first observation of V particles, there were thought to be half a dozen subcategories for V particles and a sprinkle of objects loosely called theta, tau, and K particles—all known only from a few hundred photographs and all of which might be anything or nothing. Each lab had only a few examples of every type of event, and their piecemeal publication in widely scattered journals made it impossible to gather a global picture. As a result, nobody knew what they were seeing or why it was there; they did not know how they should refer to it or even if anyone else had seen the same thing. Physics here was completely in the hands of the experimenters; the theorists who had been so confident after Shelter Island were now utterly lost, reduced to vague supposition and unsupported guesswork.

Among the many mysteries, the puzzle of the V particles stood out most sharply. Found at the beginning of the new era, these particles seemed to be abundantly produced, once there was enough energy, which suggested that the force responsible for their existence was powerful. A strong force's strength is demonstrated by the likelihood it will manifest its effects, and the plethora of newly formed particles therefore testified to the might of the force creating them. Such a powerful force should cause the V particles to decay quickly—"easy come, easy go," is the rule of thumb. By looking at cloud chamber photographs, physicists were able to make guesstimates of the time that it should take this strong force to pull the V particles apart. The calculations were somewhat awry: V particles had a lifetime *ten trillion* times longer than expected.[121] When a theoretical prediction is off by thirteen orders of magnitude, it is an indication that something strange is in the mix.

The riddle's solution lay in an old idea called *isotopic spin*. The concept of isotopic spin, although not the name, was invented by Heisenberg as an almost extraneous sidelight to his theory of the nuclear force. As a matter of convenience, Heisenberg imagined the proton and the neutron to be different states of the same particle, in somewhat the way that the atomic isobars carbon-12 and nitrogen-12 have the same mass but different charges. (By right, the name should really be "isobaric spin," not "isotopic spin," because the reigning metaphor has nothing to do with atomic isotopes, such as carbon-12 and carbon-14, which have different mass but the same charge.) To describe this relationship formally, Heisenberg proposed that each particle be thought of as spinning like a top in some imaginary space. If the axis of spin points up, the particle has a positive charge and is a proton; if the axis points down, it has no charge and is a neutron. Heisenberg said that the proton and the neutron could then be assigned the same value—½—of a hypothetical quantity later baptized isotopic spin.[122] One recalls that ordinary spin, too, is associated with a direction, and thus an electron with spin ½ has an axis of spin that is indicated by saying that its spin orientation (S_z in the diagram) is $+\frac{1}{2}$ or $-\frac{1}{2}$. Similarly, isotopic spin is associated with a direction, and thus the upward proton is said to have an isotopic spin orientation (I_z in the diagram) of $+\frac{1}{2}$; the neutron, $-\frac{1}{2}$.[123]

As put forth by Heisenberg, isotopic spin was not only complicated and confusing, but also of little real use; most of his colleagues paid little attention. Four years later, however, a single issue of the *Physical Review* con-

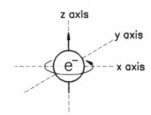

Electron rotating around z axis
has "upward" axis of spin

Ordinary Spin	Particle	Orientation of spin in magnetic field (technically known as S_z because it is the z axis on graph)
$S = \pm\frac{1}{2}$	Electron	$S_z = +\frac{1}{2}$ $S_z = -\frac{1}{2}$
Isotopic Spin	Particle	Orientation of isotopic spin in imaginary charge space (technically known as I_z, as above)
$I = \pm\frac{1}{2}$	Nucleon	$I_z = +\frac{1}{2}$ $I_z = -\frac{1}{2}$
$I = \pm 1, 0$	Yukawa particle, or pion	$I_z = +1$ $I_z = 0$ $I_z = -1$

ORDINARY SPIN AND ISOTOPIC SPIN

tained three papers that dramatically resuscitated the idea. The first gave an experimental value for the force that clamped together protons in the nucleus; the second analyzed the data to show that the force was equal between two protons and a proton and a neutron; the third used the first two as an excuse to relaunch isotopic spin, arguing that the nuclear force, whatever it was, could not "see" the difference between the proton and neutron, and that therefore they should be regarded as two states of one particle called the *nucleon*.[124] In 1938, one of Pauli's assistants, a Russian emigré named Nicholas Kemmer, extended the idea past the nucleon. He showed that if one were to take seriously Yukawa's notion of a particle that transmitted the nuclear force, one would actually have to assume the existence of three such entities—three pions, in today's language—that could be regarded as states of one particle, with three different axes of isotopic spin: up, down, and sideways.[125] (Note that, as in the case of the pion, the isotopic spin of the particle may have a value of 1, whereas its axis of spin may be some other number, such as 0.)

In an important paper, Oppenheimer and Serber then postulated that the total amount of isotopic spin stays the same in particle interactions, in the same way that the total energy, momentum, and electric charge do not change.[126] All four are said to be "conserved." The difference is that the conservation of isotopic spin is approximate; it is conserved only in interactions involving the nuclear force, whereas energy, momentum, and electric charge are conserved no matter what.

A hitherto unknown law of nature had shown its presence, but briefly, like a flash at the corner of an eye. Enter Abraham Pais, vigorously. Pais was a Dutchman then at the Institute for Advanced Study. In a remarkable talk at a 1953 conference commemorating the hundredth birthday of Hendrik Lorentz, Pais indicated the course that elementary particle physics would follow for the next two decades.[127] Pais was thirty-five; much of his twenties, ordinarily among the most productive years of a physicist's life, had been spent hiding from the Germans in Amsterdam. Eventually he was caught and put into a Gestapo prison. He was lucky enough to survive. Earlier than most Occidental theorists, he had the hunch that the new cosmic ray particles were important.[128] In the years before the Lorentz conference, Pais and several Japanese had come independently to a conclusion that now seems baldly obvious: The creation and decay of V particles are not attributable to the same agent. Unable to think of a real reason why this should occur, Pais complained about the lack of clues: "The search for ordering principles may indeed ultimately have to be likened to a chemist's attempts to build up the periodic system if he were given only a dozen odd elements."[129] Speaking at the Lorentz conference, he began by listing "a number of questions in meson physics that seem to be of outstanding interest."

The first question concerns the isotopic spin. Ever since it became clear that proton and neutron can transform into each other and in many instances are exchangeable in nuclear systems, we have been faced with the question whether a theoretical foundation could be given for the fact that these two particles seem to behave like different states of one entity now called [the] nucleon. The formal shorthand of isotopic spin takes cognizance of the situation but explains nothing of course.[130]

He then argued, in general and by example, that great discoveries were yet to be made, for new variables like isotopic spin would be the key to the study of both the nuclear forces and the cosmic ray particles. More such intrinsically conserved qualities should be sought, and their description merited the introduction of new quantum numbers, the first in twenty-five years. (Using today's language, Pais was the first to urge the search for higher symmetries.) By pushing and pulling these properties, twisting them into mathematical knots, one could rank the elementary particle interactions into a hierarchy and peer inside their inner workings.

Although encumbered by a half-dozen other ideas and a wholly incorrect model, a program for the next generation lies swaddled within the nervous evidence of Pais's creativity. The paper had little direct influence, however, because it was almost immediately superseded by a concrete example of the kind of inquiry Pais advocated, in the form of a new quantum number proposed by a twenty-two-year-old named Murray Gell-Mann, who was then beginning a meteoric ascent into prominence. A young man who gave a rich new layer of meaning to the term "brash," Gell-Mann knew almost nothing about Pais's ideas; he had been led to the same considerations independently.[131] Whereas Pais's style was more discursive and conversational, Gell-Mann was terse, oblique, clipped. One article was a hint, the other an announcement: Another era had begun.

Gell-Mann's paper—"Isotopic Spin and New Unstable Particles"—is remarkable, a pocket symphony in the modern style. It begins with a quick fanfare:

[L]et us suppose that both "ordinary particles" (nucleons and pions) and "new unstable particles" [here a list of V particles] have interactions of three kinds:
(i) *Interactions that rigorously conserve isotopic spin. (We assume these to be strong.)*
(ii) *Electromagnetic interactions. (Let us include mass-difference effects in this category.)*
(iii) *Other charge-dependent interactions, which we take to be very weak.*

This music, familiar to the cognoscenti, had never sounded so clearly. In modern terms, one speaks of a hierarchy of interactions: *electromagnetism*, which binds the electrons around the nucleus, conserves electric charge, and is fully described by a field theory, quantum electrodynamics; the *strong force*, which holds the nucleus together, conserves isotopic spin, was thought to be transmitted by pions, and had no theory to describe it whatsoever; the *weak force*, which causes particles to decay into lighter particles, does not seem to conserve anything at all, was not known to be transmitted by any particle, and was imperfectly covered by Enrico Fermi's theory of beta decay; and *gravitation*, which holds planets around suns and suns in galaxies, is fully described by the theory of general relativity, plays no role on the subatomic level, and until recently seemed unreconcilable with quantum mechanics. Today, electromagnetism, the strong and weak interactions, and gravity are thought to be the basic forces of the Universe, from which all others derive.

This ordering of forces was matched at about the same time by an ordering of the particles they affected. Particles that feel the effects of the strong force are now known as *hadrons*. Hadrons in turn are split into two subcategories, the heavy *baryons*, like the proton and neutron, and the lighter *mesons*, like the pion. Those particles, such as the neutrino and electron, which do not feel the effects of the strong force, are called *leptons*.[132] Four interactions for leptons and hadrons: the cosmos in a clause.

Having explicitly laid out the ladder of interactions, Gell-Mann considers papers by Pais and another theorist, David Peaslee.[133] In connection with their work, Gell-Mann writes, "The author would like to put forward an alternative hypothesis that he has considered for some time. . . ." The entering theme is deceptively modest: Assume that the V particles have an isotopic spin half a unit different from that of ordinary particles. If the nucleon and pions are set at ½ and 1, respectively, then the heavy and light V particles are 1 and ½, respectively.

Now Gell-Mann laces together the initial brassy chords and the theme: The stronger interaction of type (i) can produce V particles with isotopic spins of, say, $+1$ or -1, only if it makes *two* of them at once. Because $(-1) + (+1) = 0$, no new isotopic spin is created, and the net isotopic spin is unchanged. And because the heavy and light V particles are ½ a unit off the neutron and proton, there is no way the strong force can make Vs into nucleons without changing the net isotopic spin by ½—which cannot occur. Therefore, Gell-Mann notes, "[A]s far as (i) is concerned, the decay of new unstable particles into ordinary ones is forbidden. . . . Only interactions of type (iii), which do not respect isotopic spin at all, can lead to decay." Shifting the isotopic spin values neatly solved the riddle of the V particles. Simple addition and subtraction showed how particles could be created by a powerful force that was unable to make them decay.

Simultaneously, Gell-Mann wrote another paper that cut more deeply into the V particles.[134] Coming across the manuscript now is like reading the private drafts of an author, such as Mark Twain, who was chary of revealing his most ardent beliefs; widely circulated and widely influential, the article was held back by its author until it was obsolete, and eventually never published. An important idea thus slipped in quietly, through the back door, for in that paper Gell-Mann set out what is in effect a formula, good for any interaction or single particle, that reads, in its modern form,

$$Q = I_z + B/2 + S/2$$

where Q is the electric charge, I_z the axis of isotopic spin, B the number of baryons, and S a new quantum number, to which Gell-Mann attached the whimsical and provocative label of "strangeness." Particles with non-zero values of S thus became "strange particles." Strange particles are V particles, that is, those new particles with odd values of isotonic spin. (The same idea was developed independently by a Japanese physicist, Kazuhiko Nishijima.)[135]

Strange indeed. Strangeness was a wholly new phenomenon manifested only in the particles created at high energies, although the presence of Q indicated its links with old properties such as charge. The formula was couched exclusively in terms of elementary particle properties, without reference to atoms or nuclei. It accounted concisely for the observed data, but stood mute on their causes. It told physicists that they had reached a plateau, and that there was much to learn. It marked the full emergence of a new discipline: elementary particle physics.

We once asked Abraham Pais to think back to the early 1950s, when humanity first saw what at this moment appears to be the shape of the fundamental constituents of the Universe. What would the physicists of that era have thought if somebody had told them that thirty years later there would be the first attempts at a unification theory? Would that have shocked them? "Oh, absolutely!" he said. There was a nostalgic softness in his tone. "It was a *wonderful* mess at that time. Wonderful! Just great! It was so *confusing*—physics at its best, when everything is confused and you know something important lies around the corner."[136]

In the summer of 1953, cosmic ray workers from all over the world gathered in Bagnères-de-Bigorre, a small Basque town in the foothills of the Pyrenees a few thousand dry feet beneath the Pic du Midi laboratory. Compared to today's conferences, which are split up into sub-sub-categories, eagerly covered by the press, and attended by thousands of researchers and job-hunting graduate students, the gathering at Bagnères was a modest affair, a matter of a hundred and fifty cosmic ray enthusiasts, *vin ordinaire*, and cots in the town schoolroom. It was the first time these cosmic ray experimenters had been in one place since the war, the first time many of these scientists felt certain that they were not the only ones seeing peculiar stuff in their cloud chambers. Moreover, these Vs and Ks and thetas—whatever they were, and whatever they should be called—were not flukes or unimportant details, but something of vital interest.

Many of the experimenters at Bagnères remember the conference as one of the greatest ever held. They came to take stock of where they had been; they left, six days later, with the feeling that the physics of elementary particles was a scientific domain every bit as rich with implication as the study of the nucleus. Committees were organized to classify the known particles and to lay down the rules for the nomenclature of future discoveries. Greek letters were assigned to each new particle: small if light, capital if heavy. The conventions were published simultaneously in journals across the globe, and the Bagnères attendees spoke excitedly of the onslaught of new particles to come.[137]

The new particles were discovered, but not by cosmic ray physicists. At Bagnères, an experimenter named Marcel Schein created a sensation with the report that a new and enormous machine in Chicago called a cyclotron, one of the first modern particle accelerators, had successfully thrown a spray of protons at a metal plate with enough power to create a burst of V particles. (Schein's claim later turned out to be mistaken.) The conference members realized that accelerators in laboratories could easily match the energies of cosmic rays in the upper atmosphere. No longer would experimenters have to seek exotic forms of matter on mountaintops. They would be made in the tens of thousands, spat out from machines like so many cans of soup. At the conference closing, Leprince-Ringuet exhorted his fellows in the common room of the Bagnères schoolhouse. "We must hurry," he cried.

We must run without slackening our pace: we are being pursued—pursued by the machines! . . . We are—I think—a little in the position of a group of mountain climbers climbing a mountain. The mountain is very high, maybe almost indefinitely high, and we are scaling it in ever more difficult conditions. But we cannot stop to rest, for, coming from below, beneath us, surges an ocean, a flood, a deluge that keeps rising higher, forcing us ever upward. The situation is obviously uncomfortable, but isn't it marvelously lively and interesting?[138]

It was too late. To explore the new realms of matter found in cosmic rays, cosmic rays themselves would be virtually abandoned by physicists. The euphoria at Bagnères was the high-water mark of cosmic ray physics; although the town dedicated Place Hyperon in honor of a cosmic ray particle, the scientific interlude that began at the beginning of the century with leaky electroscopes had come to a close.

III

The Weak Force

10
Symmetry

AT HALF PAST FOUR IN THE AFTERNOON OF DECEMBER 10, 1979, THREE men—Sheldon Glashow, Abdus Salam, and Steven Weinberg—entered the auditorium of the Stockholm Concert Hall to a flourish of trumpets. They were to be awarded the Nobel Prize for physics, for their construction of a single theory incorporating weak and electromagnetic interactions, an achievement that began, almost unnoticed, in the years following the discovery of the strange particles. The laureates walked down the wide aisle— Glashow and Weinberg in tails and studs; Salam in full Pakistani formal regalia, including shoes with toes that curled several painful inches into the air—and onto a platform, where they were introduced and individually extolled by a middle-aged member of the Swedish Academy of Sciences. At the end of each peroration, the two thousand people in the audience applauded as the prizewinner walked to center stage to meet King Gustav XVI, who presided over the ceremony. Each physicist shook the king's hand and received a leather-bound diploma, a hefty gold medal, and a letter informing him of when and how to collect his share of the prize money.[1]

After the rite, the three physicists, the laureates in chemistry, medicine, literature, and economics, the royal family, and about half the people in the concert house were bundled into a fleet of chartered limousines and buses that conducted them through the bitter cold to Stockholm City Hall. (The Peace Prize is awarded in Oslo; the economics prize, which was established later, is technically the Nobel Memorial Prize.) A turn-of-the-century waterfront building constructed at lavish expense and frequently said by Swedes to be the Continent's finest, the City Hall features an enormous Gold Room lined with the largest single display of twenty-four-karat gold mosaics in the world. There the laureates were subjected to a series of toasts and pronunciamentoes and then led by the king onto a balcony overlooking yet another gigantic space, the Blue Room. The high bare walls of the Blue Room are composed of individually chiseled red bricks whose rough beauty convinced the original architect to change his mind about painting them blue, as had been planned. An ornate marble staircase debouches onto the gray floor of the room, which on Nobel night is jammed with students from the University of Stockholm, who have their own banquet. At the sight of the royal family and the rather stunned laureates, the students jumped to their feet and lustily bellowed a Swedish drinking song that nobody seemed to feel needed translation. Dancing commenced, and spread to the Golden Hall. After two o'clock, some students abducted the physicists for a Swedish ritual known as a *vickning*, a postprandial event featuring, mainly, more eating and drinking, the champagne and caviar now being replaced by schnapps and herring. At dawn, the revelers were released—but had to be at the bank early in the morning to pick up their checks.[2]

Winning a Nobel Prize involves a week of festivities that is the Swedish World Series, World Cup, and Olympics all in one. The laureates are shuttled back and forth among so many cities and so many elaborate dinners that the actual prize ceremony, impressive as it is, is often reduced to a splendid blur in the memory. At some point, however, each prizewinner is asked to give a formal address. The speeches by the awardees of the Peace Prize and the Prize for Literature often attract global attention; those given by the winners of the science prizes are usually ignored, because they commemorate discoveries that are already well-known in the research world. The speeches in 1979 were something of an exception, for the physics prize celebrated the most important advance in particle physics for a generation, and a step toward the full unification of nature that is the dream of every physicist.

Weinberg's speech, like those of Glashow and Salam, was a historical stroll through the path he had taken toward unification. He began in the simple, clear style that is the envy of his colleagues and the wellspring of a parallel career as a popularizer of science.

Our job in physics is to see things simply, to understand a great many complicated phenomena in a unified way, in terms of a few simple principles. At times, our efforts are illuminated by a brilliant experiment . . . But even in the dark times between experimental breakthroughs, there always continues a steady evolution of theoretical ideas, leading almost imperceptibly to changes in previous beliefs. In this talk, I want to discuss the development of two lines of thought in theoretical physics. One of them is the slow growth of our understanding of symmetry, and in particular, broken or hidden symmetry. The other is the old struggle to come to terms with the infinities in quantum field theories. To a remarkable degree, our present detailed theories of elementary particle interactions can be understood deductively, as consequences of symmetry principles and of a principle of renormalizability which is invoked to deal with the infinities.[3]

The importance of the criterion of renormalizability was appreciated in the late 1930s and early 1940s; the understanding of symmetry and its role in the Universe came afterward, and resulted in the current attempt to cover all of space and time in one theory.

Several years after his Nobel, we talked with Weinberg about renormalizability and symmetry. A tall man with wavy red hair exploding into gray curls at the back of his neck, Weinberg is now the head of the theory group of the department of physics at the University of Texas, in Austin. His dark eyes stand out in his pale, slightly flushed features. Half-moon reading glasses protrude from his shirt pocket. A certain measure of gravity, solidity, and dignity is present in his bearing; he speaks well, in slow, deep, concise sentences. On the right side of his office window is tangible evidence of the rewards of accomplishment in theoretical physics: ceremonial photographs of the Weinberg family with the king and queen of Sweden, Queen Beatrix of Holland, and Jimmy Carter, then president of the United States of America.

He hadn't been working there much, he said. Having the writer's habit of pacing while he thought, he preferred to do physics at home, where he could keep a television on his desk. It is a pleasing image: a man in his study, trying to strain sense through a screen of equations, *The Edge of Night* flickering unheeded on the desk as he asks himself, Now what is important in all this here?

We asked him how he had become involved in quantum field theory. "When I was a graduate student, that was the latter part of the fifties," he said. "It was a somewhat confused time, because there had been a great triumph of quantum field theory with the development of quantum electrodynamics—and then a complete failure at extending it beyond the area of quantum electrodynamics. I once took a look at Fermi's Silliman lectures, which were written in 1952, because I had to give the Silliman lectures in 1977, twenty-five years later. These are a lecture series at Yale. I looked back at what Fermi had said, and boy, was that depressing! Attempts made to understand the physics of nuclear forces, mostly based on conservation laws and guesswork and really weird ideas about weak interaction. It just didn't seem to be going anywhere. The one thing that seemed like a fantastic and indisputable accomplishment was the success of quantum electrodynamics. You could calculate till your eyes fell out of your head and you would get the right answer that agreed with experiment to umpteen decimal places. There was nothing else in physics like that. It was an ideal model for the way physics should be.

"By the way, I think in that respect I was not at all untypical. In most graduate schools in the 1950s, the people who were most ambitious at doing theoretical physics—not of a phenomenological kind but a fundamental kind—took quantum electrodynamics as their ideal. There was one thing, I suppose, that mattered more to me than it might have to other people. And that is that I was impressed with the *uniqueness* of quantum electrodynamics. You know, the theory had these infinities, and then it was discovered that the infinities would cancel, provided you took an appropriate definition of what you meant by mass and charge. The thing that fascinated me was the fact that the whole infinity cancellation would only work if the theory was limited to be just the original, simplest possible quantum electrodynamics. If you tried to muck around with quantum electrodynamics—

the way, for example, Pauli had suggested, adding additional terms—then the whole application of renormalization theory would break down. You just couldn't do it.

"Now, to many people, that just showed that renormalization theory was just—it *accidentally* worked in this special case, but didn't really mean that much, and you shouldn't take it that seriously, and perhaps you shouldn't take quantum field theory that seriously. But to me, that always seemed the most exciting thing there was. I never liked these arguments of simplicity. I don't think nature chooses simple equations just because they're simple. I think there are some simple principles underlying everything, and then the equations are just what they are. So merely saying that quantum electrodynamics is just the very simplest possible minimum theory never appealed to me. But here was a principle—the infinities had to cancel—that was a golden key that explained why the theory was the way it was. It was a tremendous expansion of the power of pure thought."

In other words, going from a criterion of simplicity to one of necessity?

"Right, right," Weinberg said. He went to the coffee dispenser and began pouring coffee. Beyond the window rose the hundreds of miles of Texas hill country, gentle green ridges oddly reminiscent of the foggy slopes of northern California. Weinberg said that one still looked for simple equations, but that this was not something to celebrate. "Now, Dirac had taken the attitude that—he just wrote down the equation [for the electron] that seemed to him the simplest, and that it was beautiful, and therefore that was why it was true. I've never liked that attitude. I find that very antithetic to the way I think about things. But renormalizability—obviously the theory, in order to make sense, had to have no infinities. That was a requirement that was reasonable to impose on a physical theory, and that seemed the key that would explain why the physics was the way it is."

One of Weinberg's first papers after his Ph.D. was a tough proof that dotted the last *i*'s and crossed the last *t*'s in the demonstration of the renormalizability of quantum electrodynamics.[4] Published in 1960, the article is still the only piece of Weinberg's oeuvre that he thinks would be regarded with respect by a professional mathematician. In fact, Weinberg asked a mathematically inclined physicist, Arthur Wightman of Princeton, to check through the argument; when Wightman handed the paper back, he commented that there was blood on every lemma. "I'm very proud of it, and it is often quoted," Weinberg said. An expression of amusement crossed his face. "I think it's very rarely read."

At about that time, Weinberg became explicitly interested in the question of symmetries, a subject that had been present in physics for many years but whose importance was not fully appreciated for much of that time.[5] Used loosely, symmetry means harmony or balance. Physicists and mathematicians, however, define the term more precisely. They say that something is symmetrical if one or more of its aspects is indifferent to a change. A rubber ball can be turned freely and its appearance won't be altered; therefore, as a physicist might say, the ball is "symmetric with respect to rotation." More important, space and time themselves have certain symmetries. Space, for example, has a kind of symmetry called by the clumsy name of "translational invariance," which means, simply enough, that space is exactly the same no matter where you are in it. The laws of physics are the same whether you are in New York or Nagasaki, the moon or the bottom of the ocean. Time has "time translational invariance," an even more cumbersome expression for an equally simple idea, namely, that no particular instant in time is inherently different from any other. In other words, physics is the same whether it is yesterday or tomorrow, a thousand years ago or a week from last Sunday. Because of these two symmetries of space and time, business executives in Toronto can know today that next week, when they attend a conference in Brussels, water will still flow downhill and their morning *café au lait* will not crawl out of its cup.

If translational invariance did not hold true, a rocket ship thundering through the heavens might suddenly slow down, losing its momentum, just because it had entered a different part of space. Conservation of momentum is, so to speak, wired directly into space, its presence implied by the symmetry of translational invariance. And if time translational invariance did not hold—that is, if the laws of physics could change from second to second—it would be impossible to assure that energy would be conserved. Similarly, the conservation of angular momentum is linked to the fact that it does not matter which way you turn an object around in space.

Newton implicitly based his laws on another kind of symmetry known as Galilean invariance, named after Galileo Galilei, from whose work it is derived.[6] Imagine you are a passenger in an automobile and that you are drinking a cup of coffee. The simple fact that the coffee stays in the cup when you are cruising along at a steady fifty-five miles per hour just as it does when you are stopped means that the behavior of physical laws does not depend on the velocity of the system to which they are applied. Strictly speaking, this is only true when the velocities are constant and do not change, as is shown by the fact that the coffee tends to slop out when you accelerate out of a stop light or brake to a halt.

Einstein's theory of special relativity stems from the recognition that Maxwell's equations say that the speed of light is always the same no matter how fast you are going. In other words, electrodynamical laws are symmetric with respect to the velocity of light. Unfortunately, it is mathematically impossible for the two symmetries of Maxwell and Newton to coexist. Einstein came up with a way of reconciling them, but only at the price of supposing—in flat contradiction to Galilean invariance—that the behavior of some physical properties such as mass and length *does* depend on the velocity of the system to which they are applied. The special effects predicted by relativity only show up at enormous velocities, which means that you can go right on assuming that the coffee in your morning cup will not be affected if you drive on the freeway. Because the equations were actually invented first by Hendrik Lorentz, the new kind of symmetry is called Lorentz invariance. Einstein, however, had never heard of this work by Lorentz, and was the first to understand the physical meaning of the equations.[7] But despite his use of symmetry it is not at all clear that Einstein realized then or later that the symmetries were the fundamental entities he was dealing with. This is the modern point of view expressed by the physicists who write textbooks on relativity, such as Weinberg, whose treatise on the subject is widely read.[8]

"It's sometimes hard for a physicist today to appreciate the frame of mind of the physicist of the 1930s," Weinberg said to us. "For example, take isotopic spin. Today, if a physicist who had never heard of atomic nuclei was suddenly exposed to a lot of information about them, he or she would immediately begin to ask questions about what are the symmetries of the interactions that hold them together?" He sketched the mathematical reasoning that would lead a physicist to the answer. "That's not the way people thought in the thirties," he said. "As far as I can understand it—and I wasn't doing physics then—their attitude was, what is the *dynamics* of the interactions? What is the theory of the nuclear *force?* And in the course of making a theory of the nuclear force, it might have a symmetry between protons and neutrons. If it did, you could form conclusions, like the conclusion that the nuclei form multiplets, but that would be a side issue. It would only be relevant as an incidental thing that you would use to test a particular theory of nuclear forces. The idea of separating off the question of symmetry as a completely separate subject which you study in its own right—and put off the dynamics for later—is more modern." Caught up in his own thinking, Weinberg had described a full circle around the neat expanse of his desk. He recalled that Einstein had used symmetries to develop special relativity. "The mood of 1905 was, the electron has just been discovered, the next thing to do is to make a model of the electron. And the obvious thing is that the electron is pure electromagnetic self-energy. People like Lorentz, [the great French mathematician Henri] Poincaré, [Max] Abraham, and so on made such models. . . . Einstein's real contribution was precisely to say—he didn't say it in these words, but this must have been his attitude—'We don't know what an electron is! Don't bother me with your models of the electron!' Many people—[J. J.] Thomson—thought that the electron was the only fundamental particle. Einstein said, 'This is not the time to say what the electron is. Let's just look at one aspect of it—what is the symmetry governing the laws of nature?' And he said, 'Well, it's Lorentz invariance. It's a different symmetry. It's not the one you thought it was.'

"I would say that symmetry again and again plays two roles. One is that very often symmetry is the thing to think about because it's the only way you can make progress. Einstein could only make progress by thinking about symmetry because in 1905 it was premature to make a model of the electron. Einstein understood—I don't know how!—but somehow or other he knew that that was not a time to make a model of the electron, that was the time to think about symmetry. In the beginning of the 1960s, it was not the time to make a theory of the strong interactions. The pieces were not in place. It was the time to think about the symmetry of the interactions, and Gell-Mann and [Yuval] Ne'eman did that and made tremendous progress. The other way that symmetry comes in, though, is more characteristic of Einstein in 1915 [general relativity] and [Chen Ning] Yang and [Robert] Mills in 1954 and also the development of the electroweak theory in 1967. And that is that the symmetry isn't just the only thing you can get a handle on, it is the thing that actually *drives* the dynamics. It's the central problem."

He was asked if this second kind of symmetry was what he had in mind when he talked about the simplicity of nature.

"Yeah. At the deepest level." He qualified the statement a little. "Well, I think it is. I don't know why nature has these symmetries, but Lorentz invariance, for example, the most important symmetry of all, is not only a symmetry which governs the form of the equations, it tells us what the equations are about.

"Particles are bundles of energy and momentum. What are energy and momentum but the quantum numbers defined by [time and space] translations? What is angular momentum but the quantum number which is defined by rotation? In a sense, if you have an elementary particle, and you describe how it behaves under various symmetry transformations, including translations, rotations, gauge transformations, then you've said everything there is to say about the particle. The identity of the particle is fixed by its symmetry properties. The particle is nothing else but a representation of its symmetry group." He laughed. "The Universe is an enormous direct product of representations of symmetry groups. It's hard to say it any more strongly than that."[9]

The role of symmetry in modern physics was established by Amalie Emmy Noether, one of the most important mathematicians of this century.[10] Emmy Noether was born in the German university town of Erlangen in 1882, where her father was a mathematics professor. Although university rules ordinarily prohibited women from matriculating, Noether obtained permission to attend lectures; of the 968 students at the University of

Erlangen, there were two women, neither of whom had been legally allowed to matriculate. In 1904, when the law changed, Noether took her degree; four years later, her doctoral thesis was accepted, *summa cum laude*, by the same university.[11] There was little of the rebel about her. She simply loved mathematics of the most abstract kind. From 1908 to 1915 she taught, without pay, at the Mathematical Institute of Erlangen and occasionally substituted for her father at the university; in 1915, she moved to Göttingen, where she became an important but still unpaid part of a team that worked on calculations for Einstein's formulation of general relativity.

Despite her evident brilliance and the support of the mathematicians, the Göttingen philosophical faculty, which encompassed disciplines from philology to physics, refused to accept her application for a salaryless position. The great mathematician David Hilbert, however, had her lecture under his name. (Deeply interested in the mathematical aspects of physics, Hilbert held the conviction that "physics is much too hard for physicists."[12] Indeed, he derived much of the theory of general relativity on his own.)[13] At the same time, he pressed for her promotion, angrily declaring at one meeting of the Philosophical Faculty, "I do not see that the sex of the candidate is an argument against her admission as *Privatdozent* [teaching assistant]. After all, we are a university and not a bathing establishment."[14] Evidently finding Hilbert's remark more provocative than persuasive, the assembled professors rejected Noether's application. She was finally given a teaching position in 1919. Three years later, she was promoted from teaching assistant to "unofficial associate professor" with the grudging admonition from the Prussian minister of culture that "this designation does not carry with it a change in her duties. In particular, her position as *Privatdozent* and her relation to her faculty remains unchanged; neither is she to receive the salary due an official position."[15] A year later, in 1923, she began to be paid for her work. She had been in the field full-time for over a decade.

Noether did her best work in her later years, which is rare for a mathematician. She was neither a manipulator of formulas nor a virtuoso calculator. Rather, she looked for the ideas that lay at the heart of entire disciplines, and strove to discover the conceptual bedrock upon which real mathematics could be based. Because of the abstraction and generality with which Noether worked, she became widely respected among her colleagues—which didn't save her from ejection by the Nazis. The Göttingen school of physics and mathematics was destroyed in 1933 when Noether and the other Jewish members of the Philosophical Faculty, such as Max Born and James Franck, lost their jobs. Many refugees found jobs in the United States, although not necessarily ones suited to their eminence. After fleeing Germany, Noether could only find a temporary position at Bryn Mawr College, outside of Philadelphia. In 1935, she died unexpectedly after surgery.

Noether's friend, the mathematician and physicist Hermann Weyl, himself a refugee, delivered her memorial address. Near the end, he remarked, "She was not clay, pressed by the artistic hands of God into a harmonious form, but rather a chunk of human primary rock into which he had blown his creative breath of life."[16]

Had Noether been a man, her appearance, demeanor, and classroom behavior would have been readily recognized as one of the forms that absent-minded brilliance frequently assumes in the males of the species. Fat, loud, and disorganized, Noether cared little for how she looked or what she ate. During lectures, she repeatedly withdrew the handkerchief she kept tucked in her blouse, waved it about furiously to demonstrate a point, and tucked it back in. She neither backed down from arguments nor easily relinquished the speaker's role. Weyl confessed that he and his colleagues sometimes referred to her unkindly as "der Noether," with the masculine article.[17] There is a photograph of Noether taken at the Göttingen train station just before she fled to the United States. Her humorous mien, touched with melancholy, her thick glasses and baggy rumpled overcoat, and the plain flat-brimmed hat pulled down over her forehead make her seem indeed like a female version of Einstein.[18]

The theorem on which much of today's particle physics is based forms only a small part of her total oeuvre, and is usually ignored by mathematicians more interested in her fundamental ideas on commutative ring theory and algebraic number theory. Developed for her *Habilitationschrift*, the second thesis that people in Germanic school systems had to write before joining the faculty of a university, the work was an offshoot of David Hilbert's interest in relativity.[19] As in her more mathematical papers, Noether went straight for the most fundamental statement about the kind of relations with which Einstein wrestled. Briefly, Noether's theorem demonstrates that wherever there is a symmetry in nature, there is also a conservation law, and vice versa.[20] In other words, the symmetries of space and time are not only linked with conservation of energy, momentum, and angular momentum, but each *implies* the other. Conservation laws are necessary consequences of symmetries, and symmetries necessarily entail conservation laws.

The simplicity, power, and depth of Noether's theorem only slowly became apparent. Today, it is an indispensable part of the groundwork of modern physics, and scientists have compiled a list of over a dozen important conservation laws and their associated symmetries. They have also discovered different types of symmetries, the most important of which is a *gauge* symmetry.

Gauge symmetry was introduced to physics by Noether's friend Hermann Weyl in 1918, the same year that Noether proved the theorem that now bears her name. A pupil of Hilbert, Weyl had also worked with Einstein in

Switzerland. Like Einstein, Weyl was afflicted by the urge to unify. In a series of three papers, he tried to show that electromagnetism, like gravity, is linked to a symmetry of space,' which would be a step toward showing that the two forces are aspects of the same thing. [21] He based the argument on what now has the name of "gauge invariance," a term sufficiently undescriptive that even many physicists admit they wish it could be replaced.

A gauge, quite simply, is a measuring standard. When measuring most things, the size of the standard can be changed at will, and the results remain the same. The length of a board, for example, is the same whether a carpenter's measuring tape is marked in inches and feet or centimeters and meters; as long as the proper conversion is made, identical pieces of wood can be cut in both systems. Weyl called this kind of symmetry "measuring stick invariance," then changed it to "gauge invariance," apparently thinking of the metal devices used by railroad engineers to measure the distance between train tracks. [22] Weyl also realized that gauge invariance comes in two varieties, *local* and *global*, and that local gauge invariance was an extremely important idea.

Local gauge symmetry is difficult to visualize or describe in simple terms, and may perhaps best be approached by contrasting it to global gauge symmetry. A happy example of global gauge symmetry is the earth itself, which has been divided by cartographers into longitude and latitude lines. Zero degrees is arbitrarily set as the longitude of an observatory in Greenwich, England. If the whole system were turned ninety degrees, zero would fall in the middle of Memphis, Tennessee, and the international dateline would be outside Dacca, Bangladesh. Despite the changes, pilots would be able to navigate perfectly well. They would just add ninety degrees to their charts—changing their gauge, so to speak. Mathematicians might say that navigation on earth is symmetric about a longitude change of ninety degrees. This is a *global* gauge symmetry, for all longitude readings are changed in the same way. They are all ninety degrees west of what they were before.

Another example of global symmetry can be drawn from the world of high finance, or at least that portion of it described in elementary economics textbooks. If the income of every person and the price of every commodity suddenly increased tenfold, nothing would happen. People would still spend their wages on the same things in the same way, except that now paychecks and prices would have an extra zero. The supply and demand of every commodity would be undisturbed and, in the jargon of economists, all markets would still "clear." The forces of supply and demand can be said to be globally symmetric with respect to a change in the economic measuring stick—money.

Now imagine that incomes and prices were to jump around *randomly*, some rising, some falling by different chance amounts. The situation should then change dramatically: Supply and demand would be out of whack. Some people would be clamoring to buy more than is currently offered for sale, while others would be unable to afford their previous standard of living. This state of affairs, according to classical economists, brings into play the "invisible hand" of market forces. Prices adjust, decreasing in one place, increasing in another, until once again all markets clear. Automatically compensating for each random change in the system, the invisible hand maintains the original symmetry, which in this case is a *local* one. (In fact, economists make such claims only for small stochastic perturbations, but the principle holds true.)

Given the arbitrariness of the changes in local symmetry, it is hard to believe that any physical system could have this property. And indeed, gauge symmetries cannot occur unless the random changes in one aspect of a system are precisely compensated for by a change in another aspect, so that a quantity related to both remains conserved. Such compensation cannot take place unless a force intervenes. Indeed, it presupposes the existence of one. Maintaining local gauge invariance is not an easy, passive task; the Universe must act to preserve it. Thus understanding a force means tracing it back to an originating symmetry. In physics, the forces are real entities that act in the world, whereas economic forces are the half-understood sum of many individuals' actions.[23]

Weyl indeed postulated that space does have local gauge symmetry. He argued that the behavior of space and time in fact can vary randomly from place to place and instant to instant, but these chance alterations are canceled out by the activity of the electromagnetic fields that permeate the cosmos.[24] The result of this local gauge symmetry, by an application of Noether's law, is the conservation of electric charge. Soon after receiving a draft of Weyl's paper, Einstein appeared to kill the idea by pointing out that such local gauge invariance led ultimately to the erroneous prediction that if one took a clock around the room it would not show the same time as an identical clock in the original place because of all the gauge changes.[25]

In 1927, another theorist, Fritz London, revived Weyl's work with the suggestion that he had the right idea but applied it to the wrong symmetry.[26] Electric charge is indeed conserved, he said, but the associated symmetry is not the scale of space and time, but rather a symmetry of a much more abstract property, the phase of the Schrödinger wave equation. The phase of a wave is that part of it one is looking at—foamy crest, briny trough, or something in between. When two waves are *out* of phase, crest matches with trough and they cancel out; two waves *in* phase, on the other hand, reinforce each other. One of the enduring lessons of quantum mechanics is that particles have wave characteristics, and are described by wave equations. (Recall that Bohr, Heisenberg, and Schrödinger had long arguments about exactly this point.) Among the wavelike features of any elementary particle is its phase, which is intimately tied to such traits as electric charge. Changing the phase of a wave equation therefore would usually lead to changes in the electric charge. It is known that the electric charge of, say, an electron absolutely *never* changes. As Fritz London showed, the reason the phase can shift and not affect the charge is because of the gauge symmetry, which compensates for such changes by creating virtual photons—and hence an electromagnetic field—whose actions ensure the conservation of electric charge. Canceling out the effects of any phase change, the field protects the charge with single-minded drive. Thus the gauge symmetry of the phase implies, even *creates*, the dynamic actions of the electric field. From this protective action on the subatomic level stems, incredibly enough, all the phenomena of electromagnetism that we see about us, from the small shocks of static in carpeted rooms to the jagged arc of lightning across a storm sky. A single symmetry gives birth to them all.

Fittingly, it was Weyl himself who put the final formulation on this idea; it is too bad that he retained the name "gauge symmetry" for what was now a symmetry of phase.[27]

Just as someone who had never played chess would have to watch for a while to pick out the usually untouched king as the most important piece on the board, it often takes scientists a while to put their fingers on the central aspect of any phenomenon. Although physicists were well aware that electromagnetism could be described as a local gauge field, they did not know if this was particularly significant. Indeed, the concept stayed nearly dormant for twenty-five years.[28]

Modern gauge field theory was largely created at Brookhaven National Laboratory in early 1954 by Chen Ning Yang and Robert Mills. In a single short paper published that year in the *Physical Review*, "Gauge Invariance and Isotopic Spin," Yang and Mills built the frame upon which modern quantum field theory is woven. At first their ideas were treated with skepticism and even derided as being pure mathematics, without any physical significance; now, theories of Yang-Mills gauge symmetries are powerful enough to account for almost every particle and force in existence.

The thrust of the paper occurred to Yang in 1948, when he was a graduate student at the University of Chicago. He had recently left war-torn China, where his father was a mathematician, to study with Enrico Fermi.[29] Under the theory that his given name, Chen Ning, might be too difficult for Americans to say, he awarded himself the nickname "Frank," after Benjamin Franklin, a man whom the youthful Yang admired. He was impressed with the idea that gauge symmetries could be used as a base to construct the entire theory of quantum electrodynamics. "When I was a graduate student," he said to us, "people talked a lot about gauge invariance. People would say that a calculation was not right, for example, because the result was not gauge invariant. It was a kind of check on calculations. But it was not appreciated that gauge invariance is a principle that can *generate* forces. Today we recognize it as the only principle that can generate forces. It had been understood that way for electromagnetism in the 1920s, only somehow people were not paying attention. They also didn't realize the principle could be moved to new situations."[30]

Yang began to suspect, in other words, not only that such a description might be essential to an understanding of electromagnetism, but that there might be other kinds of local gauge fields. He was unable to carry his postulate further than this. Theorists with incomplete ideas are like people with songs in their heads that they can't identify; they can't stop trying to place the tune. Yang finished his thesis and went to the Institute for Advanced Study, in Princeton, New Jersey. He kept picking up and dropping gauge invariance. In 1953, when he spent some time at Brookhaven, the idea still bothered him, and he told Mills, his office mate, about it. "There was no other, more immediate motivation," Mills said. "He and I just asked ourselves, 'Here is something that occurs once. Why not again?' "[31]

To find a conserved quantity like electric charge, Yang and Mills seized upon isotopic spin. The paper begins with celerity:

The conservation of isotopic spin is a much discussed concept in recent years. Historically an isotopic spin parameter was first used by Heisenberg in 1932 to describe the two charge states (namely neutron and proton) of a nucleon. The idea that the neutron and proton correspond to two states of the same particle was suggested at that time by the fact that their masses are nearly equal, and that [many] nuclei contain equal numbers of them.

They note that the strong force does not care whether the axis of isotopic spin is up (proton) or down (neutron).

Under such an assumption one arrives at the concept of a total isotopic spin which is conserved in nucleon-nucleon interactions. Experiments in recent years on the energy levels of light nuclei strongly suggest that this assumption is indeed correct.

Saying that the strong force conserves isotopic spin is the same as arguing that it can not see the difference be . ween neutrons and protons. Yang and Mills said that therefore, as far as the strong force is concerned, "the differentiation between a neutron and a proton is then a purely arbitrary process."

The point made, the two men then busied themselves with the argument of the paper. Reminding the reader that in electromagnetism the phase of the wave function can be shifted arbitrarily in space and time because the action of the electromagnetic field will invariably cancel out the alteration, they proposed to do the same thing with isotopic spin. After randomly spinning the third axes of isotopic spin—that is, after imagining that they had haphazardly rearranged the labels "proton" and "neutron" throughout the Universe—they hypothesized the existence of a "*B* field" to counteract the change. Just as the *raison d'être* for the electromagnetic field is to ensure the gauge symmetry of electromagnetic interactions about the phase, so the *B* field maintains the gauge symmetry of strong interactions about the orientation of isotopic spin.

Yang and Mills then spent three pages working out what this *B* field would be like if it exists, and what the characteristics would be of the virtual *B* particles that transmit it. The paper concludes:

The quanta of the B *field clearly have spin unity [i.e., one] and isotopic spin unity. We know their electric charge too because all the interactions that we proposed must satisfy the law of conservation of electric charge, which is exact. The two states of the nucleon, namely proton and neutron, differ by charge unity. Since they can transform into each other through the emission or absorption of a* B *quantum, the latter must have three charge states with charges ±1 and 0.*

Today, the *B* particle is called a *vector boson*, which is jargon for a particle with a spin of one. (The name is redundant: A boson is a particle with integral spin, whereas any particle with a spin of one is called a vector particle.) Yang and Mills hoped that the equations would work out in such a way that this purely speculative *B* particle would convey a force identical to the strong force. They would thus have an explanation for strong interactions that would be in the same language as quantum electrodynamics, the model theory.

Unfortunately, when Yang and Mills did the mathematics, the figuring seemed to suggest that their conjectured vector boson had electric charge but no mass. This was most discouraging: Massless particles, such as photons, are easy to make in experiments, and charged particles, such as electrons, are easy to detect in them. The two men could not understand why, if charged, massless vector bosons existed, they would not already have been discovered. "We do not have a satisfactory answer," they admitted.[32]

On February 23, 1954, Yang spoke to a small audience at Princeton about the work he and Mills were doing. The audience included luminaries such as Pauli and Oppenheimer, who was now in the midst of his battle over his security clearance. Although Yang did not know it, Pauli had explored the possibility of generalizing gauge invariance a year before, but had decided not to publish because of the mass problem.[33] Yang recently recalled the occasion:[34]

Soon after my seminar began, . . . Pauli asked, "What is the mass of this [vector boson]?" I said we did not know. Then I resumed my presentation, but soon Pauli asked the same question again. I said something to the effect that this was a complicated problem, we had worked on it, and come to no definite conclusions. I still remember his repartee: "That is not sufficient excuse." I was so taken aback

that I decided, after a few moments' hesitation, to sit down. There was general
embarrassment. Finally Oppenheimer said, "We should let Frank proceed." I then
resumed, and Pauli did not ask any more questions during the seminar. I don't
remember what happened at the end of the seminar. But the next day I found the
following message:

> Dear Yang:
> I regret that you made it almost impossible for me to talk to you after the
> seminar. All good wishes.
>
>   Sincerely yours,
>   W. Pauli

Despite the unresolved question about the mass of the vector boson, Yang
and Mills decided to publish their work. "We did not know how to make the
theory fit experiment," Yang said. "It was our judgment, however, that the
beauty of the idea alone merited attention."[35]

Although it would not be fully realized for fifteen years, Yang and Mills
had transformed the role of symmetry in quantum physics. They had given
symmetry muscle, shown how it could create forces and set particles in
motion. In essence, their message was this: "Look at particle properties. See
if you can find any unchanging qualities and try to imagine a gauge field that
might account for them. Work out the properties of that fictional field and its
associated virtual particles. Are they like anything in the real world?" If they
are, then it would be a major step toward unification, toward the day when
scientists would be able to trace out the laws of the Universe from a single
principle, one that reaches out to cover the whole of nature in a single
explanatory web. Yang and Mills had raised the hope that the clue to the
diversity of the world lay in symmetry.

In the meantime, however, physicists had to occupy themselves with
the question of how to use this new insight. Yang and Mills, and many of
their colleagues, chose to look at the strong force and the heavy particles it
affected, about which a wealth of data was pouring in from the first particle
accelerators. Although they couldn't know it, that was a blind alley. Most of
the speculation about Yang-Mills gauge fields—the term was soon applied to
any local gauge field—stayed shut in physicists' notebooks, and gauge symmetries lay dormant until years later, when Steven Weinberg, Abdus Salam,
and Sheldon Glashow applied them to the weak interaction.

11
Weakness

WHAT ARE NOW KNOWN AS THE WEAK INTERACTIONS HAVE HISTORICALLY been associated with a striking degree of experimental confusion ever since their accidental discovery by Becquerel in 1896. The *rayonnements invisibles* that elided through the heavy black paper in which he wrapped his photographic plates largely consisted, we say today, of the electrons emitted by the beta decay of neutrons in his uranium crystal. The process of beta decay was named and distinguished from its confreres by Rutherford; Becquerel himself made the hypothesis that the "beta ray" was a speeding electron, although this universally accepted belief was in fact not completely proven until much later, in 1948, by an experiment performed by Gertrude and Maurice Goldhaber, in which they showed the identity of beta rays and atomic electrons.[1] No one had any idea why every now and then an atom should spit out an electron. After the alpha-scattering experiments of Rutherford, Geiger, and Marsden, the location of the phenomenon was thought to be the nucleus, but beta decay was in no way less mysterious.

The most puzzling aspect of beta decay was that the electrons came out of the atom with a dizzying variety of speeds. Radioactive atoms generated electrons in a continual rain, but there seemed to be no preferred speed or direction to the flow.[2] From the law of conservation of energy, one would expect that the energy of the emitted electrons would be equal to the energy lost by the nuclei that ejected them. Measurements showed that this was not the case.[3] In 1927, two experimenters at the Cavendish totted up the full heat energy of a small amount of radioactive material and compared it to the energy of the electrons it created. There was little connection between the two figures.[4] Other, more precise experiments discovered the same result.[5] Niels Bohr said that these experiments represented irrefutable proof that the law of conservation of energy must not apply to radioactive decay.[6]

Rather than give up conservation of energy, Wolfgang Pauli proposed—reluctantly, perhaps, and certainly with less than usual vigor—that beta radiation actually consisted of *two* particles, the already known electron and a previously unimagined entity later dubbed the "neutrino." Pauli had an ambivalent attitude toward publishing radical ideas; he announced his guess at the end of 1930 through a letter to some colleagues—"Dear radioactive ladies and gentlemen," as he called them—begging off from a conference.

(He wanted to go to a dance contest instead.)[7] The neutrino was small, light, electrically uncharged, very hard to detect; it carried off the missing energy.

At the same time, there was the puzzle of where, exactly, the electrons came from. It was generally assumed that they must somehow be in the nucleus until the question was reopened by the development of Heisenberg's uncertainty relation. Recalling that one formulation of this principle links the uncertainty in the position of a particle to that of its momentum, it is a simple matter to show that pinning down an electron's location to the nucleus of an atom, which is extremely small, implies an enormous uncertainty as to the particle's momentum, and the corresponding likelihood that the latter quantity is extremely large. A large momentum means a high velocity. If electrons indeed whiz around inside the nucleus at great speed, why do they not shake loose anyway, long before beta decay has a chance to kick them out? On the other hand, where else could the beta decay electrons come from if not the nucleus?

(Heisenberg doubted that the electrons were in the nucleus at all, a skepticism that he shared with Victor Weisskopf on a hot day in the spring of 1931 as the two men sat outside the entrance to a swimming pool. In Weisskopf's recollection, Heisenberg said, "These people go in and out all very nicely dressed. Do you conclude from this that they swim dressed?")[8]

On the theoretical side, the first measure of order was established in 1933 by Enrico Fermi, who took advantage of the recent discovery of the neutron and Pauli's neutrino suggestion to create the modern picture of beta decay: neutrons turning into protons, giving out electrons and neutrinos as they do so.[9] Later, when it was more clearly realized that the creation of a particle is always balanced by the creation of an antiparticle, the picture would be amended again to say that the newly manufactured electron is accompanied by an *anti*neutrino.

Even so, the picture discomfited some physicists. It was odd to imagine that the electron and neutrino were not originally in the nucleus at all, just as a bubble is not in a bubble pipe until one blows through the mouthpiece.[10] Hard, too, to picture an elementary particle, one of the bricks that comprise the Universe, falling apart. Maurice Goldhaber, who was at the Cavendish at the time of the first experiments with the neutron, one day described his alarm. "We found that the neutron was definitely heavier than the hydrogen atom, which is a proton and an electron," he said. "So I could easily see that a neutron could decay into a proton and electron." In fact, as was found later, the average time before beta decay hits a free neutron is about eighteen minutes. "We decided, here is a fundamental particle which should be radioactive. I remember being shocked by that idea—that a fundamental particle could *decay*. This is now taken for granted. But such a basic concept, that a particle can decay— One knew of course that nuclei decay, that particles in the nuclei change, but elementary particles seemed to me . . ."

He put his finger to his upper lip, considering. A cold wind keened at the window. His office is at Brookhaven National Laboratory, where Goldhaber is involved in such things as examining the unification theories that sprang from such earlier work as the measurement of the neutron lifetime. Once dismayed at the end of a particle, Goldhaber found himself joining experiments testing unification theories that implied the eventual end of all things. "Maybe I shouldn't have been shocked. But it seemed to me a shocking idea then, that a fundamental particle could *decay*."[11]

The theory of the weak interaction was first exposed to the world between Christmas and New Year's Eve of 1933 from a small, bare, cold hotel room in the minute Italian hamlet of Selva, nestled in the Alps about twenty miles east of Bolzano, where four physicists from the University of Rome— Enrico Fermi, Franco Rasetti, Edoardo Amaldi, and Emilio Segrè—had decided to spend their Christmas vacation. One evening, after a full day of skiing, Fermi invited the others into his hotel room to talk about the paper he had just sent off for publication. There were not enough chairs, so the four men crowded together on the edge of the bed. Fermi bent over, scribbling calculations on a sheet of paper he had propped against his thighs, while Rasetti, Amaldi, and Segrè craned their necks to read his drawing. Sore from repeated tumbles on the icy snow, Segrè kept changing his position restlessly. But the three were impressed, although a bit confused, by the bold theory that was emerging on Fermi's knees. Beta decay was not only associated with the unheard-of transmutation of neutrons and a never-seen neutrino, Fermi told them, it was the handiwork of an entirely new force of nature.[12]

When theorists are confronted with a phenomenon they don't understand, they often cast about for something similar they *do* understand, borrow its theory, and squeeze as much of it as possible into the new phenomenon. This had been precisely Fermi's strategy. The most promising similar theory was quantum electrodynamics, despite the infinities then driving theorists to distraction. Fermi therefore set out to model beta decay on electromagnetism.

An electron emits a photon when it changes from one state to another. This change of state is brought about by a familiar force of nature: electromagnetism. In the process of electron scattering, for example, two electrons approach each other, exchange a virtual photon, and fly away like so many billiard balls. This interaction is said nowadays to be composed of two "currents," which might loosely be described as the two halves of the diagram below, that is, the two V-shaped electron lines. Fermi pictured the process of beta decay likewise as a combination of two currents, a neutron-proton current (*n* to *p* in the diagram) and an electron-neutrino current (*e* to *v*)—again, the two halves of the diagram. Avoiding all speculation as to what, exactly, happened at the point where the two currents approach, he argued merely that the changes of state are brought about by an extremely feeble force, trillions of times weaker than either electromagnetism or the nuclear force responsible for strong interactions. Its effects are almost always swamped by the other two forces, like the political course of a tiny nation caught between two superpowers. Nevertheless, the actions of this small force can sometimes make a decisive difference when the effects of the other two offset each other, as in unstable nuclei. The result is beta decay. (There is also a form of inverse beta decay, in which a proton changes to a neutron, absorbing a neutrino and releasing a positron.)

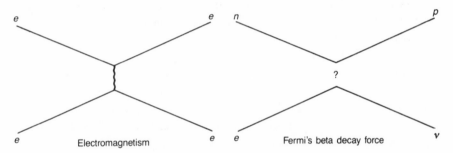

The strength of the electromagnetic interaction between two objects is described by its coupling constant, alpha, the infamous 1/137, which in a sense indicates the absolute value of the force. Fermi wrote down his theory of beta decay to include such a coupling constant, and was able to deduce its value from the tables collected by experimenters. This Fermi constant is extremely small, as befits such a feeble interaction. Although its exact value depends on the system of notation employed, physicists sometimes say the strength of the weak force relative to electromagnetism is on the order of 10^{-13}, a number whose tininess is best appreciated when written out in full: 1/10,000,000,000,000.

Using the analogy with electromagnetism and the value of the beta decay coupling constant, Fermi was able to calculate the range of energies the emitted electrons should have, when beta decay could take place, and many other features of the process. His rapid series of papers on the subject is extraordinary: The theory was his first and only work on beta decay and was so complete that its description of the phenomenon has remained essentially unchanged to this day. At a stroke, Fermi first accurately described beta decay, first identified the separate nature of its cause, and first predicted most of its essential features. "Seldom was a physical theory born in such definitive form," Rasetti has written.[13]

The editor of *Nature*, to whom Fermi sent his first paper, was less appreciative. Fermi's note was bounced immediately, with the remark that its speculations were too remote from physical reality to be of interest to the practicing scientists who made up the audience of the journal. The idea was therefore published first in a small Italian review, *Ricerca Scientifica*; follow-up articles appeared in Italian and German.[14] (Maurice Goldhaber, in fact, explained Fermi's Italian paper to physicists at the Cavendish.) By the time they appeared, Fermi, the last twentieth-century physicist to make great contributions to both theory and experiment, had turned his attention to the experiments on which much of his fame rests—bombarding the nuclei of all the elements with neutrons, creating radioactive forms of some of them in the process. These experiments were eventually to make him one of the fathers of nuclear physics and the atom bomb. Certain British experimenters found this to be a more promising route to good science; at the end of April the following year, after Fermi had published some preliminary data, he received a letter from Rutherford whose parting salutation is almost a conditioned reflex: "I congratulate you on your successful escape from the sphere of theoretical physics! You seem to have struck a good line to start with. You may be interested to hear that Professor Dirac also is doing some experiments. This seems to be a good augury for the future of theoretical physics!"[15] (Needless to say, experiments did not form a major part of Dirac's oeuvre.)

One small theoretical adjustment was made to Fermi's thesis. For mathematical reasons, Fermi argued that beta decay would take place rarely unless the summed spins of the constituent protons and neutrons in the original and the final nucleus remained the same.[16] (Here the discussion is of ordinary spin, not isotopic spin.) Two theorists, George Gamow and Edward Teller, showed that beta decay was also mathematically possible if the spins differed by one, and developed the appropriate equations.[17] Ever since, beta decays where the spin of the nucleus remains unchanged have been called Fermi transitions; those where the spin of the nucleus changes by one are called Gamow-Teller transitions. Although Gamow and Teller developed their equations only as a theoretical possibility, actual instances of Gamow-Teller transitions were soon discovered.

Despite its theoretical success, Fermi's theory soon ran afoul of experiment. His articles predicted that the emitted electrons would come out with varying energies, but that if experimenters plotted the number of beta electrons against their velocity, the resultant curve would have a particular shape, somewhat like the famous bell curve used to grade students in large classes.[18] Two British and three Soviet experimenters promptly measured this curve with the electrons emitted from radium and radioactive phosphorus. They found many too many slow electrons.[19] To fit the data, George Uhlenbeck, the co-discoverer of spin, and Emil Jan Konopinsky, both of whom had recently moved to the University of Michigan, formulated a brand-new, modified theory, which became known as the K-U theory after their initials.[20] Discerning the velocity of the electrons was no easy task, and the work was made even harder by the fact that the weak force is so feeble that beta decay does not happen very often. In order to scare up enough electrons to measure, the British and Soviets had used thick chunks of radioactive material. Unfortunately, if the source is thick, electrons emitted in the middle of the stuff must travel through many layers of atoms, ricocheting as they go, before emerging into the open air. The resultant deflection and loss of energy distorted the readings and the graphs based on them. When thinner slivers of radioactive material were used, the experimenters found fewer slow electrons than before.[21]

Although the new set of experiments invalidated the K-U theory, the curve for low-energy electrons still did not agree with Fermi, leaving the understanding of beta decay in what was beginning to be a customary state of confusion. The war intervened, and most of the players in nuclear physics and electrodynamics became preoccupied with building weapons or hiding from Fascism. An exception was Willis Lamb at Columbia, who thought about measuring the fine structure of hydrogen; another was Chien-Shiung Wu, who spent most of the war years teaching physics at Smith College and Princeton University. Born in China in 1912, Wu came to the United States twenty-four years later to win her doctorate. Although the skill and care with which she worked were immediately apparent, her sex guaranteed her anonymity. At the end of the war, she was finally taken in at Columbia University's Division of War Research. The hostilities concluded, she began to consider the question of the low-energy electrons from beta decay. Slow electrons are particularly susceptible to scattering effects, the way, for instance, slight ripples on a putting green may make a barely moving golf ball swerve round the hole but have no effect on a faster one. Even the thinnest of radioactive sources would throw out slow-moving electrons in odd ways if its surface was uneven; the electrons could also be deflected by the backing material and the detector itself. In 1949 Wu experimented with an intensely radioactive film of copper—the technology to make this kind of thing was a legacy of Los Alamos—and was able to show with convincing thoroughness that Fermi's theory held true through the gamut from fast to slow electrons.[22]

During the sixteen years experimenters tried to sort out the electron end of beta decay, they attempted in addition to discover whether the neutrino existed.[23] At the time Fermi proposed his theory, the plausibility of the scheme depended on the near-impossibility of detecting the neutrino, which explained why it had not yet been observed. It is unlikely that Fermi realized at the time just how elusive this particle was going to be. Hans Bethe and Rudolf Peierls calculated that neutrinos of average energy could pass through fifty billion miles of water without hitting a single atom.[24] Unless a particle interacts with matter, it cannot be found. As the decades went on, the undetectability of the neutrino became an embarrassment. Massless, chargeless, hardly present but supremely necessary, the neutrino was an enormous lacuna in the physicist's image of matter. By the beginning of the 1950s, scientists were beginning to regard it as a sort of epistemological boondoggle, a poltergeist particle that belonged in its own special category of being.[25]

In 1952, a thirty-three-year-old Los Alamos scientist named Frederick Reines took a leave of absence to think about questions that would be worth devoting a lifetime to answer. After some months of staring at his empty office, he said later, "all I could dredge up out of the subconscious" was the recollection that the frightful by-products of an atomic explosion include a big blast of neutrinos—if they exist. Given enough neutrinos passing through a detector, some of them might strike a proton, causing inverse beta decay, that is, producing a neutron and positron.[26]

Reines talked about blowing up an atom bomb to find neutrinos with a friend, Clyde Cowan, and together the two men realized that if the detector was sensitive enough the experiment could be done in the considerably less hazardous surroundings of a nuclear reactor. (When informed that Reines had decided to try the reactor, Fermi drily commented that the new experiment was much better if only because it did not require setting off an A-bomb every time somebody wanted to check the results.)[27] The idea was that the neutrinos from the reactor would pour into a vat of dilute cadmium dichloride. Reverse beta decay would take place. Positrons would quickly strike and annihilate electrons, releasing photons; neutrons would be absorbed by cadmium nuclei, which acknowledge the action by emitting more photons. The two parallel bursts of photons would travel outward, passing into a layer of highly flammable triethylbenzene, a smelly substance somewhat like household cleaning fluid that has the happy property of flashing when struck by high-energy light. Amplified, the light is picked up by photomultipliers, devices like the sophisticated uptown cousins of the phototubes that keep open elevator doors. The task of the experimenters was thus to find and measure those twinned sparks of light.

The first run, at Hanford Engineering Works, in Hanford, Washington, was a nightmare of stacking and restacking lead shielding, cleaning dirty pipes, and trying to prevent the liquid detector from making the paint peel. Nonetheless, Reines and Cowan thought they saw a hint of the neutrino's presence.[28] Nearly three years later, using a much larger system that weighed ten tons even before it was surrounded by lead blocks, Reines and Cowan found the signals that enabled them to wire Wolfgang Pauli that the neutrino had been discovered three and a half decades after its invention.[29] (It took another eight years before the find was independently confirmed.)[30] Reines set to work on another problem, that of trying to see if a neutrino could scatter off an electron. Twenty years later, he proved it could.[31]

Millennia of intermittent thought went into deciphering the character of the gravitational force, into realizing that whatever pulled objects to the ground was the same as whatever bound the moon to the earth. Lightning, static electricity, and magnetism were known for centuries before humanity became aware that they were actually different aspects of one thing—electromagnetism. The weak force, too, was recognized only after scientists learned that what they thought were several different phenomena were in fact part and parcel of just one. But this new force, its actions mostly hidden from our eyes but no less fundamental than the other two, came to light in a time of frenzied activity that telescoped the scientific process. The weak interactions were discovered, named, and incorporated into a unified theory within the span of a single generation, but only after the most appalling experimental confusion as to exactly what they were.

Linking together the various aspects of the weak force began when clever investigators suspected that the Fermi beta decay force also played a role in the emission and absorption of muons.[32] (Muons, one recalls, are the "ten-year joke" particles in cosmic rays that were mistaken for pions.) In 1948 and 1949, four groups of theorists independently claimed that the coupling constants of the force responsible for muon decay, muon capture, and beta decay were identical, and thus all three phenomena were controlled by the same force.[33] The implication of each paper was that beta decay was merely one example of a whole class of actions in the subatomic world, a thesis which became known as the "Universal Fermi Interaction." Strange particle decays, too, were soon assigned to the same class. If the Universal Fermi Interaction hypothesis was true, several disparate interactions would be unified as the doings of one force; if it was not, particle physics would be more hopelessly complicated than anyone had ever dreamed.

The Universal Fermi Interaction would have to survive two tests: First, this new force must be shown to have its own unique *properties*; second, the *form* of its alleged manifestations must be proven identical. Remarkably, the Universal Fermi Interaction—what physicists now call the weak interactions—survived both tests in a single year.

As it happens, the property that characterizes the weak interactions is related to a symmetry called *parity*. Like symmetry, parity has a loose, general meaning and a much more restricted scientific definition. Ordinarily, parity means equality; two individuals are on a par or have parity if they are equal in some respect like position or stature. To scientists, however, parity means the kind of equality that results under a particular kind of transformation—when all spatial coordinates are reversed, and right becomes left, down up, and backward forward. This complicated-sounding process is quite simply what happens when you look in a mirror, except that up and down are not reversed in a mirror. The sameness that results is a special form of symmetry called reflection symmetry or parity.

Physical laws were assumed to be indifferent to reflection transformation. What works for left, up, and forward works equally well for right, down, and backward. Apparent asymmetries in nature, such as the right-spiraling shells of crustaceans or the left-sided position of the human heart, are the result of historical evolution rather than physics; there is no reason why shells could not spiral left, or why a "mirror-image human" could not exist whose heart is on the right. The heart may have its reasons for its left-sided position, scientists thought, but physics does not.

When physicists speak of the parity of elementary particles, they have in mind what happens when minus signs are placed in front of the spatial variables (the Xs, Ys, and Zs of position), for the Schrödinger wave function of a stationary particle, an operation equivalent to flipping the equation in the parity mirror. If the wave function is not changed, the particle is said to have even parity and given a quantum number of $+1$; if the wave function reverses sign when its orientation in space is reversed, the particle is said to have odd parity and given a quantum number of -1. (If the particle is moving, a few extra complications enter.) Over the years, experimenters observed that if one multiplied together the individual parities of all particles in a system, the total parity, like electric charge, was conserved.

The parity of elementary particles came into sudden prominence as a result of the decays of the strange particles, which were candidate members of the Universal Fermi Interaction. While grouping together the various particles found in cosmic rays, the Bagnères conference attendees took the time to note a small peculiarity concerning two strange particles, the theta and the tau. These had almost identical masses and lifetimes and could only be told apart by the manner in which they decayed: The theta particle changed into two pions, whereas the tau became three. A stationary pion has a parity of -1. Ignoring other complicating factors, the two-pion decay has a total parity of $-1 \times -1 = +1$. The three-pion decay has a parity of $-1 \times -1 \times -1 = -1$. In this way the tau and theta were distinguished. As Australian physicist Richard Dalitz showed after the conference, such calculations left physicists in a terrible bind.[34] On the one hand, describing two different particles with almost identical physical properties required a theory so complicated that nothing remotely like it had ever been seen in nature. On the other, asserting that the tau and theta were the same thing meant that this particle could blithely decay as it pleased into products of even or odd total parity—in short, that parity was not conserved in its decay.

In April 1956, Frank Yang attended the 1956 Rochester Conference, in Rochester, New York. Organized by Robert Marshak to carry on the Shelter Island tradition, the Rochester Conferences were the principal meeting places of physicists in the United States, yearly landmarks when the luminaries of the field would gather together to discuss where they were. Unlike Shelter Island, the Rochester Conferences had equal mixes of theorists and experimenters. Some even roomed together; Richard Feynman's roommate at the 1956 conference, for instance, was a Duke University experimenter named Martin Block. Feynman was notorious for arguing heatedly with whomever he was in closed quarters, and he and Block began quarreling about the tau-theta puzzle. At one point Block suggested that parity was not conserved. Block has recalled his roommate's reaction: "Feynman, in his usual gracious way, was ready to say to me, 'How *stupid* you are!' And by the time his jaw got over to the word 'stupid' he had thought a little bit—I'm quoting Dick on this—and he said, 'Well, maybe you have something there.' We spent one week, every night, arguing until the wee hours of the morning."[35]

The last day of the conference began with a session on the "Theoretical Interpretation of New Particles." Oppenheimer held the chair; Yang gave an introductory review.[36] Several talks followed, each wrestling with the problems involved in making the tau and theta one particle, or separate particles. Delphically, Oppenheimer opined, "The tau meson will have either domestic or foreign complications. It will not be simple on both fronts."[37] A few minutes later, Gell-Mann inventoried his own possible approaches to the problem, but he, too, was perplexed. He was unable to bring himself to advocate his own ideas of dealing with the problem—only to mention that he had some. The proceedings then report: "Pursuing the open mind approach, *Feynman* brought up a question of Block's: Could it be that the theta and tau are different parity states of the same particle which has no definite parity, *i.e.*, that parity is not conserved. That is, does nature have a way of defining right or left-handedness uniquely?"[38] After Yang said that he had looked at parity violation and several other possibilities without reaching a conclusion, Oppenheimer remarked that the time had come for everyone to close their minds. Afterward, Block asked Feynman what Yang had answered. "I don't know," Feynman said. "I couldn't understand it."[39]

Two weeks after the conference, Yang moved to Brookhaven for the summer.[40] Nearby at Columbia was Tsung-Dao (T. D.) Lee, another Chinese emigré, frequent collaborator, and long-time friend. The two met twice a week to work together, with the tau-theta puzzle occupying most of their attention. One morning in late April or early May, Yang made the long drive in from Brookhaven and picked up Lee at his office. The two parked near 125th and Broadway, where there were, until recently, several fine Chinese restaurants. Upon discovering they were not yet open, Lee and Yang retired to the nearby White Rose Cafe. Over coffee, they discussed the possibility that parity could be *conserved* in the strong interaction that produced the tau and theta, yet *violated* in the weak interaction by which they decayed. Although they couldn't imagine why parity might be violated, they agreed that this possibility should be further explored.

Lee went to the office of the local expert on beta decay, Chien-Shiung Wu, on the thirteenth floor of Pupin Hall, and asked whether she knew of any experiments that had definitely shown anything one way or the other about parity in the weak interactions. Wu said that Lee would have to look through the literature. She lent him the "literature," the recently printed *Beta- and Gamma-Ray Spectroscopy*, a massive tome of almost a thousand pages.[41] Crammed with graphs and tables in tiny type, the book summarized more than forty years of work by hundreds of physicists. All Lee and Yang had to do was make sure that not one of the cited experiments had directly tested for the conservation of parity.

Several weeks of intense work later, they knew the answer. Incredibly, nobody had ever proved that parity conservation was valid for the weak interaction. On June 22, 1956, the results of their labors, "Is Parity Conserved in Weak Interactions?" arrived at the offices of the *Physical Review*. The paper was published under the title of "Question of Parity Conservation in Weak Interactions," because Samuel Goudsmit, the editor of the journal, believed that question marks in titles were a blot on the escutcheon of physics. The work is an exceptional document. It begins simply, with a confident statement of the tau-theta puzzle: The tau particle and the theta particle are known to have nearly identical properties except for their parities, and thus are considered to be separate particles.

One way out of the difficulty is to assume that parity is not strictly conserved, so that theta and tau are two different decay modes of the same particle, which necessarily has a single mass value and a single lifetime. We wish to analyze this possibility in the present paper against the background of the existing experimental evidence of parity conservation.

What is this evidence?

At first sight it might appear that the numerous experiments related to beta decay would provide a verification that the weak beta interaction does conserve parity. We have examined this question in detail and found this to be not so.

Lee and Yang did not argue that parity was actually violated in weak interactions. Pointing to a surprising gap in our knowledge of the world, they merely tried to convince experimenters to settle the question.

To decide unequivocally whether parity is conserved in weak interactions, one must perform an experiment to determine whether weak interactions differentiate the right from the left.[42]

Few experimenters were willing to drop everything to jump into the notoriously difficult area of beta decay to check an absurd idea. Some theorists, too, tended to dismiss the paper as clever but irrelevant—scientific virtuosity displayed to little apparent purpose. Freeman Dyson, for example, read the article twice, thought "this is interesting," and filed the journal back on the shelf.[43] Gerald Feinberg, who got his Ph.D. from Columbia that year, heard about Lee's struggles with parity during Tuesday night theory seminars. He was interested by the idea of parity violation, but not convinced. "People didn't know what to make of it," he recalled. "That's true of any genuinely new idea—you don't know what to think of it at first. Lee had gone through a lot of different ideas. This paper did not come with an arrow pointing to it, saying, 'This is it!' "[44]

Wu had been the first to hear about Lee and Yang's work, and was the first to consider testing it. For months she and her husband, Chia-Liu Yuan, had planned to visit the Far East on the twentieth anniversary of their exodus from China. With passage booked on the *Queen Elizabeth*, Wu abruptly canceled and left her spouse to make the sentimental journey alone. She wanted to start an experiment before anyone else realized the paper's importance. By the beginning of June, three weeks before Lee and Yang were ready to submit the paper, Wu was already lining up her collaborators.

She was fully aware of the difficulty of the experiment, because she herself had informed Lee and Yang of the possibility of its execution. Briefly, the task was to align the spins of the separate particles in an atomic nucleus, thus making them, so to speak, face the same way. The experimenters would then watch for beta decay. If the Fermi force respected parity, the electrons could have any orientation in space whatsoever and would come out in a random, symmetrical spray. If beta decay violated parity, the emission would be asymmetrical. The neutrons needed to be lined up so that their own random motions would not hide any asymmetry.

Aligning the neutrons is known as "polarizing" the nucleus. At the time, polarizing nuclei was, to understate the case, something of a challenge. The only known technique involved cooling a substance to almost absolute zero. With the nuclei nearly still, a carefully applied magnetic field could then rearrange each atom's cloak of electrons in such a way that they in turn pushed and pulled the frozen protons and neutrons into a sort of lockstep. Even at incredibly cold temperatures, the alignment could not be maintained very long because the atoms would slowly bang into one another. With luck, the polarization might last a few minutes. Wu intended to use Co^{60}, an intensely radioactive form of cobalt that has sixty protons and neutrons and the right configuration of electrons.

Not only the cobalt but the electron detectors would have to be cooled, for the electrons emitted in beta decay could not possibly penetrate the heavy insulation around the equipment. These detectors were thin anthracene crystals that gave off tiny flashes of light whenever struck by electrons; the flashes would be carried through the insulation by glass windows and a four-foot Lucite pipe to a bank of photomultipliers. Knowing that the small low-temperature laboratory at Columbia could not do the job, Wu called Ernest Ambler, a pioneer in nuclear alignment, at the National Bureau of Standards in Washington, D.C., in early June to propose a collaboration. He agreed; they began to construct the equipment. By September, Wu commuted regularly to Washington to set up the apparatus. She and Ambler also recruited Ambler's division head, R. P. Hudson, and two nuclear physicists, R. W. Hayward and D. D. Hoppes.

The cobalt had to be layered in a thin film—Wu knew all about the importance of thin films—onto a small crystal of cerium magnesium nitrate (CMN), which was then shielded in a housing made from larger crystals of the same material. Halfway into the experiment, the team discovered that nobody knew how to make the big crystals. Consulting a huge, dusty, forgotten nineteenth-century German text on the crystalline properties of CMN, they set up beakers under an array of heat lamps and dumped in as much ground CMN as the water would hold. They slowly lowered the temperature and watched the CMN crystallize on the bottom of the glass like so much rock candy. Once the crystals were glued together, Wu and her colleagues froze the mass of CMN and Co^{60} to a fraction of a degree above absolute zero. The whole ensemble fell apart. In this manner they learned that Du Pont cement loses its sticking power at cryogenic temperatures.

Even after they tied together the crystals with nylon thread, they could only keep the nuclei aligned for fifteen minutes at a time. Nonetheless, at the end of six months' work the team was almost certain that more electrons were being emitted in one direction than any other. A period of intensive checks of the equipment followed; Hoppes slept on the floor next to the equipment in a sleeping bag for fear of something going awry. About two o'clock in the morning of January 9, 1957, all checks were finally complete. Hudson uncorked a bottle of Chateau Lafite-Rothschild 1949, ceremoniously poured it into paper cups, and everyone drank to the overthrow of parity.[45]

Before New Year's Eve, Wu had told Lee and Yang of the asymmetry, but said the team was not ready to go public. But the word leaked out anyway. Someone informed Leon Lederman, another Columbia experimenter whom Lee and Yang had unsuccessfully urged to investigate parity. Lederman had a colleague, Richard Garwin, who was creating beams of muons in order to repeat the Conversi-Pancini-Piccioni effect on his own. Over a meal at the Shanghai Cafe on January 4, Lee, Lederman, Garwin, and half the Columbia physics faculty discussed the muon beam. If parity nonconservation was a general feature of weak interactions, the electrons created by the decaying muons would also emerge more in one direction than another. The experiment was a gamble, because the process of stopping the muons might wreck the alignment of their spins. Despite all the difficulties, Lederman and Garwin went ahead. Lederman telephoned Lee before breakfast on Tuesday, January 8, with the words, "Parity is dead." The effect was huge, unmistakable, even more pronounced than Wu's. Incredible that nobody had seen it before. As a matter of course, Lederman's team waited for Wu's group to finish before sending off their paper.[46]

Valentine Telegdi, an experimenter at the University of Chicago, also had an assistant making muon beams. Having run across a preprint of the Lee and Yang paper in August, Telegdi, unlike most of his colleagues, was immediately interested. Not knowing of Wu's work, he and the student began an experiment that was similar in many respects to Lederman's. He might have finished it first, but illness in the family delayed the run. His parity paper arrived two days after the other two at the offices of the *Physical Review*. It was bounced. A rewritten article was published two weeks after the others.[47] By then the news had already thrown the physics world into turmoil.

Typical of the reactions was that of Wolfgang Pauli, who with amazingly bad timing put the following remarks in a letter to Victor Weisskopf just three days before he was told the news:

I do not believe that the Lord is a weak left-hander, and I am ready to bet a very large sum that the experiments will give symmetric results.[48]

Ten days later, he wrote Weisskopf again.

Now, after the first shock is over, I begin to collect myself again (as one says in Munich). Yes, it was very dramatic. On Monday 21st, at 8:15 P.M. I was to give a lecture on "the past and recent history" of the neutrino. At 5 P.M. the mail brought me three experimental papers. . . . Now, where shall I start? It is good that I did not make a bet. It would have resulted in a heavy loss of money (which I cannot afford); I did make a fool of myself, however (which I think I can afford to do)—incidentally, only in letters or orally and not in anything that was printed. But the others now have the right to laugh at me. What shocks me is not the fact that "God is just left-handed" but the fact that in spite of this He exhibits Himself as left/right symmetric when He expresses Himself strongly. . . . How can the strength of an interaction produce or create symmetry groups, invariances or conservation laws?. . . . Many questions, no answers![49]

Pauli was not alone. Yang informed Oppenheimer with a telegram to his vacation retreat in the Virgin Islands. Oppenheimer cabled back, "Walked through door." I. I. Rabi explained to the press, "In a certain sense, a rather complete theoretical structure has been shattered at its base and we are not sure how the pieces will be put together."[50] That fall, the Nobel Committee awarded Lee and Yang the Nobel Prize barely a year after the work which it canonized.

Revolutions are seldom accomplished without some legacy of bitterness. Telegdi was sufficiently angered at the rejection of his paper by the *Physical Review* that he resigned from the American Physical Society.[51] Later, several experimenters were chagrined to realize that they had observed parity violation as far back as the 1920s but had not understood what they saw.[52] Block believes that his question at the Rochester Conference started the entire process, and that the physics community does not recognize his contribution.[53] The most lasting divisions came between Lee and Yang, whose friendship was a casualty of their increasing celebrity. In 1962 they formally ended their decade-long collaboration. Physicists took care to invite one only when the other was absent, and watched in distress when these two remarkable thinkers used the handsomely printed volumes of their collected works to attack each other.[54] Science is ordinarily a collective enterprise, with the flux of ideas as impossible to monitor as it is to measure. Thoughts and concepts flow freely, and the most solitary of thinkers is influenced by the talk in the corridors and the rumors from experiment. When the stakes are high, however, worries about priority and reputation creep in; the relationship between Lee and Yang could not withstand their own success.

Parity nonconservation was soon found to characterize all manifestations of the proposed Universal Fermi Interaction. But to clinch the case for its existence, physicists would have to do more than show that the interactions—beta decay, muon decay and capture, and strange particle decay—had similar properties and strengths. ("Similar strength" means that the various decays had apparently identical coupling constants.) A sleigh can be pulled with an equal force by, say, a small pickup truck, several Clydesdale horses, or a large pack of Siberian Huskies, but that does not imply that the nature of the entity pulling the sleigh is identical. To show that, experimenters would have to demonstrate that the *form* of the interactions is in all cases the same.

Physicists have classified elementary particle interactions into five kinds, which bear the labels scalar, vector, tensor, pseudoscalar, and axial vector. These imposing names—customarily abbreviated as S, V, T, P and A—designate the only ways that one particle's wave function can transform into another during an interaction (as the wave functions for the neutron and proton do in Fermi's beta decay theory) and satisfy the demands of both relativity and quantum mechanics. Physicists can learn much about an interaction by discovering its form. Quantum electrodynamics, for instance, is a vector (V) interaction, which implies that the transition from one wave function to another is always accompanied by the creation of a virtual particle, the photon, that has a spin of one and negative parity. For this reason, any particle with a spin of one and negative parity is called a "vector" particle. Similarly, an axial vector (A) interaction is associated with a spin-one, positive-parity particle. And a scalar (S) interaction is one mediated by a particle with a spin of zero and positive parity.

Copying electromagnetism, Fermi had hypothesized that beta decay is a vector interaction. But the beta decays of Gamow and Teller could not be described in this manner. From measuring the velocities of the emitted electrons, Fermi-style beta decays were known to be either V or S, whereas Gamow-Teller decays were either A or T. This left only four possibilities for a Universal Fermi Interaction: VA, VT, SA or ST. And there the matter stood when Wu started aligning frozen cobalt nuclei.

The Seventh Rochester Conference opened on April 15, 1957, just three months after the discovery of parity violation. A feeling of elation was still in the air. One of the most eagerly awaited sessions was the discussion on weak interactions, in which the chair had been given over to Yang and the introductory remarks to Lee. Lee quickly threw cold water on the hopes of the crowd by pointing out that although "there exists a large class of interactions characterized by coupling constants of order 10^{-13}," and that "all weak interactions seem to have striking features [such as parity violation] in common," there was no reason to believe that they were all caused by one entity, and indeed some reason to think that they were not.[55]

He drew a triangle representing the particles affected by the proposed Universal Fermi Interaction, with each vertex representing a current—again, the top or bottom half of the appropriate Feynman diagram—and each leg the interaction between the two currents. The left-side leg (proton, neutron—electron, neutrino) was beta decay, which one remembers has two different forms, Fermi and Gamow-Teller transitions. Each of the two had to be examined separately to determine which letter applied to it.

Two years earlier, experiments had been done on neutrons, Ne^{19} (an isotope of neon with 19 particles in the nucleus) and He^6 to determine their

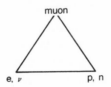

form of beta decay, with the He^6 especially noteworthy because it was performed under the aegis of Chien-Shiung Wu. He^6 pointed unambiguously toward T for Gamow-Teller transitions, whereas the other two indicated S for Fermi transitions.[56]

All that was fine, Lee said. ST fitted perfectly well into the Fermi theory. But the parity violation experiments of Telegdi, Lederman, and Garwin indicated that muon decay—the bottom leg of the triangle—was V. A single force could not be scalar and tensor in one case and vector in another; beta decay was not the same as muon decay. The similarity in coupling constants appeared to be one of nature's nasty coincidences, like the similarity between the masses of the muon and pion that led the two particles to be confused for a decade.

The case against the Universal Fermi Interaction was hammered home a few minutes later by Wu, who described some as-yet unpublished experiments with another type of radioactive cobalt. The new data contradicted the two experiments that indicated S for Fermi decays. Wu said she believed that they were vector. Combining this with the He^6 results for the Gamow-Teller transitions, she argued that beta decay was a combination of V and T, which was as compatible with Fermi as S and T. This removed the contradiction with the experiments by Telegdi, Lederman, and Garwin and should have been pleasing to advocates of the Universal Fermi Interaction.[57]

Instead, her speech threw the conference into confusion. V and T behave completely differently under parity violation. In practical terms, parity violation means that the neutrino emitted during beta decay always has a spin that went in one direction. A neutrino with "left-handed" spin rotates clockwise to the direction of motion; "right-handed" neutrinos go counterclockwise. (The names come from the act of sticking a thumb out and curling the fingers into the palm; the left hand curls clockwise, the right counterclockwise.) When one plugged V and T into the now parity-violating equations, the vector terms predicted a left-handed neutrino whereas the tensor terms came up with a right-handed particle. In other words, if Wu was correct, the neutrinos emitted in the two types of beta decay were completely different. Not only might there be no Universal Fermi Interaction, but beta decay itself might be caused by a mess of different forces.

One of the more startled conference attendees was its organizer, Robert Marshak.[58] His graduate student, E. C. G. Sudarshan, had reviewed all available experimental data about the "class of interactions characterized by coupling constants of order 10^{-13}." If there really were a Universal Fermi Interaction, Sudarshan and Marshak concluded, it had to be a mixture of V and A, specifically V minus A $(V - A)$. This was a vector interaction with some axial vector thrown in to account for parity violation. Sudarshan and Marshak had intended to present their ideas, but Wu's remarks threw them completely off kilter; moreover, there were rumors floating around that two other experiments were coming out that would back her up. To Sudarshan's annoyance, Marshak, the senior collaborator, said that they should keep quiet until the new data appeared.

One of the rumored experiments was done by Felix Boehm at Caltech. By chance, Marshak ran into Boehm's colleague Gell-Mann in late June at the RAND Corporation in Santa Monica, where both were consultants, and asked him to arrange a meeting. Over lunch, Marshak and Sudarshan outlined the $V - A$ idea to Gell-Mann and Boehm and asked the latter about his experiments; Boehm replied that his data contradicted V and T, but were compatible with $V - A$. Reassured, Sudarshan and Marshak completed their paper in a matter of days and decided to unveil $V - A$ in September at a conference in Padua, Italy.[59] In Padua, Marshak boldly declared that the weak interactions must all be caused by the same thing, and that therefore $V - A$ was the only possible way to go. "I said," Marshak recalled, "that it had to be $V - A$ for a universal interaction, and that to be so will require that four experiments be murdered—I used the word *murdered*—including the He[6] experiment. One physicist then said to Lederman during a coffee break, 'Marshak's *crazy*. How can He[6] be wrong?' "[60]

Gell-Mann, too, had been toying with $V - A$. At the Boehm luncheon, he had been interested to hear Sudarshan and Marshak also pushing the idea.[61] Gell-Mann then talked up $V - A$ with a colleague, who in turn described it to Richard Feynman. (Such is the operation of the physics grapevine.) Feynman had been thinking about weak interactions for some time, but had been unable to understand how nucleon beta decay could be S and T and fit in with anything else. He spent that summer in Brazil. When he returned to Caltech, he asked some experimenters to fill him in on beta decay.

Finally they get all this stuff into me, and they say, "The situation is so mixed up that even some of the things they've established for years are being questioned— such as the beta decay of the neutron is S and T. Murray says it might even be V and A, it's so messed up." I jump up from the stool and say, "Then I understand EVVVVVERYTHING!" They thought I was joking. But the thing I had trouble with at the Rochester meeting—the neutron and proton disintegration: Everything fit but that, and if it was V and A instead of S and T, that would fit too. Therefore I had the whole theory!

Delighted, he went home to work.

I went on and checked some other things, which fit, and new things fit, new things fit and I was very excited. It was the first time, and the only time, in my career that I knew a law of nature that nobody else knew. The other things I had done before were to take somebody else's theory and improve the method of calculating. . . . I thought about Dirac, who had his equation for a while—a new equation which told how an electron behaved—and I had this new equation for beta decay, which wasn't as vital as the Dirac equation, but it was good. It's the only time I ever discovered a new law.[62]

Feynman wrote a paper on $V - A$ with Gell-Mann, and they sent it to the *Physical Review*. It had some features not considered by Marshak and Sudarshan, but, like theirs, claimed that the He^6 experiment must be wrong. Due to the vicissitudes of publishing, the Sudarshan-Marshak article appeared months after the Feynman–Gell-Mann article. (Independently, a young Cornell graduate student named J. J. Sakurai also postulated $V - A$.)

Within a few months of the Padua conference, experimental evidence began to collect that $V - A$ was indeed correct.[63] The question was settled when Maurice Goldhaber and two collaborators performed an elegant and simple demonstration showing that the neutrino was left-handed and the antineutrino right-handed.[64] After this experiment, $V - A$ was established as the form not only of beta decay but all other manifestations of the weak interaction as well. The Fermi force that had grown into the Universal Fermi Interaction was now an accepted fundamental force of nature: the weak force.

All participants in the $V - A$ theory had many reasons to be pleased, but Gell-Mann in particular felt as if the plums had at last fallen into his lap. There had been years of experimental wrestling with the phenomena, but at last theorists had enough solid ground to stand on; the new picture of the weak interactions represented, at least in part, a hope that physics could make giant strides indeed. The vector and axial vector nature of weak interactions was deeply compatible with the vector form of electromagnetism. Two vector interactions—could they somehow be the same thing? Gell-Mann spent a fair amount of time in 1958 and 1959 thinking about a principle that would account for the existence of the weak force, something like the way Yang and Mills had shown how fields can be generated through symmetries. He never came up with a workable model, and so didn't publish any of the scribbling on his sketch pad, but he was sure that a true understanding of the weak interactions would entwine and embrace them with electromagnetism.[65]

12
Steps to Unification

ONE CASUALTY OF THE CONFUSION OVER THE FORM OF THE WEAK INTERAC-
tion was Julian Schwinger, the prodigy who facilitated the renormalization of
quantum electrodynamics. During the 1950s, he built up a school of quantum field theory at Harvard that drew its inspiration from his extraordinary
thrice-weekly lectures on the subject. He drove, in the recollection of his
students, the largest and most elegant Cadillac in Cambridge, a purring
baby-blue sedan that pulled into the lot by Lyman Hall moments before class
began.[1] "Cast in a language more powerful and general than any of his
listeners had ever encountered," one of Schwinger's advisees recalled,
"these ceremonial gatherings had some sacrificial overtones—interruptions
were discouraged and since the sermons usually lasted past the lunch hour,
fasting was required."[2] After a short break, students would gather in a clump
outside Schwinger's office in the hope of a private meeting with the master,
whose fundamental insights were written in a wholly idiosyncratic system of
symbols that avoided all use of the Feynman diagrams that by then pervaded
theoretical physics.

Schwinger navigated a solitary course that paralleled the path of the
rest of the field, independently duplicating when he did not anticipate the
summed efforts of every other person in particle physics. His great physical
intuition and enormous breadth of interest made him into that rarest of
creatures, a one-man community of thought—a forbidding role that eventually isolated him from his fellows. Like Yang and Mills, Schwinger had
been impressed by the way that the theory of quantum electrodynamics
could be associated with a symmetry of phase. Characteristically, he derived
a mechanism somewhat like that of Yang and Mills in his own way, and with
his own notation. (He tried a global, rather than a local gauge symmetry.) But
instead of using the idea to look only at strong interactions, as most of his
colleagues were doing, Schwinger turned also to the weak. And in October
and November 1956, while Sudarshan and Marshak at Rochester were putting together their first notions about the $V - A$ form of the weak interactions, Schwinger dazzled his students at Harvard with a series of lectures
attempting to derive the very existence of the weak force from the symmetries of its interactions—an extension of Noether's theorem and a much more
ambitious project. As if this were not enough, Schwinger introduced, for the

first time anywhere, the idea that the weak and electromagnetic interactions might be two facets of the same phenomenon.

Schwinger had been thinking about unification since before the Second World War, when he worked with Oppenheimer at Caltech. "In 1941," Schwinger once told us, "when I was Oppenheimer's assistant, we talked about all parts of physics. And Fermi's theory of beta decay was one of the hot theoretical topics because obviously it had divergences that were even worse than those we were already conscious of [in electrodynamics]. It was really messy. And I had a thought. Fermi's theory was then expressed in terms of a basic coupling constant. The basic coupling constant in electrodynamics is 1/137. It's a pure number, which is why the whole thing works. Now, Fermi's coupling constant is a dimensional quantity, and since it arose in the days when—well, it's concerned with electrons, so the electron mass was used as a unit of length." (Oddly, by fiddling with the ergs, centimeters, and so on used in the equations, it is possible to use a mass to express an associated length.) "It turns out to be some absurdly low number, 10^{-13}, something or other. Now, I was then a nuclear physicist. And I knew that while the electron was involved in beta decay, it was something that the proton and neutron also engaged in. So I said, 'First of all, why isn't the proton or the nucleon mass the important thing?' When I introduced that, the coupling constant became 10^{-5}, a more reasonable number." Using the much larger mass of the nucleon as a standard pushed the coupling constant up by a factor of a hundred million. "Then, being very aware of electrodynamics, I said, 'Hey, could it be that if the relative mass [of the virtual particle in beta decay] was a couple of hundred proton masses, the coupling constant would become the fine structure constant—1/137?'

"It was numerology," Schwinger said dismissively. "Numerology. But—that's the whole idea. I mentioned this to Oppenheimer, and he took it very coldly, because, after all, it was an outrageous speculation."

We asked if this had all happened in the 1940s.

He nodded. "In 1941. But it's the idea of the intermediate particle, with the mass coming from making the coupling constant to be the fine structure constant. And so that had always been on my mind. In 1956, when I was hit with parity—it was a blow to me, parity violation—I began to rethink the whole subject. And at that point, I introduced the idea explicitly, that there was a vector particle which carried weak interactions, and all the rest of it."[3]

Parity violation was indeed a shock to Schwinger. When the rumors about Wu's experiment began seeping out of the Bureau of Standards, Schwinger urged his Cantabrigian colleagues not to jump to the idea of parity violation before the results were known. In one of the discussions that took place everywhere in the physics world in the first days of 1956, Schwinger told his fellows that parity was too lovely a symmetry to be easily abandoned. The argument was interrupted by a call from Schwinger's mentor, I. I. Rabi. Schwinger returned with the sad words, "Gentlemen, we must bow to nature."[4]

Like many elementary particle physicists, Schwinger was pushed by parity violation into rethinking the nature of the weak interactions. The thoughts grew into a series of lectures, and out of the lectures came a paper, provocatively entitled, "A Theory of the Fundamental Interactions," which was published in the November 1957 issue of the *Annals of Physics*.[5] An expansive piece of writing, it is filled with Schwinger's trademarks: an evident love of mathematical ingenuity, an utterly eccentric notation, and a bold, imperious style that makes almost no reference to the work of other physicists. (This last has offended many of Schwinger's colleagues, whose reputations depend on such citations.)

The most important part of the paper is a section at the end, on the class of particles called leptons, which are unique in that they respond only to the weak force and not to the strong. In 1957, there were three known types of lepton: the familiar electron; the neutrino; and the muon—the "who ordered *that*?" particle—which is exactly like the electron except that it is about two hundred times heavier. All are as impervious to the strong force as a tree stump is to an electromagnet. In his paper, Schwinger simply assumed the existence of two types of neutrino, one associated with the electron and one with the muon. True to form, he did not identify his conjecture as such, but wove it into the logic of his argument.

("Schwinger always pretended that everything he wrote was a direct revelation from God," one of his colleagues told us. "You'd be reading along and suddenly there's two neutrinos—a direct revelation from the Lord!"[6] On the other hand, other physicists had independently made the same guess beforehand.)[7]

Schwinger needed this first speculation to make a second bit of speculation come out right. He postulated that the weak force is carried among the leptons by three particles—the photon and two hypothetical vector bosons, which are now called W^+ and W^-, and are analogous to the B particles in the paper by Yang and Mills.

There had to be *two* vector bosons to account for the way that weak interactions convey electric charge. When the weak force transmutes an uncharged neutron into a positive proton, a virtual particle is emitted, and in its decay an electron and an antineutrino are created. Schwinger argued that when the weak force transfers a positive charge, that charge is carried by a positive version of the vector boson, now called the W^+, whereas when the weak force transfers a negative charge, it is carried by a negative version of the vector boson, the W^-. The ordinary photon would be the intermediary when no charge is involved.

Schwinger told his classes that this picture of the weak force had far-reaching consequences. Because the photon and the Ws make up a triplet, Schwinger said, electromagnetism and the weak interaction should be treated as part and parcel of the same phenomenon; that is, he had unified them. But he was more cautious in print, as is proper. He never used the word "unification" in the paper, referring instead to the weak force as "a partner of the electromagnetic field."[8]

There was no compelling reason for his supposition; Schwinger merely wanted to demonstrate that his pet notion could be used to construct a coherent theory. In addition, he found the result aesthetically pleasing. His colleagues did not. During a visit to New York City, Rabi told him bluntly, "They hate it."[9] Upset by the reaction, Schwinger was further dismayed to learn that the experimental data he was relying on were wrong. By artful juggling, he had managed to make the description of the combined weak and electromagnetic interactions fit the data on the then-believed V and T form. (Schwinger, too, had been at Rochester when Wu argued in favor of that interpretation of the experimental findings.)[10] Between the time he submitted the paper and the time it was published, $V - A$ was in. By the time Goldhaber and others proved that $V - A$ was right, Schwinger had thrown up his hands and turned to other areas of physics.

Before he did, however, he asked a student, Sheldon Glashow, to look at the possibility of a connection between the weak and electromagnetic forces—to see if a connection could still somehow be established. "It was a vague request," Glashow recalled later. "He said, 'Think about intermediate vector bosons as agents of weak interactions.' My problem was to think about intermediate vector bosons. And that's what I did for two years—think about it."[11]

Sheldon Glashow's path toward unification began at the extreme northern tip of Manhattan, where a little side street, Payson Avenue, runs uphill for a long block beside Inwood Park. Facing the grass and trees is a row of small two-story houses, one of which—number sixty-five—was Glashow's boyhood home. Glashow's father, Lewis Gluchovsky, had come from the town of Bobruysk, in White Russia, where the family had owned a Turkish-bath house ("*the* public baths of the city," according to Glashow). When he arrived here, the immigration authorities changed his name; shortly thereafter, Gluchovsky/Glashow became a plumber in Manhattan. He married another immigrant, Bella Rubin. They had two boys, aged fourteen and eighteen, when the third, Sheldon Lee, was born. One of Glashow's brothers became a doctor, the other a dentist. Sheldon decided at the age of eight or nine that he wanted to be a scientist.

"It was the beginning of the Second World War," Glashow said to us. "People were interested in those days in identifying airplanes and understanding how bombs are dropped, and I got curious about the ballistic problem. My brother explained to me that if a plane drops a bomb and doesn't take some evasive action, the bomb is going to blow up right underneath where the plane is *going* to be. That of course struck me as a very odd fact, because the assumption is that when the plane drops the bomb, the plane is moving forward and the bomb is not.

"I had a little notebook, this black-and-white speckled notebook that kids still have. My brother would write little lessons—two or three or five such lessons—about things he had learned. Not long before, Harold Urey had discovered heavy water, and Urey was his professor in school, so he was a little turned on about it, and he communicated that to me."

Lewis Glashow was not entirely pleased by his son's interest in science. "My father told me that science would be all right, but why didn't I do it in my spare time and go into medicine, like my brothers. I differed with him slightly about that. He relented when my brothers came back safely from the war. When I entered high school, he helped me build a chemistry lab in the basement, where I loved to perform long and dangerous experiments."[12]

Glashow's passion for science was not directed to physics until 1947, when he entered the Bronx High School of Science. After classes, the boys in the Science Fiction Club—there were no girls—would cluster around an old laboratory table littered with surplus hospital Bunsen burners and the tracework of test-tube racks. The preferred topics of conversation included the contents of a magazine called *Astounding Science Fiction* and the ideas of L. Ron Hubbard and Immanuel Velikovsky. Two of Glashow's best friends were Gerald Feinberg and Steven Weinberg. (Feinberg became a distinguished physicist and has written several respected books; Glashow and Weinberg shared the Nobel Prize in physics in 1979 for their contributions to a theory incorporating the weak and electromagnetic interactions.) As boys, all three contributed to the Science Fiction Club fan magazine, *Etaoin Shrdlu*; the name is a term from typography.

The founder of Bronx Science, Morris Meister, was a pioneer of fast-track education and a tireless advocate of the notion that if bright, science-oriented students are brought together, certain ill-defined but nonetheless valuable learning processes will occur. As far as Bronx Science is concerned, he seems to have been right. Since its founding, in 1938, it has produced three winners of the Nobel Prize for physics—more than any other high school in the world ever has, and as many in those forty-eight years as most countries, including France and Italy.

Considering that statistic, it is surprising to learn that the physics taught to Glashow, Weinberg, and Feinberg was of dubious merit.[13] In Feinberg's recollection, the instructor who taught his introductory course, "for all we could tell, did not believe in the existence of atoms."[14] The textbook was written by a man named Charles Dull, who seems to have lived up to his name. Students interested in pursuing advanced physics were offered a choice between "automotive physics," in which the kids soberly took apart and put together an old Curtiss-Wright airplane engine each semester (a good course "for those who believe the automobile is here to stay," the 1950 yearbook states); and "radio technology," in which students put together shortwave kits of the sort advertised on the back of *Boy's Life*.[15] Glashow took neither.

Despite the modest course offerings, a kind of *samizdat* science took place outside the classroom. Glashow persuaded his mother to buy him college outline texts whenever she dragged him downtown to shop; at lunch Feinberg often led excited conversations about quantum mechanics. Most important, however, was the give-and-take among the Science Fiction Club members; the boys competed with each other to reach the most complete understanding of the latest physics discoveries they had encountered, but they also helped stragglers to catch up. Feinberg said, "After the meetings, Shelly and I would spend hours on the phone talking about science. An hour, hour-and-a-half a day—it drove our parents crazy. We were sure we were going to find wonderful things."

After graduation, Feinberg chose to stay in New York City and go to Columbia. Glashow and Weinberg went upstate, to Cornell. Although Cornell has become an internationally important center for physics, in the early 1950s it had the nickname "Moo U" and was known chiefly for agricultural science. Cows grazed on the lawns, and campus lore had it that professors were not allowed to eject dogs that might stray into their classrooms. Although some prominent physicists—Richard Feynman, for example—were at Cornell when Glashow and Weinberg entered, in the class of '54, their courses were restricted to graduate students. Those who taught undergraduates regarded most forms of collaboration among students as dishonest, and after Bronx Science Glashow found that attitude hard to take. "There was a style mismatch. Some of us were thrown out of school for cheating. We would collaborate on their dumb problem sets and they didn't like that, for example. One of us would show up with six copies of homework—sort of identical homework—and the teacher would simply ask if one of us wanted the 100 percent credit or if we wanted to divide it up. It got to the point where I flunked a course taught by a solid state professor. I just found the problem sets absurd and refused to do them."[16]

In his senior year, Glashow was exposed for the first time to quantum field theory. He was the only undergraduate in the class, and the subject baffled him. At exam time, the professor, Silvan Schweber, was nonplussed to learn that he could give all the graduate students pass-fail grades, but had to give an undergraduate a numerical one. Glashow recalled: "So he said, 'How 'bout an eighty-five?' Which was not too low, not too high. I wanted to try for something better, so I said, 'No, give me an exam.' Schweber asked me a bunch of questions and I got every single one of them wrong. At the end, he said, 'How 'bout an eighty-five?' I said, 'Great!' "[17]

("I still remember giving him the oral," said Schweber, who went on to Brandeis University, in Cambridge, Massachusetts. "It was clear that he was a very unusual undergraduate. It was also clear that he thought he could do the work with his left hand. My memory plays tricks on me—I can't be sure now why I did it—but by giving him that [very hard] test I may have been trying to tell him you can't do physics that way.")[18]

Despite his trials, Glashow learned some physics, most of it in the same way he had at Bronx Science—by shooting the breeze with his friends. He argued with Weinberg, dashed off letters to Feinberg, and dropped down to New York City on weekends to hear lectures; back in Cornell, he latched on to classmates who could teach him something. Perhaps the most important material he absorbed came from a graduate student, Harold V. McIntosh, who had decided to pursue physics after getting a degree from the Colorado School of Mines. Brainy and eccentric, McIntosh had a little coterie of enthralled undergraduates whom he introduced to his current obsession, the mathematical theory of groups. Glashow said, "He was always coming up to me with problems involving vibrating bedsprings and jiggling ozone atoms and things like that which he claimed group theory was capable of solving. Actually, what I learned from him was as relevant as any course I took there."[19] (The last Glashow heard, McIntosh was teaching physics in Mexico.)

Glashow chose to do his graduate work at Harvard, where he might be able to write his thesis with Schwinger, one of the heroes of the effort to renormalize quantum electrodynamics. At first, Glashow found Harvard graduate school little better than undergraduate life at Cornell. "Nothing is quite as dull as being here," he wrote to Feinberg in November of 1954.[20] He saw Schwinger rarely, and considered the classes insufferably tedious.

By Glashow's second year, however, he was taking courses directly from the master, an experience that he found both electrifying and peculiar. A charismatic teacher, Schwinger spoke at lightning speed, covering the blackboard with equations, many of which were expressed in his own "Schwingerese." Regardless of the description in the catalogue, Schwinger's advanced graduate courses were always about whatever he was working on at the moment. He sometimes began class by saying that he had just realized that what he had told them the day before was utterly wrong. Glashow loved him. A shy man, Schwinger had a reputation for aloofness outside the classroom; his students claimed that all they ever saw of him was the flash of his florid silk tie as he ducked into the men's room to avoid them. ("He suffers from an uncertainty principle," Glashow said.)[21] Nevertheless, at the beginning of his third year, Glashow asked Schwinger to be his thesis adviser. So did eleven other students. To Schwinger's consternation, he found all of them waiting in his office one day. Thinking to test their abilities and avoid dealing with them as a pack, he devised an enormously complicated problem and instructed the multitude to work on it and return, one by one, with their solutions.

"So *of course* we collaborated," Glashow recalled. "The twelve of us came back a few days later, all at the same time, having solved the problem. He was happy with the solution—it was very elegant—but not so happy with the complete failure of his scheme to sort out the masses. So he then came up with a problem for each of us to work on individually, and in that way, against his will, got us all started on our theses."[22]

The question Schwinger awarded Glashow was one close to his current concerns. He asked Glashow to explore the behavior of the virtual particles Schwinger had hypothesized to be the carriers of the unified weak and electromagnetic forces—but this time to look at them in the light of the new experimental data on weak interactions. Glashow did not know he was being put on the track that would eventually lead him to a grand unified theory; he knew only that Schwinger was asking him to look at the sort of wild idea that was dear to his heart.

Today, Glashow's thesis—the product of those two years of thought— sits on his bookcase, stuck in a black spring binder on a neglected top shelf. He had to hunt for it, wading through the stacks of preprints and journals that clutter his office, when he wanted to show it to us. On a bulletin board beside the door were many strata of notes and memos, including a letter from physics students in Beijing claiming to have discovered antigravity ("Is *most* important result!"), newspaper advertisements showing how Transcendental Meditation can harness the power of the Grand Unified Field, and a number of clippings about the Soviet physicist Andrei Sakharov, in whose defense Glashow has been active. "Here it is!" Glashow said, standing on a wobbly stool. He hefted the volume appreciatively. "These things are always long, to show you have lots of bright ideas, and filled with tons of calculations—student showboating. Mine is a complete parade of crazy digressions. I haven't looked at this in *years*. But there's one part I'm still proud of. Here, wait, it's in the appendix."

This was no surprise. Graduate students of all stripes traditionally hide their most radical statements in footnotes and appendices, where they can be disavowed if necessary. Glashow was now sitting at his desk, wreathed by cigar smoke and shaking his head at the maunderings of his youthful self. When he found the page, he stuck an admonitory finger in the air.

"*Yes.* 'It is of little value to have a potentially renormalizable theory of beta processes [i.e., weak interactions] without the possibility of a renormalizable electrodynamics.' That's because the Ws are charged, so you had to get into quantum electrodynamics to talk about them. Here we go. 'We should care to suggest [his voice rises here; his pleasure in the younger Sheldon Glashow is apparent] that a fully acceptable theory of these interactions may *only* be achieved if they are treated together.' "

He snapped the binder shut, kicking up a little cloud of dust. "That's the basic idea of the electroweak theory right there, that the weak interaction'd never make any sense by itself because it's always mixed up with electromagnetism. So you have to look at them together. It sounds simple, but that kind of realization is a big deal. Before Newton could work out his theory of gravitation, he first had to realize that the force that moved the planets around the sun and the force that dropped apples on people's heads were the same thing. So, in a certain sense that sentence won me the Nobel Prize. It took thirteen years more to show that I was right, but damn it, I was."[23]

While working on his thesis. Glashow won a National Science Foundation fellowship, a prestigious award that entitles its recipient to study abroad. Glashow spent two years, from 1958 to 1960, at the Niels Bohr Institute of Theoretical Physics, which invites a small number of physicists every year to participate in the discussions held there. Glashow attended lectures, gave talks, met the great Bohr, and in general lived the life of which he had dreamed years before at Bronx Science.

At the Institute, Glashow expanded the appendix of his thesis into a full-fledged paper, "The Renormalizability of Vector Meson Interactions," which he submitted to the journal *Nuclear Physics* in the fall of 1958.[24] (Maddeningly, physicists for years used "vector meson" and "vector boson" interchangeably in this context.) Unlike Schwinger, Glashow thought that a Yang-Mills theory—that is, a local gauge theory—could be the basis for a unified theory. Moreover, Glashow wanted to prove that such a unified Yang-Mills theory of the weak and electromagnetic interactions was indeed renormalizable, as he had alleged. By November, when the paper was done, he thought he had it.

Pleased with himself, he went to England in the spring of 1959 to present his work. "I lectured in London or somewhere. They all sat there smiling very politely, told me that they liked what I was doing, and went home to tear it apart."[25]

"They" in this case refers mostly to Abdus Salam, who later shared the Nobel Prize with Glashow and Weinberg. Salam and his collaborator, John Ward, had also been inspired by Schwinger's paper to look into the question of linking the weak and electromagnetic interactions. Although the two men were skilled at the techniques of renormalization, they had been unable to rid the theory of infinities. When Glashow claimed he had, they were appalled. Salam said to us, "Ward and I were in London, and he was just passing through. To our surprise, he was working on the same thing—having been put on it by Schwinger. He had the W^+, W^-, and the photon in one group—that part was common. But then he also did a quite *amazing thing!* He said the theory was renormalizable! My God! this young boy was claiming that this theory was renormalizable! It cut me to the quick! Both of us considered ourselves *the* experts on renormalization. So there we were, the two great experts on renormalizability, wrestling for months with the problem—and here was this slip of a boy who claimed he had renormalized the whole thing!" Salam laughed. "Naturally, I wanted to show he was wrong— which he was. He was completely wrong. As a consequence, I never read anything else by Glashow, which of course turned out to be a mistake."[26]

Glashow had made mathematical and physical errors; his theory was not, after all, renormalizable. In fact, as Salam delightedly pointed out, if one did the math correctly one actually proved that the model of two Ws and a photon was not and could not be renormalizable, exactly the opposite of what Glashow had asserted.[27] For Glashow, the episode was embarrassing and annoying—a physicist's nightmare. ("They don't make graduate students that dumb any more," he said, groaning.)[28]

He was now faced with a choice more dependent on his personality than on the rules of scientific inquiry. Young theoreticians have to come up with bold ideas if they are to acquire a reputation, yet their papers cannot be half-baked. Glashow's idea was bold, but it was wrong. If he kept going, he risked being labeled an eccentric; if he went on to something else, it would be an admission of defeat. Glashow chose not to abandon unification; instead, he went further out on a limb, advocating an even wilder, more speculative extension of his original position. He threw himself into a sequel, dashing off notes to Feinberg and Weinberg and receiving encouraging criticism, turning over his ideas with the bemused care of a raccoon washing a shiny ring.

In March 1960, after Glashow gave a talk in Paris, Gell-Mann, whom Glashow had never met, invited him to lunch. Gell-Mann was at that time a visiting professor at the Collège de France. As he was then in the middle of an extraordinary decade in which he totally dominated particle physics, Glashow was more than happy to be his guest. The two went to a seafood restaurant. Glashow mentioned an aversion to fish; his dining companion explained that serious people liked seafood, that Glashow was a serious person, and that therefore Gell-Mann would order fish for both of them. Over the entrée, the two men talked about vector bosons. Loathing the odd notation, Gell-Mann had not read Schwinger's paper, but he had himself been thinking for several years about ways to reconcile the weak and electromagnetic forces. He urged Glashow to press on, and invited Glashow to join him at the California Institute of Technology in the fall to work on field theory. "What you're doing is good," Gell-Mann told Glashow. "But people will be very stupid about it."[29]

"Partial-Symmetries of Weak Interactions," the paper that established Glashow's claim to a Nobel Prize, was published in the November 1961 issue of *Nuclear Physics*—one month before its author's thirty-first birthday.[30] The paper has the dry, confident tone that characterizes the best scientific prose. "At first sight," Glashow begins,

there may be little or no similarity between electromagnetic effects and the phenomena associated with weak interactions. Yet certain remarkable parallels emerge with the supposition that the weak interactions are mediated by unstable bosons.

Next, Glashow squares up to a problem that had arisen first in the work of Yang and Mills and was still unsolvable: how much these W bosons weigh.

The mass of the charged intermediaries must be greater than [zero], but the photon mass is zero—surely this is the principal stumbling block in any pursuit of the analogy between hypothetical vector [bosons] and photons. It is a stumbling block we must overlook.

Glashow simply flags the question in passing with a "don't know" sign, and turns to the rest of the theory. This time, he argues that the theoretical problems can be resolved if one assumes not three virtual particles, but *four*.

In order to achieve a . . . theory of weak and electromagnetic interactions, we must go beyond the hypothesis of only a triplet of vector bosons, and introduce an additional, neutral vector boson, Z_S.

As he confesses,

The reader may wonder what has been gained by the introduction of another neutral vector [boson].[31]

What had been gained was this: The Z_S allowed Glashow to separate the weak and electromagnetic interactions, and to deal mathematically with each in turn. (Today a very similar entity is called Z^0—"zee-zero"—and without getting into unnecessary complications we shall henceforth identify the Z_S as the Z^0.)[32] In the first Schwinger-Glashow scheme, the photon had worked overtime—both for electromagnetism and the weak force—with the result that the two functions interfered with each other in the theory. The new model was far neater.

Unfortunately, the neatness came at the expense of unification. Glashow could only resolve the difficulties with renormalization by "detaching" the weak and electromagnetic forces; they now worked hand in hand, but were no longer fully unified. In this model, the weak and electromagnetic forces within the atom are like two children around an elaborate Lionel train set; each has a separate control panel, and frantically switches the gates, toots the whistle, and twists the throttle without consulting the other. The motion of the train is the result of the actions of both, and depends moment by moment on what each is doing. So it is for subatomic particles. Surrounded by a haze of photons, Ws, and Zs, their movements are a synthesis of the effects from all of them.

The contribution of each force to the total interaction is expressed in a ratio, so much weak to so much electromagnetism, that Glashow describes in the paper. It is customarily known as a "mixing angle" because of a principle from high school trigonometry that most students immediately forget: Any ratio or percentage can be expressed in terms of a trigonometric function of some angle. Today, to Glashow's chagrin, the mixing angle is sometimes called the "Weinberg angle."[33]

It is the Weinberg angle because seven years after Glashow's paper was published Weinberg reinvented it. No one had paid much attention to Glashow's paper: It had appeared in an unknown journal and, if that had not been enough to guarantee its obscurity, there was the fact that it was written in Schwinger's odd notation.

A deeper reason was the sheer ungainliness of the model. Glashow had jammed together two gauge theories. The first was ordinary quantum electrodynamics. The second was a theory of the weak force based on the use of isotopic spin by Yang and Mills. But instead of using ordinary isotopic spin, as they had, Glashow postulated the existence of an isotopic spin-like quantity that applied to electrons and neutrinos; its conservation generated the weak force. But because parity is conserved by electromagnetism and violated by weak interactions, half of the model applied equally to left- and right-spinning particles, whereas the other part applied only to left-spinning particles—which meant the model looked like an unholy mess.

"When $V - A$ was established," Gell-Mann said, "everybody began working with electromagnetism, the weak force, and Yang-Mills, trying to unify. But it never worked. The crucial missing step was Shelly's idea. He pinned down the screwy, ugly, disgusting, stupid, horribly asymmetrical model nature had picked out to work with." Indeed, Glashow's paper was the first rough sketch of a triumph of modern physics, a theory tying together the weak and electromagnetic forces. This electroweak theory, as it is called today, does not fully unify the two forces. Nevertheless, it ties them together so firmly that most scientists refer to it as unified. Furthermore, it became a crucial step on the road to the more speculative fully unified theories that dominate the field today.

Intrigued by Glashow's work, Gell-Mann took it before publication to the 1960 Rochester Conference.[34] "I gave the first clear description of his ideas," Gell-Mann said, meaning that he translated the theory into a form his colleagues could understand. "The reaction was very unenthusiastic."[35]

One reason for the lack of enthusiasm was that in addition to the theoretical peculiarity of Glashow's model, it had run into problems with the experimental data. By adding a third vector boson, the Z^0, to make the model look like a gauge theory, Glashow had created a new problem having to do with what are called weak neutral currents. In general, particle interactions are described in terms of currents; when the neutron of beta decay turns into a proton, this is known as a "weak current," because the current is the result of a weak interaction. In Glashow's hypothesis, the weak current occurs when the neutron emits a vector boson. The neutron-proton current can be further typified as a "charged weak current," because the uncharged neutron becomes a positively charged proton—a change of charge occurs. In terms of Glashow's electroweak model, the neutron emits a negatively charged vector boson, the W^-. If there were a weak interaction in which the neutron did not change identity, this would be a "weak neutral current." Such an interaction would occur, Glashow argued, if the neutron emitted a Z^0. The problem was that experimentalists had looked at millions of weak interactions and were certain that weak neutral currents did not exist. Not once, in hundreds of careful experiments, had they ever observed a weak interaction that might be the passing of an uncharged vector boson.

Unfortunately for Glashow, in this case nature was playing a trick on experimenters. Because weak interactions are usually swamped by those of the strong and electromagnetic forces, experimenters had done almost all of their work in this area with the particles in which the effects of the weak force were easy to see—strange particles. Strange particles only decay via the weak force, and physicists had used them since their discovery as a laboratory to study that interaction. For reasons that Glashow would only unravel a decade later, strange particles are just about the only type of particle in which weak neutral currents of the kind experimenters were looking for do not exist, and thus their years of research did not bear on Glashow's theory.

At the time, of course, he did not know this. Like his colleagues, he believed that the data on strange particles held true for all forms of matter. To save his idea, Glashow invented a reason why the Z^0 had never shown its face. He asserted that it must be a lot heavier than its brethren, the W^+ and W^- To create a Z^0, experimenters would therefore have to bring to bear much more energy than any existing accelerator could generate. No wonder the particle had never been seen. (Such theoretical gambits infuriate experimenters, by the way.)

Alas, this strategy threw the baby out with the bathwater: In order to explain how weak neutral currents could exist but not be observed, Glashow had hypothesized a Z^0 so heavy that it would hardly ever put in an appearance. The paper concludes, "Unfortunately our considerations seem without decisive experimental consequence."[36]

The problem of the mass of the vector bosons and the question of weak neutral currents dogged the electroweak theory from the outset, and these difficulties were not resolved for years.

In 1960, after Glashow finished his paper on weak interactions, he moved to Pasadena, where he became part of a team scrambling to keep up with Murray Gell-Mann. One of the true *enfants terribles* of twentieth-century science, Gell-Mann entered Yale at fifteen and received his doctorate at the age of twenty-one. A polymath, he is sometimes said by his colleagues to have no particular talent for physics, but to be so smart that he is a great physicist anyway. Although only three years older than Glashow, he had been on the scene almost a decade longer. By 1960 he had the run of the California Institute of Technology and was spewing out ideas at a dizzying rate. He is a stocky man, whose short gray hair grows in tight curls; he moves with bulletlike determination, his shoulders thrust forward, his gaze fixed ahead. Notorious for his drive and asperity, he is also quick and generous with praise. Like Glashow, he loves conversation. The two men enjoyed each other's company and soon began to work together on an ambitious article that they hoped would pave the road to unification.

The result—"Gauge Theories of Vector Particles"—appeared in the *Annals of Physics* in September 1961.[37] It is a curious document; Gell-Mann has described the first half as a "classic, a landmark," and the second part as a "mess."[38] The first half sets up a kind of grammar for quantum field theory—a catalogue of possibilities. Drawing upon group theory (Glashow's introduction to it at Cornell, under the tutelage of Harold McIntosh, had not been a waste of time), the two physicists noted that the sets of hypothetical virtual particles described in Yang-Mills theories corresponded to the generators of a particular type of group.

A group is a set of mathematical objects related by a symmetry. An example of a group is the six sides of a die, which has the same shape no matter what side you are looking at. The symmetry is shown when the cube is rotated ninety degrees in any direction. Other groups are the sets of all ways one can turn a round rubber ball or a book. Interestingly, the latter two groups are quite different. It makes no difference whether a ball is turned to the side then flipped upward or the other way round. On the other hand, if this book were to be turned to a side and then flipped up the cover would be in a different place than if these movements were done in the other order. Mathematicians say that the second group, the set of book rotations, is not commutative, because the outcome of several successive rotations depends on the order in which they are performed. Because, as Heisenberg discovered in the 1920s, quantum variables are often noncommutative, this kind of group might be expected to be of interest to physicists, and indeed it is. Moreover, as Gell-Mann and Glashow discovered, all possible variations of such groups had already been laid out nearly thirty years before by the great French mathematician Élie-Joseph Cartan. Cartan had given the groups names. For example, Cartan's group U(1) corresponded to the gauge field for quantum electrodynamics, and his group SU(2) × U(1) to Glashow's electroweak model. (It is usually pronounced "ess-you-two-cross-you-one." SU stands for "special unitary" group and U for "unitary." SU(2) is the name for the group defined by all possible ways of turning something like a book; it is also the group for all possible rotations of the axis of isotopic spin.)

In the first half of their paper, Glashow and Gell-Mann showed that every one of the groups Cartan defined corresponded to a Yang-Mills gauge field, and that if one constructed a Cartan-style group of virtual particles there would always be a Yang-Mills field associated with it. The difficulties arose in the second half of the paper, in which Glashow and Gell-Mann tried to do something practical with the classifications borrowed from Cartan. They proposed a theory of the strong force to go along with Glashow's electroweak theory, and expressed them both in terms of Cartan's groups. With this trick, they sought to treat the strong and electroweak forces as Yang-Mills gauge fields. In this way, the strong and electroweak forces would be explained in the same language as electromagnetism, and thus would be linked to the only proven theory in elementary particle physics: quantum electrodynamics. But the attempt was like a game of basketball and a game of hockey played in the same arena at the same time. Gell-Mann explained, "We wanted to construct a Yang-Mills theory for the weak interaction and one for the strong interaction. But to do this, we had to have them both in the same space. And they *fought* each other. It was *terrible*. They would get mixed up with each other in some ghastly way, and the weak interactions would turn strong—it was *bad*, it was *sick*. Finally, I gave up on it. I said the hell with it. This one, I said, we're not going to solve now."

Glashow abandoned the effort, too. He was discouraged not only by their failure to put the electroweak and strong forces into a single theoretical framework, but also by the still unresolved problem of the mass of the W and

Z particles and by the lack of experimental evidence for weak neutral currents. For almost a decade, he turned his attention to the strong force, consigning all his earlier work to one of the dusty piles of paper in his office.

He returned to unification only once during this time, in the summer of 1962, when he spent a few months in Turkey. Roberts College (now the University of the Bosphorus), where he taught a seminar with Abdus Salam, is in the town of Bebek, which overlooks the Bosphorus. It is not far from Istanbul, where Glashow journeyed on the young American's ritual pursuit of hashish. Salam, who disapproved, accompanied Glashow to Iki-Bucuk Street, where a criminal named Ceri sold them a considerable quantity of greenish-brown Bursa hashish. After smoking much of it, Glashow is reported to have babbled incoherently about unification into the early hours of the morning. According to participants, Glashow resisted all efforts to quiet him, and pontificated with the volume and conviction of a Biblical prophet. [39]

As a testament to his concern, "The Secrets of S. L. Glashow," a notebook he kept during that summer, has a page labeled "For Fun." On the top is the question, "How does finite mass come about?" There was no answer. But the bottom half of the page and part of the next contains a sketch of an $SU(2) \times U(1)$ model that is, in retrospect, almost exactly correct. At the end, the mass of the Z^0 is predicted to be twice what it actually is ("I didn't do my arithmetic carefully"). The notes conclude with an almost audible sigh: "So much for this model—rather offbeat—just intended as a curiosity."

13
Broken Symmetry

GLASHOW WAS BY NO MEANS THE ONLY PHYSICIST TOYING WITH GAUGE theories, nor was he the only theorist with a model of how one might work. From 1957 to 1964, he executed an unknowing intellectual *pas de deux* with the man who later joined him in Stockholm, Abdus Salam. Often physicists happen more or less independently upon the same conclusions, but the temporary harmony between the separate styles of Glashow and Salam is as striking as the sudden, simultaneous appearance of African motifs in the work of Picasso and Matisse. For years Salam ran the International Centre for Theoretical Physics in Trieste, a stark modern cluster of cement-and-glass slabs pocketed in a hillside with a view of the Adriatic Sea. The Centre— Salam follows the British spelling—is funded by two United Nations agencies, the International Atomic Energy Agency and the United Nations Educational, Scientific, and Cultural Organization. Its aim, purely and proudly, is to promote basic research in theoretical physics, to foster thinking unfettered by the demands of practicality, in the belief that in the long run such scientific work is just as valuable to any nation, no matter how poor, as education to fit narrow and immediate needs, no matter how pressing. Salam suggested the idea of creating the Centre in 1960, and fought relentlessly for its creation at United Nations fora. He has given much of his life to the cause of Third World education, and, as shall be seen, some of his science. The Centre won new life after he received a Nobel Prize for the $SU(2) \times U(1)$ model of electromagnetic and weak interactions.

He has a position there and another at Imperial College, in London; he is on the board of directors for a score of Third World and other committees; he advises and consults and in between finds time to do physics. When we first met him, he had just flown to Trieste from the United States and was soon to go to Geneva for a meeting of the CERN Science Policy Committee. A consul from Yugoslavia was expected to arrive with a plaque announcing Salam's induction into that country's Academy of Sciences. Despite the frantic schedule, Salam himself seemed as calm as the surface of a summer lake. Beneath his chair was a white shag rug, a small space heater, and a favorite pair of worn slippers. His face was thoughtful behind twin barriers of beard and glasses. There was a Dickensian sense of coziness in the room. The feeling of comfort was heightened, perhaps, by the racket outside of the gelid *bora*, the notorious Triestino winter wind which gusts so fiercely that ropes are strung along the sidewalks to prevent pedestrians from being blown into traffic. Salam had on a thick brown suit of the sort that looks incomplete without a watch fob. A blackboard filled one wall with the calligraphy of grand unification. On another wall at eye level: a framed quotation from the Koran.

Salam was born and raised in what was then the Punjab province of British India. A gifted child, he created a sensation in his hometown of Jhang by having the highest marks ever recorded in the entrance examination for the Punjab University. Through a series of lucky coincidences, he was enabled after graduation in 1946 to leave the riot-torn subcontinent for England a year before partition would have irrevocably wrecked such an arrangement.[1] He finished an undergraduate degree in mathematics at Cambridge but found himself at the end with another year's scholarship and a hankering to try some physics. He went to his supervisor, the astronomer Fred Hoyle, explaining that he wanted to be a theoretical physicist and had one more year of scholarship money. Hoyle informed him loftily that proper physicists had to take actual experimental courses; otherwise, he told Salam, they could not look their colleagues in the eye. "I said, 'Look, first of all, this physics course which you are asking me to take lasts two years. I have funds for one year only. Secondly, I have not done physics for about six years—proper experimental physics. How the *hell* do you expect me to cope with the lab work?' And he said, 'Never mind, it's a challenge to you.' " Salam went to see his college adviser, a geologist, who rubbed his hands in glee at the prospect of learning whether someone who had won a first-class undergraduate mathematics degree in two years could make a first in the undergraduate physics course in one. "You see," he said to Salam, "[N. F.] Mott and G. P. Thomson [both later Nobel winners] tried, but they failed."

"So he just wrote my name down for physics," Salam said. "I went to the lab in physics and *by God* it was *hard!*" His voice is soft, rather quiet, the vowels held in the throat as if he enjoyed them; he has the English habit of whispering for emphasis. We had come upon him in another place—Geneva, as it happened—six months later and asked him what it was like as an undergraduate in the Cavendish. Salam had just given a talk on Kaluza-Klein theories, the type of unification ideas that then interested him most, and folded, ten-dimensional spaces made a chiaroscuro of equations on the blackboard behind his head. By coincidence, he was wearing the same suit; it had a watch fob, or something like one. "In the Cavendish, there was the old equipment, ancient equipment, and nothing but. Rutherford's own equipment—and you were supposed to make it work. You had to blow glass tubes yourself and carry them three flights of steps. It was a torture. They wanted it to be torture, and they succeeded. I remember the first experiment which I was given. It took me four full days to complete. Basically rather simple, the experiment—you had to measure the difference of two sodium D spectral lines, the wavelength difference, by an interferometer method. I took three days to set up the equipment, set it up properly, and then I took three readings. Three readings on the principle that I wanted to get a straight line—two points to determine a straight line, and the third to prove it. I took this piece of work to Sir Denys Wilkinson, who is now vice chancellor of Sussex University and one of the brightest experimental physicists in the U.K. He was one of the supervisors who awarded you marks on your write-up. He looked at this with a quizzical look on his face. He said, 'What's your background?' I said I came from mathematics. He said, 'Oh, I can see that. You realize that you have to take *one thousand* readings before you have a straight line. This is just not worth grading. Go back!' " Salam laughed. He had not gone back. He never showed Wilkinson his lab work again. In another experiment, he said, "I had these Poiseuille's tubes, through which you look for the laminar flow of water. But I couldn't get the laminar flow. My tubes, which I had blown myself, were probably all clogged up with God knows what. So I invented a new theory to fit my facts." He almost failed the experimental exam, but did so well on theory that he got a first. "The day the results came out, I met Wilkinson in the Cavendish. Wilkinson called me and asked me what class I had gotten. I said, 'I got a first'—rather modestly. He was facing me. He turned full round in a circle. You know, like this—" he stood up, miming his teacher's little dance of astonishment, the vents of his wide suit a-flap "—and then he said, 'Shows you how wrong you can be about people!' "[2]

At Cambridge it was the tradition that first-class graduate students became experimenters, whereas lesser lights went into theory. Because he had done so well, Salam's reward was a promotion to the experimental halls he detested. He was put to work under Samuel Devons, firing hydrogen isotopes at each other and measuring the resultant collisions. He loathed it. "It was hard for me," he told us in his office in Trieste. "I lacked the sublime quality of patience with experimental equipment, which—certainly in the Cavendish—never functioned."

Salam asked Devons if he could leave to join a theoretical department. Devons agreed, provided he could find a supervisor. Salam approached Nicholas Kemmer. After Yukawa had invented the theory that a virtual particle—then called the mesotron, now called the pion—was responsible for the strong and weak interactions, Kemmer had described all possible forms of interactions that could be mediated by mesotrons. After the discovery of the pion and the renormalization of quantum electrodynamics, the natural question was if Yukawa's meson theory could also be renormalized. "The question was, was the pion a renormalizable theory?" Salam said. "At our first interview, Kemmer said to me, 'Well, all the problems have either been solved by

Feynman, Schwinger, Tomonaga, and Dyson, and [P. T.] Matthews is going to solve all the problems concerning mesons. So go to Matthews and ask if he has any problems left.' I went to Matthews, and I said, 'What have you been up to?' He said he had spent two and a half years trying to renormalize meson theories. He had found that only spin-zero mesons would work." This was encouraging; the pion has a spin of zero. "He had done the calculations to one-loop order and shown that the theory of spin zero was renormalizable up to the second order. Matthews said to me that I should read Dyson and look at the general problem of renormalization of meson theories. Following Dyson, in a couple of days I produced a general scheme of renormalizing all meson theories of spin zero."

That quickly?

"Very quickly," Salam said, chuckling. "I went to Matthews and I said, 'Look, is this the problem?' He laughed and he said, 'This is *not* the problem. You've done dimensional analysis, which only shows that various factors fit and everything would be fine—*if* one can show that the infinities *really* can be removed, each to its proper place. That's the problem. That problem devolves to a rather obscure point in Dyson's papers about overlapping infinities. It's so obscure: we really have to work on that.' " In complicated interactions, the reaction can go more than one way along the tangle of loops and vertexes in the graphs. The question was whether each subinfinity had to be removed individually, as if the others were not present. Dyson claimed to have solved this for quantum electrodynamics, but had not given his proof. In the meson theory, Matthews had encountered even more virulent snarls of infinities, many of which overlapped in a vicious manner. Would Dyson's claim hold for these also? "Matthews said, 'I'm taking a vacation. You can have this problem until I come back to work in October. If you don't solve it till then, I'll take it back.' That was the sort of gentlemen's agreement we had.

"This was probably April or May [1950]. I rang up Dyson at Birmingham, where he then was, and said, 'I would like to talk to you. I am busy renormalizing meson theories and there is this problem of overlapping divergences which you claim to have solved.' He said, 'I'm leaving tomorrow for the U.S.A. If you want to talk, come tonight to Birmingham.' "[3] Salam took the train to Birmingham and early the next morning asked Dyson to show him the solution to the overlapping infinities.

"I have no solution," Dyson said. "I only made a conjecture."

"It was as if the earth had opened and I had sunk into it," Salam said on another occasion. "Here was my hero, Dyson, blandly confessing to me that he had no proof and that he had just made a conjecture and gone on!" Salam laughed heartily. "It turned out to be right, of course." Using some hints from Dyson, Salam soon showed that the overlapping infinities indeed could be accounted for; he then wrote a long, yet in some ways oddly sketchy demonstration that the theory of the strong interaction with spin-zero particles could be renormalized to all orders. (Indeed, the complete method was not published until 1983.)[4] "We believed at that time in Yukawa's theory, which said that pions were the final secret of nuclear forces, and we went on believing it till 1951 or 1952. The expectation was that meson theory, once renormalized, would completely solve the strong force problem. And that's why I became famous, because I had proved that the theory was renormalizable. In 1950, when my proof was completed, we believed this was the end. We expected it to be *the* theory. The end! And we lived in that fool's paradise, euphoric for a year or so, until the excited state of the nucleon, the so-called delta, was discovered. Then the bottom fell out of the business. At the same time, the strange particles started to be discovered. So the situation became even more chaotic."

Although Salam was not permitted by Cavendish regulations to submit the renormalization as a thesis for another few years, he was immediately offered a position at the Institute for Advanced Study. Upon going, he discovered his reputation had preceded him, and was horrified to learn that he was regarded as an expert theorist. "All I had was one year of research," he said to us. "The result was disastrous for me. I knew *nothing* of physics except what I had done myself." He was afraid to reveal his ignorance. As a result, he said, "I could learn nothing. In learning, you have to ask questions which are sometimes exceedingly foolish if you don't know the subject. That's why people stop producing when they become very famous, because they cannot bear to write papers that may prove to be trivial. Once that happens, that's the end. You are no longer able to make stupid mistakes. Personally, since then I have never subscribed to this view that I should stop writing articles that might be trivial."

Before finishing his doctorate, Salam returned to Pakistan, where he began teaching at the University of Lahore. To his dismay, he learned that he was the only practicing theoretical particle physicist in the entire nation. No one cared whether he did any research. Worse, he was expected to look after the college soccer team as his major duty besides teaching undergraduates. Just before Christmas of 1952, he received a cable from Wolfgang Pauli, who was on a tour of India. Overjoyed, Salam flew from Karachi to Bombay. When he knocked on Pauli's door at the Tata Institute after traveling all night, Pauli let him in and without a word of greeting snapped, "Schwinger's wrong! I can prove it!"[5] Upon returning from Bombay, Salam was "chargesheeted" for leaving Pakistan without permission.

Kemmer kept in touch with Salam. When Kemmer decided to leave Cambridge for Edinburgh, he managed to get Salam appointed in his stead. Salam went with his wife and daughter to England early in 1954. One of his first graduate students was a moody young man named Ronald Shaw, who, while working for Salam, independently invented Yang-Mills theories for his Ph.D. thesis.[6] In Trieste, we asked Salam what had been his reaction to the difficulty of assigning mass to the vector bosons. "Well," he said, "I think the . . . " He stopped and pushed around the espresso cup on his desk for a few seconds. "Mass is a very important thing, something whose origin we must explain," he said at last. "But it's also the most difficult problem in physics. Just look at the range of masses. You start with the electron and go to the Planck mass, which is twenty-two orders of magnitude larger. My reaction has always been, don't worry about the mass problem—take it the last. So although Pauli shouted Yang down for ignoring mass, I personally have never let myself be put off by it, especially when the primary interaction and its basic symmetry is yet to be determined." Sipping his coffee appreciatively, he added, "To me at that time, the establishing of the symmetries was the primary motivation."

Like many other particle physicists, Salam was drawn into the field of weak interactions by the tau-theta puzzle. Unlike most others, he was enthusiastic about Lee and Yang's proposal that parity violation might be at work. Appalled at this violation of symmetry, Salam could not understand how nature could prefer left to right. Then he was the first to realize that parity violation could be tied to the existence of a neutrino that spins in just one direction—what is called, in the jargon, a "two-component" neutrino.[7]

When Salam proposed this idea in 1956, before experiments confirmed the left-handedness of the neutrino, Pauli was scornful. "Give my regards to Salam and tell him to think of something better," was his message. And when Salam went on to claim the existence of the Universal Fermi Interaction, he was treated to a second blast from Pauli:

For quite a while I have [made] for myself the rule: "If a theoretician says 'universal,' it just means 'pure nonsense.' This holds particularly in connection with [the] Fermi interaction, but otherwise, too." And now you, my dear Brutus, come with this word! . . . So I have not seen your paper, but I have some small hope . . . that you have already withdrawn it.[8]

We asked Salam why Pauli was so dead set against universality. "Pauli was the oracle for my generation," he said. "So it was fun to see how wrong he was on both counts. But regarding universality, I think he was so angered by Einstein's claims of the unification of electromagnetism and gravity that he disliked anything that resembled it. We, the young people, liked universality, but our seniors did not. They—I'm speaking of the generation of Pauli and Dirac—had also wanted in their time to do everything with one stroke. You see, the greater physicists are, the greater their desire to finish off the subject in their own lifetimes. For them, the proton was going to be the antiparticle of the electron, and this was to be the end. Everything wrapped up in one tidy package. When those ideas didn't work, when new physics came, they became discouraged with the whole idea of grandiose unifying schemes.

"What really shocked them was the muon. There was no reason for it to be there. I remember telling my seniors it was just a heavy electron and getting indignant denials. 'How dare you! It weighs two hundred times more than the electron!' The mass syndrome again! In that sense, the muon was a giant step backward for particle physics. There is the famous remark of Rabi about the muon: 'Who ordered *that*?' Instead of being seen as a replicated electron, it made them think that everything was lost and there was no chance of ever coming back to a unifying picture. It took decades for us to recover our ambition."[9]

Curiously, Pauli's wish was granted. After writing his paper on the Universal Fermi Interaction in the beginning of 1957, Salam went to the seventh Rochester Conference. Like Marshak, he, too, was alarmed by Wu's talk of V and T. Salam's version of the Universal Fermi Interaction was premised on the supposition that the weak interactions dealt only with left-handed particles, as is the case for $V - A$. To his later chagrin, he stopped at the offices of the *Physical Review* on his way back from Rochester, and pulled his article from the editor's hands.[10]

Schwinger's hypothesis later in the year that the weak and electromagnetic interactions might be mediated by a triplet consisting of two vector bosons and the photon delighted Salam, who has a fondness for the imaginative leap. He had already imagined the possibility of a virtual particle for the weak interaction. Why not add in electromagnetism for an elegant and unified theory? With John Ward, a theorist who had renormalized the meson theory at about the same time, Salam proposed a new unified model that directly linked Yang and Mills, universality, and unification.[11] The paper describes the same model as in Glashow's first paper, and is contemporaneous in publication.

"Ward and I were trying to make Schwinger's idea into a gauge theory. We estimated that the mass of the W would be at least thirty times the proton mass—huge, it seemed to us. We were very ashamed of calling this heavy object an elementary particle. And we tried to diminish it as much as possible. We tried our best to keep the mass down." Salam laughed again, swinging the chain on his glasses. "We found reasons to insert the square root of two in the appropriate places—a standard theoretical ploy! We wrote a second paper in '64." He rummaged in a drawer. "This is the manuscript of that one," he said, displaying a tired mass of yellow legal paper. He flipped through it. The pages were covered with scratched-out phrases, canceled equations, inserts, additions, and erasures. Hidden in the mire of second thoughts was the complete SU(2) $\times$ U(1) model; like Glashow, they had found the Z^0. Also like Glashow, they had not resolved the mass question or the difficulties with neutral currents. Salam dropped the scribbled pages on the table with a gesture of disgust. "I did not know until '72 that Glashow had done this already."[12]

After 1964, Salam, too, set aside unification, partly because he was distressed by a general change in the theoretical climate. Frustrated by their decade-long inability to formulate a theory of the strong interactions and disheartened by the phenomenological character of the $V - A$ theory, a prominent group of scientists began to argue that quantum field theory had failed, and that new ideas would have to take its place. But there was nothing ready to supplant it, and most physicists kept right on using field theory to do calculations even as they decried its shortcomings. "Many of our colleagues tried to throw quantum field theory out, keep only relativistic quantum mechanics, and confront its predictions with experiment," Salam said. "I mean, all those people who believed in field theory were badgered. I didn't doubt the thing. But there were others, like Geoffrey Chew, who believed that quantum field theory was dead." Salam was reminded of the only time he had ever met the Soviet physicist Lev Landau, at the 1959 Rochester Conference, which was held in Kiev as a gesture toward Sputnik-era détente. "He was flamboyant, wearing a shirt of extraordinarily vivid colors. He saw me from a distance in the crowd and said, 'Oh, you're Salam! The man who scooped me in inventing the two-component theory of the neutrino!' This was flattering, so I said yes. He said, 'Come here, come here!' I went to him. He said, 'Aren't you *ashamed* of yourself?' This was a shock. I said what for. He said, 'Aren't you a believer in field theory?' I said yes. He said, 'I have just shown that the Hamiltonian in field theory is zero. Aren't you ashamed of yourself?' This was the so-called Landau ghost which he had discovered in quantum electrodynamics, portending that the theory must possess a ghost. Basically, a ghost is a particle that violates the laws of cause and effect. Awful things, those ghosts. Some years later, however, Landau was shown to be wrong. There are no Landau ghosts in local gauge theories." Shaking his head, Salam said, "But in 1959 there were so many people who thought that field theory was rubbish, and only *fools*—" he pointed to himself "—talked about something like *gauge* field theory."[13]

Part of the reason Salam was unworried by the mass problem in his $SU(2) \times U(1)$ model was that he had at least a rough notion of how to solve it. In a lengthy footnote to the $V - A$ paper he yanked from the *Physical Review* is a sketch of an idea now called spontaneous symmetry breaking, which shows how asymmetrical systems can end up being described by symmetrical equations.[14] One of the first and most well-known examples of broken symmetry is the theory of ferromagnetism, which was worked out by

Werner Heisenberg in 1928. Magnets have north and south poles, which means that they have a preferred orientation in space; they are not symmetric, as physicists use the term. If a bar-shaped iron magnet is heated in a forge, it loses its magnetic properties and becomes symmetric—there is no way of distinguishing one end from the other. When the metal cools down, however, it spontaneously recovers its magnetization and again acquires a north and south pole. The symmetry, physicists say, is "broken" or "hidden"; it is manifest when the bar is red-hot, concealed when the magnet is cold.

In an intuitive way, Salam thought the real, asymmetrical world might be described by a theory that was itself perfectly symmetrical. He hoped that he might be able to construct a gauge invariant theory, that is, a theory with the same kind of symmetry as quantum electrodynamics, which described the asymmetric real world in which vector bosons had to have masses if they existed at all. Salam's thoughts were worked out for electrons, not for W or Z bosons, but he had the hunch he was on the track. Although he can be very careful in his formalism, he cares most about broad ideas and general concepts, about the gist of things; after he has skimmed the cream, someone else can look after the mathematical niceties. "I am an intuitive physicist," he says. "Give me the basic thrust and let me get on to the next problem."[15]

In the summer of 1961, Salam attended a conference in Madison, Wisconsin, with Steven Weinberg. The two men had met eighteen months before, when Weinberg, then a junior member of the faculty at Berkeley, told Salam about his proof of the last loopholes in the work on overlapping infinities in quantum electrodynamics. They had hit it off immediately. In Wisconsin, they were both excited and appalled by a series of conversations about broken symmetry with a young Cantabrigian theorist, Jeffrey Goldstone.

Goldstone, too, had studied field theory, but had become interested in superconductivity, the phenomenon in which certain substances lose all resistance to electricity when they are cooled to a few degrees above absolute zero.[16] Discovered at the beginning of the century, superconductivity resisted explanation until 1957, when John Bardeen, Leon N. Cooper, and John R. Schrieffer of the University of Illinois developed what is called the BCS theory of superconductivity.[17] (Cooper, like Glashow and Weinberg, is a graduate of the Bronx High School of Science; Bardeen, Cooper, and Schrieffer won the Nobel Prize in 1972.) When first formulated, the BCS theory was attacked in many quarters because it did not seem to be gauge invariant; by 1960, however, at least some physicists realized that superconductivity was in reality an example of the spontaneous breaking of the gauge symmetry of electromagnetism. When the temperature gets cold enough, the symmetry breaks, and strange things happen.[18] The few particle physicists who understood superconductivity wondered if perhaps the trick of breaking the symmetry could be carried over into some other area of inquiry. Jeffrey Goldstone was one of these physicists.

The most obvious asymmetry in particle physics is the difference in mass of the various elementary particles. Goldstone himself was intrigued by the electron and muon, which are identical except for their masses. From his work in superconductivity, Goldstone, like Salam, hoped that some kind of symmetry breaking could account for the difference. Unlike Salam, Goldstone actually tried to construct such a model during the summer of 1960. He sketched out a simple unrealistic model of particles somewhat like pions interacting with particles like protons (the hot magnet, so to speak), broke the symmetry, and tried to figure out what would happen (the cold magnet). To drive the symmetry breaking, Goldstone hypothesized the existence of a special field permeating all of space that alters the conditions under which the symmetries hold. In Goldstone's conception, the vacuum of outer space is, in short, not really empty; it is everywhere saturated by the symmetry-breaking field. Assuming the existence of this field broke the symmetry, all right, but *en passant*, it created a bizarre problem. The equations with symmetry and the equations without symmetry were equivalent only if he threw in another, massless particle into the stew. In other words,

$$\text{symmetry} = \text{broken symmetry} + \text{extra particle}$$

This was not good. The new particle would presumably be strongly interacting. "Massless particles of that type, like the photon, lead to long-range interactions, like electromagnetism," Goldstone said. "The strong force has a range of 10^{-14} centimeters or so. So you were in trouble."[19]

Goldstone spent that summer in the theory wing of CERN, where his office was near that of Sheldon Glashow. The two men talked over symmetry breaking, but could not see how it fit into anything, including and especially Glashow's own work on $SU(2) \times U(1)$. Goldstone told Glashow that he was completely confused about what this extra particle meant. In Goldstone's recollection, Glashow raised his eyes heavenward and said, "Oh, come on! Publish it anyway!" Somewhat reluctantly, Goldstone did.[20] "It's a ragbag of remarks and techniques," Goldstone told us, "including, thrown in at the end—on Glashow's encouragement because it was unsupported—the statement that you *always* got zero mass particles [if you broke a gauge symmetry]. I would have sworn it was true, but I certainly didn't have any kind of proof at that point."[21]

Despite the lack of proof, this "ragbag of remarks" snapped Salam and Weinberg to attention when Goldstone explained it the next year in Wisconsin. Salam had just shown that the one in Glashow's first theory, in which the masses of the Ws were put in "by hand," were unrenormalizable. Spontaneously broken theories just possibly might run around this problem. They might start with massless particles, and end up with massive ones. Weinberg, too, was intrigued. He had recently come across another paper, by University of Chicago theorist Yoichiro Nambu, that talked about the same analogy with superconductivity, but had not really understood it. "I think that week with Jeffrey Goldstone in Madison was really the first time I began to think about these things seriously," Weinberg told us. "I fell in love with symmetry breaking. It was clear that the old style of looking for symmetries was pretty well played out. It just seemed that it was going nowhere." The way Goldstone thought about symmetry struck a chord in Weinberg's heart. "That I thought was just beautiful, and it gave me the feeling that it wasn't just a bright idea of Nambu about an analogy with superconductivity, but that there really was a whole body of theory that one could master and apply in various ways."[22]

On another occasion, Weinberg returned to the subject of spontaneous symmetry breaking. "The point is," he said, "there clearly weren't a lot of symmetries left to be discovered, like isotopic spin, which were clearly manifest in nature. The wonderful thing about broken symmetry is that there still may be a lot of deeper symmetries left that are still hidden. It wasn't the *breaking* that excited me, but the potentiality that there are still symmetries left for us to discover. It's as if you were mining for some kind of metal—gold, say—and you'd done all the surface mining and you can't find any more, and then you find out that gold is also found underground in certain kinds of rock formations. You get excited not because it's deep underground, but because there's still some gold you haven't found yet."[23]

Thinking this, it is no wonder that Weinberg and Salam were dismayed by Goldstone's conclusion that symmetry breaking must always produce particles of a type that do not exist. Salam couldn't use it to save SU(2) × U(1); Weinberg couldn't find more symmetries. "Both Weinberg and I disbelieved Goldstone," Salam said. "We christened his particle 'the snake in the grass'—always ready to strike. After the talk, we argued a little bit with him—a good physics discussion. Weinberg and I came away feeling it *had* to be wrong. There *couldn't* be this particle." That fall, Weinberg went to Imperial College, where the two continued to look for the flaw in what is now called the Goldstone theorem. Not only were they unable to find any, they demonstrated the theorem was true in no less than three separate ways. "We started working and ended up proving Goldstone was right," Salam said. "More elegantly, or so we claimed. Then, after the paper was written, we began to wonder if maybe some of the ideas in it had somehow not really come from our talk with Goldstone. He was at MIT at the time, and we cabled to ask him if he wanted to be in on the collaboration."[24]

Goldstone telegrammatically agreed. "Basically, I didn't write a single word of that paper. They wrote it. But some of the results in fact were things we had discussed at length in Wisconsin that summer."[25]

The Goldstone-Salam-Weinberg paper left no doubts about the unwelcome massless particle; it was there, and everyone concerned was exceedingly glum about it.[26] The three men had invented the mathematics for studying the possibility that the vacuum might be saturated with special fields—the prospect that nothingness might have, as physicists say, structure—but were unable to do anything but produce unwanted particles with their idea. Eighteen years later, when Weinberg won his Nobel Prize, he recalled the dismay he felt then: "I remember being so discouraged by these zero masses that when we wrote our joint paper on the subject, I added an epigraph to the paper to underscore the futility [of trying to make the vacuum beget mass]: it was Lear's retort to Cordelia, 'Nothing will come of nothing: speak again.' Of course, the *Physical Review* protected the purity of the physics literature, and removed the quote."[27]

We once asked Glashow why nobody involved had thought of what seems obvious in retrospect, making a connection between the massless boson in SU(2) × U(1) and the massless bosons in spontaneous symmetry breaking. You have to recall the time, he said. "Goldstone showed that if a symmetry broke down by itself, there would necessarily exist certain massless particles," he said. "Now, there are none of these massless particles. They don't exist. This was just a little exercise in abstract quantum field theory. What Yang and Mills showed was that if you had this kind of crazy gauge theory, then the particles that would carry the force would be massless. So here we have two types of theories, quite independent—one [being] gauge theories, the other spontaneous symmetry breaking—both of which produce massless particles. Two different types of massless particles, neither of which exists in nature. Two sets of guys doing crazy stuff. How could you know any of this would be relevant to any other part?"[28]

In the early 1960s, evading the Goldstone theorem and, as it became known, the Goldstone boson, became a sort of cottage industry in the coterie of physicists still interested in field theory. In the United States, Great Britain, and Belgium, half a dozen theorists tried their hand at it, proceeding less from the conviction that the theorem was empirically wrong than the belief that spontaneous symmetry breaking was too pretty a concept to be fouled with such ugly consequences. The first to find a loophole was Julian Schwinger, who pointed out that broken local symmetries and broken global symmetries might be different animals in a paper that few people read and even fewer understood.[29] One of those who tried to comprehend was Philip

Anderson, a condensed matter physicist at Bell Laboratories, then in Murray Hill, New Jersey. (Condensed matter physics is the somewhat grandiose name for the study of the behavior of atoms in solids.) Anderson had taken courses from Schwinger at Harvard and tended to follow his papers. He picked up on Schwinger's main thrust and argued from it that Goldstone, Salam, and Weinberg were obviously wrong.[30] Superconductivity is an example of broken symmetry in which no extra massless particles appear, and hence is an exception to the Goldstone theorem. Anderson suggested that if electromagnetism can escape Goldstone, then other locally symmetric gauge fields might do the same thing.

It is likely, then, considering the superconducting analog, that the way is now open [for a theory] without any difficulties involving either zero-mass Yang-Mills gauge bosons or zero-mass Goldstone bosons. These two types of bosons seem capable of "canceling each other out" and leaving finite mass bosons only.

What happened, he said, is that in a gauge-invariant theory the Yang-Mills bosons "become tangled up" with the Goldstone bosons and end up with a mass. The massive particles could exist because they might be heavy enough not to be seen.

We conclude, then, that the Goldstone zero-mass difficulty is not a serious one, because we could probably cancel it off against an equal Yang-Mills zero-mass problem.[31]

In retrospect, Anderson's suggestion that these two theoretical embarrassments would take care of each other was incredibly prescient. His idea is precisely what is now believed, which makes it all the more remarkable that almost nobody looked at his paper although it was printed in the *Physical Review*, then the most important journal in the world, and was written in anything but an obscure style. Anderson's article went unread because he was a condensed matter physicist and therefore in some sense not a member of the club.

"Everyone in condensed matter physics knew about Goldstone bosons," Anderson said to us. "If you break symmetries you get what we call 'collective excitations,' which are the equivalent. All Goldstone did was put it into a theorem. That sort of thing—" he chuckled scornfully "—impresses particle physicists."[32]

Among the few to read Anderson were Walter Gilbert of Harvard, and Abraham Klein and Benjamin Lee of the University of Pennsylvania. As often happens in theoretical physics, reasoning from the same base of data led to opposite conclusions. Klein and Lee supported Anderson; Gilbert said that superconductivity and relativistic quantum field theory were totally separate domains.[33] The dispute helped trigger a spate of more or less independent activity by two Belgians, Roger Brout and François Englert; and a

Scot, Peter Higgs. (In addition, Goldstone himself had done some work on the topic a bit earlier.) Each discovered that, in a sense, all the conflicting claims about symmetry breaking were correct. All relativistic theories with broken symmetry had massless particles except—the *sole* exception!—Yang-Mills theories, that is, theories based on local gauge symmetry. Exactly as Anderson had suggested, the massless particles that were a problem for theories such as Glashow's SU(2) × U(1) do become "tangled up" with the massless particles created by spontaneous symmetry breaking. In a sort of mathematical fratricide, the two sets of particles eat each other up and the vector bosons that transmit forces magically acquire mass. Left over at the end is another, as yet unseen particle called the Higgs boson. The entire process is a theoretical Rube Goldberg device that only works for a very small class of theories. Some physicists, such as Glashow, think the hurlyburly of spontaneous symmetry breaking is so awkward and ugly that it is a gravy stain on the tie of physics. Others, such as Weinberg, regard symmetry breaking as a wonderful limiting device, for it means that nature and humanity are forced to work with local gauge invariance. It is a necessary condition; the symmetries again constrain us.

In a perfect world, a world in which science moved logically and scientists were not swayed by fashion or blinded by intellectual prejudice, the discovery that spontaneous symmetry breaking can endow vector bosons with mass would have produced a flurry of theoretical activity.[34] As it was, Brout and Englert were too identified with condensed matter physics for anyone to pay attention; Goldstone was told by Gilbert that this ludicrous scheme could not possibly be correct and therefore, to his regret, did not write up his ideas; and Higgs had trouble even getting people to publish his papers. His first piece showed how to evade the Goldstone theorem; the second, which gave a very simple example, was bounced by the European journal *Physics Letters*, on the grounds that it was irrelevant to particle physics.[35]

Brout phoned Salam and tried to explain the exception to the Goldstone theorem, and discussed it in his solid-state textbook, *Phase Transitions*. He knew that a vector boson with mass spoils the renormalizability of weak interaction theories, and with Englert began to sniff around the question of whether symmetry breaking would be the answer. They were sure it was, but could not prove it, although they published the suggestion in 1966. As Brout recalled:

Perhaps the most thrilling moment of my life as a physicist was when, in conversation, we conjectured that the photon and $W^{\pm}$ were gauge bosons of a broken isotopic spin group, thereby unifying weak and electromagnetic interactions into a common renormalizable theory. We were conversant enough with the weak interactions to realize that one isotopic spin group alone would not do. So *we worked—and very hard indeed—to get a good group. And we failed to find one [i.e., SU(2) × U(1)], and so we dropped the matter.*

After working on symmetry breaking, Weinberg joined a number of other physicists in an attempt to develop chiral symmetry, which treated the isotopic spin of left- and right-handed particles independently. (*Chiral* comes from the Greek word for "hand.") The symmetry is broken by the weak force, which only interacts with left-handed particles; the massless particle associated with the broken symmetry is the pion. Or, rather, because chiral symmetry is only approximate, the pion is approximately a Goldstone boson, and has a small mass. "The fact that there was a way of avoiding Goldstone bosons simply no longer seemed very exciting," Weinberg said. "We didn't need to avoid them. There they were—it was the pion. I didn't pay a great deal of attention to the work of Higgs at all for that reason. I think most theoretical physicists didn't pay much attention to it for that reason. The Higgs mechanism seemed like a peculiarity that didn't have any relevance to the real world. In the real world, there were Goldstone bosons, and pions the example. The general idea of using symmetry breaking, though, captured the attention of a large number of physicists in the mid-1960s."[36]

Weinberg then tried to develop a theory of the strong interaction based on the combined effects of unlimited numbers of low-energy pions. "It wasn't successful. I wrote some papers about [large numbers of low-energy] pions which are *totally* ignored, and justifiably so. They have *no* importance. I remember I gave a talk about it at a conference in Amherst, and nobody paid the slightest attention. Very often my colleagues show good taste like that."[37]

Doggedly, Weinberg tried another tack, constructing a model in which the intermediate particles were a triplet of mesons. "I tried to make a Yang-Mills theory for pi, rho, and A1 mesons, with the rho and A1 being like photons but getting their mass through broken symmetry and the pion being the Goldstone boson. It was not a beautiful theory, and I was not very happy with it. And then, at a certain point [in the fall of 1967]—I have some memory of driving to my office at MIT—I suddenly understood that I'd been applying the right idea to the wrong problem. I realized that the whole mathematical apparatus I was constructing for the strong interactions was *just* what I needed to understand something I hadn't really been thinking about, namely, weak interactions and intermediate vector bosons. Like everyone else at the time, I knew that the weak interactions were carried by intermediate vector bosons. I suddenly realized, 'My God, this is the answer to the weak interaction!' Now I was starting out with an exact gauge theory, for the Goldstone boson had been eliminated by this mathematical formalism in which I had re-created the Higgs mechanism. I knew that Yang-Mills theories with the mass added by hand were not renormalizable—everyone knew that. Because it was gauge invariant like electromagnetism and the massive particles got their mass by spontaneous symmetry breaking, I thought the theory just might be renormalizable."[38]

Weinberg's paper came quickly.[39] Just three pages long, the article reads as if written in a single surge of confident thought. With quick, sure strokes, Glashow's $SU(2) \times U(1)$ model is reinvented and twined together with symmetry breaking. The arguments of Yang and Mills are sketched; at long last, the virtual particles have mass. With massive vector bosons, Weinberg now had an answer to the first of the two difficulties that had defeated Glashow. Next came the problem of neutral currents, the second of the two puzzles. Weinberg had no idea of what to do about it. Nobody had ever seen anything that might be construed as being caused by a Z^0. Like Glashow, Weinberg dodged the snag. He restricted the model to leptons—electrons, muons, and neutrinos—by fiat. Having banged his head against the strong force for years, he didn't want to complicate his model with the hadrons that could feel its influence. Moreover, he knew that neutrinos were so hard to work with that no experimenter could possibly have seen a neutral current involving them, and that therefore the purely leptonic neutral currents in his model might still exist. If $SU(2) \times U(1)$ could be somehow renormalized, he would have taken an enormous step forward. The last words of the paper convey the special sense of scientific excitement as clearly as anything in the literature:

Is this model renormalizable? We do not usually expect [Yang-Mills] gauge theories to be renormalizable if the vector boson mass is not zero, but our Z and W bosons get their mass from the spontaneous breaking of symmetry, not from a mass term put in the beginning. Indeed, the [equation] we start from is probably renormalizable. . . . And if this model is renormalizable, then what happens when we extend it to include the . . . hadrons?[40]

What happens is the biggest stride forward in elementary particle physics since 1947. Full of excitement, Weinberg explained his electroweak theory to the Solvay Conference of 1967—and was greeted by a resounding "So what?" Faced for the first time with the model that now forms the backbone of high-energy physics, the assembled theorists could not have been less interested. It did not fit in with the aesthetic of the time; it was restricted to leptons; it did not account for neutral currents: For all these reasons, Weinberg, like an accepted painter whose first works in a new style are ignored by the art world, was left to tinker alone with his sketch pad as long as he continued in this line.

In the early months of 1967, Salam was tutored in the techniques of evading the Goldstone boson by a fellow theorist at Imperial College, Thomas Kibble, who had examined the Higgs mechanism in considerable detail.[41] Salam saw immediately the application to the SU(2) × U(1) model he had developed three years before with Ward. Every fall, Salam gave a series of lectures at Imperial College. The subject was always a rough presentation of whatever he was working on at the moment. In the fall of 1967, he presented his electroweak theory. Like Weinberg, he did not know what to do about neutral currents or renormalization, but he didn't mind. He had the general idea, and thought nobody else cared enough to worry much about the details. Toward the end of the year, a colleague told him about Weinberg's preprint. Salam checked Weinberg's paper. The two models were identical. "It was something I had been working toward little by little for many years," he said. "With the new stimulus of spontaneous symmetry breaking, all these things I had worked on for years—local gauge theory, chiral symmetry, $V - A$, unification—even possible renormalizability—just came together. To me, it was mother's milk."[42]

Nonetheless, he decided against publishing. Instead, he planned to extend the model to baryons and mesons, thus superseding Weinberg, who had only considered leptons. Unfortunately, he couldn't do it. He tried a dozen schemes, but could not avoid predicting neutral currents. His exasperation was increased by having no time to think. The International Centre for Theoretical Physics was moving out of temporary quarters to a permanent location in Trieste, and needed transfusions of money on an almost weekly basis. Salam spent most of his day on fund-raising. In addition, he was organizing a conference to commemorate the establishment of the Centre, and had managed to convince twelve Nobel laureates, including Bethe, Dirac, Heisenberg, and Schwinger, to attend. There was no room for physics.

He presented the SU(2) × U(1) model at a small congress held outside Göteburg, Sweden.[43] Impressively titled the Nobel Symposium despite its relative insignificance, the gathering was postponed twice. It took place only two weeks before the large Trieste conference, and Salam barely had time to attend. He left in the middle of the Symposium for London to pick up a grant for the Centre, then returned to Göteburg to give his talk. As a result, the speech was half-prepared to the point of near-incomprehensibility. SU(2) × U(1) appears in the middle, flickers into visibility, and then is shrouded by a screen of other notions. Nobody paid attention; few understood it; some were critical.[44] Overwhelmed by preparation for the forthcoming Trieste conference, Salam never wrote out a paper; the only record of his thought is the published transcript of the talk. Salam did not go back to unification for four years.

The resistance to tying together the weak and electromagnetic forces was in large measure aesthetic. Not only was quantum field theory as a whole regarded as old hat, as passé and overtaken by time as Lawrence Welk, but

SU(2) × U(1) was in some views a startlingly nasty-looking model. Thomas Kibble, for one, recalls reading Weinberg's article and thinking that "it was very intriguing that one could do something like this, but it was such an extraordinarily *ad hoc* and ugly theory that it was clearly nonsense."[45] His colleagues evidently agreed: In the next two years, the only physicist to mention Weinberg's work in print was Salam.[46]

Consider what Glashow, Salam, and Weinberg were offering to the community of physics. The weak and electromagnetic interactions are said to be tied together. The photon, two W particles, and the Z^0 form a family, even though the photon has no mass or charge and the W particles weigh eighty times as much as a proton and have charge. In addition, the huge masses of the W and Z come from an incomprehensible and suspect mechanism involving unseen Higgs bosons. The weak and electromagnetic interactions are said to be tied together despite the fact that the weak interactions violate parity and the electromagnetic interactions do not. Perhaps the strangest twist of all is that the ordinary isotopic spin upon which Yang and Mills based their theory is for leptons replaced by "weak isospin." Before, the proton and neutron were treated on an equal footing; in weak isospin, it is the electron and neutrino. The weak portion of the electroweak force is supposed to be generated by the symmetry between these two wildly dissimilar particles. Incredible! Weinberg and Salam asked the community to believe that the weak interactions were based on a Yang-Mills symmetry that had never been seen; that the force was carried by three vector bosons that had never been observed; that these bosons formed a family with the photon; and that the whole business could in some way be made renormalizable. "Rarely has so great an accomplishment been so widely ignored," Sidney Coleman of Harvard wrote later. It is easy to see why; in 1971, Weinberg called his own SU(2) × U(1) model "repulsive."[47]

On the other hand, the flagrant awkwardness of the scheme could be seen as a testament to its significance. The world around us is not symmetrical, and the untidyness of SU(2) × U(1) is a reflection of the asymmetry of nature itself. Behind the clumsy jamming together of unlike objects in the model are the simple, powerful rules of creation—a source of satisfaction, elegance, and beauty that the best theorists look for in their work.

Weinberg took us to lunch one day at the faculty club of the University of Texas. The afternoon was crisp, dry, almost cold; wind made students huddle into their sweaters. There were trees, tall buildings, sloping asphalt paths: the usual impedimenta of modern higher education. The faculty club rang with conversation; we had to lean forward a little to hear Weinberg. He talked about what he had thought when he had proposed the spontaneously broken gauge theory of SU(2) × U(1). "I was tremendously excited," he said. "I do remember very clearly that my point of view at the time was, 'This is the way that theories of the weak interactions must be.' But not, 'I know that SU(2) × U(1) is right.' I regarded that theory as illustrative, and I still—I can't quite get over the fact that it turned out to be right. I was sure that was the *way* the theory had to be. It had to be a spontaneously broken gauge theory." It did not have to be that particular model, he said.

"You can't test general ideas experimentally. You can only test concrete realizations of general ideas. If you have a general idea—broken symmetries are important, or we should reformulate physics in terms of dispersion relations, or something like that—you can't test that. All you can do is test concrete, specific theories. So when it comes time for an experimentalist to prove a theory is right, it's always a concrete, specific model that he is proving right. And that's where all the attention is focused on. That's what is regarded as the greatest success. To me, the underlying ideas are much more important, and often they are the real breakthrough. In that particular case, I regarded the real breakthrough as this understanding that these spontaneously broken gauge symmetries would work. And that they would be a natural way of understanding the weak interactions. And the specific SU(2) × U(1)—" he spread his hands in amazement "—it's just fantastically lucky that that was the right theory."[48]

IV

The Strong Force

14
The Eightfold Way

BEFORE THE TWENTIETH CENTURY, PHYSICISTS GENERALLY WORKED BY themselves, presented their ideas before small societies or clubs, and had them promptly printed in journals that mixed scientific papers, minutes from meetings, and birthday greetings to distinguished colleagues. Many physicists were gentlemen amateurs; there wasn't much money in science. In England, the first university physics laboratories appeared only in the middle of the nineteenth century, when the invention of the telegraph made it desirable to have people around who knew something about electricity.[1] These early laboratories were dusty, high-ceilinged rooms whose brick walls were lined with bottles of chemicals. A spiderweb of primitive wiring connected thumping vacuum pumps, smelly wet batteries, and glowing cathode ray tubes. Scientists wore wing collars and frock coats on the job; they spent much of their time trying to get the vacuum pumps to work. If the equipment did function, a physicist could perform an experiment in a few days. James Clerk Maxwell, the eminent Scottish physicist who showed that electricity and magnetism were two aspects of a single phenomenon—electromagnetism—did his thinking by himself, in the time (according to legend) he was able to snatch away from the social obligations of his demanding wife. Einstein, too, fashioned the theory of relativity single-handedly, one of the reasons he has often been called the last of the classical physicists.

All this has changed. The construction of the standard model and the unified theories that rest on it has gone hand-in-hand with the transformation of physics itself. A tapestry woven by many hands, as Glashow once put it, the standard model is the work of hundreds of people, an act of sustained, collective creation which recalls the effort that produced the great Gothic cathedrals.[2] It required the construction of enormous machines—the "atomsmashers" of the Sunday supplements—and huge centers to house them. Whether they work in blue jeans or business suits or white laboratory coats, physicists now work *together*, chewing over scientific problems with their colleagues. Experiments are done by scores or even hundreds of Ph.D.'s in concert, designing, building, and operating million-dollar machines run by the fastest computers on the planet.

Brookhaven National Laboratory, in Upton, Long Island, has a campus of more than 5,000 acres and a staff of 3,200—all for research. When we went there one freezing winter afternoon, the campus stretched for miles

253

in splendid, snowy isolation. A ring that could encircle Monaco had been dug out of the woods, and earth-moving equipment lay in the ice about it. Fine cold rain pelted the snow. Their luncheon concluded, a throng of researchers streamed in their winter clothing through the fields around the cafeteria building. Some were going to offices where every available inch was covered with blackboards and bookcases; these scientists had spent the morning scribbling equations about fermions and bosons, leptons and quarks, and all the other arcanae of the subatomic world. Other workers were about to enter quite a different world—an experimental hall the size of an aircraft hangar but crammed so densely with computers, cables, and concrete shielding that only one small pathway was available for people to go from one end to the other. Earlier that day, they had supervised the operations of the yellow bridge crane overhead, which spanned the entire width of the room and carried giant spools of cable and electromagnets the size of a human being.

Brookhaven is one of three national particle physics laboratories in the United States: the other two are the Enrico Fermi National Accelerator Laboratory (Fermilab), in Batavia, Illinois, and the Stanford Linear Accelerator Facility (SLAC), in Palo Alto, California. Other high-energy physics laboratories include CERN, in Switzerland; DESY, in Germany; KEK, in Japan; and UNK, in the Soviet Union. Each is dominated by one or more particle accelerators, machines that break atoms apart and drive the fragments to high speeds. The stream of particle "bullets" thus created is shunted out of the accelerator through various channels into special components known as detectors. There the protons or electrons slam into other particles or into the nuclei of atoms. Pieces fly all over the place, and the detector tracks their trajectories for clues to what is taking place. (The whole process has been likened to firing a gun at a watch to see what is inside.) In the early 1950s, accelerators supplanted cosmic rays as the chief means of studying particles; they could reach higher energies, and could produce many more particles in the same amount of time. Today, these machines are the biggest pieces of scientific equipment in the world.

We had come to Brookhaven to talk about the era of accelerators with Nicholas Samios, the director of the laboratory and a participant in several of the chief discoveries of the past three decades. The discoveries concerned the nature of the strong force and the particles that feel its influence. They took place at about the same time as the unraveling of the weak force that has just been chronicled, but completely overshadowed the evolution of the electroweak theory. Little wonder: The strong interactions are the most powerful in the universe, holding sway over the subatomic world like an inflexible dictator. The desire to understand the strong interactions mandated the accelerators among which Samios's generation of physicists has spent its active years; the successful outcome is at once a capstone of contemporary physics and a landmark on the road to unification. It was early February, budget time on the Potomac, and Samios met us between conferences with federal officials who wished to know why the taxpayers should pay hundreds of millions of dollars to make new forms of matter that have no apparent use. During allocation season, he frequently takes the shuttle to the Capitol, pleading the cause of science before half a dozen oversight committees. His job had recently been made especially difficult: In addition to lobbying for his own laboratory, he was one of many particle physicists promoting the construction of a gigantic new accelerator called the Superconducting Supercollider, a machine some fifty miles in circumference that was to be built in the Texas prairie until it was canceled by Congress in the fall of 1993. If built, the multibillion-dollar ring would have collided protons together at energies that have not been reached since the first fractions of an instant after the Big Bang.

Samios invited us to sit down at a round table in his office that faced a blackboard covered with scrawled Lagrangians and budget figures. He has longish black hair combed straight back over his head that nonetheless tends to fall over his face when he speaks vehemently, which he often does. Samios loves to talk physics, and enthusiastically rattles off particle masses, meson lifetimes, and quantum numbers the way a baseball fan might talk of batting averages, ERAs, and RBIs. Simply put, he relishes the game; like a baseball fan, a certain nostalgic softness filters into his voice when he talks of the 1960s, when there was money for all, nobody had heard of the deficit, and the scale of science was not so grand. The United States had just introduced a new, aggressive style to the sport and American physicists were snatching discoveries right and left. We asked him about the American style of physics.

"America is a place you *do* things," he said. "It comes from the West, okay? It's important to get on with the job instead of setting up committees to study the best possible way. So the style in American physics in the late forties and early fifties was—after the war many places wanted to get into science again. And depending on the drive of different places, each person wanted their thing, to get on with it. And the thing at that time was [a type of particle accelerator called] a cyclotron. Cyclotrons were at Chicago, at Columbia, Carnegie-Mellon—called Carnegie Tech at the time—at Caltech, and at Harvard. People were just interested in *doing* things—in great contrast to today. Before you do anything, you need five blue-ribbon panels to review things. And so as a result you *study* things to death. In those days, you didn't justify it to the utmost, you just *did* it. For instance, there's a document which is the justification for the AGS." The Alternating Gradient Synchrotron, built in 1960, is still the largest accelerator at Brookhaven. "It's a five-page letter, written by the then Brookhaven director, Leland Haworth, to the Atomic Energy Commission, and *that* was the AGS proposal. It's a five-page letter![3] There was no detailed theoretical justification. No theorist is mentioned. He says, 'We believe that as you raise the pion energy interesting things will happen, and this machine can do that.' And the AGS turned out to be one of the most productive accelerators ever built."

Cyclotrons began in the early 1920s, when a few experimenters wanted to find a way to split atoms without relying on radioactive materials. The first to succeed even partly in artificially kicking electrons to high energies was Rolf Wideröe, a Norwegian-born engineer at a university in Germany. Wideröe wanted to put a lot of electrons into a disc-shaped container permeated by an electric field. The positive end of the field would naturally attract the negatively charged electrons. By moving the field in a circle, one could induce the electrons to rotate faster and faster, until they were traveling nearly the speed of light. Unfortunately, when Wideröe actually built the device, he could not prevent the electrons from flying off to the side.[4]

Ernest Orlando Lawrence, an American of Norwegian descent teaching at Berkeley, accidentally picked up Wideröe's paper. Although unable to read German, he studied the drawings and built his own accelerator, blissfully ignorant of Wideröe's problems. Lawrence thought up his own variant of the design, a sort of circular metal sandwich. The top and bottom layers consisted of two coils of wire: electromagnets. The middle layer resembled a steel and brass pillbox and was made of two electrically charged metal semicircles—or *D*s, as they were called—with a small space in between. If electrons were squirted into the middle separation, they would rush toward the positively charged semicircle. Passing through a port in the bar of the *D*, they would feel the influence of the electromagnets, which snapped the particles around the semicircle and back toward the second *D*. This time, the other *D* was positively charged. Picking up speed, the electrons raced toward it, passed through, were turned around, and so on and so forth, until they received so many small "kicks" that they traveled at huge velocity. Drawn on, disappointed, and attracted again, the particles spiraled out until they reached the edge of the machine, where they flew through an open window and smacked into a target, ripping its constituent nuclei asunder. The thing seemed too complicated to build, but Lawrence and a student went ahead and did it anyway. The electromagnet was four inches across; the whole device fitted easily on a tabletop.[5]

Lawrence was a bull of a man, a driven experimenter whose temperament was volcanic and habits were erratic. He was well aware that John Cockroft and E. T. S. Walton were working on a rival accelerator under the doubtful and somewhat disapproving eye of Ernest Rutherford. Despite their chief's attitude toward spending the necessary money, Cockroft and Walton were the first to succeed in smashing atoms with a machine, prodding the reluctant Rutherford to admit that the whole business "seems to be worth the expense and the effort."[6] It was Lawrence who built the largest machine. He literally danced with glee when the speeding protons broke the one-million-volt mark.[7]

The measure of energy used by particle physicist is not the ordinary volt, but the electron volt. An electron volt is a unit, like a calorie.[8] A cup of chocolate mousse contains eight hundred diet-shattering calories, which means that every mousse molecule has an energy of a few electron volts. The molecules in a candle flame, by contrast, burn with a few hundred electron volts. When the enormous particle accelerator in Fermilab reached a trillion electron volts in 1984, the particles inside, small but furious entities, had the energy of hundreds of billions of candle flames. Lawrence's 1932 machine went to one million electron volts, or 1 MeV, but he wasn't satisfied: He wanted 5 MeV.

The Berkeley laboratory where Lawrence worked had no money, no professional engineers, no tools more complicated than a lathe. Accelerator parts were scavenged from radios, bought secondhand, or made on-site. Nothing worked exactly the way it was supposed to, and nobody really understood what was going on. The electrons or protons to accelerate came from a tank of hydrogen that was subjected to an intense electrostatic field that pulled the protons one way and the electrons the other. Sometimes the flammable hydrogen leaked into the accelerator, where it blew up. If the machine was on when physicists descended into the bowels of the enormous magnet, it created electric currents in their brains—a form of self-induced shock therapy. The physicists saw fireworks and strange colors. After discovering they were giving themselves jolts of the sort used to "treat" the clinically insane, they stopped going down if the power was on. When the particles hit the target, they knocked out a dangerous spray of neutrons. To protect themselves, Lawrence and the Berkeley physicists filled dozens of jerry cans with water and stacked them around the target area. They leaked, dripping hot water on the physicists' heads. There were no gauges. To find out how much electricity was loose inside the accelerator, physicists taped a nail to a stick and gingerly approached the machinery, waiting for a spark to leap across to the nail. If the voltage was high, the electricity sometimes jumped from the nail to the hand of the physicist holding the stick; if the voltage was very high, the stick holder went to the hospital.[9]

The war changed everything. Before Hiroshima and Nagasaki, Lawrence had begged for private money to build his machines; he used his Nobel Prize, which he won in 1939, to convince the Ford Foundation to give him a million dollars to build an accelerator that could go up to 100 MeV.[10] He ended up working on the Manhattan Project instead. After the war, atomic energy and physics were big news. People touted fission as the energy source that would power the nation into the future, producing electricity so cheap that it would not be cost-effective to meter it. Physicists convinced the new Atomic Energy Commission, the Manhattan Project's successor bureaucracy, to pay for some accelerators as well as reactors.[11] In addition, the U.S. military, which had just beaten Japan by exploiting the practical consequences of the previously arcane study of nuclear structure, was willing to fund more research that would lead to newer and better weapons. A passionate anticommunist, Lawrence argued before congressional committees that new weapons like particle-beam death rays and radiation bombs capable of killing the populations of cities while leaving the buildings intact could best be devised by building particle accelerators.[12] Although the efficacy of such testimony is difficult to assess, it is certain that the Office of Naval Research, the Atomic Energy Commission, and other federal agencies spent hundreds of millions on the construction of ever-larger particle accelerators in the 1950s and 1960s, initially unaware that they were going to have to keep them running at a cost of *more* hundreds of millions. Some physicists think that when Congress discovers that accelerators will not produce death rays and superbombs, Congress will stop funding particle physics. It is the task of Samios and his colleagues today to make the case for pure science, basic research untroubled by application to war.

In 1952, the first multibillion-volt accelerator, the Cosmotron, was dedicated at Brookhaven. A year later, Nick Samios got his B.S. from Columbia University, one of the universities associated with Brookhaven. Blissfully unaware of the hoped-for military benefits from experimental physics, he received his Ph.D. from the same school.

"Columbia was really a center of physics at the time," Samios said, "because of the people who were there. T. D. Lee was there, Feinberg, Weinberg, Steinberger, Lederman, Rubbia—and then people came by for a year, or to give a lecture series, people like Gell-Mann and Pais. Lots of people around, lots of excitement." The influence of Lawrence and Fermi was strong, he said. Although the accelerators kept costing more, the cost per MeV went down. Samios took off his sports coat, pushed up the sleeves of his white turtleneck. "Strange particles had been found in cosmic rays, but I came in at a time when the machines were just starting to operate. In fact, my early work was done at the Cosmotron. We worked on lots of games—finding particles, looking at their decays, their spins, the beta decays. There were lots of interesting questions because everything was a puzzlement. Today, in the texts, everything is nice and neat, but at that time the whole thing was wide open. The first great breakthrough was the Gell-Mann–Nishijima scheme for strangeness. But then, after that, the particles kept increasing, both mesons and baryons."

Particles, particles everywhere. Neutrinos, pions, *K* particles, lambdas, sigmas, and xis, all with their own spins and masses and lifetimes and typical decays and isotopic spin and strangeness, different states, different resonances—a wilderness of data that nobody could interpret. Most of these particles are seen infrequently in cosmic rays, because the flux of cosmic rays is not enough to produce many instances of the rare processes studied by particle physicists. In addition, cosmic rays are not under experimenters' control, making them difficult to use systematically. Instead, the accelerators would fire a burst of protons at a small metal strip; energy would change to mass, weak and strong interactions would take place, and in a fraction of an instant a spray of exotic particles would be produced. As physicists increased the force of the collisions, they discovered that at certain energy levels the numbers of pions, K particles, and so forth that appeared in the detector suddenly increased. Just as more patrons pour out of a crowded concert hall when another exit is opened, the jump in particle production showed that a new channel for interactions had opened up. The channel was an additional way that particles could decay; physicists realized that this extra path indicated the existence of a novel type of short-lived particle which quickly changed into the floods of pions and K particles that were observed in detectors. The first particle to be found in this way is now called the delta. It was discovered in 1952, when Fermi and his colleagues slammed together protons with just enough power to create a few deltas.

The energy needed to make the particle can be employed to calculate its mass, through the famous relation between energy and mass discovered by Einstein; as a result, physicists tend to speak of particles having a mass of a certain number of electron volts. The proton, for example, weighs in at 938 MeV, the pion at 135 MeV, and the light electron at 0.5 MeV. Similarly, the quantum numbers of new particles could be inferred by studying the quantum numbers of the decay products. In this fashion, experimenters discovered and named dozens of new entities, which they variously and not always consistently referred to as "states," "resonances," and "new unstable particles." The work was not easy, and errors were frequent. A table of meson resonances was published in the *Review of Modern Physics* in the spring of 1963. Twenty-six were listed, of which nineteen are now no longer believed to exist.[13]

Nobody knew what to do with the particle population explosion, or how to relate the new discoveries to each other. Berkeley published a handbook with information about the sixteen known particles in 1958; two years later, so many more had been found that a second issue, with a handy wallet card, was printed. Samios has the latest edition in his office: the size of a small telephone book, it catalogues some two hundred particles, not counting all of their various avatars, resonances, and states.[14]

"You could go back in the literature [of the '50s and early '60s] and find dozens of classificatory schemes," Samios said. He flipped idly through the pages of the *Review of Particle Properties*; the summed work of hundreds of experimental physicists sped through his fingers in the form of reproduced computer tables. There was an amused lilt to Samios's voice. "They'll explain *lots* of things. They'll give you a whole spectrum of particles. They're all ambitious," he said, shutting the book. "And they're all wrong."

☐ ☐ ☐ ☐ ☐

Taxonomy is the study of the principles of classification, especially the classification of objects based on common characteristics. A proper taxonomy of the subject material is essential for the progress of a science. Chemistry was a jumble of cookbook recipes and unrelated facts until Dmitri Mendeleev put together the periodic table in the nineteenth century, thus ordering the elements and their study at a stroke. Natural history began to change into biology with Carl Linnaeus's lifelong attempt to classify and name every species of plant and animal in the world. Darwin's *Origin of Species* can be regarded an attempt to explain the relationships established in Linnaeus's *Systema naturae*.

The diversity of chemical compounds and biological species was always apparent; the variety of subatomic particles was an unwelcome surprise. Particle classification was at first regarded as an uninteresting chore. Looking back, however, taxonomy was the key to understanding both the fundamental constituents of matter and the strong force. A successful scheme for categorizing the diverse forces of matter is also a prerequisite for unification, which seeks to weave together both forces and particles into a seamless whole. Theorists today stand on the solid base of the taxonomy of matter established in the 1960s.

The earliest classification of elementary particles assigned them to three "weight classes," leptons, mesons, and baryons. Admittedly crude, this scheme did have things to recommend it. For instance, all the leptons were immune to the strong force, so it made sense to lump them together. Moreover, there was believed to be a conservation law called "conservation of baryon number," which meant that the same number of baryons entered and exited any interaction, although the species of baryons might change. (Stueckelberg, once again, had first proposed this idea, and, once again, had been ignored.)[15] Everything that was neither a lepton nor a baryon fell into the class of mesons. This classification scheme was tidy but arbitrary. Like separating dogs into breeds by weight, it revealed nothing about their nature and the genetic relationships among them.

As particles proliferated, the lepton-meson-baryon triad showed its limitations. Physicists turned to quantum numbers, the handful of numbers that describe a particle's dynamics. By the 1950s, they had grown to six, and included a particle's spin, parity, electric charge, baryon number, strange-

ness, and isotopic spin. Some quantum numbers, like electric charge and
baryon number, are conserved in all known types of elementary particle
interactions, whereas others, like isotopic spin and strangeness, are conserved in only some. As Emmy Noether had shown, each conservation law is
connected with a symmetry, and particles can therefore be arranged in
groups according to the various symmetries that resulted from the way they
conserved quantum numbers. For instance, isotopic spin is conserved in all
strong interactions, which are therefore said to possess isotopic spin symmetry. Scientists put particles with the same isotopic spin into small groups
called *multiplets*. An example is the nucleon, which is the multiplet formed
by the proton and neutron. Each particle in a multiplet has roughly the same
mass and is identical in all other features except electric charge, which is
described, one recalls, as the orientation of the third axis of isotopic spin.
This "isospin" scheme had helped Gell-Mann and Nishijima develop the
notion of strangeness.

A second attempt to order the particle zoo involved trying to conceive
of some of the new particles as composites of the old ones. The first to try out
this idea were Enrico Fermi and Frank Yang, who hypothesized that the
recently discovered pions might be made up of nucleon-antinucleon pairs.[16]
Figuring that "the probability that all [the new] particles should be really
elementary becomes less and less as their number increases," they offered
their calculation "more as an illustration of a possible program . . . than in
the hope that what we suggest may actually correspond to reality." Although
at first blush it may seem absurd to suppose that a combination of two heavy
particles (the proton and antiproton, in this case) could produce a third (the
pion) that weighs one-sixth as much as either of its two purported constituents, Fermi and Yang pointed out that the excess mass could turn into the
energy that bound together the proton and antiproton. They were unable to
show how this would actually occur, and their paper is today noteworthy
principally for introducing the notion that some subatomic particles might be
made from others.

In the early 1950s, several physicists expanded on the Fermi-Yang
approach.[17] The most noteworthy of these attempts was performed by an
unusual group of Marxist-inspired Japanese physicists led by Shoichi Sakata
of Nagoya University. While a high school student, Sakata came across *Dialectics of Nature*, a posthumously published work by Karl Marx's collaborator
Friedrich Engels that attempts to extend the methods of dialectical materialism to the realm of natural phenomena.[18] Just as dialectical materialists
perceived society as an endless sequence of interacting social levels, so
Engels viewed nature as an endless sequence of physical strata. The task of
the dialectical scientist is to discover the interconnections of these levels,
recognizing that any particular stratum of matter, such as atoms or particles,

is but one in a teeming panoply of material forms. Sakata kept a quote to this effect from *Dialectics of Nature* on his desk as a "precious stone" to guide his work; until his death in 1970, he regarded Marx, Lenin, and Mao as the founders of a "true science" of society upon which physics should be modeled.[19] An extremely capable theorist, Sakata used dialectical materialism to great effect for many years—which most physicists today would take as an example of how far one can go with a nutty idea. Sakata took the Fermi-Yang nucleon-antinucleon idea and, because strange particles had just been discovered, added a third constituent, the lambda (the post-Bagnères name for the V particles discovered by Rochester and Butler).[20] By combining protons, neutrons, and lambdas with their respective antiparticles, Sakata was able to account in an approximate way for all of the particles then known.

The population explosion of hadrons—particles affected by the strong force—brought both good news and bad news to the Sakata group. In 1959, three younger colleagues—Mineo Ikeda, Shuzo Ogawa, and Yoshio Ohnuki—and, independently, Yoshio Yamaguchi, demonstrated that the mesons, whose number had grown to seven, could be constructed by a slight reworking of Sakata's model.[21] They also predicted the existence of an eighth, which was duly discovered.[22] Simultaneously they tried to build up the baryons, supposing that each was made from two of the basic triplet, together with the antimatter version of the third. They ended up with a mess, a loose gaggle of fifteen possible baryons, with particles like the sigma, xi, and proton, which have similar masses and the same spin and parity, treated as totally different entities. And there the composite approach of Sakata seemed to stall.

In retrospect, the separate searches for symmetries and fundamental constituents were two sides of a coin, but they were not recognized as such at the time. Sakata warned against the "inverted viewpoint of believing the ultimate aim to be a discovery of the symmetry properties" as a type of "theology," whereas his critics ridiculed the Japanese group's sometimes dogmatic insistence that proton, neutron, and lambda begat everything else.[23] In any case, the question was resolved with unexpected speed, largely through the efforts of Murray Gell-Mann, whose role in strong interaction physics is something like that of Linnaeus and Darwin combined.

Murray Gell-Mann has published less in refereed journals than his influence might indicate. He suffers now and then from writer's block brought about, at least in part, by the belief, uncommon among theorists, that publishing a wrong idea leaves an indelible stain upon one's career. A theorist's insight, he has said, "should be gauged by the number right minus the number wrong, or even the number right minus twice the number wrong."[24] Even by this lofty standard, Gell-Mann's score is remarkably high, for he has a gift for recognizing symmetries and patterns in places where they are not apparent to his fellows. Time and time again, he has plucked from the chaotic tables of data about particle properties a connection that illuminates the entire discipline and sends his colleagues scurrying to follow him. When he received what he calls "one of those Swedish prizes," in 1969, the Academy cited just this aspect of his work: The award was given in recognition of Gell-Mann's "contributions and discoveries concerning the classification of elementary particles and their interactions."

Gell-Mann is a busy man these days. In addition to his position at Caltech, he runs the Santa Fe Institute, devoted to the new science of complexity, and assists the MacArthur Foundation, the enormously wealthy endowment that funds many charities, including what have become famous as the "genius" awards. Had the awards existed in the 1950s, it would be hard to imagine a better candidate than Gell-Mann himself, a brash, forceful, sometimes intolerant, and remarkably cultivated man who rapidly rose to prominence in a difficult and competitive field. His father was a teacher of languages, and Gell-Mann has retained the habit of pronouncing foreign names in an impeccable accent. He is an alarmingly knowledgeable amateur birdwatcher and collector of pre-Columbian pottery. The range of his interests is sufficiently large that some of his colleagues have remarked that he is the world's greatest physics dabbler—an image that Gell-Mann has fostered by proclaiming that he only entered physics after his father informed him he would starve as an ornithologist.[25]

When we first spoke with him a dozen years ago, he was on a lightning cross-country tour and could only speak to nonphysicists on the plane between stops. We found him on the aircraft already buckled in and tapping his fingers with impatience; the only trace of the exhausting series of lectures and seminars that he had just delivered in half a dozen cities was a slight darkening under his eyes. His California tan had not faded from the gray spring skies of New York City. "It's a long flight," he said. "I hope you're interesting." We asked him a question. He groaned, looked out the window. "People talk a great deal of nonsense about science," he said. "And you have *obviously* reaped the consequences of that." The cabin doors clicked shut, and we flew from New York to Los Angeles.

In the air, we asked him about his introduction to physics. He went to Yale at fifteen, he said, but hadn't learned much there. The real physics came at MIT, where his adviser was Victor Weisskopf. "There's a joint Harvard-MIT seminar," he said. "I attended that, but I didn't have any idea what a theoretical seminar was. I didn't have any idea what research was, really. I thought it was some sort of class, and that the young physicists who spoke there were trying to make an impression on the older ones. At college, what goes on mainly is pleasing the teacher, not really understanding anything for its own sake." The stewardess was bustling about with a cart of tea and coffee.

"I didn't know what it meant to have a critical understanding of the theory and to know what was going on at the frontier. And so at an early Harvard-MIT theoretical seminar that I attended, I listened to a student who was just receiving his Ph.D. at Harvard talk about a calculation he had done on the ground state of a certain nucleus, boron-10." B^{10} is a boron nucleus with a total of ten neutrons and protons; like all nuclei, it has a certain intrinsic spin which is related to the summed spins of its constituent neutrons and protons. "Everybody knew that the spin of boron-10 was one unit. Of course he had to find that the ground state had spin one. He did, and he told how. While he was talking, I was speculating whether some of the more experienced physicists sitting in the front row would think it was good or bad. At the end of the talk, the people in the front row said nothing. But a grubby little man with a three days' growth of beard sitting next to me, who had crawled out of some lab in the basement at MIT, got up and said in not very cultivated English, 'Say, dey mehsured da spin of boron-10, and it ain't one, it's t'ree.' And suddenly I realized that the whole point of theoretical science was to understand what was *really happening*. It was not to please the teacher or older colleagues, but this grubby man from the basement who was reporting somebody's measurement."[26]

Gell-Mann's Ph.D. was delayed for six months because he was slow about writing his thesis. After a short stay at the Institute for Advanced Study, he spent a few years at the Institute of Nuclear Studies in Chicago, working under the aegis of Enrico Fermi. The weather in the South Side of Chicago is appalling, frigid in winter and searing in summer; Gell-Mann was distressed to discover that the university regarded machinery, not human brains, as the proper target of air-conditioning. He had a few thoughts about the V particles which he had trouble writing up in the heat. His concentration was helped wonderfully by the arrival of a draft notice; his occupational deferment form had not been properly completed. Facing the prospect of KP in Seoul, he began to write furiously. A year earlier, at the age of twenty-two, he had solved the riddle of the V particles by inventing strangeness but only now was writing it up. In 1955 he moved to Caltech, where he decided to stay; there, two years later, he and Richard Feynman, a fellow Caltech physicist, helped to straighten out the $V - A$ form of the weak interaction. Fascinated by Yang-Mills fields as soon as the idea was presented, Gell-Mann was one of those who, like Glashow and Salam, spent the end of the 1950s trying to extend the idea to the weak force. He called the vector boson the X. He still does, although everyone else in the field refers to it as the W.

While working on the forces, he also tried to figure out a taxonomy for the ever-growing multiplicity of subatomic particles. He first tried working with the symmetries, proposing, then quickly withdrawing, a global symmetry for the strong interaction much like the suggestion floated by Schwinger that started Glashow on his work.[27] As did the Japanese, Gell-Mann then switched to the composite approach. He, too, guessed that the proton, neutron, and lambda were somehow the basic entities of the Universe, and that all other observed particles were created simply by mixing and matching those three. "It was a big mess," he recalled. The xi, sigma, and nucleon were not in the same family. "I didn't like it, and I didn't publish it."[28]

In 1959, he spent a year at the Collège de France, where he tried to use Yang-Mills theory to account for the strong interaction and the strongly interacting particles. As it turned out, he was hampered by his ignorance of mathematics, especially group theory. Although he knew that electromagnetism demonstrated one kind of gauge invariance and Yang and Mills had come up with another for isotopic spin, he did not know that the two symmetries were described respectively by the groups U(1) and SU(2). A group, one remembers, is a collection of items related by specific mathematical rules; to take a simple example, all the integers,

$$\ldots -3, \ -2, \ -1, \ 0, \ 1, \ 2, \ 3 \ldots$$

comprise a group. Every group has one or more "generators," which are entities that give rise to the group when certain mathematical operations are performed. The integers, for instance, are generated by the number 1, because if you add or subtract 1 any number of times from any integer you can generate the entire group. The generator of the U(1) group of quantum electrodynamics is electric charge, with the photon the agent that ensures the gauge symmetry. Similarly, the three generators of the Yang-Mills SU(2) group imply the existence of three vector bosons. In modern terminology, Gell-Mann was trying to find a larger, more complex structure formed from a greater number of generators that could generate the full range of hadrons; at the time, knowing no group theory, never having heard of terms like SU(2) and U(1), he put it in terms of looking for what he called "another Yang-Mills trick," only bigger.[29]

A worrier, a fretter, Gell-Mann slogged through calculations every morning at the Collège de France until he felt as if he were banging his head on a wall. Discouraged, he met his French friends for a good lunch and a couple of bottles of wine, and returned to his office in a much better frame of mind but with his computational facilities impaired by alcohol. Glashow visited Paris in March of 1960 and showed his model of SU(2) × U(1), one Yang-Mills and one electromagnetism. Seeing a concrete example of how one could push Yang-Mills further, Gell-Mann was enthusiastic. Unfortunately, SU(2) × U(1) was bigger, with four generators (three vector bosons and a photon), but not *different*. As he later put it,

I worked through the cases of three operators, four operators, five operators, six operators, and seven operators, trying to find algebras that did not correspond to

what we would now call products of SU(2) factors and U(1) factors. I got all the way up to seven dimensions and found none. . . . At that point, I said, "That's enough!" I did not have the strength after drinking all that wine to try eight dimensions. Unfortunately I did not pay sufficient attention to the identity of one of my regular companions at lunch. It was Professor Serre, one of the world's greatest experts on Lie algebras [the kind of mathematics for the groups with which Gell-Mann was working].[30]

After his return to Caltech that September, he combined forces with Glashow, worked out the examples again, and again got nowhere. Glashow went to the East Coast for Christmas vacation, leaving Gell-Mann on his own. By chance, he talked to Richard Block, an assistant professor of mathematics at Caltech, who informed Gell-Mann that he had spent the last six months reinventing the wheel. Gell-Mann looked at Cartan's table of Lie groups and felt like a fool. Eight generators, the one that he had not bothered to work out, corresponded to SU(3), the simplest group that was not a composite of SU(2) and U(1).

"Of course," he has said, "if I'd had my mathematical wits about me, I would have realized it much earlier. It's trivial." He had actually sat through courses and lecture series on group theory. "But, you know, the way they teach math is so abstract and peculiar, it's very hard for a student to know what's going on. Mathematicians tend to present things in such a—what is the word?—such a *nonconstructive* way. They like to prove that there is something, but not actually *show* you what it is. When they give examples, they are so trivial that you don't learn anything from them."[31]

What Gell-Mann perceived about SU(3) has the simplicity that is the retrospective hallmark of good physical ideas; fitting the thought to the data was a matter of detail work, of algebra, of tidy manipulation and setting up tables. The group SU(3) has eight generators loosely corresponding to a collection of Yang-Mills vector bosons. Two of the generators represent isotopic spin and strangeness; the other six are rules for changing the value of the first two.[32] Metaphorically, the arrangement of the generators can be thought of as being like a chessboard, with the rows and columns representing different values of strangeness and isotopic spin. The six other generators are like the rules that dictate where a piece can move on the chessboard. A knight in the center of an empty board, for example, can move to any one of eight possible squares. In this case, the generators are rules like "Move the knight up two rows and over one column to the left." And the "representation" of the group is the set of squares on which the knight can land plus the one that it sits on. For SU(3), the six generators alter isotopic spin and strangeness during elementary particle interactions. Two of them change isotopic spin by one unit more or less without changing the strangeness; the other four alter the isotopic spin by one-half and the strangeness by one. In

Gell-Mann's original, Yang-Mills–style formulation, the changes in isotopic spin and strangeness were the work of no less than eight vector bosons.

The simplest and smallest representation of SU(3) has just three members (see page 268), and the six generators are the six possible transitions from one member to another. Other, larger representations are created from aggregates of the smallest, which is why the latter is called the "fundamental" representation of the group. Sakata and his followers argued that the three objects in the fundamental representation are the proton, neutron, and lambda, and in fact produced an SU(3) model based on this idea. Gell-Mann did not believe in the primacy of the three particles, but after considerable study was unable to decide what should be in this representation.

Gell-Mann was dismayed, but he had no choice but to continue if he wanted to use SU(3). He simply skipped over this representation, and turned to the next simplest. There he saw a fit with the quantum numbers for the baryons with the lowest mass: the familiar neutron and proton, the lambda, and five other particles—the positive, negative, and neutral sigma, and the recently discovered xi minus and neutral xi. All have the same spin (½) and the same parity (+1).

The eight particles can be plotted neatly; six of the particles are at the points of a hexagon, and the other two particles are at the center. Seven mesons with the same spin and parity fit neatly into another hexagon; to fill the empty place Gell-Mann predicted the existence of a new particle, the eta. Gell-Mann ran into more serious difficulties, however, when it came to the next largest representation, which has ten members and is called a decimet; only four baryons with the right spin and parity were known. Perhaps the other six indeed existed, but Gell-Mann thought that he was getting into dangerous terrain with the decimet; he decided for the moment not to talk much about this representation and stick to the octets.

Working furiously in the quiet of a campus deserted by Christmas vacation, Gell-Mann finished his particle periodic table in the first days of 1961. With some gaps and a few problems, he had successfully assigned most of the known particles to families. He honored the family of eight baryons, the representation that fit SU(3) most exactly, by calling his idea the "eightfold way," in joking homage to the teaching of the Buddha.[33]

Now this, O monks, is noble truth that leads to the cessation of pain: this is the noble Eightfold Way: *namely, right views, right intention, right speech, right action, right living, right effort, right mindfulness, right concentration.*

To Gell-Mann's considerable annoyance, his jest has fed the notion that quantum physics has something to do with the mysteries of Eastern mysticism. The only mystery to Gell-Mann was *why* the thing worked.

After typing up the eightfold way as a Caltech preprint, he began to

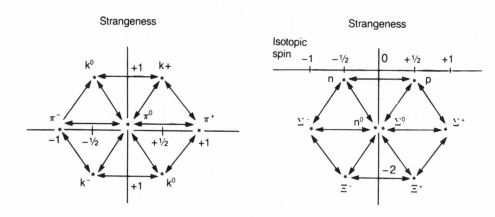

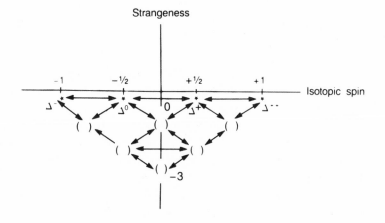

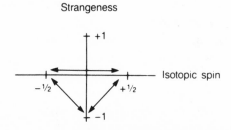

The eightfold way. Here the arrows represent the operators of the group SU(3). The smallest representation (*bottom*) is the fundamental representation, from which the other three are built. Particles in parentheses had not been discovered at that time.

worry about his inability to explain the fundamental representation. The inexplicable disparity in the masses of the particles in the octets was also troubling. In addition, he was at the same time collaborating with Glashow on their attempt to derive both weak and strong interactions from Yang-Mills theories. When that collapsed, he withdrew the Yang-Mills theory that had inspired the eightfold way, and finally presented it as an abstract symmetry whose origin he did not know. Just as Mendeleev had discovered the regularities of the periodic table without the faintest notion of their source, Gell-Mann had invented a taxonomy for the hadrons but had no answer to the obvious question: What caused it?

A similar model was presented independently by Yuval Ne'eman, a colonel in the Israeli army and an amateur physicist who was, during the Begin government, a right-wing member of the Cabinet in Tel Aviv; he was once in the Knesset, where TV cameras sometimes zoomed in to catch him doodling equations during dull speeches.[34] A living refutation of the belief that scientists are absent-minded dreamers, Ne'eman may have had the most unorthodox career of any physicist in this century. He is a *sabra*, a Jew whose family lived in Israel before the partition of Palestine in 1948. Ne'eman joined the Haganah, the Jewish underground, at the age of fifteen; about half his four years as a college student at the Israel Institute of Technology, in Haifa, were spent outside the classroom, fighting for the creation of a Zionist state. A child prodigy, he has grown into a short, pugnacious man with thinning hair and intense, slightly squashed features hidden behind the thick black frames of his glasses. We met him in a conference at Fermilab, where he moved around the expansively gesturing physicists with his briefcase clamped securely at his side, watching the proceedings with an intelligence officer's shrewd eye. By 1958, Ne'eman said, he had set up the national defense policy, risen to the rank of colonel, commanded an infantry battalion during some of the heaviest fighting in the 1948 war, and served as the acting head of the Israeli secret service. He was also tired of not working with his first love: physics. Ne'eman asked his friend and superior, General Moshe Dayan, for two years' leave to study physics in Haifa. Dayan agreed, provided Ne'eman did the work between appointments at his new job—defense attaché in London. Ne'eman went to London, armed with yellowing twelve-year-old recommendations from his teachers in Israel and a crisp new note from Dayan.

"When I decided to do physics, I was going to do relativity," Ne'eman said. "But because of the London traffic, I ended up doing particle physics. It turned out that general relativity was centered on King's College. The embassy was in Kensington. It was impossible to be in both places because London traffic is so jammed that I could never get from one to the other. I went looking for something else closer. I found Imperial College." Ne'eman found a professor in the Imperial College catalogue under "theoretical physics," and astonished him by walking into the office in an Israeli colonel's uniform and asking where people studied unified field theories. "He said, 'I don't know about *unified* field theory, but Salam over there is doing *field* theory.' " It was indeed fortunate that Ne'man was directed to Salam, one of the few physicists interested in unification at the time. Salam laughed at the letter from Dayan and told Ne'eman to bring a recommendation from a physicist. Ne'eman never did, but Salam accepted him anyway—partly, he has said, to repay a debt incurred by Islamic science, which in its medieval heyday owed much to Jewish scholars.

We met Ne'eman late one moony night on the steps of the central administration building at Fermilab, a tall curvilinear structure that rises fifteen stories above the empty Illinois prairie. The conference was over for the evening, the blackboards wiped clean, and physicists trickled from the building in animated knots of two and three, talking of symmetries and string theories. A bus was late; waiting scientists dotted the steps. Ne'eman stood erect in the slight breeze, one foot on a higher step, talking in long, proud, harshly rolling sentences. Oddly romantic moonlight played on his square face and heavy glasses. "I worked on evenings and weekends," he said. "And I taught myself quantum mechanics properly from books. After some time, Salam gave me some calculations to make, and I made the calculations, so I learned field theory. Then there was a gap of a year, because in July of 1958— the end of that school year—there was a revolution in Iraq. The Iraqis assassinated their king and their prime minister, and there was a feeling that the whole Middle East was going to collapse and fall in the hands of [President Gamal Abdel] Nasser [of Egypt] and the Russians. The British wanted to overfly Israel with two battalions of paratroopers to protect King Hussein [of Jordan]. And I found myself having to negotiate with them what they would pay for that overflying of Israel. We ended up with the possibility of purchasing two submarines and fifty Centurion tanks. So between the summer of fifty-eight and the end of fifty-nine I barely had time to listen to a seminar. I was completely busy organizing the training of two submarine crews that had never seen a submarine in their lives and had to be taught from the beginning." He described with some amusement the process of driving in his uniform from the base to a seminar, changing his clothes in the Imperial College parking lot. "It was very bad for my physics. I kept sending letters to the chief of staff—Dayan had meanwhile been replaced—telling him that this was not the original deal I had made with Dayan."

A van pulled up. Ne'eman threw the door open and found a seat without interrupting his story. "He regarded all of this as some kind of intellectual escapade. He wanted me to create the national defense college in Israel. So I did him a favor. I prepared a draft for the national defense college, which was afterward created, but I told him that I was going to stay in physics because I liked it. We agreed, finally, that he would send me a replacement for a year, and the replacement arrived in the spring of 1960. On May 1, 1960, I was liberated from my army responsibilities and was given the possibility to stay in London until the next summer at the expense of the state of Israel." The former child prodigy was thirty-five and just beginning his career in the world of science.

Ne'eman heard about Yang and Mills from Thomas Kibble, a theoretician who later helped develop spontaneous symmetry breaking; at the same time, he learned the rudiments of group theory from Salam. Like Gell-Mann, Ne'eman thought of combining group theory and Yang-Mills to produce a taxonomy of the hadrons. Unlike Gell-Mann, he had the advantage of knowing some group theory, but he had the disadvantage of having to learn the rest of physics. "I started coming up to Salam every time with a new model, trying to produce a model for the particles that were known at that time. He would look at it and say, 'Oh, you've again been wasting your time on these things. Why don't you do this calculation I gave you? What you've just done is something that was suggested by [J. C.] Polkinghorne and me ten years ago.' Then I would come up the next time and he would say, 'Oh, that's global symmetry. That was invented four years ago, five years ago, by Gell-Mann and Schwinger.' So he was becoming impatient and saying, 'You know, you told me you just had one year, and you're wasting it on these things. Why don't you start working on a proper calculation that you know where you start and where you finish?'

"I, on the contrary, was gaining confidence, because if I was reproducing serious things that others had thought about then I was not on such a bad track. I told him, 'After all, I'm doing physics only because I love the material, so I want to do what I love to do.' He said, 'Okay, but you're—' he used these words '—you're embarking on a *highly speculative search*. But if you do it, do it properly. Don't be satisfied with the group theory I taught you here. That's nothing. Go in depth, study the thing properly, and then come up with something.' " Salam mentioned that he had heard of a Soviet mathematician named Dynkin who had supposedly done something useful with groups. After searching with some confusion through both the *American Mathematical Society Transactions* and the *American Mathematical Society Translations*, Ne'eman found a thesis by one E. B. Dynkin in the *Translations* that consisted of a modified extension of Cartan's catalogue of groups. Ne'eman was looking for a group big enough to contain strangeness and isotopic spin but small enough to avoid extra complications. From the list, he worked out a set of five groups that satisfied his criteria: SO(5), SO(4), Sp(4),

SU(3), and G_2, the last being one of five "exceptional" groups for which Cartan could not find mates. "Emotionally, I felt that maybe G_2 should do it," Ne'eman said. "It was one of the exceptionals, and I didn't know what that could be. I worked, and I drew the diagram, and it came out to be a Star of David. I said, 'That must be the finger of God!' " He laughed. "But then it turned out that I didn't like the transitions that G_2 gave. I had myself an entire series of criteria, and they didn't obey my criteria. Whereas SU(3) did obey them."

Delighted, Ne'eman waited for Salam to return from the Rochester Conference in October 1960. Salam was impressed, but had heard Sakata disciple Yoshio Ohnuki expound on the Japanese SU(3) model. Disliking the proton-neutron-lambda triplet for the same reasons as Gell-Mann, Ne'eman wanted to publish his own SU(3) anyway. In February, he asked his former secretary at the embassy to type up the article.

A month later, he found Salam in the office at Imperial College with a fat mimeograph from Caltech in his hands and an astonished look on his face. He admitted sheepishly that seeing Gell-Mann's name attached to SU(3) made the whole thing more plausible. At about the same time, Ne'eman's paper was bounced by the editor of *Nuclear Physics* because the Israeli embassy had single-spaced the manuscript, making the welter of equations nearly impossible to read. After retyping, the article was published in July of 1961. In the meantime, Ne'eman began to worry about the same thing that bothered Gell-Mann. If SU(3) was indeed a symmetry of nature, what was the fundamental representation? Could he really pretend it was not there?

The van arrived at the hotel, a bright oasis of conviviality in the midwestern night. Jukeboxes hammered in the bar; the restaurant bustled with the last dinner seating. Physicists thronged the lobby, stumbled over piles of luggage. Ne'eman made directly for a vinyl-covered chair and sat down. He had only a few minutes more to talk before his midnight telephone appointment with an Israeli journalist doing a story about Ne'eman's periodic abdication of his responsibilities as a member of the Knesset to play at theoretical physics. Although he had a special dispensation from the Israeli supreme court to leave the country for up to a month every year on physics-related matters, Ne'eman attracted considerable media suspicion that he was somehow pulling a fast one. "They always find me," he said, checking his watch. He sat hunched over his briefcase in the middle of the room, a small and fiercely solitary man in a rumpled suit, utterly untouched by the roar and sentiment and laughter echoing off the plastic walls about him.

Theoretically, the eightfold way was beautiful: One simple group gave rise to octets and decuplets that contained all the known particles and could be presented in simple diagrams. Experimentally, SU(3) was a disaster.[35]

Gell-Mann had started with two octets and been forced to add a third when a new batch of mesons was discovered. Even so, there were more holes in Gell-Mann's multiplets—almost a third of the total—than there had been particles ten years before. Years later, when we visited Brookhaven, Nicholas Samios recalled the time from the privileged view of hindsight. "There were many people who had many schemes," he said. "In fact, there were some people who said it was a crock of shit, SU(3) couldn't be correct. People tend to say, 'Oh, my God, Murray came up and showed it, and everybody said "fantastic!" ' But that's simply not true."

We asked why SU(3) was disbelieved. Samios produced a blank sheet of paper and quickly put a set of dots into an inverted pyramid; four dots in the top row, followed by rows of three, two, and one—ten dots in all. The decimet he said, is very neat. The particles in the top level have a strangeness of 0, in the second a strangeness of -1, in the third a strangeness of -2, and the single particle at the bottom, a strangeness of -3. "The mass difference between the different levels is the same, about 150 MeV," he said. "It works out magnificently! But it was incomplete, okay? Murray had *this*"—pointing to the top layer of the diagram, the quartet of delta particles—"but nothing else. There was not enough information to say whether he was right or wrong. He needed more data, okay?"

That additional data came soon enough.[36] In June of 1962, Gell-Mann, Ne'eman, Samios, and about a thousand other physicists attended a Rochester conference held, in a gesture of international amity, at CERN. Some CERN experimenters had measured the by-products of proton-antiproton annihilations. The data contradicted the Sakata model, whereas they fitted the eightfold way. Elated, Ne'eman asked to speak, only to discover that the chairman of the symmetry session was Yoshio Yamaguchi, an ardent partisan of the Sakata SU(3) model. Yamaguchi refused to let Ne'eman talk about the rival model, having at first misunderstood the calculation. "There were all kinds of funny papers at that session," Ne'eman recalled. "There was another Japanese who explained 137—things like that!—but the octet was not allowed."

In addition, at the conference another team of European experimenters announced the discovery of two new particles: the xi-star, by Samios and collaborators, and the sigma-star. (The star refers to the particles' similarity to other sigma and xi particles.) Although little was yet known about them, their approximate masses fit the decimet. Ne'eman guessed that the sigma-star and the xi-star would fill in the middle two levels of the pyramid, leaving one hole at the bottom. Below the xi-star should be a single particle that is about 150 MeV heavier and has a strangeness of -3.[37]

The plenary session on strange particle physics took place in one of the rather barren auditoriums in the main CERN complex. The rapporteur,

G. A. Snow of the University of Maryland, cautiously went through the still-confusing tangle of data about the competing SU(3) models, then turned the session open for comments. Full of his prediction, Ne'eman raised his hand impatiently. The chairman looked directly at him and said, "Professor Gell-Mann."

Gell-Mann, too, had understood the significance of the experimental findings. He strode to the platform and took the microphone. If the spin and parity of the xi-star and sigma-star are "really right," he said, "then our speculation [about the decimet] might have some value and we should look for the last particle, called, say, omega minus." Because the masses of the new particles fitted the eightfold way, Gell-Mann also predicted the mass of this new particle: 1685 MeV. On the way up the stairs to his seat, he noticed Ne'eman's name tag; the two proponents of the eightfold way met for the first time.

Minutes afterward, the convention broke for lunch in the CERN cafeteria, a sunny L-shaped room surrounded by picture windows that open out onto a view of patios, rusty lawn furniture, and, very occasionally, the Alps that surround cloudy Geneva. Samios and another Brookhaven experimenter, Jack Leitner, approached Gell-Mann and asked him to predict how the omega minus would decay. Gell-Mann jotted down a prediction on a paper napkin in his small neat handwriting. The three talked about how one would find the omega, if it existed, and Samios put the napkin in his pocket to show to Maurice Goldhaber, then the head of Brookhaven.[38]

For about a year Samios had been planning to do an experiment on Brookhaven's spanking new accelerator, the AGS. The AGS speeded up protons to nearly the speed of light, and smashed them into a dense target material. By using magnets, the various fragments of this collision—muons, pions, K particles, and the like—could be sorted out, and split into secondary beams which were in turn directed into particle detectors. Early in his career, Samios had decided that the future lay in beams of strange particles, because nobody had ever seen what happened when a lot of strange matter interacted with ordinary matter. "Since the name of the game was strangeness," Samios said, "it's best to try to make a beam made out of Ks [a type of strange particle], because if you start with strangeness you could leap one more strangeness, or two more strangeness to begin with."[39] The beam of K particles would be directed into a vat of liquid hydrogen, which was inside a particle detector called a bubble chamber.

Bubble chambers were then the most sensitive detectors available. Invented by Berkeley physicist Donald Glaser in 1953, bubble chambers are large and somewhat dangerous devices whose care and feeding preoccupied experimental physicists for years. We once asked Robert Palmer, Samios's longtime collaborator and an accelerator expert, to describe how a bubble chamber worked. A British subject who has been at Brookhaven for thirty-five years, he still works on designs for accelerators that he hopes will be built next century. "I like to say a bubble chamber is like a pressure cooker," Palmer said. "In a pressure cooker, you can heat up the water inside to well above the boiling point, and it won't boil, because the pressure inside is so high. Now, you take the lid off, and what happens? Funnily enough, nothing—not right away. Then it starts boiling. Where does it start boiling? On the surface. It doesn't start boiling in the middle of the liquid—it doesn't know how.

"But in a bubble chamber—in which liquid hydrogen is used, to make it more sensitive—it can. If a charged particle races through the middle of the liquid, the friction heats it up. The liquid hydrogen starts to boil in the wake of the particle, and a string of tiny bubbles forms.

"Then you wait about a millisecond to let the bubbles grow. All of a sudden you *slam* the lid back on the pressure cooker—in the bubble chamber, it's a big piston—flash your lights, and *click!* the cameras peering through the portholes snap a picture of all the little strings of bubbles. Then you advance the film, reset the chamber, and do it again about a second later."[40]

Pouring through the thousands of photographs taken every day in a bubble chamber is a tedious process that is usually assigned to lowly graduate students. Knowing that the AGS experiments would produce tens and even hundreds of thousands of photographs, Brookhaven physicists hired a team of people from the outside. Mostly Long Island homemakers who needed extra income, they worked around the clock, three shifts a day, seven days a week, with special projectors that blew up the seventy-millimeter film onto white tabletops. The scanners, as the women were known, copied the tracks of the events onto magazine-sized sketch books, and then measured the various angles and lengths.

When the mass of tracks is blown up onto a tabletop screen, the result is astonishingly lovely—a thicket of white lines skirls across a dark gray field, intersecting, brachiating, looping like the lines in a Ptolemaic drawing of the heavens. Each type of particle leaves a characteristic track called a signature, and the scanners were taught how to recognize them. Electrons, extremely light particles, produce faint, wiggly tracks, for they are easily knocked about amid the heavy protons of the hydrogen nucleus, like a ping-pong ball attempting to fly through a roomful of baseballs. Protons leave thick, straight lines that run like pencil strokes across the slide. As in a cloud chamber, a magnetic field is placed in the bubble chamber to help identify the particles; it causes the positively charged particles to bend one way and the negatively charged particles to bend the other. The amount of deflection indicates their mass, for the paths of lighter particles bend more than heavier ones. Neutral particles leave no tracks.

"When I was a student," Samios said, "we started off with a chamber that big." He held his hands a few inches apart in front of his face. "Then we went this big." His hands spread about a foot. "Okay? But this was *big*! A thousand liters. So here we were, with a new beam and a new chamber, and we were going to push both to the limit. And along comes Murray with his prediction of the omega minus. That provided us with a focus. We wanted to look for it."

Samios and Palmer built a pipeline from the target that was hundreds of feet long and lined by huge gray magnets that pulled and pushed at the stream of particles until everything but the kaons was filtered out. At the end of the tunnel, the K particles struck the hydrogen in the bubble chamber, producing showers of new particles that had to be precisely analyzed. The experiment started in November of 1963, and promptly ran into trouble. The particles going into the chamber acted nothing like K particles. After a month of figuring, the team discovered that the kaons were being swamped by pions. They spent weeks adjusting the magnets used to select out the Ks—and got nowhere. Early one winter morning when Samios and Palmer were on shift together, they decided that the Ks had to be hitting something somewhere inside the beam line, creating a secondary burst of pions. From the geometry of the situation, the two men were able to calculate approximately where the obstruction must be. They turned everything off and walked down the line of magnets. A machine part called a collimator was sticking out close to the target region, a small piece of metal that had cost them two months' time.

Samios and Palmer were not the only ones seeking the omega minus. Gell-Mann had made his prediction at CERN, which had its own accelerator. A British team built a K beam just like the one at Brookhaven, except six months earlier. Unluckily, their bubble chamber didn't work. A French group did have an operating bubble chamber, but the English refused to use their detector unless the experiment remained entirely British, an idea that the French group rejected. The British kept on struggling with their own detector in proud isolation.[41]

The Brookhaven bubble chamber, too, had trouble. A set of black strips about an inch wide hung from the wall of the chamber like a vertical venetian blind to cut down the reflective images from the camera and lights. Soon after the beam was introduced into the chamber, the repeated jolting of the piston knocked ten of the slats loose. They fell forward and came to rest against the glass window.

"There we were, in the middle of the night, [bubble chamber designer

Ralph] Shutt, myself, W. Fowler, Palmer," Samios said. "And we looked at the window. The question is, Did you damage the glass? Because if you damage the glass, and put pressure on it, then the glass breaks, and you have a thousand liters of liquid hydrogen coming out—you have a real catastrophe. So there we were, with ten of the hangers down, and Shutt looking in. He asked all our opinions, but he had to make the decision. He said, 'Expand.'" The three-foot piston thudded into the liquid hydrogen. "Nothing happened. And he made the decision to continue.

"The other option was to dump the chamber, open it up, fix the slats, and lose a month. The logical thing would have been to stop, to do it right. But we wanted to get on with it. My feeling is, you get five of these things, then you've got to stop. But if you always stop at the first fix, there are usually two or three problems further on. If you fix the first one right, when you get started again, you get the next one. But you've wasted a month before you even *knew* about the other ones! So you go as far as you can without jeopardizing things. We ran, and we kept taking data."

During the months at the end of 1963 and the beginning of 1964 that the Samios-Palmer group could not get their apparatus to work, they were the target of the wrath of other Brookhaven experimenters, who wanted the time and the equipment that were being granted to Samios's team. "We had first priority for a certain window, okay?" Samios said. "But whatever time we took, other people didn't get, so we didn't have carte blanche. We had to perform. If we made mistakes, people would say, 'Why don't you put these guys off for a while and let us do our things, and when they're in better shape, have them come back.' We had big debates. And in fact, I lost some."

In January 1964, Maurice Goldhaber left for a conference in Coral Gables, Florida, where he told the physicists present, including Gell-Mann and Ne'eman, that nothing interesting had been found in fifty thousand bubble chamber exposures; theorists began giving up on the eightfold way, and the Sakata group took new hope.[42]

In the meantime, Samios, Palmer and their team continued to analyze what eventually became more than a million feet of film. "We arranged the scanning so that there would be a physicist on shift at all times," Samios said. "In case any candidate events or something queer showed up, a physicist would judge whether it was good or not. We took data, December, early part of January . . ." He waved his hands to indicate the passage of time. "The physicists would scan, and then, as you're scanning, if someone had questions, you'd stop and you'd go over and look at what they had and tell them yes or no. And then you'd continue, day after day."

Samios was on duty, spooling through bubble chamber film, measuring angles and particle tracks, when a striking negative slid across the white surface of his scanning table. The *K*s usually swept in lines from left to right across the frame, slowly spreading out to either side like sheaves of wheat held in an unseen hand. In this frame, one stalk was snapped off. A K particle had come in, hit something, and created a new particle that veered violently downward. A foot away, a thin V appeared in mid-air, the two arms crossing lazily after a few inches.

An old bubble chamber rule of thumb is that the line of flight of the invisible particle that created the V can be estimated by laying a ruler between the two intersections. If it misses the point where the K struck the hydrogen atom in the chamber, that is a sign that some interesting, albeit invisible physics took place in the interval. "So I said, 'This is a real candidate. Looks interesting.' But I didn't do very much with it. I said, 'Measure it.' Early the next morning—I think I was on the night shift—they were measuring it. And while it was on the measuring table, there was a group around looking at it, and somebody noticed a gamma ray."

More exactly, they noticed that an electron-positron pair suddenly burst into existence a few inches away from the candidate interaction. The speeding K had struck a hydrogen atom, creating a minute cornucopia of particles. One of the products was a photon, which then turned into an electron and positron that spiraled away from each other like watch springs. This was a stroke of luck: By measuring the curvature of the tracks of the electron and positron, the experimenters could determine their mass, and hence the energy of the originating photon. Subtracting that from the energy of the K they would know how much energy was available to create particles, such as an omega minus.

"They noticed one gamma ray first. And then someone said, 'Hey, could there be a second?' This was on the measuring table. They said, 'Hey, there *is* a second!' I said, 'That's *crazy!*' " The likelihood that *two* photons had come out and turned into easily measurable pairs of electrons and positrons was quite literally a million to one. "I said, 'Measure it quickly! Get it off and put it on a regular scanning table!' We had templates—" calibrated French curves for particles "—so I could measure these two gamma rays by hand." In about two hours, Samios was almost ready to declare that they had found an omega minus.

Some thirteen particles were involved in the event. The only reason that the physicists could be certain they had an omega minus was the sheer dumb luck that all the final products—the two particles in the V and the two electron-positron pairs—left easily measurable tracks. This good fortune allowed them to work out the mass exactly. Samios and his team finished a letter to the *Physical Review* and sent it out.[43] By the time it was printed they had come across a second. The Europeans published a year later.

Samios called up Gell-Mann, who gracefully pretended to be surprised by the news, although he had already heard it through the scientific grapevine, which travels at nearly the speed of light. The Brookhaven group later sent him a picture that still hangs on the wall of his office. The following winter, the laboratory mailed out a Christmas card bearing an artist's rendering of the first omega minus event.

The omega minus discovery clinched the case for Gell-Mann's SU(3) scheme. The eightfold way not only arranged particles into patterns, but it successfully predicted where hitherto unknown particles should be found. Unfortunately, nobody had the faintest idea *why* it worked. The person who figured this out was, once again, Murray Gell-Mann.

15

The King and His Quarks

THROUGHOUT THE LAST CENTURY *QUARK* WAS A RARE, POETIC TERM FOR A particular type of animal call, the cry of a heron or gull. Nobody is ever likely to use the word in this sense again, for during the past two decades it made an abrupt transition in meaning from bird caw to subatomic particle fragment, surely one of the most bizarre etymological twists in the history of language. This movement began in 1939 with the publication of James Joyce's last novel, *Finnegans Wake*. Its hero is a Dublin pub owner named, variously, H. C. Earwicker, Here Comes Everybody, and even "Heinz cans everywhere." He is asleep throughout the book; his dreams, which are recounted in its pages, are the expressions of a collective unconscious, reenacting myths, historical incidents, and even the rise and fall of civilizations. As befits a dream, however, the events and scenes of the book do not unfold with the sequential logic of a television miniseries, but through a dense collage of puns, repetitions, misspellings (which begin with the deliberate omission of an apostrophe from the work's title), allusions, and other linguistic highjinks.

The fourth episode in the second section of *Finnegans Wake* begins with the hoots and babbles of four old men, who represent the four authors of the Gospels, the four ancient historians of Ireland, and anything else that comes in fours, as they chortle over the old Celtic romance of Tristan, a young nobleman, and Iseult, the wife of Tristan's uncle Mark, King of Cornwall. The sequence opens with the huzzah:

> *Three quarks for Muster Mark!*
> *Sure he hasn't got much of a bark.*
> *And sure any he has it's all beside the mark.*

Like everything else in *Finnegans Wake*, *quark* has many meanings, all of which have been fiercely argued by scholars. The term is said to be a kind of Bronx cheer for the poor cuckolded king: Three raspberries for the pigeon! It is also a play on a demand that one might have heard in Earwicker's tavern for three hearty Irish quarts of ale: Buy that man a drink! In German, *quark* is a variety of particularly runny cheese: The four men describe Mark as beleaguered by screaming gulls, and the sentence therefore also refers to a deluge of bird droppings—bombs away!

Quark might have passed into the etymological limbo reserved for the obsolete words resurrected by modernist writers had not Murray Gell-Mann accidentally revived the word in connection with his thoughts about hadron taxonomy. On leave from Caltech, Gell-Mann passed the winter of 1962–63 at MIT, from where he traveled, one blustery March day, to present a lecture at Columbia. A guest speaker at Columbia is customarily treated to lunch at the faculty club by a member of the department, and this time the honor fell upon Robert Serber. Walking up Broadway in a freezing wind, Serber, Gell-Mann, and several other physicists began talking about the eightfold way. Serber asked Gell-Mann about the different representations of SU(3), not all of which can be matched with elementary particles. In particular, the scheme had no room for the smallest SU(3) family, which has just three members. Mathematically, such an absence is peculiar, because the larger representations are built up by fitting together an array of the smallest. For this reason, the three-member representation of SU(3) is known as its "fundamental" representation. Serber did not understand how nature could use the octet and decuplet without taking some account of their basic constituent, the fundamental representation.

The physicists entered the faculty club dining room, a large, elegant space in a building overlooking Morningside Park and South Harlem. In Serber's recollection, he asked Gell-Mann why he had not considered suggesting that the SU(3) families were indeed formed from the fundamental representation, which was equivalent to saying that the particles in the octets were made out of littler particles. The eightfold way might therefore be viewed as the set of all possible means of mixing and matching several of these subunits, analogous to the way the more than one hundred chemical elements are formed out of protons, neutrons, and electrons. In this way, taxonomy would be explained by structure. "I pointed out that you could take three pieces and make protons and neutrons," Serber said. "Pieces and antipieces could make mesons. So I said, 'Why don't you consider that?' "[1]

"So I *showed* him why I hadn't considered it," Gell-Mann told us later. "It was a crazy idea. I grabbed the back of a napkin and did the necessary calculations to show that to do this would mean that the particles would have to have fractional electric charges— $-\frac{1}{3}$, $+\frac{2}{3}$, like so—in order to add up to a proton or neutron with a charge of plus or zero. [No such fractional charges had ever been seen.] And he said, 'Oh, I see why you don't do that.' But then, thinking about it the next day or so I said, 'So what if they have fractional charges? Maybe they're permanently trapped inside. Maybe they don't ever come out and they don't ever cause any problems.' I said, 'What the hell, why not?' Speaking the next day at Columbia I may have mentioned it." The talk was in a seminar room at Pupin Hall. Afterward, over coffee, Gell-Mann used an odd word for these subunits: *quork*, to rhyme with *pork*.

"It seemed somehow appropriate," Gell-Mann said. "A strange sound for something peculiar. When I was going to publish the idea eight months later or whenever it was, late in sixty-three, I was paging through *Finnegans Wake* as I often do, trying to understand bits and pieces—you know how you read *Finnegans Wake*—and I came across 'Three quarks for Muster Mark.' I said, 'That's it! Three quarks make a neutron or a proton!' Joyce's word rhymes with *bark*, but it was close enough to my funny sound. Besides, I told myself, in one of its meanings it is also supposed to rhyme with *quart*. So that was the name I chose. The whole thing is just a gag. It's a reaction against pretentious scientific language."[2]

Gell-Mann took almost a year to write a short paper about quarks. Entitled "A Schematic Model of Baryons and Mesons," the article appeared in February 1964, the same month that the Brookhaven team discovered the omega minus and clinched the case for the eightfold way.[3] A landmark in contemporary physics, the eight-paragraph note is a model of scientific prose: brief, logical, achingly clear, so tightly and modestly drawn that its full scope may elude the reader. In the first line, the author sets forth his intention: "If we assume that the strong interactions of baryons and mesons are correctly described in terms of the broken eightfold way, we are tempted to look for some fundamental explanation of the situation."

After pointing out the algebraic necessity of the fundamental representation in the second paragraph, Gell-Mann begins to state the case for subunits in the third. Guessing the reaction to fractional charges, he at first avoids them; he therefore proposes only that all hadrons are composed of three subunits, which he calls "up," "down," and "strange," with charges of +1, 0, and 0, respectively. In addition to this triplet, there is an oppositely charged antimatter "anti-triplet" of "antiup," "antidown," and "antistrange." Mesons are made from one member of the first triplet and one of the second; the heavier baryons consist of a meson plus another neutral particle, which Gell-Mann calls *b*. First suggested in the earlier Sakata model, the neutral *b* particle has the job of maintaining the conservation of baryon number. Passed along through interactions, it prevents the ups, downs, and stranges in the baryons from all recombining into mesons.

The bit with the *b* is awkward, as Gell-Mann was the first to admit.

A simpler and more elegant scheme can be constructed if we allow non-integral values for the charges. We can dispense entirely with the basic baryon b. . . . *We then refer to the members* u$^{+2/3}$ *[up with a charge of* +2/3*]*, d$^{-1/3}$ *and* s$^{-1/3}$ *of the triplet as "quarks"* q *and the members of the anti-triplet as antiquarks* q̄.

Exactly as Serber suggested, baryons are now made of three quarks; a proton, for instance, consists of two ups and a down clamped together by the strong force. The lighter mesons are composed of one member of the first trio (a quark) and one of the second (an antiquark).

How did this explain why the eightfold way successfully ordered particles? Gell-Mann argued that the quarks and the antiquarks that make up baryons and mesons divvy up each particle's electric charge. Thus the proton—two ups and a down—has a total charge of $(+2/3) + (+2/3) + (-1/3)$ $= +1$. Similarly, the neutron, which Gell-Mann supposed to be made of an up and two downs, has a total charge of $(+2/3) + (-1/3) + (-1/3)$. The particles in the SU(3) baryon families consist of every combination of three units of $+2/3$ or $-1/3$; the result is always an integer.

Quarks not only explain SU(3)'s usefulness, they make sense out of strangeness and isotopic spin. A particle's strangeness is given by the number of strange quarks inside it; a strangeness of 0, for instance, means that no strange quarks are inside. Easy![4] And isotopic spin is determined by the number of up and down quarks in the particle: A proton, which has more ups than downs, has an upward axis of isotopic spin; the neutron, with more downs than ups, spins on a downward axis. Values of strangeness and isotopic spin are constant in strong interactions because, as it happens, the strong force cannot change one kind of quark into another. It can shuffle them around like cards in three-card monte, or hold them together in an iron grip, but it cannot alter the proportions of up, down, and strange.

Gell-Mann suggested that although the weak force cannot hold quarks together, it might get them to change types, or "flavors." And that trait would account for the long lives of strange particles, which can decay into ordinary particles only when the feeble weak force finally asserts itself and turns a strange quark into an up or down.

Up, down, strange: Just three quarks explained phenomena that theorists had been puzzling over for years. Nevertheless, Gell-Mann said, "Quarks went over like a lead balloon." The reason was that the fractional electric charges $+2/3$ and $-1/3$ contradicted one of the most thoroughly established rules in physics. Over the half-century during which anyone had been looking, no particle with, say, two-thirds the charge of a proton had ever been spotted. Such particles simply could not exist.

Gell-Mann asserted that quarks had never been seen because they cannot be seen. For reasons he could not explain, quarks must always be chained together, like the hero and heroine of Alfred Hitchcock's *The Thirty-Nine Steps*. ("You can't even pull one out with a quarkscrew," Glashow has said.)[5] This, too, did not go over well. Experimenters did not like the notion of something that *in principle* could not be found with the aid of their machines. Their frustration was not assuaged by Gell-Mann's oracular pronouncements that the failure to see fractionally charged quarks meant that "they exist but are not 'real.' "

(Gell-Mann explained once that he used those terms "because I dreaded philosophical discussions about whether particles could be considered real if they were permanently confined. While a colleague of mine falsely claims to have a doctor's prescription forbidding him to engage in philosophical debates, I really do have one, given to me by a physician who was a student in one of my extension courses at UCLA.")[6]

To others, at least, it seemed that Gell-Mann hedged when it came to the question of the ontological status of quarks—whether they were actually *there*. Although he closed the paper by proposing experiments "to reassure us of the nonexistence of real quarks," Gell-Mann did not expect that experimenters might find them. "Even I thought the idea of observable fractionally charged particles was crank," he said. "My intuitive idea was that maybe they were permanently stuck inside and couldn't get out and were therefore what I called 'mathematical' quarks—that was very hard to explain to people. They thought it was some sort of cop-out. So I asked, 'Why don't you look for them experimentally if you want them to be real?' Then people said, 'That's crazy, too. Everybody knows that there won't be any experimental particles that have fractional charges." About the time he finished the paper, he called his former teacher, Victor Weisskopf, who had been appointed the director of CERN. "I was in Caltech, he was in Geneva. In the course of the conversation, I said, 'By the way, Viki, I have a very good idea for how to account for all the mesons and baryons; they are made of particles of charge $-1/3$ and $+2/3$, three kinds of them.' And he said, 'Oh, nonsense, Murray, don't waste time on a transatlantic call talking about stuff like that.' And I said, 'Well, look, maybe they exist as observable particles. Do you think it would be worth trying to look for them at CERN? How about a search for observable particles of this kind?' And he said, 'Murray, let's talk about something important.' "

The quark paper appeared in the CERN journal *Physics Letters*, whose editor at the time was the Franco-Polish theorist Jacques Prentki. "We were just starting out," he said, "and nobody from the States was submitting papers. I said, 'Murray, why don't you submit something?' So I got the quark paper, which he did not think would be accepted by the *Physical Review*. Now, everybody knew that if you took the hadrons and got them broken up, you ended with fractional charges. That is arithmetic. But it took Gell-Mann to make a very beautiful paper from it. I accepted it immediately. The blame for such a crazy paper would fall on Gell-Mann, not the editor of *Physics Letters*."[7]

Many physicists *were* thinking about triplets. Any theorist with a rudimentary knowledge of group theory could wonder about the fundamental representation of SU(3). Yuval Ne'eman had worried about the problem from the beginning, finally deciding that the regularities of the eightfold way, like those of the periodic table, would be understood fifty years later.[8] Unable to bear the thought of waiting half a century, Ne'eman took a stab at explaining the structure within a few months. There were three known leptons;

Ne'eman asked Salam if a baryon could be made by combining neutrinos, electrons, and muons. Salam laughed outright, then said that no idea is too stupid if it works.[9] With another Israeli theorist, Haim Goldberg, Ne'eman worked out the mathematics of the quark model, but shied away from hypothesizing the existence of quarks themselves. Their hedging, awkwardly written paper was sent to *Nuovo Cimento* in February 1962; it passed almost unnoticed, and is still hard to read.[10]

Somewhat later, Julian Schwinger and T. D. Lee independently came up with fundamental triplets of their own. Respecting fifty years of experimental practice, they, too, were leery of fractional charges. Schwinger scoffed at Gell-Mann's up, down, and strange; in his article, mailed the month after Gell-Mann's article appeared, he said they could presumably be detected solely by "their palpitant piping, chirrup, croak, and quark."[11]

The saddest story belongs to another Caltech theorist, George Zweig, who duplicated the quark model exactly at almost the same time. He called the quarks "aces," and strenuously insisted that they could be detected. Unfortunately, he was at CERN during the year of the quark, and resisted a laboratory regulation that he had to publish his papers in a particular form in the laboratory journal, *Physics Letters*. Zweig had worked out an elaborate scheme of diagrams for the aces, but the editors would hear none of it. Despite heated argument, Zweig never published his long, careful article on aces.[12] (One man's meat: Gell-Mann sent his article to *Physics Letters* voluntarily because he was sure he would be hassled by the American *Physical Review Letters*.)[13] Word got around, though, and the reaction to aces was, in Zweig's words, "not benign." It was all right for someone of Gell-Mann's stature to advocate the lunatic notion that most of matter was made up of ineffable entities that were invisible to experiment; having no reputation to protect him, Zweig was denied an appointment at a major university because the head of the department thought he was a "charlatan."[14]

Nevertheless, there was a certain elegance to the idea of three quarks. For one thing, the three quarks matched up neatly with the three leptons, and physicists were beginning to discover that symmetry begets unification. But no sooner was this pattern pointed out than it was disrupted again through the discovery of a fourth lepton, in what amounted to the last hurrah of Columbia University as the dominant center of particle physics in the United States, and thus the world. Whereas the eightfold way arose as a frontal assault on a discipline-wide problem, the fourth lepton was first suspected by theorists worrying over small paradoxes, the sort of tiny irritants whose slow consideration can with luck produce pearls of insight. There is a dictum known as the totalitarian theorem—Gell-Mann took it from a line in *1984*—which states that every particle interaction not forbidden is compulsory; it is up to experiment merely to determine the rates. By 1960, physicists believed that the electron and muon were exactly, precisely, absolutely identical, except that the latter was more than two hundred times heavier than the former. By the totalitarian theorem, muons should be able to turn into electrons through the simple expedient of emitting gamma rays that take away the extra mass in the form of the photons' energy—unless some rule forbade it. Experimenters had duly searched for evidence of such decays, and found they occurred, if at all, less than one ten-thousandth as often as ordinary decays. Theorists had a little puzzle.

The late 1950s was also when many physicists began to seriously consider the possibility of an intermediate vector boson that plays a role in weak interactions.[15] If the vector boson is admitted to the theory, then the muon-electron decay should occur as follows: the muon turns into a W^- and a neutrino, the W^- releases a photon, and the neutrino and W^- recombine into an electron. Having set up the chain of interactions, Gary Feinberg calculated the probability that it would take place: It was five times more than the experimental limit. Therefore either the boson did not exist or the decay was expressly forbidden for some other reason. Feinberg guessed the other reason might be that the neutrino from the muon simply could not become an electron because it has what one might call an essential "muness." Thus, two neutrinos exist, one each for the electron and muon. Although such ideas had been bruited about before, they were not respectable, and Feinberg, then a young postdoc at Brookhaven, was sufficiently mindful of the climate of opinion that he tucked his solution into an obliquely written footnote.[16]

Despite Feinberg's hesitation, the footnote stirred discussion, which subsequently foundered on the absence of satisfactory means of studying neutrinos. Because neutrinos experience only the weak force, they rarely interact with matter. Floods of neutrinos from the sun shoot as easily through the earth as if the planet were a sheet of tissue paper. Reines and Cowan had barely been able to discover the neutrino by examining the trillions given off every second by a nuclear reactor; although most physicists believed that the particles exist, a smaller number found their demonstration convincing.[17] In the next five years not a single direct measurement of neutrinos was performed anywhere in the world. For this reason, Feinberg did not even consider the possibility that the question of two neutrinos could be settled by experiment.

The answer to the dilemma came in a roundabout way, a product of the afternoon coffee breaks on the eighth floor of Pupin Lab at Columbia University. During these hour-long respites, students and faculty alike gathered before the blackboard, tossing out pet ideas and watching T. D. Lee run through their consequences with chalk in hand. One day in November 1959, the *koffee klatsch* was attended by a Columbia experimenter named Mel Schwartz, who by chance found himself on campus without a class to teach. The subject of the coffee break was the experimental study of weak interactions. Schwartz is a square-shouldered man with the brash, tough manner that is the hallmark of what Europeans mean by the American style. Infuriated by bureaucracy and committees, he left high energy physics for twenty years to work in Silicon Valley. He returned in the 1990s, after winning the Nobel Prize, and ended up at Columbia. There are certain people who love to talk about their passionate interests. Schwartz is one of them; when we asked about the discussion that day in Pupin Hall, the phrases tumbled out. "Everybody was sort of gathering around, throwing around crazy ideas on how to measure the weak interactions at high energies," he said. "Somebody was discussing how to scatter electrons at very, very high energies and looking at the deviations—you know, the cross-section as a function of angle. There are a whole lot of other ways you can look for weak interactions, but they're all ugly, because they're all terribly masked by other things that are taking place." The much more powerful effects of the strong and electromagnetic forces tend to swamp weak interactions. "At the end of the hour there was no conclusion. I think T. D. and Gary Feinberg and a few other people were there. I went home, and then that evening I was pondering the thing, and it became suddenly obvious that the right way to do that would be to make use of neutrinos. The reason being, of course, that neutrinos have *only* got the weak interactions. So if you wanted to study the weak interactions, the question became, could you make enough neutrinos? Once you think to yourself, in a very simple-minded way, that you can make a neutrino beam, it's all a matter of calculating how many you can possibly get, and whether there'd be enough to do something."

By the end of February 1960, Schwartz worked out what would be necessary to produce and detect a beam of neutrinos. (In Europe, Glashow had just worked out his $SU(2) \times U(1)$ model.) At the same time, Lee and Yang, whose partnership was not yet defunct, produced a list of theoretical questions about neutrinos that should be resolved. First in the series was whether there are two neutrinos. The pair of two-neutrino papers were published back to back in the *Physical Review Letters*.[18] As difficult in practice as it is simple in theory, Schwartz's method for studying neutrinos involved shooting protons into a chunk of metal, creating pions, which then decay into floods of muons and the neutrinos that accompany their creation. All slam into a thick steel wall. Only the neutrinos make it through the wall and into the detector on the other side. To discover whether two types of neutrinos exist, the experimenters must identify the particles produced in the detector when the neutrinos hit something. If they include muons but no electrons, then there is some conserved quality of "mu-ness," and both muon and electron neutrinos exist. If muons and electrons are created in equal abundance, then only one type of neutrino exists, and theorists had some rethinking to do. Schwartz's Letter, published in March 1960, claimed neutrino experiments "should be possible within the next decade." In fact, the first design work began within months.[19]

"There were two machines being built at that time," Schwartz said, "both of which had the capability of doing this experiment: the AGS and the CERN machine. Now, the CERN machine was about six months ahead of the AGS. In the spring of 1961, a group at CERN began looking into the experiment, and we began looking into the experiment at Brookhaven." The rivalry between the two teams was overt and intense from the first: Both were composed of ambitious young men aching to make names for themselves in what looked like wide-open territory—the completely unexplored terrain of neutrino interactions. Moreover, the Brookhaven physicists were led by Schwartz and his Columbia colleague, Leon Lederman, and the CERN group included yet another Columbia experimenter, Jack Steinberger. "There was a certain period of resentment because of the fact that we were in competition with somebody from Columbia who had decided to go off and do the experiment elsewhere. In any case, one day we got the exciting news that the experiment at CERN was a disaster, because Jack had made a mistake in his calculations. Somebody by the name of von Dardel had gone over Jack's calculations and discovered an error and found that you couldn't do the experiment operating out of a five-foot straight section." The straight section was the segment of pipe between the ring of the accelerator and the target. CERN physicist Guy von Dardel discovered that the straight section was too short and that the magnets on the ring were therefore close enough to defocus the pions to one side, which would diminish the neutrino intensity to the point where the experiment became impossible. Physicists are rarely kind to one another, and experimenters in competition rejoice over their opponents' misfortune. "That became of course very exciting to us. The so-called von Dardel effect meant that we were going to be there first. Number one, we had a ten-foot straight section; number two, there was no opposition at Brookhaven toward doing the experiment, whereas at CERN there was very strong opposition."

We asked Schwartz why CERN hadn't installed a longer section, which was, after all, only a matter of five feet of pipe. "Well, the place at that time— I'm not sure how it is now—was not overly full of cooperation. Different teams from different countries were always at each other's throats. If one guy got screwed, the other guy would jump up and down with joy." At a different point in the conversation, he came back to the subject. "Everybody else who was not doing the experiment was in competition with the people trying to do the experiment. And so the minute that von Dardel discovered this problem, the first reaction should have been, 'Well, let's go and switch the experiment to a ten-foot straight section'—right? In fact, the reaction was, 'Can the experiment.' Which made us very pleased. Of course it was about as stupid a thing as CERN could have done, in retrospect."

Discouraged at the nationalistic battling within the international lab, Steinberger returned to Brookhaven, where the hatchet was buried and Steinberger invited to join the experiment. The team spent the rest of 1960 and all of 1961 building the apparatus. The plan was to send pulses of protons from the AGS down the ten-foot straight section into a beryllium target, creating a cloud of pions, which were in turn directed into a seventy-foot pipe. About one-tenth of the pions decayed to produce, among other things, billions of neutrinos. Everything then smashed into a pile of steel forty-two feet thick that only the neutrinos could penetrate. The steel was scrap, rusty chunks cut from the sides of old Navy cruisers and deemed worthless enough to be wasted on science. Piling it around the detector took Lederman, Schwartz, and Steinberger months of hard labor. On the other side of the steel was a ten-ton detector consisting of many thin aluminum plates set close together; the plates were electrically charged, and the passage of charged particles created sparks. By photographing the sparks, experimenters could obtain approximate pictures of the particles' trajectories. Once a day, they calculated, one out of the billions of neutrinos passing through the steel would strike a proton or neutron in the detector, producing an interaction.

The experiment began in early 1962 and, as is the case with all genuinely new techniques, something went wrong immediately. In this case, it was the AGS itself, which leaked protons and gave rise to contaminating radiation. Because the team was expecting only one neutrino event a day, it would not take much to hide what they were looking for. After some time, Lederman recalled later, "we traced it down to part of the beam escaping the target and hitting the wall of the vacuum chamber so that neutrons would come down into the concrete and underneath and scatter from the concrete up into the detector." The team figured they could stop the leak if they piled lead right along the wall of the brand new accelerator. "At the time there was a line beyond which Mr. G. K. Green, who is in charge [there], said we could put shielding only over his dead body; this would have made a small but unsightly lump in the shielding. We compromised."[20] Seven days a week, twenty-four hours a day, physicists sat outside the detector, watching the dials, consuming sandwiches and coffee, playing chess, and sometimes napping. The experiment ran eight months until July 1962, during which time over a hundred trillion neutrinos passed through the detector. Just fifty-one of them hit something on their way. Not a single electron was found in the fragments of these collisions, whereas muons were found in all fifty-one.

The American team was elated. With pluck and luck, they had soundly beaten CERN, and they popped open the bubbly to celebrate. We asked Schwartz about the pleasure of being first, of having clear priority to the discovery. (At the time, he was still in self-imposed exile from physics.) "You know, now I'm in a business [computer systems] where the measure is very simple. It's how many bucks can you bring in, right? If your company makes enough profit, then you're a big man. If it makes a little profit, you're a small man. If it makes no profit, you're miniscule. So it's very simple to make a measure of a man. In physics, the only measure you make is general recognition. In that situation, you have an awful lot of people fighting for the only money that exists, which is the money in recognition. It's a big poker game, with a certain amount of zero-sum, so to speak. In other words, if I win, you lose. If I get the priority for that particular thing, you haven't got it." Experimenters often ask the question: Who was the *second* person to say "$E = mc^2$?" (Lederman, Schwartz, and Steinberger shared the Nobel in 1988.)

The tally now stood: Leptons 4, Quarks 3. The symmetry between them could be restored by tinkering with the number of quarks, but this seemed pointless to most physicists, who regarded them as Gell-Mann's fractionally charged folly. Not Sheldon Glashow. Hypothetical particles perpetually locked inside all the particles in the nucleus was just the sort of loony supposition he thrived on, and he was determined to play with it. Glashow quit Caltech in a fit of annoyance in 1961 when the college bookstore refused to cash one of his checks. Exasperated at the lowly status awarded to postdoctoral fellows, he wangled a junior faculty position at Berkeley and moved upstate, where he spent some years energetically propagandizing for the eightfold way.[21]

In the spring of 1964, Glashow returned to the Bohr Institute. Bohr had died, but little else had changed. The Institute was still full of bright young theorists talking up their ideas in tiny offices. One of the other newcomers was James Bjorken, a tall, sandy-haired theorist who had arrived from Stanford a few months earlier. The two men had met in northern California a little while before. Soon Glashow hit Bjorken with his latest pet notion, namely, that there might be a *fourth* quark. He called it *charm*.

Charm excited Glashow, because it arranged the Universe back into an elegant pattern. With the discovery of the second neutrino, the existence of four leptons was universally accepted. If a fourth quark could be summoned into existence, Glashow reasoned, then the subject matter of physics could be divided into two families. The first family makes up ordinary matter; the second occurs in high energy processes.

quarks	leptons
up	electron
down	neutrino

ORDINARY MATTER

quarks	leptons
charm	muon
strange	neutrino

WEIRD MATTER

Unfortunately, he was unable to find any justification for charm other than his certainty that God could not have been stupid enough to decree four leptons and only three quarks. Nonetheless, he managed to persuade the reluctant Bjorken to send the idea off anyway.

It is hard to imagine two theorists with more divergent styles. Bjorken has long arms and long legs and the slight stoop of a very tall man. There is a certain outdoorsy ranginess to the way he occupies a chair; one suspects immediately that he has spent a fair amount of time on hiking trails. His voice is a quiet, almost avuncular tenor. Glashow is tall, too, but with thick graying hair that falls in a tangle over his forehead, needing less to be combed than to be subdued; his glasses are often askew; he speaks with a pure New York inflection untouched by years in Harvard Yard. Glashow has spent his career producing ideas in a hot-brained hurry, tossing one aside the instant another occurs to him, whereas Bjorken is slower, more methodical, closely tied to the phenomena. Bjorken has spent his career working closely with experimenters, first at SLAC and now at Fermilab; he likes trying to guess what the machines will find before they are turned on. The article with Glashow is perhaps the most speculative in his oeuvre.

"At the time, it was not a particularly earth-shattering paper," he said. "The idea of a fourth quark had been floating around among many people, who also wrote papers about it at the same time as we did. You have to remember the flavor of the time. It was an easygoing time for models. Models came and went. Salam was particularly prolific; he would explore every idea, build a model for it, and most all of them were wrong. A mountain of papers proposing various models were published by many authors, including Shelly, that went straight into the trashcan. There was no reason to think that this paper would fare any better. The idea turned out to be right. But then, I've seen Shelly equally enthusiastic about things that didn't pan out. We made a lucky hit, but we weren't the only ones. The one thing unique about that paper was the name we gave the fourth quark: charm. That is what stuck."[22]

When they submitted the charm article, Bjorken indicated the spirit with which he took the whole enterprise by signing a fictionalized version of his name: "B. J. Bjørken," "B. J." being the monicker he is known by, followed by the Danish spelling of his Swedish father's name. In the indexes of scientific publications, there is a mysterious "B. J. Bjørken" who has written only one paper. It appeared in *Physics Letters* in August 1964—just seven months after Gell-Mann's original quark article.[23]

As it turned out, charm worked more magic than either Glashow or Bjorken knew. When incorporated into SU(2) × U(1), charm explained why experimenters had never observed weak neutral currents.

In many cases, the greater the number of ways in which a particle can decay, the more readily its decay will occur. This much is a familiar part of our world. For example, if two doors on a bus are open rather than one, twice as many people can exit at any given time. But in the quantum world such "exits" can interfere with one another, and opening an additional "door" may have the paradoxical effect of reducing rather than increasing the frequency with which particles decay.

So it is with strange particles—the kind that physicists had always used to study the weak interactions. According to the SU(2) × U(1) theory, weak neutral currents exist, which means that strange particles should often emit virtual Z^0s and turn into ordinary particles without any change in electric charge. But experimenters had never seen any such event, let alone the number likely if SU(2) × U(1) were correct. Charm reconciled Glashow's lovely theoretical model to this ugly experimental fact. By a quirk of mathematics (and nature), the probability that a strange particle will emit a Z^0 and become an ordinary particle and the probability that it will emit a Z^0 and become a charmed particle are almost equal. It is as if the strange particle stands irresolute between two equally tempting options and cannot bring itself to choose either. (Glashow wrote later, "As it happened, a sign in the equation that defines [the charmed reaction] is negative, and the two interactions cancel each other.")[24] Thus, although weak neutral currents should be able to occur in strange particles, they never do. Experimenters would have to look at other particles for evidence of such currents.

Unfortunately, when Glashow and Bjorken wrote their paper they didn't see the power of charm. "There it is in black and white, and we missed it!" Glashow said recently, smiting his brow to theatrical effect. "If you read my 1961 paper with Gell-Mann, you'll say, 'These people say the theory doesn't work and something is missing.' And if you read my 1964 paper with Bjorken, you'll exclaim, 'By God, that's it! A fourth quark!' But did we notice it? No, not for six years."[25]

In 1964 Glashow was not even thinking about weak neutral currents. He was preoccupied by how poorly quarks fitted into field theory. If no theorist could come up with a field that would hold quarks together in a way that would explain why they were never detected in experiments, it was difficult to understand how adding a fourth quark would help matters. Troubled, Glashow gave up on charm, as he had earlier given up on SU(2) × U(1), because he didn't know how to show that it was right. For the next six years he published inconsequential papers, threw others away, and grew cranky. In 1966, he became a full professor at Harvard, and arrived on campus to find other members of the physics department in the same funk.

□ □ □ □ □

"There's a long tradition of theoretical physics, which by no means affected everyone but certainly affected me, that said the strong interactions are too complicated for the human mind." The speaker was Steven Weinberg. We were in his living room, and he had been inveigled into speaking, in his synoptic way, about the strong force and the many attempts made by physicists to understand it. "When I was a graduate student at Princeton, I worked with Sam Treiman, who was my thesis adviser. I don't know remember whether Sam said it to me or we said it to each other, but that was the feeling. You just couldn't get anywhere trying to understand them. What you could do was sort of walk around the outside of them—" look for small effects and simple situations "—which gave you such a feeble interaction that you really could study it mathematically and learn about its symmetry properties." Other people, of course, had waded into the fray, directly confronted the strong force. "I never got involved in that," Weinberg said. "I thought, that's not—maybe their minds can do that, but mine couldn't."[26]

Although both Gell-Mann and Ne'eman based SU(3) on Yang-Mills, the inability of local gauge field theory to account for the eightfold way eventually made them both shy away. The subsequent explication of the representations in terms of quarks removed Yang-Mills theories one step further; by 1965, gauge field theory, which today dominates particle physics, was sufficiently out of favor that even Weinberg, one of its more ardent proponents, could remark that "no one would ever have dreamed of extended gauge invariance if he did not already know Maxwell's theory."[27]

Between 1964 and 1969, there were no major experimental discoveries concerning the strong interactions. Without pace-setting experiments, theoretical ideas tend to branch out, wide, ramose, and shallow, splitting, like the roots of a palm tree, into endlessly fine distinctions, ready to be torn out by the first strong wind. In a repeated, faddish pattern, new wrinkles and clever techniques spark frenzies of speculation that subside as physicists run into blind alleys. Considering the theoretical work of the 1960s, one wants to retreat before the barrage of calculation in the journals, the thousands of articles that nibble at the corners of large, poorly understood questions, and come to small conclusions.

An attempt to evoke the prevailing mood runs into the historiographic difficulty that many if not most high energy physicists spent their time on pursuits that seem remote from what are now considered central problems. Some theorists, for example, put SU(3) and the two orientations of spin together to form SU(6); after two years of frenetic calculation, it was demonstrated that the theory was not Lorentz invariant.[28] When that failed, theorists turned to symmetries like U(6), SU(12) and E_7, another one of the "exceptional" groups. All were eagerly embraced and readily discarded. Gell-Mann made a habit of building new techniques on field theory but removing the particular wrong version of field theory from the final product, likening the process to the French chef's trick of cooking pheasant between two slices of veal, then throwing away the veal. He had done this with the eightfold way; he did it again with "current algebra," which treated the currents of the interactions as the fundamental entities, rather than the particles comprising them.[29] Other researchers tried to link the spins and masses of elementary particles with what were called "Regge poles," after the Italian theorist Tullio Regge. Some Regge enthusiasts believed that their poles, together with calculations of scattering probabilities, would displace quantum field entirely, leading eventually to the reign of "nuclear democracy," in which no particles would be fundamental and every entity in the subatomic world was a transitional form of every other.[30]

In the huggermugger of currents and Regge poles and groups, good ideas vanished if not advertised. Even as the majority of theorists cast about for some clue to the strong interaction and the quarks it held together, if it held them together, the later key to success lay entombed in the back issues of the *Physical Review*. If ideas are not picked up quickly, they are generally forgotten unless their authors continue to promulgate them energetically. In this case, the author was Yoichiro Nambu, a soft-spoken, quiet man who has earned the dubious distinction of being described as the John the Baptist of elementary particle physics—"there in the wilderness before everyone else," Bjorken once put it.

Nambu's manner is self-effacing, almost retiring; his features seem to recede from view, his words to fall into the back of his throat. He doesn't parade his accomplishments.[31] Born in 1921, he is one of the last members of the great brief flowering of Japanese theoretical physics that began with Nishina and Yukawa in the glory days of quantum electrodynamics. At the beginning of the 1940s, Nambu attended the University of Tokyo, where the curriculum did not include particle physics. He taught himself about cosmic rays, relativity, and quantum field theory with a group of friends. When he could, he attended the seminars at Riken given by Tomonaga and Nishina, including one that discussed a letter from Sakata proposing for the first time the two-meson hypothesis. The university was exceedingly formal, comatose with tradition and bureaucracy, the last legacy of an aristocratic style of education that was about everything but learning. Students did not question professors. Professors did not fraternize with students. Hints of change came from the direction of the physics department, but only small indications.

Within a year after Pearl Harbor, Nambu was drafted into the Imperial Army and put to work in Osaka researching the microwaves used in Japan's recently developed land-based radar. One of Nambu's tasks was to steal a document from Tomonaga, working for the Imperial Navy; the two branches of the Japanese military had as little to do with each other as possible, and the Army got word that Tomonaga had written a paper in which he used some of Heisenberg's ideas to understand microwaves. Through a professor who worked for both sides, Nambu managed to get a copy of Tomonaga's work.

After Hiroshima and Nagasaki, Nambu was demobilized. After a brief stint at the University of Tokyo as a research assistant, he was appointed to a post at Osaka City University, a new school that was trying to rise up from the wreck of postwar Japan. Living in the confused desperation of a beaten nation and with their leaders under suspicion, the Osaka group nonetheless managed to keep abreast and even ahead of their victorious American colleagues. With two other Osaka theorists, Nambu wrote one of the first papers on the production of strange particles.

Nambu is now an emeritus professor at the Enrico Fermi Institute of Nuclear Studies at the University of Chicago. One of the long row of neoclassical buildings at the school, the institute is directly across from the site of the old football stadium, where, in a basement squash court, Enrico Fermi supervised the creation of the first man-made nuclear chain reaction. These days the football stadium is gone; in its place is a set of tennis courts and a tall, rounded monument to atomic energy by Henry Moore. Nambu's window overlooked the courts and the sculpture, which were dimpled with snow on the cold February day we visited him; he told us that he and the rest of the particle and nuclear physicists had watched the princes and potentates at the dedication ceremony from their office windows.

We asked him how it was that Japanese science had been so strong despite so many obstacles.

"That's a curious thing," he said, and then waited a while with his hands flat on a long table in his office. "Yukawa had this great vision. He was a self-made man, in the sense that he never went to study abroad. He was a very highly independent person. And Sakata and Taketani, who were his collaborators, they had their own ideas. . . . I don't know how they arrived at those views. I just don't know. I mean, because they were based on Hegelian, on Marxist philosophy."

By all accounts Sakata was a powerful man whose smooth, sociable demeanor masked an overbearing streak. He rebelled against the authoritarian tradition of Japanese academics only to use it to his own benefit when he became established. Holding that real but unseen particles ran counter to the tenets of existence discovered by Engels and Lenin, Sakata refused to consider quarks as a serious possibility. Nambu subscribed to this dialectical view, and to some extent still does. It proved fruitful for him, but unfortunately for most others in his native land the ideological pressure to conform was too high. "No great independently minded physicists have appeared since," he said.

After the war Oppenheimer invited Yukawa to Princeton one year, and Tomonaga the next; Nambu followed in 1952. Somewhat to his surprise, he never returned to the country of his birth except for visits. He learned English quickly and well, and went to the University of Chicago a few years later. His style is one of reasoning from first principles, much as a coloratura might constantly return to the primary scales. A good portion of his time has been occupied with the task of attempting to reconcile currently fashionable theoretical notions to fundamental physical principles. In the late 1950s, for example, Nambu listened to a lecture by a graduate student, John Schrieffer, on the "BCS" theory of superconductivity that he had recently helped to develop. In the middle of the lecture, Nambu realized that the BCS theory violated gauge invariance. Schrieffer came soon afterward to the University of Chicago as an assistant professor, and Nambu talked with him often. He soon showed that superconductivity was an example of broken symmetry, that the U(1) of quantum electrodynamics did very strange things at low temperatures. Then, with a student, Nambu developed a method of symmetry breaking for strong interactions.[32] The audacity of ascribing to the vacuum—empty space!—a structure was remarkable; just as bold was their assertion that when SU(3) was broken, a massless particle popped out. They claimed it was the pion, even though the pion is not massless, and said that it was not "the primary agent of strong interactions, but only a secondary effect." Since Yukawa's invention of the meson a generation before, the pion had been assumed to be the agent of strong interactions. Now it was relegated to the role of a by-product. The argument later became the basis for the surge of work on chiral symmetry in the 1960s.

We asked Nambu whether he had been attracted to the quark model. "Oh, yes. Except for the fact that these quarks had to have these funny charges—a strange fractional charge. *That* I didn't like. My belief in general was that what you assume as constituents cannot be fictitious, but must be real." His reaction, he admitted, was perhaps influenced by the Sakata school.

We said that the Marxist method did not sound too different from the traditional way of doing physics. "Maybe it's a matter of abstraction," Nambu said. Then, quickly: "But when you start building the particles out of quarks, it didn't seem realistic or correct to me just to regard these quarks as symbols." He made a rare gesture by tracing a little symbol in the air. "I had some difficulty with physicists who could easily swallow that—you know, quarks as just another symbol."

Did he have other objections? Nambu nodded. "This again had to do with one of the basic principles of physics, the Pauli exclusion principle. In a very simplified language, it says that two particles with spin one-half in the same state cannot occupy the same place at the same time. The same state—actually, to be rigorous, the electron has to be specified by its position and spin direction. So if you put two electrons in the same place with the spins pointing in the same direction, that is just not allowed. And this quark model just ignored it completely!"

Gell-Mann had said the quarks had to have spins of plus or minus one-half if three of them were to make a baryon with a spin of one-half or three-halves. (The plus or minus indicates the direction of spin.) To make, for example, a delta, which has a spin of 3/2, requires three up quarks, and 1/2 + 1/2 + 1/2 = 3/2. Three quarks with the same spin are thus packed together inside a delta, which flagrantly violates the Pauli principle. The problem was referred to as a problem with "the statistics." Oscar Greenberg from the University of Maryland worked out an alternate type of statistics—what he called *parastatistics*—in which the spin one-half particles could avoid the Pauli principle. He applied it to Gell-Mann's quark model, but not in a way most physicists understood.[33] Nambu was not satisfied with that route, either; he thought he could fix the statistics and have integer charged quarks to boot.

"You cannot deny that the quark model can explain a lot of things," Nambu said. "I tried to save two things: one, the statistics and two, to make them observable." He was slowly pushed in the direction of making the quarks different from each other. "There is always a hesitancy to increase the number of fundamental constituents. But I was slowly driven to this idea: If there is one kind of quark, why not two?[34] In other words, two types of ups, two types of downs, two types of strange. I tried that first. Then just for the heck of it I tried three—three kinds of quarks. And I wrote a preprint in which the possibility of three was mentioned in passing."

In 1965, Nambu was invited to give a lecture at Carnegie-Mellon, called the Carnegie Institute of Technology in those days, and there he looked at a letter sent to him by a young graduate at the University of Syracuse, Moo-Young Han. Han had used group theory to elaborate on Nambu's three-triplet idea, and, like a jazz player pulling unexpected riches from a popular tune, had discovered treasures within. Through the mail, Han and Nambu put together a paper proposing three types of quarks; it was published in the *Physical Review* without its authors having met face-to-face.[35]

Nambu hypothesized at the start that every quark carries one of three charges, in much the same way that a pion can bear a positive, negative, or neutral electric charge. The quark charge and electric charge, however, have nothing whatsoever to do with each other. Rather, quarks have a new, never-before-seen charge that years later was christened *color*, although it is not ordinary color. There are three "color charges," today called red, green, and blue.[36] (Confusingly, Nambu called his new charge *charm*, even though Glashow and Bjorken had already used the term for the fourth quark.) Baryons are composed of three differently colored quarks; a proton, for instance, might be a red up, a blue up, and a green down. Mesons, on the other hand, would be colored quarks and "anticolored" antiquarks; a pion might be a red up and an "antired" down.

Exactly as positive and negative electric charges attract one another, so do red-blue-green and color-anticolor combinations. Furthermore, like the photon associated with electric charge, the color charge has eight particles—"gluons," in current parlance—that hold colored quarks together. The common proton, in this picture, appears as an extraordinarily complicated object: three differently colored quarks crowded into each other like people on a rush-hour bus, surrounded and permeated by a swarm of gluons.

An odd view, perhaps, but one compatible with integrally charged quarks. Three colors also avoid the statistics problem, for the quarks inside the proton are no longer in the same state; they are differently colored, hence in different states, and can happily coexist.

There was more. Implicit in the scheme—and here was Han's contribution—is the startling suggestion that there are two types of SU(3): one, the eightfold way, is a symmetry of particles that feel the strong force; the other is a color symmetry that pertains to the force itself. Moreover, the intense strong interaction that holds together protons and neutrons in the nucleus is merely a secondary by-product of the second SU(3), the enormously powerful "color force" that grips the quarks. The exchange of pions that clamps together the constituents of the nucleus, and thus every atom, is a feeble reflection of the forces among quarks; the pions should be regarded, Nambu wrote later, "as perturbing forces rather than the decisive factors in the physics of hadrons."

Up, down, and strange quarks were far out enough for most physicists; the Han-Nambu triplets of each seemed even more outrageous, especially when the two men were forced to admit at the end of their paper that at the energies available to experimenters it was "difficult to distinguish" their model from the regular quark model. Unhappily, Nambu realized that arguing that the proton is really a collection of colored quarks and gluons was equivalent to saying that the strong interactions could best be explained *entirely in terms of particles that had never been seen*, and that the particles that could be observed were nothing but secondary manifestations of these invisible entities. Although he was predisposed to believe that matter was composed of a series of levels, such dealing with ineffable entities bordered on the metaphysical, if not the crank. Nambu did not know if his guess was profound or simply a clever new way to be wrong. In any case, nobody seemed to be interested. Convinced that the *Physical Review* would never accept further work on color, Nambu turned over the idea a little more, briefly considering making a Yang-Mills theory out of it, and then gave up. He let the paper with Han slide into the morass of failed speculation on the strong interaction, where it remained for another five years.[37]

□   □   □   □   □

"It really was an extraordinarily boring time in particle physics," Howard Georgi once said of the late 1960s. He was in his office at Harvard, leaning back in his chair, in a good-humored, discursive mood. "Hardly anybody knew what they were doing." Small bursts of laughter punctuated his sentences, as if thinking over the confusion of the past inevitably threatened to spill into simple guffaws of amusement. "Witness the fact that the $SU(2) \times U(1)$ model—Weinberg's model of leptons—was written down and completely ignored by everyone. Him, too, really. No one even blinked at it in 1967. I remember looking at it as a graduate student, and I said, 'Well, it doesn't look renormalizable to *me*.' And dismissed it.

"The reports of friends of mine who were here—" he swept out a hand, indicating Harvard "—at the time were not encouraging, because this place . . ." He paused. "Well, Shelly [Glashow] tends to get depressed during boring periods of physics, and when he gets depressed he sort of gets grumpy and wanders around not doing much of anything for long periods of time. And that tended to happen during this period." He paused again, reconsidered. "Actually, I should say it wasn't obvious that anything was going on, although in fact there was a ground swell of what turned out eventually to be physics coming from the experimenters on the West Coast." We asked Georgi what had been so unexpected about that experimental work. He laughed. "Really, who would have ever thought then that somebody would show that *quarks* existed?"[38]

The ground swell came from the Stanford Linear Accelerator Center, a brand new national laboratory south of San Francisco in the Santa Clara Valley. Begun in 1962, SLAC is centered on almost five hundred acres of countryside—ridged, verdant, marked by forest—that belongs to nearby Stanford University. The heart of the facility is a tunnel a dozen feet wide and two miles long that cuts from the campus laser-straight toward the Pacific: the linear accelerator. Controversy over the size and use of the machine accompanied the laboratory's parturition. For the years before its construction, the accelerator was known as Project M—*M* for monster—and regularly derided by East Coast experimenters. The two-mile length of the machine, they whispered, had been chosen only because that was the longest straight path the planners could make on Stanford property (which was true); electrons, they fairly shouted, were thoroughly understood by quantum electrodynamics and could never produce much new physics (which turned out not to be true). What could be less interesting than slamming electrons down a big sewer pipe into a vat of hydrogen?

The monster now lies rigid across the Santa Cruz foothills like the backbone of a huge extinct beast. Tunneling relentlessly through sharp loess hillocks, under orchards, and even under Interstate 280, the accelerator starts a quarter mile from the San Andreas Fault. (The laboratory has an earthquake committee that reviews all experiments and supervises construction; even the cafeteria vending machines are bolted to the wall.) Although the beam itself is forty feet below the grass, its path is marked by a sort of corrugated iron shed that runs the entire length of the line and contains almost two hundred and fifty man-sized devices known as klystrons. A klystron is an amplifier for microwaves; low-power microwaves come in one end and high-power microwaves go out the other. The high-power microwaves push the electrons down the pipe in the tunnel. When we visited the klystron gallery, as the long shed is called, the monster was on, and small flashing red lights played in mock alarm over the walls and ceiling. It was an unseasonably warm day in February, and the faces of the joggers who regularly use the gallery as a track were glossy with sweat. The klystrons buzzed amiably, noisy but not uncomfortably loud, sending short bursts of microwaves down precisely engineered copper tubes that disappeared into the earth; each tube connected to the beam pipe, and the quick pulse was timed to meet with a bunch of electrons as they shot toward the target, whipping them to greater vigor.

The electrons come from a piece of hot metal oxide subjected to an electric field of intensity sufficient to rip the light little electrons free of their nuclei; shooting out one end of the "gun," they enter the accelerator and are almost immediately pushed to within a whisker of the speed of light. All the further "kicks" they receive from the klystrons do not significantly increase their velocity. They go at almost the same speed, but accumulate energy, storing it like misers, ready to slam into a target with minute, awful violence. Joggers need twenty minutes to run the course of the gallery; the electrons whip through it in about one hundred-thousandth of a second. At that speed, they could travel 'round the world seven times in a single second.

SLAC is open to anyone who cares to enter, a rule that Wolfgang Panofsky, the founder and first director of the laboratory, fought for with considerable vigor. Short, plump, bespectacled, and articulate, Panofsky, who is nicknamed 'Pief' after a German cartoon character, has been involved in political struggles over arms control and scientific budgets since the advent of fission. He is said to be in many ways responsible for the ban on atmospheric testing of nuclear weapons, a treaty that has helped reduce the level of radioactivity in the air. To create SLAC, Panofsky had to negotiate a course through both the bureaucracy in Washington and the unpredictable politics of the San Francisco Bay area. The laboratory was designed so that the base of the accelerator faces a small grassy hill from whose summit we could see the earthmoving equipment for the next SLAC machine, the linear collider. Designed to take one beam of electrons and one beam of positrons from the original linear accelerator and send them down opposite sides of a circular track until they hit each other head on, the collider is an attempt to explore the ramifications of the electroweak theory quickly and cheaply. Most of the new work will be underground; the largest structures on the surface are the old experimental halls, factory-sized concrete and metal boxes where the first evidence of quarks was found.

When the long accelerator switched on in 1967, one of the first experiments was a SLAC-MIT-Caltech collaboration that measured what happened when electrons smacked into protons at high speeds. (Actually, the electron and proton don't directly hit each other, but interact through virtual photons.) The expectation was that the paths of most electrons would be only slightly bent by coming near a proton, but that a few would come close and be bounced off at a wide angle, the way a few of Ernest Rutherford's alpha particles had been knocked back by the atomic nucleus. If there were something hard inside the proton, something with its own charge and mass—something like, say, a quark—the electron would recoil, much as Rutherford's alpha particles had recoiled from the concentrated electric charge of the hard little nucleus in the center of the atom.

"The Rutherford thing was self-evident," James Bjorken recalled later. He was a student at Stanford when the new machine was proposed. "I certainly remember Leonard Schiff [the head of the Stanford physics department] saying something about it the first time that Project M was talked about at the colloquium at Stanford." Bjorken had been sitting in the seminar room, still marveling at the concept of a machine with dimensions described in terms of miles. "Something two miles long was really *off scale*. It was just *enormous*. The first real geographic machine.

"But anyway, it was announced there'd been some planning done by the senior people and then there was a general colloquium in the physics department to explain this to the community there. And Schiff was the guy who talked about the theory. I remember him talking in those terms, that Project M would give enough energy to look at instantaneous charge distributions inside of a proton." At least to Schiff, that much was clear: One could hunt for constituents in this way. The problem was going to be interpreting the results. "Nobody knew at all the right descriptive language for that, because this was clearly a very relativistic situation, whereas all the previous applications were nonrelativistic in nature. [So the barrier was not the generalization of the Rutherford idea to looking inside of a proton.] That much was self-evident to anyone who thought at all about it." Later, after his return from Copenhagen and charm, he joined the theory group at SLAC and quickly became preoccupied with the experiments about to be run on the machine. "The real problem was how to handle the problem that this was a very new situation, because whatever the explanation was, it had to include relativistic motion. Extreme relativistic motion, in fact. In a very essential way. Whatever the constituents were inside of a proton, they were moving around at the speed of light, sort of, and the old-fashioned way of doing things just didn't work. That was the barrier that was so hard to overcome."

From the point of view of an electron in the SLAC tunnel, the protons in the hydrogen target rush toward it at close to the speed of light. At such speeds, as Einstein showed, peculiar things happen; from the vantage of the electrons, the protons are flattened out into pancake-shaped objects that move with nearly infinite momentum and seem to be frozen in time. The bizarre relativistic environment made for circumstances that many theorists thought to be both hard to predict and of little intrinsic interest: hard work, no profit. The experimenters were sure that they would come across odd things, but nobody quite knew how they should interpret the results when they came in. Bjorken had many friends on the MIT-SLAC-Caltech experiment; as the beam line neared completion, he started making calculations. He didn't look up for more than two years. "I got really hooked on the problem," he said ruefully. "Worked relatively obsessively on it. I can remember saying at some point or another, 'There's all sorts of other physics around, why the heck am I stuck on *this* thing? I can't get loose from it.' "[39]

He had been trained to approach physics linearly, to build up a deductive chain of reasoning, a series of "ifs" leading to a small number of "thens." Calculating electron-proton scattering didn't work like that at all. On the research frontier, all the half-certain theoretical notions that were jumbled in the foreground had to be applied at once, a little dubiously, but none could be relied upon. The yardsticks were quantum mechanics and relativity. Gell-Mann's current algebra gave a solid foundation, but one of limited extent. Bjorken soon got down to the slippery precepts of common sense. Don't get stuck on a model; strip away assumptions and see what you can believe if you start off believing nothing.

The first batch of low-energy trial runs started in January of 1967. They went well. So well, in fact, that the Caltech members of the group dropped out of the experiment because the results had been completely in accord with expectations. The electrons ricocheted off the protons; a few were sent off at wide angles. We told you so, the Caltech experimenters were informed by their colleagues on the East Coast. The remaining experimenters—Henry Kendall and Jerome Friedman of MIT, and Richard Taylor of SLAC—pressed on in the second half of the year, looking for cases where the electrons had scored more direct hits on the protons, creating a splash of new particles, and considerably complicating the situation. By the beginning of 1968, the group was ready to start looking at the data.

Here a distinction must be introduced. The first set of runs, in which electrons and protons simply bounced off each other, measured what is known as *elastic* scattering, interactions in which the particles knock each other about like so many billiard balls. When, in the later runs, the electrons had lost enough energy in the collisions to create new particles, this was *inelastic* scattering. "The whole East Coast establishment didn't believe that electron scattering of any kind was interesting at all," Richard Taylor recalled. "Ever! They *still* don't." Inelastic scattering was regarded merely as an uninteresting way of producing resonances and measuring their properties, work that had in most cases already been done; elastic scattering, in which particles retained their identities, was thought of as the way to study the structure of a proton. The electron came in, bounced off whatever was inside, and left, without confusing the issue by making more particles. Several elastic scattering experiments had already been performed, and Taylor liked to quote a Harvard physicist's remark: "The peach didn't seem to have a pit."

When the later, inelastic data was processed by the computers, however, the experimenters found themselves staring at an odd batch of numbers. A relatively large number of electrons seemed to fly off to the side even when the proton exploded in the collision. Theorists had predicted rather vaguely that the likelihood of such wide-angle interactions would be small; instead, the probability of an electron careening off to the side was a hundred times what they had imagined. It was unclear what this meant.

Bjorken thought he knew—"sort of," as he put it. From half a dozen sources—principally Gell-Mann's current algebra, but also from work on neutrino scattering using various mathematical rules—he had come to the conclusion that the possibility of scattering at the large angles would depend on the ratio of two simple quantities: the energy loss of the electron and the momentum transferred from the electron to the proton. Above a particular energy threshold, the electrons would stop behaving as if they were hitting big protons and start acting as if they were ricocheting off something as small as another electron. That small something might be a quark, although Bjorken was anything but certain about it.

A model of caution, the experimental group was afraid that the photons radiated by speeding electrons sufficiently complicated the interaction to make any calculation dubious.[40] They did their analysis two different ways and released their data in slow bits and pieces. As the numbers dribbled out, Bjorken made his own rough graphs. The curves more or less did what he thought they would. Although he was young and leery of pressing his views, he showed his graphs to the puzzled experimenters. They were interested in the fit, but baffled by his abstract style of exposition.

Nonetheless, they felt they had enough to go on. In September 1968, Panofsky told a large conference in Vienna about what was becoming known as "scaling."[41] The presentation was lengthy, evenly phrased, and sufficiently encyclopedic that no member of the audience seems to have noticed that Panofsky ended the talk by mentioning "the possibility that these data might give evidence on the behavior of point-like, charged structures within the nucleon."

At that point, Richard Feynman paid a short visit to SLAC. After working on the weak interaction, he drifted out of elementary particle physics temporarily, because he thought there were not yet enough experimental clues to make good guesses, and into the quantum theory of gravity. Little research had been done into quantum gravity since the flowering of interest in the subject created by Einstein's theory of general relativity. Feynman struggled for a few years and made some small headway, but decided to return to high energy physics.[42] Tall and sparely built, with thick long gray hair combed straight back over the skull, he had features marked by years of vivid internal experience and a bold, precise, intense, and almost imperious manner of speaking. His office in Caltech was by Murray Gell-Mann's; one overworked secretary served them both. An enthusiastic amateur percussionist, Feynman announced his presence on the floor by rapping drum patterns on the corridor walls with his knuckles as he approached. He walked with the jaunty, hips-out stride of a young blade checking out a bar the day after payday.

Sunlight filled Feynman's large, cluttered office. A pair of glasses rested at an angle in his shirt pocket. His dislike of interviews had increased since the Nobel Prize made him willy-nilly into a media figure; his voice was edgy, suspicious, ridged with theatrical intonation. He is scornful of history and philosophy. We asked how he had become interested in electron-proton interactions. Staring straight into his interlocutors' eyes, he waited for three long seconds before answering. "I had not been in physics much," he said. He spoke with an urban snap, the fast emphatic cadence of a city dweller. "I mean, I'd been doing gravity, waiting for phenomena to grow so that it was more interesting. Then I thought I'd better get back into hadronic physics. I asked some friends where a problem was and they said that high energy collisions are peculiar between protons. About the way the cross-section behaved and some other things." In the jargon, cross-sections are measures of the likelihood that particular types of interactions will take place. "I worked it all out and didn't see anything peculiar. And then I got interested when I realized that the machines were going to be for high energies, so I started to work on the theory of what happens at relativistic energies, trying to guess what the behavior would be."

He didn't want to imagine that Gell-Mann's quarks were inside the protons, and that this would affect the interaction between two protons. Instead, he guessed agnostically that there was an indeterminate number of unspecified objects he called *partons* inside each proton, and tried to proceed from there. "Partons were nothing but the field quanta of some future relativistic field theory, whatever it may be. The method I had of thinking was to not decide yet which it is but to see what would happen in general. I gave it a word, so that meant I didn't have to say quarks or whatever. I could say, 'Whatever they're made out of, when we find out later, they'll behave like this.' We can find out what they are and how they're distributed by experiments. I was noticing I could deduce something irrespective of what they were made out of." He went to SLAC to find out what the experiments were doing. We asked him if he had heard of scaling beforehand.

"No. They said that Bjorken had proven by some way that the data should be plotted a certain way and that it fitted pretty nicely, that it behaved like he said it would behave. They asked me why did it behave like that. He [Bjorken] at the moment was out of town. I said I didn't know a cause, why don't you ask him? They said something about that he had tried to explain it, but it wasn't very clear or something. They showed me the data. They showed me that it fitted this thing, this function business that he had. And then I went home after they showed it to me—" *home* being his motel room "—and I began to think about it." Abruptly he comprehended that scaling was caused when an electron was going fast enough to interact with an individual parton, not the proton as a whole. The complicated, messy interaction with the proton is then suddenly replaced by a much cleaner collision between two pointlike objects. "I suddenly realized that I was very *stupid*, that I'd been working on a doubly hard problem, that the [collision of] two protons is harder [to understand] than one proton being hit by an electron. I thought I was rather dumb. I had not thought of the easy experiment to analyze, I had always tried the harder one. I had some half-assed ideas about it, but when they showed me this experiment, I started to analyze it from my other point of view, which was the partons. And believing that that was what Bjorken had done, I returned to them and said, '*This* is why Bjorken says there should be scaling.'"

It wasn't, exactly. Bjorken returned the next day and discovered that Feynman, whom he had never met, was in a state of great and alarming enthusiasm over scaling. "We got together," Bjorken recalled, "and he said, '*Of course* you know this,' '*Of course* you know that.' And I said, 'Uh, sort of.' Some of the things I knew very well and other things I didn't and other things I knew better than he did, but in any case he just went away."

The laboratory was thrown into a state of excitement by Feynman's visit. His analysis was intuitive and transparent, and, as Henry Kendall said, "Feynman being Feynman, if Feynman says that you fellows are observing pointlike constituents in the nucleus, then you pay attention."[43] Whereas Bjorken expressed his ideas in the language of current algebra, Feynman had a simple picture: The proton actually consists of a swarming mass of particles. We don't know what they are now, but they move around freely inside the proton. When an electron is going fast enough, it sees the partons as almost motionless because of the slowing of time associated with relativity. Whereas under ordinary circumstances, the electron interacts with many partons— the proton as a whole—it now goes by fast enough to hit just one. Thus by carefully watching how the electrons rebound, you could deduce something about the nature and distribution of these partons. Chiefly, what was learned was that the partons moved around inside the proton as freely as balloons at a birthday party.

Pleased, Bjorken, too, went to work on partons. With a colleague, E. A. Paschos, Bjorken started working on trying to understand other aspects of the phenomenon: what happened when photons or neutrons hit partons, and then how the struck parton turned into a lot of hadrons. In the middle of the work, they ran into a minor question of etiquette: Feynman never got around to publishing anything about partons. It seemed *comme il faut* for the SLAC team to wait for Feynman to announce the idea himself before printing work based on it. Bjorken was too shy to get on the phone and ask Feynman. Eventually, they published the parton model first, with a prominent and somewhat embarrassed acknowledgment to Feynman.[44]

A more serious disagreement arose when Paschos asked Feynman if, as seemed plausible, the partons were quarks. No, Feynman said. "I did not feel too sure about this, basically because I was worried by the fact that my theory was for partons, which if hit hard enough would come out with only little further interaction. Quarks, the theory went, couldn't escape the nucleus—if they went too far they would find large forces bringing them back." Feynman told Paschos not to jump to conclusions.[45]

Down the hall, Gell-Mann thought this was ridiculous, and said so. Frequently. "The whole idea of saying that they weren't quarks and antiquarks but some new thing called 'put-ons' seemed to me an insult to the whole idea that we had developed. We'd put in so much effort inventing quarks and antiquarks and so on, and here Feynman and other people were talking about these put-ons as if they were something different." A coolness developed between the two men. "It made me furious, all this talk about put-ons. The implication was that there was something fundamental about the scaling behavior of put-ons, and this was not true. It's only a relatively good approximation to the correct theory of quarks and gluons." Gell-Mann paused, reflected; his voice grew amiable. It was getting late in the afternoon. Soon he would startle us and a waiter in a Chinese restaurant by ordering dinner in what was evidently serviceable Mandarin. "But there was a very important physical point that was made by Bjorken," he said. "And then explained, I guess, in a somewhat more—ah, what shall I say?—popular manner by Feynman. And that was that deep in the interior of the nucleon the quarks were almost free."[46]

In the long row of candles that had to be lighted before the strong force emerged out of darkness, the scaling experiments set the match to the one closest to the center of the mystery. Scaling emerged slowly, through a series of meticulous experiments that, one by one, checked and cross-checked various aspects of the phenomenon until the early 1970s.[47] In 1990, Kendall, Friedman, and Taylor shared a richly-deserved Nobel Prize. In their quiet, tenacious, matter-of-fact way, the experimenters overturned previous dogma and established with icy rigidity that there were freely moving *somethings* inside the neutron and proton, little wheels inside bigger ones.

Somewhat paradoxically, scaling did not create a conversion en masse to the quark model. A curious dilemma: The simplest, most naïve quark model, the picture of quarks as little objects floating inside the particles we see, was amazingly good at accounting for many of the peculiarities of the hadrons and was backed by an impressive number of experiments at SLAC. The sole feature the quark model failed to account for was the quarks themselves, which stubbornly persisted in their improbability. How could they bobble helter-skelter inside a proton yet be stuck so firmly together they could never be seen as individual entities? Weinberg, for one, was extremely skeptical. "The only way you could understand why we weren't seeing physical quarks directly was to imagine that quarks were really very heavy, and somehow inside the nucleon the binding energy almost exactly canceled their rest mass, so that the neutron and proton weighed much, much less than the quarks inside them. That seemed absurd dynamically; I couldn't understand how that could be. And if that were true, then the success of the quark model wouldn't be explained. You know, the successes of the quark model had to do with, for example, calculating the magnetic [behavior] of the neutron and proton on the basis that they're just bound states of three fairly independent quarks. And if the quarks really were very heavy and very tightly bound, then their existence wouldn't account for the success of the quark model." He laughed good-humoredly. "It was a paradoxical situation. In fact, I remember Murray Gell-Mann at a conference saying that the discovery of [observable] quarks would be very exciting and it would be very revealing and very important, but the one thing it wouldn't cast any light on is the success of the quark model."[48]

V

The Great Synthesis

16
Killing the Hydra (Part II)

IN THE FALL OF 1969, HARVARD ACQUIRED TWO NEW POSTDOCTORAL FEL-
lows, Luciano Maiani and John Iliopoulos, from Rome and Paris, respectively. The campus was quiet, resting between spells of activism; police and
petitioners were gone from the streets, replaced by the usual Cambridge
panhandlers. Lyman Laboratory, which houses the physics department, is a
few minutes' walk north of Harvard Square, tucked into a corner by Oxford
Street. A brick building undistinguished by any particular style, Lyman
emanates the decaying Victorian gentility that suits academia. Harvard's
physics department was then among the second rank, but near the head of
the second rank; Maiani had chosen to go there, rather than a top university,
because he feared he would never have access to the physicists in a place like
Caltech. Iliopoulos came because he knew Glashow. Both of the new men
were field theorists; both were interested in weak interactions; and both
knew that their approach to physics made them members of a vulnerable
minority. Like young men everywhere, they had hopes of getting into something big, something important. And—it cannot be put more exactly—their
dreams came true.

That Iliopoulos and Maiani would fire the first volley in a barrage which
would rapidly transform particle physics seems all the more remarkable
when one considers the state of play as they entered Lyman Laboratory that
fall. Physicists spoke of four interactions: gravitation, electromagnetism, and
the weak and strong forces. Each was described by a separate body of theory
expressed in a distinct mathematical language; passing from one to another
was like shutting a book of Balinese folk tales and picking up a juridical code
in medieval French. Gravity had been brilliantly accounted for by Einstein
but remained resolutely mismatched with quantum mechanics. Quantum
electrodynamics, the first and most successful quantum field theory, explained electromagnetic phenomena to umpteen decimal places, but its insights did not seem readily applicable to any of the other interactions. Although a $V - A$ theory described the weak force, the physical mechanism of
the interactions was not understood—did they occur via virtual W particles?—and the equations remained stubbornly unrenormalizable. Finally,
the study of the strong forces was in approximately the same state as that of
contemporary art: a tidal wash of fads and short-lived movements, band-
wagons driven by briefly dominant personalities, an uncertain jittery market

311

of anxious theories with little agreement on the simplest assumptions. Such confusion did not seem unusual to Iliopoulos or Maiani; it was the normal state of the science they had grown up with. They hoped to see a few cobwebs swept away in the course of their careers. The amazing fact is that within three years of their arrival in Harvard the first modern unification theory was proposed.

Iliopoulos is Greek by origin, a man whose thick, wavy hair is black enough to make his skin seem pale.[1] He did his undergraduate work in the early 1960s. There being no place in Greece to obtain a good graduate physics education, he moved to France, where tuition was free. He did not speak a word of the language. Annoyed with the relentlessly practical engineering classes he had been forced to take in Greece, he opted for theory, rather than experiment. His faculty advisers were Claude Bouchiat and Phillipe Meyer, two of the more important figures in the slow postwar resuscitation of French physics. The French school was known for being mathematically oriented—excessively so, perhaps. Bouchiat and Meyer were more interested than most in applications; Bouchiat's wife, Marie-Anne, is a careful experimenter with the European sense of craft.[2] Bouchiat taught Iliopoulos to sail close to the solid shore of the experimental data. He also interested his student in the weak interactions, which the successes of the eightfold way and current algebra had reduced to a minority interest. Because foreigners then could not hold tenured positions in French universities, Iliopoulos went to CERN for a couple of years before getting his *doctorat d'état* in 1968.

At CERN, he worked with Jacques Prentki, the house theorist, who was also interested in weak interactions. A nervous, fidgety man with a heap of white hair and an extraordinary Franco-Polish accent, Prentki shared Iliopoulos's lack of interest in Regge poles and enthusiasm for field theory. They decided that something should be done about weak interactions. In Prentki's small, bare, smoky office, they set to work. "Until that time, weak interaction theory was very phenomenological," Iliopoulos remembered in the course of a long conversation. He was speaking in an office at Rockefeller University, where he was paying a two-months' visit. It was early afternoon; Iliopoulos, a late riser, was sipping his morning cup of coffee. He wore his sweater buttoned Gallic-style along the shoulder. "You had this Fermi theory, which you could use to compute to the first order, and then you ask questions about what happens. And we had learned all these things about symmetries and current algebra and all that. A small number of people felt it was the right time to really ask the right questions about what is the *theory*. First of all, we had run out of processes to compute. All the weak interaction processes were more or less computed already—this was the late sixties— and people really felt that we now had enough experience to go beyond.

Since we had no idea what would be the right theory, this involved a kind of step-by-step approach."

("We believed physicists should have their heads hanged in shame," Prentki said. "Forty, fifty years since Fermi, and what have they learned fundamental about weak interactions? *Nothing!* V − A is wonderful, beautiful, but that is really just making Fermi violate parity." What about the SU(2) × U(1) of Glashow, Salam, and Weinberg? "We knew about that. Shelly did his work in Europe, some here in CERN. I remember seeing Weinberg's paper the first time. Whoof! I not only read the paper, I lectured on it in Paris. But these papers, they do not come with ribbons saying, 'Take me, I am correct.' I saw there was something very fine about it, but the situation was also not very satisfactory. He had a nice way of calculating SU(2) × U(1), but it was not renormalized. There's something at the end about hoping it can be renormalized, but that is not physics, that is *hoping*. Iliopoulos and I said, 'We are going to do physics. We are not going to hope.' ")[3]

The blessing and the curse of the Fermi theory and the V − A theory erected on its back was that they artfully avoided the question of how, exactly, the weak force did its work. As soon as V − A had been announced, Gell-Mann and others had realized that the form of the interaction was compatible with the existence of a positive and negative vector boson, but this had not led to substantial progress, still less a resolution of the endless infinities. Iliopoulos, Prentki, and Bouchiat decided a first step would be to sort them out.

"We were trying to see if we could arrange those divergences," Iliopoulos said. He spoke quickly, sentences pouring out in a soft urgent rapid flow. "We classified them into leading divergences, next-to-leading divergences, next-next-to-leading, and so on and so forth." Others had taken the road before them.[4] The idea was to set up an energy cutoff, and see what happened if they said the theory was only good to a certain energy level. "From a mathematical point of view, it didn't make much sense, because all these things were infinite anyway, but they were less infinite when we put a cutoff. So you have the things that would diverge most strongly with the cutoff, then the things that would diverge with a power less." Strange things started to happen. "It was a phenomenological theory. You would say, suppose that the cutoff tells you up to which energy you could trust the theory. You can't trust it at energies that are much more than the cutoff. Then you try to guess the order of magnitude of the cutoff. If you do nothing to the [V − A] theory, the cutoff comes out to be very, very small—so small that you can't trust the theory *anywhere*." What happened when they just went ahead nonetheless was that the equations worked out so that the weak interactions became much more powerful. "We looked at that and we said, 'Well, suppose that they *are* strong interactions.' " In other words, what if the weak and strong forces were somehow the same thing? "What kind of bad effects would that have? You would produce parity-violating strong interactions, strangeness-violating strong interactions, which don't exist." They discovered that if they put in a clever symmetry-breaking mechanism, they could sweep the parity violation under the rug. This also got rid of the effects of the leading divergences.[5] "We were left with the next-to-leading divergences. They are equally bad, but you sort of say, 'Well, it's one power less, who cares?' " But even that didn't work: The next-to-leading divergences, although much smaller, made rare processes occur frequently. "So you had to get rid of those, too, hoping that somehow this would get inside, let you see the right thing." The whole procedure was intellectually untidy, mathematically unrigorous, but physics at the edge is frequently that way.

Glashow was visiting CERN, and Iliopoulos showed the half-complete work to him. Glashow, too, was pushing around the infinities in the weak interaction, working by brute force, trying to squeeze them into the edges where they wouldn't show. The two men decided to collaborate when Iliopoulos came to Harvard, and agreed that they could use the criterion of renormalizability to try to figure out a real theory. At the beginning of November, they were joined by Luciano Maiani, a young Italian field theorist. The three men's interests meshed immediately: All were working with the same level of ignorance on similar phenomena. Maiani brought some suggestions from Europe; Glashow and Iliopoulos rejected his ideas with the bluntness customary among physicists. Within two months, they had come up with a pivotal piece of the standard model.

The atmosphere around the University of Rome had the stillness of shell shock; the days when tear gas and terrorism accompanied students to class remained in memory, and the gateway to the school on Piazzale Aldo Moro was flanked by sullen *carabinieri*. At the time, tanks had only recently stopped being part of campus life. The university was large, desolate, creakingly underfinanced; the heavy neoclassic buildings of the science wings were sprayed with political graffiti and surrounded by indifferently tended islands of grass. Luciano Maiani's office was on the second floor, in the middle of a twist of dusty corridors. A graduate student guided us through, silent as Charon, turning imperturbably this way and that in the dim light.

Maiani was waiting for us, a cigarette burning in a sixteen-millimeter film canister that served as an ashtray. He has a large head, expressive dark blue eyes, and black hair that is swept back from a high forehead sticking out to the side like the cartoon image of an orchestra conductor. His voice is deep, penetrating, an unmistakable peninsular bass. "Want one of these?" he asked, pushing the cigarette box across the desk. They were MS, the state-owned brand. "MS, *morte sicura*," he said. "They are disgusting."

Maiani's family is from San Marino, the minute city-state in northern

Italy. He was born, however, in Rome, and studied there as an experimenter. In the midst of writing a thesis on solid-state detectors, he decided to move to theory, and in particular the theory of weak interactions. There was something of an Italian tradition of weak interactions, a chain of work that started with Enrico Fermi, was interrupted by Mussolini, and reimported by Raoul Gatto from the United States. In 1963, Nicola Cabibbo, an Italian then at CERN, figured out a general formulation for hadronic weak decays. The strange particles, as usual, did not behave exactly like the ordinary particles, and Cabibbo was obliged to introduce a parameter that, speaking crudely, related the probability the weak force would change the strangeness of a particle to the likelihood that it would not. This parameter became known as the "Cabibbo angle," because of the trigonometric law that ratios can be expressed in terms of some angle. (The measured value of the angle is about thirteen degrees.) The angle was absolutely necessary to calculate weak interactions, but its existence was one of the mysteries surrounding them.[6]

Five years later, Cabibbo and Maiani decided that there should be some way to compute this angle, rather than simply letting experimenters find it. "We thought we had a solution," Maiani said. "It was a very strange idea, which was connected with some ideas that Gell-Mann had. That is, that in fact the weak interactions are strong." He had his feet on the desk, his head resting against the cool metal wall of a file cabinet. In unknowing parallel to Iliopoulos, Bouchiat, and Prentki, hundreds of miles north in Geneva, Cabibbo and Maiani had hit upon the notion that the divergences in some way tied the weak and strong interactions together. "Now, you worry whether the parity violation which is in the weak interactions would propagate to the strong interaction improperly. But this can be exorcised. We came out with a theory of the Cabibbo angle which gave a good result.[7] But this theory had a problem, which we became aware of just in the summer of sixty-nine, while I was packing and moving to Harvard." The problem was that a consequence of their ideas was the prediction that a long-lived K meson should decay often and easily into two muons. It does not. Maiani arrived in Cambridge and was delighted to find Glashow and Iliopoulos had similar interests—so similar, in fact, that they immediately informed him that his ideas could not possibly be correct.

"We started discussing furiously," Maiani said. "Because I was defending my work, and they were attacking it, there was always a lot of discussion in which two people were attacking a third one." He laughed. "I was a fool."[8]

"We finally convinced him that what he had done with Cabibbo was not relevant," Iliopoulos recalled. "I don't know how long it took, but he finally admitted it." Nonetheless, Iliopoulos and Glashow liked the essential idea of computing the Cabibbo angle by getting rid of divergences. From his thesis, Glashow also knew that using a Yang-Mills formulation also got rid of infinities; he had even claimed, erroneously, that local gauge symmetry eliminated all of them. "Every day," Iliopoulos said, "invariably, one of us would come up with an idea. Then the other two would prove to him he was wrong. We tried all sorts of different recipes, and nothing worked. There were long calculations and trying to—I mean, there was this angle that we tried to untangle—it was impossible. This took us some time, and we got frustrated. And then we really convinced ourselves—we didn't have any rigorous proof—that there was no solution in the then-standard model of three quarks and four leptons. We couldn't renormalize it. So then we started to change things in other ways. The first thing we tried to do was put in more leptons."

After playing with leptons for a while, they gave up and plugged in a fourth quark, which Glashow, remembering his 1964 paper, again called charm. Almost at once they realized they were onto something. All three men liked the obvious tidiness of four quarks and four leptons. Moreover, adding in an extra quark to the equations made many more divergences vanish; the theory became much neater. Finally, Glashow at last realized that charm canceled the possibility of neutral currents for strange particles, neatly removing one of the major problems of his $SU(2) \times U(1)$ model. He told us, "What I don't understand is why I forgot that charm was relevant for so long. But how often can you forget something for six years and have no one take you up on it? The next paper, with Iliopoulos and Maiani, uses that observation to solve a major problem of physics. But nobody else was aware of the *problem*, let alone finding the solution."[9]

The *problem* came from Maiani's work with Cabibbo. Their Lagrangians predicted K decays that didn't happen, but, the three men realized with astonishment, *so did the regular* V − A *theory*.[10] Every now and then, strange particles, such as Ks, should emit both a W^+ and a W^- and through a complicated interaction decay into two leptons. Such an interaction would mimic the final result of a weak neutral current. (The weak neutral current would arise when the K emits a Z^0 that in turn decays into a lepton pair.) Glashow, Iliopoulos, and Maiani calculated that experimenters should already have seen such decays, and took their absence as an indication that charm was on the job. They were hugely pleased; physicists routinely predicted new particles, but it was not often that one got a chance to postulate the existence of an entirely new state of matter comprised of whole families of charmed particles.

Excited, the three theorists took their work to the office of MIT theorist Francis Low, and talked it over with anyone who came in. They mentioned that charm might fit into an $SU(2) \times U(1)$ model. One of the physicists who wandered by was Steven Weinberg. Glashow laid out the case for charm, mentioning that it might fit into some kind of Yang-Mills scheme. A spectacular failure in communication ensued; Weinberg did not see the relevance of charm to his work three years before. Indeed, in Maiani's recollection, he reacted with instinctive dislike to the extreme lack of economy implicit in introducing whole families of unknown particles to shrink the reaction rate of an obscure class of particle decays. His momentary lapse—together with those of Glashow and Iliopoulos, who knew of Weinberg's paper—meant that the three men continued to put the masses in by hand, instead of employing symmetry breaking. Which meant in turn that the first draft of their paper, which contained the statement that an $SU(2) \times U(1)$ Yang-Mills theory with four quarks might be renormalizable, was bounced by the referee at the *Physical Review*. They excised the offending sentence, inserting instead the remark that the "more daring speculation" of a Yang-Mills theory at least "does not make the theory more divergent." Although the new phrasing was almost as unsupported as the old, it was allowed, and "Weak Interactions with Lepton-Hadron Symmetry" appeared on October 1, 1970.[11]

All three were convinced of the existence of charm, but Glashow's certainty was the most ebullient. Glashow, Maiani, and Maiani's wife went to a Cambridge hangout with the curious name of Legal Sea Food. Over the meal, Glashow informed Maiani's wife that he was extremely pleased with what has come to be known as the GIM mechanism for avoiding strangeness-changing neutral currents. She asked if it was important. Glashow replied, "It will be in the textbooks." They gave a seminar on charm to the experimenters at the Cambridge Electron Accelerator, a small machine run by Harvard and MIT. Glashow flatly informed the audience, "From now on, everything is chemistry. We understand everything now. All you have to do is go out and find neutral currents and charm, and then everything is wrapped up."

The experimenters didn't see it that way. Charm seemed like the worst sort of theoretical fantasizing: In order to make sense out of an as-yet unformulated theory of the weak force, the GIM paper reached into the domain of the strong force and created a fourth quark—when few people felt sure of the existence of the first three. Nonetheless, charm was to Glashow's taste; it matched his physical intuition. When Maiani decided to return to Italy after his wife became sick, he was able to obtain passage for the family on a luxury ship from Boston to Rome. They decided to have a farewell party. Glashow arrived with a trash basket full of ice and champagne, a brace of cigars in his pocket. Toasts were drunk. Talk got loud; it was another one of the expansive collegial gatherings that make the science go round. At a certain point in the proceedings, in Maiani's memory, Glashow, Iliopoulos, and he cornered MIT experimenter Sam Ting. They told him about charm and liquidly informed him why it was important. If electrons and positrons collided with sufficient energy, they said, the burst of energy from matter-antimatter annihilation should produce charmed particles. Ting smiled skeptically and said nothing. Maiani told him charm was the next principal item on the experimental agenda; the words "Nobel Prize" seem to have been mentioned. And Ting smiled and said nothing. Experimental physics is a serious business, and he had no interest whatsoever in spending years chasing after some lunatic speculation that would be disowned by its theoretical parents a few weeks later.[12]

After Maiani's departure, Glashow and Iliopoulos determined to show the referee that the divergences canceled even in a Yang-Mills theory. They spent the first six months of 1971 matching infinities against counter-infinities. The work was arduous, painstaking, deeply mathematical, exactly the kind of highfalutin abstract work Glashow and Iliopoulos loathed. They proceeded without a clear plan, trying to match up each individual tangle in the thicket with another: the head-banging approach. Eventually they showed that the situation was not automatically worsened by going to a Yang-Mills theory.[13] "It was the most intelligent paper I ever wrote," Iliopoulos told us. "The most intelligent that either Shelly or I have ever written. It's a completely obscure paper. Nobody ever read that paper and nobody ever will read it because it's so difficult."

In June, Glashow presented the work thus far to a small conference on renormalization at the Centre de Physique Théorique, in Marseille.[14] During the meeting he received a rude shock. "There we are, laboriously canceling out infinities, when this guy Veltman comes up to us. And he says, 'My young student 't Hooft has solved the problem of renormalizability of gauge theories. All the work you've done for the past year is a waste of time!' And he was right. His student had spontaneously re-created the whole SU(2) × U(1) theory, and renormalized it to boot."[15]

□   □   □   □   □

The Dutch physicist Martinus Veltman was one of the few who had clung unswervingly to quantum field theory throughout the 1960s. Now at the University of Michigan in Ann Arbor, stubborn, independent, isolated "Tini" Veltman labored on renormalizing Yang-Mills theories at the University of Utrecht for five years despite the almost complete indifference of his fellows. Veltman is a proud man with a full rack of dark hair and a direct manner who identifies whatever he doesn't like as "baloney" or "crap." We met him in Robert Serber's office after he gave a seminar about alternatives to the Higgs boson. Unlike some of his colleagues, Veltman thinks that spontaneous symmetry breaking, the Higgs mechanism, and the rest of it is, not to put it too plainly, a lot of hooey—at best an approximation to some more elegant truth. When we entered the office, Veltman was slumped in Serber's old red leather chair, half glasses riding down on his nose, a *toscano*-style cigar nearly finished in his mouth. His beard was trimmed in the

European manner around the adam's apple, patches of white marking the jawline. His hair rose back from a large sloping forehead. He spoke in an English transfigured by a Dutch accent, shimmeringly pure vowels shining like brass through his beard. His affect was melancholy; he began by regretting that he had agreed to speak with us at all.

"I don't like to talk about this. It opens up old wounds. I just want to close it off and continue working. You say it and say it and nobody listens, so after a while. . . ." He looked distractedly about the office. *"No matter what I say*, it sounds like sour grapes," he said, earnestly, believably. "You try to solve a problem and sometimes you don't even know that you've solved it. In your innocence, you do things and you aren't even aware of what are doing, and nobody—" Veltman waved the cigar disgustedly. "Ach, here I am, talking about it again.

"I started this business in the spring of 1968. I spent a month at Rockefeller University, doing nothing but thinking about the problem of weak interactions. Where did you go for the theory?"[16] He had arrived at that position in a highly idiosyncratic manner. After writing his thesis on formal properties of field theories, Veltman became interested in current algebra, which Gell-Mann had abstracted from field theory and then thrown away the field theory. (One recalls Gell-Mann's use of a metaphor from haute cuisine: cooking a slice of pheasant between two slices of veal, then discarding the veal.) The residue was a set of relations among currents that supposedly could be manipulated to produce many of the rules and predictions of field theory without the attendant ambiguities and mathematical difficulties. In fact, however, field theory is so beset with technical complexities that careful calculations often bring unwelcome surprises; the current algebra taken from field theory had problems of its own. Years before, Julian Schwinger showed with characteristic force that two methods of doing the same somewhat obscure calculation in field theory produced different answers unless additional restrictions were factored in.[17] The same problem showed up in current algebra in the form of extra terms called "Schwinger terms." Mathematics being a consistent subject, one should not get two different answers to the same question. No matter how you add, two plus two should always equal four, not three or five. In 1966, Veltman showed that adding in two extra fields to the current equations canceled out the Schwinger terms and restored mathematical sanity, albeit at the cost of sacrificing some of the useful features of current algebra.[18] That was not the end of it, however. When Veltman sent a preprint to the theorist John Bell, a friend at CERN, Bell realized almost immediately that Veltman had proceeded by "avoiding as much as possible the creaking machinery of field theory and simply *imposing* gauge invariance" on the equations.[19] Although Veltman hadn't known it, the fields that he had introduced as abstract technical devices were in fact Yang-

Mills fields applicable to weak and electromagnetic interactions.[20] Gell-Mann had based current algebra on Yang-Mills and then discarded the Yang-Mills; coming round the other way, Veltman had deduced the veal from the flavor of the pheasant without knowing how it had been cooked.

Easy, in retrospect, to state with a single phrase. In fact, Veltman didn't know what to make of Bell's paper for almost a year. Confused, Veltman left Utrecht to spend a month at Rockefeller, where he decided to take advantage of the peace and quiet to figure out the article. One day in his Rockefeller office, while staring out the window at the 59th Street Bridge over the East River, Veltman suddenly grasped the heart of Bell's argument: that local gauge invariance was *the* key to a successful theory of the weak interactions. Current algebra, Bell was saying, is just a mask for a more fundamental Yang-Mills theory.[21]

Why hadn't other people realized the significance of Yang-Mills to the weak interactions? Because, Veltman said, nobody could show the theory was renormalizable. Why couldn't they show that? Several reasons. One was that in addition to the ordinary self-energy-type infinities, the intermediate vector bosons had electric charge, and thus could interact with each other. This meant that calculating involved horribly complex Feynman diagrams with three and four bosons looping and intertwining. Self-coupling was not the only cause for despair. The intermediate vector bosons must have mass, and they do not in a pure Yang-Mills theory. When one simply awarded them a mass, as Glashow had done, the picture became even darker.

Veltman decided that if nature had chosen to express herself in terms of local gauge invariance, then Yang-Mills theories could not truly contain infinities. At this point he was faced with the choice of examining pure, massless Yang-Mills theories or theories with extra mass terms. Massless theories were obviously not realistic, so Veltman elected the latter. He didn't know at the time that Salam and Komar had shown seven years before that massive Yang-Mills theories were nonrenormalizable; Veltman's ignorance was fortunate, for their proof was incomplete.

In a local gauge theory many of the infinities match up neatly with ones of the opposite sign. The trick is to show that in the vast sea of possible diagrams not one divergence is left uncanceled, and that no hidden infinities spoil the game. Veltman realized that pairing infinities the conventional way—lining up leading divergences, next-to-leading divergences, and so on—was going to be fruitless. Radically new methods were needed. He found them, but at the price of making himself incomprehensible to almost every other physicist in the business.

Progress in physics often depending on discerning the right hair to split, Veltman first drew a vital distinction. In the diagram pictured above—a "one-loop diagram," as it is called—an electron emits and reabsorbs a virtual photon. In the brief interval of the photon's life, it can have any momentum whatsoever. Its existence is ruled by the uncertainty principle; standard rules of cause and effect do not apply; calculations seem to produce an infinity. Such a virtual photon is said in the jargon to be "off the mass-shell." Before and after the virtual interaction, the electron is on the mass-shell. "You try to play this renormalization game," Veltman said, "and you realize these diagrams have infinities that disappear on the mass-shell, but not off of it. These off-shell divergences are utterly hopeless, unless you change the rules."

We asked how one changed the rules.

"It's complicated," Veltman said. He stood up, moved heavily to the blackboard. "I think in diagrams, not equations. You have a loop, like so—"

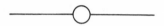

"—you subtract a negative infinity—"

"—the result appears as mass. You don't follow. In gauge theory, you have a freedom in the choice of variables—yes, you know this. Good.

"In classical physics, a gauge transformation takes you from one formulation of the theory to another. That is gauge invariance, in a way. All the formulations are the same, they give you the same result. The question was, How does this take place in quantum field theory? This people could not do. People thought that the diagrams were slightly different for each gauge transformation, that you just fiddled it, but really they are totally different diagrams." Revolution, not reform. "You can't just adjust the Feynman diagrams for gauge theory. The rules are different. Each transformation means that you have to have a totally different set of rules. You had to derive a technique to go from unitary rules, where every line corresponds to physical particles, to others, where diagrams have fewer divergences off the mass-shell." *Unitary rules* are the rules of the real world, where particles are physical objects and all probabilities add up to one hundred percent. Veltman went from unitary rules to a new gauge, a gauge of his choice, where the divergences matched neatly off the mass-shell but very odd things happened in the Feynman diagrams. Chief among the strange things was the appearance of objects of negative probability, "ghost" particles with the impossible quality of having less than zero chance of being around.

In the early 1960s, the apparent prediction of such entities in quantum electrodynamics contributed to the climate of disbelief in field theory. Here, however, Veltman decided that they didn't matter if they disappeared by the end of the calculations. If the ghosts always stayed off the mass-shell and never made it into the final state of an interaction, it didn't matter whether they appeared in the calculation. In the early summer of 1968, he wrote a paper in which he used these ghosts to get renormalized rules for the simplest infinities in massive Yang-Mills theories—one-loop diagrams.[22]

"The immediate reaction was essentially zilch. I was told I was nuts. My colleagues told me I was disappearing into a black hole. Sidney Coleman [of Harvard] told me, 'Tini, you're sweeping an odd corner of the weak interactions.' Most other people were doing fancy things—even Mr. Weinberg was doing no work on gauge theory. They were busying themselves with Regge poles, phenomenology, all kinds of funny things." What made him stick to field theory? "Field theory was my own belief. I have always gone my own way. It is my natural habitat, it is as if you ask a fish why he likes water."

We once asked David Politzer, a Caltech physicist who was a first-year graduate student at Harvard at the time of GIM paper, if the thought of constructing gauge field theories had really been as dead as those in the discipline recalled. In addition to being an active physicist, Politzer actively watches the field, and has many enthusiastically expressed notions about what has happened to elementary particle physics. "I took a course from Glashow in 1970 on weak interactions," he said. "Aside from it being quite inspiring, and aside from the fact that I didn't understand very much, I do remember that in a whole year course on weak interactions, he briefly addressed the question of renormalizability—he mentioned an absolutely disgusting model put together by Gell-Mann, Low, someone, someone, someone, and someone—and did not even *mention* the work of Weinberg. This was 1970, and this was Shelly Glashow! He didn't even mention his own work, for which he supposedly got the Nobel Prize. What had happened was that people had been sort of burned by their lack of understanding of gauge theories." He named a few seared physicists. "They wouldn't touch it with a ten-foot pole! And it is true that Tini Veltman was the only one who kept slugging away at it seriously. Then things turned around."[23]

In August 1968, Veltman went to the Laboratoire de Physique Théorique et Hautes Energies, at Orsay, outside of Paris. The city was still reeling from the student revolution, and Veltman found the intellectual climate interesting if remote from his own views. He was told in Orsay that several Soviet physicists had managed to renormalize massless Yang-Mills to any order—two loops, twenty loops, one hundred loops, whatever they pleased. Although Veltman had chosen to ignore such theories as physically irrelevant, the Russians' resounding success raised his hopes for the massive case.[24]

Veltman revamped his one-loop renormalization proof using the Russians' methods, and commenced working on the next level—loops within loops, two-loop diagrams. These were much more complicated, and Veltman found himself engaged in the sort of sprawling computation that induces math anxiety. He became an expert in little-known mathematical techniques, especially a trick called "path integrals." He prepared computer programs to work on two-loop diagrams and let the machine take over—but the printouts showed that at two loops the massive theory was damningly unrenormalizable. His hopes were evidently dashed; years of work seemed to have reached a dead end; he had indeed swept clean an odd corner of the weak interactions.[25]

Events from this juncture were shaped more by the bent of Tini Veltman's character than compelling scientific logic. This supremely stubborn man kept going, his hopes resting now on a paradox. When he tried the massive theory with an intermediate vector boson of zero mass, the answer was not the same as in the massless theory. One equation had a particle of no mass, the other a massless particle. Zero is zero, and surely the two cases should be identical. They weren't. One was full of infinity, and the other wasn't.

By this time, he had acquired a graduate student named Gerard 't Hooft, a trim, slightly built man whose finely cut youthful features were later filled in by a prominent, barely reined-in moustache. In his inexpensive gray suit, 't Hooft might easily pass for a salesman of electrical appliances, an impression dispelled the instant he opens his mouth: 't Hooft is a very clever fellow. (The 't is an abbreviation for *het*, which corresponds loosely to the French *le*.) When 't Hooft completed his undergraduate studies at Utrecht, Veltman was the only theoretical particle physicist around, and 't Hooft asked him to be his adviser. Veltman agreed, and gave him lecture notes to write up.

Gerardus 't Hooft was born in 1946 and raised in The Hague. Physics ran in the family: His great-uncle, Frits Zernike, won the Nobel Prize in 1953 for inventing the phase contrast microscope, and his uncle, Nicholas van Kampen, was a well-known professor of theoretical physics at the University of Utrecht. While in high school, 't Hooft often asked his uncle questions about physics. "He usually forced me to rephrase them many times until I was more precise," 't Hooft later recalled. "In this way, I learned more about physics than if he had given me straight answers." (On the other hand, being related to van Kampen had its disadvantages; 't Hooft first had to convince Veltman that he wasn't just his uncle's nephew.)

In college, he learned as much particle physics as he could; a classmate did an undergraduate thesis on spontaneous symmetry breaking, and through him 't Hooft picked up a few vague notions on the subject. In the summer of 1970, 't Hooft went to a physics summer school in Corsica where with impressive celerity he learned a great deal about renormalization and spontaneous symmetry breaking from the theorists Benjamin Lee and Kurt Symanzik. "I was very shy in those days, but I did approach both Lee and Symanzik with one question: How do you do this in a Yang-Mills theory? They both said they knew nothing about Yang-Mills theories."[26]

In the meantime, 't Hooft contemplated thesis topics. Veltman made several suggestions, none of which 't Hooft found particularly exciting. He was intrigued, however, by what Veltman himself was working on: Yang-Mills theories. In a first paper, 't Hooft tackled the massless case. He believed the Soviets had only argued that the theory was renormalizable, but not actually done it. This he set out to do, and finished early in 1971.[27] The calculation was the first explicit statement of the renormalizability of massless Yang-Mills theories, although the result was already accepted by the cognoscenti.

"When I had my first draft ready," 't Hooft told us, "Veltman was in Paris for the year. I thought I had done something important, and it was difficult to get his attention at that time. He came back several times and, you know, we had big fights even about this first paper, because he did not agree with the way I formulated things. In fact, my first version did not contain anything about the unitarity proof." That is, 't Hooft had not shown that he ended up with only physical particles and not ghosts.

Once he had done that, Veltman was still suspicious. There was yet another hurdle. Working together, John Bell and Roman Jackiw had discovered, in 1967, what was to be called an "anomaly" in pion decay. An anomaly shows up when, again, two different means of calculations produce two different answers. The anomaly in pion decay was discovered independently by Stephen L. Adler, and the resultant trouble is still called, alphabetically, the Adler-Bell-Jackiw anomaly.[28] Basically, any theory in which something like the Adler-Bell-Jackiw anomaly could occur is not gauge invariant—that is, it is wrong. Veltman had also come across the anomaly himself, but had not clearly understood its significance for a while. By 1969 he had realized its importance to the point of assigning it to 't Hooft as a subject for an undergraduate thesis. Previously, 't Hooft had known the anomalies were there, but thought they were of secondary importance. Now Veltman informed him they were paramount. There could be no anomalies hiding in the computations. Once 't Hooft had shown that, Veltman was ready to be impressed. The massless case would be licked. Of course, the theory would still have to be massive to apply to the real world.

Early in 1971, Veltman had a conversation with 't Hooft that he has never forgotten. In Veltman's recollection, the interchange went as follows:

M.V.: *I do not care what and how, but what we must have is at least one renormalizable theory with massive charged vector bosons, and whether that looks like Nature is of no concern, those are details that will be fixed later by some model freak. . . .*
G.'t H.: *I can do that.*
M.V.: *What do you say?*
G.'t H.: *I can do that.*[29]

"And this he could not believe," 't Hooft said years later, "because he had been working on the massive case himself for so long, and he was convinced that it couldn't work. He may disagree, but I think for me this was the very important moment. He was certain the massive case wouldn't work. And I said, 'Yes, but you need this extra particle. There will be an extra scalar particle around, and then it will work. I am convinced I can do it.' " The extra particle is the Higgs boson; *scalar* is a technical expression meaning that it has a spin of zero. On the surface, the suggestion was foolish. In most situations, tossing in a scalar particle ends up by producing ghosts. "He said, 'No, no, no, it doesn't work because you will get the wrong statistics.' Then I said, 'You *can* do it. I am convinced one can do it.' And so he said, 'Write down the Feynman rules, and I'll check.' By that time he had his computer program ready to check such things. He said, 'Actually, it's rather easy for me to check, if you have a scalar particle.' I wrote down the Lagrangian—the Higgs Lagrangian—but to make it acceptable to him I just wrote down that the mass was there, the interaction, and so on. It looks rather strange. It's not really recognizable as coming from a spontaneous symmetry breakdown. So it just looked like a lengthy Lagrangian with a bunch of extra terms in there which looked rather arbitrary and crazy."

To Veltman, 't Hooft had just shown the result, not the derivation? He had hidden the Higgs mechanism?

"The point was that he didn't want to hear about the Higgs mechanism. So he said, 'Just forget about the Higgs mechanism. Just give me the Lagrangian that you think works.' I ignored where it came from. I just wrote down the whole theory in the broken representation with the extra scalar particle and its self-interactions." Among the many odd-looking terms that flowed from 't Hooft's pen was a factor of four that came from the Higgs boson. "He said, 'Well, I'll take this Lagrangian. I don't believe your factors of four, they look crazy, but I'll take this Lagrangian and put in it my program.' "

Veltman traveled to Geneva to put the paper into the CERN computer. He went down in a skeptical frame of mind, worried that his student did not seem to understand the consequences of making an error in a paper that was sure to be widely discussed. In 't Hooft's recollection, Veltman soon called up in a state of excitement and said, " 'It nearly works! You just have some factors of two wrong.' But that was because he had not copied my factors of four, because he didn't believe them. So then he realized that even the factor of four was right, and that all canceled in a beautiful way. By that time he was as excited about it as I had been."

A few weeks later, David Politzer was at a summer school in Erice, Sicily, run by the Italian experimenter Antonino Zichichi. The school is on the rugged western coast, and the delights of the Sicilian countryside, the sea breeze, and the excellent cuisine have ensured its popularity. In quest of the perfect snorkel, Politzer one day went rowing with Sheldon Glashow. In the middle of the ocean, Glashow suddenly told Politzer that some Dutch student claimed he had renormalized Yang-Mills theories. Politzer asked if Glashow understood the work. He still recalls the response. "No," Glashow said. "Either this guy's a total idiot or he's the biggest genius to hit physics in years."[30]

Initially, 't Hooft had thought of Yang-Mills in terms of a theory of the strong interactions, and had actually played with a model in which the rho mesons, the lightest mesons with a spin of one, played the role of intermediate vector bosons. The rho's interacted—"mixed" is the term of art—with the photon. The relevant group was, by rank coincidence, $SU(2) \times U(1)$.

When Veltman took 't Hooft's proof to the CERN computer center, he also asked a resident theorist, Bruno Zumino, if anybody had done anything else with an $SU(2) \times U(1)$ model. Zumino is known for being one of the rare physicists who actually reads all the journals; he recalled Weinberg's 1967 paper, and Veltman phoned Utrecht with the citation. A small communications failure transpired, the wrong reference was copied down, and 't Hooft couldn't find the right paper, although he did find another Weinberg article that he thought was pretty good. Having been provided with the basic idea, he sat down and started to duplicate the model. He was fairly well along when Veltman returned with a photocopy. Turning through the pages, the two men realized they were in possession of something lovely: a fully renormalizable model of the weak and electromagnetic forces.[31]

As luck would have it, a big conference was scheduled in Amsterdam that August, and Veltman happened to have been given the task of arranging the theoretical talks. It pleased him no end to be able to schedule a special session devoted to renormalization. "Talk number one," Veltman said, "was given by Salam, who was trying to remove the infinities by using gravitation. Talk number two was T. D. Lee, who was introducing physical particles with bad properties. So first I let Salam talk about his baloney, then T. D. Lee about his attempts, and then came the real thing. I got up and I remember announcing to the audience that here was a renormalizable theory 'every bit as good as quantum electrodynamics.' That phrase is not in the proceedings, but I remember the moment very well."[32]

Alas, Veltman's pleasure was quickly to sour. His relationship with 't Hooft did not survive their joint rise to prominence; the teacher and student ultimately could not be together as equals. A sensitive, private man with strong views about the nature of physical truth, Veltman became irritated that the specific $SU(2) \times U(1)$ model became the focus of attention rather than the proof that an entire genre of theories was useful; unwilling to advertise that the student's success was based on his master's techniques, Veltman drifted into bitterness. He withdrew from the limelight, moving across the ocean to the University of Michigan, where with unaltered tenacity he followed his intuition that the Higgs mechanism is an ugly tear in the otherwise unmarred tapestry of field theory.

To Steven Weinberg, 't Hooft's proof just seemed like hand-waving. Then he heard that his friend, the Korean-American physicist Benjamin Lee, was working on it. Lee had taken Veltman's course on his path integral techniques in Paris three years before, and was one of the few people in a position to understand what 't Hooft was doing. Besides lending 't Hooft's work his considerable prestige, Lee spent most of August translating it into a form other theorists could comprehend.[33] "I was really impressed with that," Weinberg recalled. "I thought if Ben Lee takes this seriously, I've got to take it seriously." Working on the problem himself, he slowly grasped that an extraordinary period in the history of elementary particle physics had begun, and that in the middle of the tumult and celebration was nothing other than his old $SU(2) \times U(1)$ model.[34]

□　□　□　□　□

Gell-Mann, in the meantime, was thinking of anomalies. If properly understood, the Adler-Bell-Jackiw study of pion decay might at last make some progress on a small but interesting problem that had defeated theorists for some time. According to the $V - A$ theory and its extensions, the neutral pion ought never to decay into two photons. It does. Adler, Bell, and Jackiw argued that this discrepancy appeared because of their anomaly. Gell-Mann hoped that properly disentangling the anomaly would lead to understanding pion decay.

In late 1971, he teamed up with a young, enthusiastic colleague from Germany, Harald Fritzsch. The collaboration began auspiciously, several hours after a predawn earthquake knocked all the books off the shelves in the Caltech library and set askew the paintings in Gell-Mann's office. They were soon joined by William Bardeen of Stanford (who is not the Bardeen of the BCS theory). Bardeen had recently worked at Princeton with Stephen Adler (who *is* the Adler of Adler-Bell-Jackiw) on a further calculation of pion decay, and brought with him a piece of news from New Jersey.[35] In a talk at Columbia, Adler had shown that pion decay could be computed well enough to test various models of hadron constituents. Three fractionally charged quarks came off very badly, predicting low by a factor of three. Such disagreement, Adler said, means that *"the quark hypothesis is strongly excluded."* (Emphasis in original.) On the other hand, Han-Nambu quarks with three colors did reasonably well.[36]

Reluctant to have quarks strongly excluded just as scaling seemed to give evidence of their existence, Gell-Mann, Fritzsch, and Bardeen realized they could fix things up by adding color to the original fractionally charged quarks; that is, by making each of the three "flavors" of quark come in three additional types or "colors." Why not? "We gradually saw that that variable was going to do *everything* for us!" Gell-Mann said. "It fixed the statistics, and it could do that without involving us in crazy new particles. Then we realized that it could also fix the dynamics, because we could build an SU(3) gauge theory, a Yang-Mills theory on it."[37] The new idea was similar to Nambu's old model of color, except with Yang-Mills and without integral charges; it had, Gell-Mann thought, the feel of something very promising.

Just as quarks resolved many of the problems with the strong force but raised the question of their own existence, so color answered questions about quarks, but in turn raised the question of *its* existence. Although every hadron in the Universe was allegedly composed of colored building blocks, nobody had ever seen anything like color. The three men took recourse in Gell-Mann's old solution: Color is not observed because it *cannot* be observed. Only particles in which red, blue, and green or color and anticolor neutralize each other can exist; the world is made of uncolored, neutral, "white" composites. The tidiness of the scheme enchanted Gell-Mann. "I assumed that everybody was working on that," he said. "It turned out they weren't."[38]

In September of 1972, at a conference celebrating Fermilab's opening, Gell-Mann presented an almost complete picture of the strong interaction.[39] Hadons, he said, are composed of three types, or "flavors," of quarks, each of which can come in three colors. Up, down, and strange can be red, blue, or green. The quarks are held together by a gauge field whose quanta, gluons, are the vector bosons that mediate the strong force. Whereas the quantum of electromagnetism, the photon, comes in only one type and is uncharged, gluons come in eight types and are colored, like the quarks they interact with. Gell-Mann called the new theory "quantum chromodynamics"—*chromos* is the Greek word for "color"—in an obvious but loose analogy with quantum electrodynamics.[40]

Although Gell-Mann presented quantum chromodynamics in a bold and forthright manner to the Fermilab conferees, he had acquired a case of cold feet by the time the proceedings were printed, and the new Yang-Mills theory of color remains only as a kind of shimmer over the printed article, a promised theory that is never quite fully described.[41] Quantum chromodynamics was very new, and Gell-Mann was not certain he understood all the wrinkles. Oddly, however, he was not perturbed by the main cause of his colleagues' skepticism, that the quarks, colored or uncolored, could rattle around the proton like marbles in a bag and yet somehow be stuck together in perpetuity. Something that is loose can be dislodged; something that cannot be dislodged must not be loose.

One can say the words *asymptotic freedom* any number of times without causing them to discharge their informational content. The term stands as a kind of perfect emblem of the gap between scientists and non-scientists; rich with association and historical resonance to physicists of a certain age, the words are blank and impenetrable to the lay public. For a little while in 1972 and 1973, a select coterie of mathematically inclined theorists tossed about the notion as a panacea for all the ailments of strong interaction physics, telling seminars and conferences that asymptotic freedom was the wave of the future. All one had to do was prove it possible; when three physicists did, they gained renown. The standard model came together like a card game in which all the early play is but a setup for a rain of trumps at the end; asymptotic freedom was the last trick in the shuffle, when the hidden face cards came out and the round abruptly snapped into its final shape. It was discovered in 1973, after a complicated, quick little bout of competition and near-collaboration.

When we visited David Politzer, one of the three godfathers of asymptotic freedom, it was entirely natural to ask him what it was. Politzer is now at Caltech, a relaxed, loquacious, slightly cherubic man in a small office with a tiny window and a pair of brown slippers under the desk. He answered with a promptitude indicative of the number of times he had been asked this question. "Roughly speaking," Politzer said, "—I'm lying a little—it means that there is a unique class of forces that gets systematically weaker as the separation between particles gets littler. That allows you to have quarks when they're close together to be weakly interacting, and as they get farther apart, their influence on each other gets stronger instead of weaker. The 'asymptotic' means getting things close; 'asymptotic' means in the limit that there's no separation at all. But in the limit that quark separation goes to zero they are free particles. When quarks are sitting right on top of each other, they don't see each other, each doesn't feel the presence of the other one. It's only when you pull them apart that they feel the influence of their neighbors. And there's only this very small, well-defined class of gauge

theories which are asymptotically free among all possible theories that anyone has ever been able to imagine that could have that property." He fiddled
with his red vest. "Yes," he said firmly. "That's one way of saying it." Another
way is to say that asymptotically free theories, and only asymptotically free
theories, have negative coupling constants.

He was asked how he became involved with this problem.

"Ah, that's an interesting example of how science works," he said. "In
1972, I was looking for something constructive to do, having been in graduate
school already three years, passed my exams, done all sorts of things, but
really had nothing to work on." That summer, he went to a summer program
in Sicily where he learned about the Callan-Symanzik analysis, a field-the-
oretic attempt to understand scaling simultaneously invented by Curtis
Callan, Jr., of the Institute for Advanced Study and Kurt Symanzik from the
DESY theory group in Hamburg.[42] Its details need not concern us here;
Politzer merely thought it clever and inventive enough to apply to symmetry
breaking, which was not—is not—truly understood. "I was trying to do the
Callan-Symanzik analysis specifically on gauge theories to understand their
long-distance behavior or low energy behavior, hoping in the long run to
address the question of spontaneous symmetry breaking. I went down to
visit my adviser at Princeton, Sidney Coleman, to ask him what he thought
about it. [Coleman, a member of the Harvard department, spent the year at
Princeton.] He thought it was a good idea. And I asked him, had anybody
done it? He said not to his knowledge, but let's ask [Princeton theorist]
David Gross. So we go next door to ask David Gross. Gross says no, nobody's
done it. And then I discussed briefly with Gross why it wouldn't be so hard
to do. Even though it used to seem terribly complicated, if you have your
wits about you it's pretty straightforward."

David Gross's thoughts were far from symmetry breaking, although he,
too, was interested in Callan-Symanzik. The Callan-Symanzik analysis
sprang from a curiously powerful yet curiously neglected aspect of field
theory called the renormalization group, a mathematical technique originally
developed to relate the structure of the theory of quantum electrodynamics
at high energies to its predictions at low energies. The renormalization group
was first concocted by Stueckelberg in conjunction with a student, André
Petermann, but nearly all theorists know it through an article written by
Murray Gell-Mann with Francis Low in 1954.[43] "[O]ne of the most important
[papers] ever published in quantum field theory," Steve Weinberg recently
called it,

*This paper has a strange quality. It gives conclusions which are enormously
powerful; it's really quite surprising when you read it that anyone could reach
such conclusions: The input seems incommensurate with the output. The paper
seems to violate what one might call the First Law of Progress in Theoretical*

Physics, the Conservation of Information. (Another way of expressing this law is:
You will get nowhere by churning equations. . . .)[44]

As the theorists of the 1930s knew full well, when two electric charges
approach each other very closely, quantum effects like vacuum polarization
must be taken into account. Another way of saying this is that different
aspects of quantum electrodynamics come into play according to the distance
scale; because high energies are needed to push particles very close together,
the energy scale one works with requires one to treat different parts of the
theory. The procedure of renormalization can make this process more complicated and even seem to lead to inconsistencies unless one realizes that the
coupling constant—the famous alpha, 1/137—is *not* constant, but depends
on the distance and energy. Thus, the coupling constant is a "running"
constant, and at terribly short distances and terribly high energies the value
of alpha changes. The renormalization group describes the way this odd
behavior works.

With the general lack of faith in field theory that characterized the
theoretical climate of the 1960s, an idea that treated the extremes of distance—distances like 10^{-291} centimeters, arguably the smallest number ever
to appear in a serious physics equation—was not going to be discussed fully.
An independent-minded physicist named Kenneth Wilson took up the technique, however, and chewed it over in his idiosyncratic way for many years.
Wilson worked slowly; he took his doctorate in 1959, and in the next decade
published exactly six papers. In 1971, he revealed the first fruits of his
thoughts about the renormalization group.[45] Much if not most of his approach
concerned solid-state physics, but he also gave a sample calculation of how
one could apply the renormalization group to strong interactions. With assistance from Sidney Coleman of Harvard, Callan and Symanzik gave a
general prescription for such an application.

Gross had been trying to come up with field theory that would explain
scaling, but could only find theories that predicted violations of scaling,
which experimenters were not finding. Gross hoped that the renormalization
group would help him find a form of field theory that did not predict scaling
violations—in vain. By the beginning of 1972, repeated failure had driven
Gross to the conclusion that field theory was not adequate to explain strong
interactions. "I sort of decided that I was really going to *kill* quantum field
theory," he told us. "I was going to prove (1) that scaling really required
asymptotic freedom, and (2) I was going to show that there weren't any
asymptotically free theories. I started to work on both of those problems in
the fall of seventy-two. The first problem I was working on with Curtis
Callan. We tried to show that if you assumed ordinary quantum field theory
and you assume you have scaling, within the framework of the renormalization group it must be asymptotically free." The first was successful. "We

managed to show that for all nongauge theories, if you have scaling, you had to have asymptotic freedom. So that was nail number one. Nail number two was going to be that there aren't any asymptotically free theories."

The second nail was harder to strike. There were many kinds of field theories of varying degrees of difficulty. Local gauge theories were the hardest to work with, so Gross first tackled all the others and showed that they weren't asymptotically free. "The one hole left in this thing was gauge theories, which didn't fit into the same line of proof. So that hole I was going to close with Frank Wilczek, who had started to work with me as a graduate student."[46]

For some time, Wilczek, too, had been fascinated by the renormalization group, partly because so little had been done on it; the standard textbooks on field theory had brief sections on the renormalization group, but then "they just sort of *stopped*." In the months before his collaboration with Gross began, Wilczek tried to calculate renormalization corrections to weak interaction processes. This led him to wade into the technicalities of the renormalization group; from there, it was a short hop to join Gross on his quest to kill off field theory. Asymptotic freedom became Wilczek's thesis topic.

Working through the fall of 1972, Gross and Wilczek thought they were almost the only physicists concerned with asymptotic freedom. They kept committing and catching mistakes in the long, tedious figuring, but they began to wonder if Yang-Mills theories might not be asymptotically free after all. Then, a shock: Soon after Christmas vacation, Wilczek came across a preprint by Symanzik in which he spoke about asymptotic freedom. Wilczek said to us, "Now, the theory he actually used to illustrate it was a diseased theory. It's asymptotically free, all right, but also unstable. It doesn't really exist. But at the end, the very last sentence of this preprint said something like, 'Gee, it would really be interesting to know if Yang-Mills gauge theories were asymptotically free.' I was terrified. I saw my thesis going down the drain."[47]

He should not have worried about Symanzik. He should have worried about 't Hooft. In June of 1972, 't Hooft attended a congress on gauge theories in Marseille, a follow-up meeting to the one where Veltman had first presented 't Hooft's renormalization the year before. Upon descending from his plane in the Marseille airport, 't Hooft recognized Symanzik, who was scheduled to give a talk on field theories of the strong interaction. Symanzik said he had been trying to figure out if it was possible to have a force with a negative coupling constant, that is, a field that got weaker when you went closer to its source. Something like that was probably happening with the quarks. Symanzik had been discouraged, however, by various "no-go" theorems purporting to show that this was impossible. Well, 't Hooft said, he had looked into the question, and there is one class of theory in which negative coupling constants can exist: Yang-Mills gauge theories. The two men commenced to argue, and went at it from the airport to the university. Symanzik told 't Hooft that he had probably made a sign mistake somewhere, an easy thing to do given the complexity of equations in quantum field theory; 't Hooft said he didn't think so. The debate continued until Symanzik went to the dais to deliver his talk.

Symanzik told his audience that he knew just what kind of theory one needed for the strong interactions, but he didn't have the slightest idea of how to put it together. It had to be asymptotically free. But as far as physicists knew, such peculiar fields could not be; if, for example, gravity had a negative coupling constant, people would float about the surface of the earth and only acquire weight in space. At the end of the talk, 't Hooft got up and announced that he had done the requisite calculations, and that Yang-Mills theories could have a negative coupling constant. Then he sat down. Nobody there, including 't Hooft himself, quite grasped the significance of the remark. Unluckily for 't Hooft, he was still embroiled in dotting the i's and crossing the t's of his renormalization proof. To demonstrate asymptotic freedom, he would have had to write a paper explaining his technique, and another using his techniques to explain scaling, and then another. . . . He did so, but by then it was too late.[48]

One of the many mistakes Gross and Wilczek made while sweating through the calculations was a sign error in the overall result. For a few days, they thought that gauge fields, like all other fields, invariably became stronger at short distances. During that time, they happened to describe their results to Sidney Coleman. They soon corrected their goof and made some more, but they eventually proved to Gross's amazement that Yang-Mills theories could be asymptotically free. "It was," Gross said, "like you're sure there's no God and you prove every way that there's no God and as the last proof, you go up on the mountain—and there He appears in front of you." (The nail, so to speak, had turned into a door.) But in the interim, Coleman had talked to Politzer, who had run smack into a dead end with the Callan-Symanzik analysis as a means of understanding symmetry breaking.

"Okay, so I'm working on it," Politzer said, miming the action of a wizened savant scribbling field theoretical calculations. "It comes out in the end, after all the dust settles, that it's clearly totally useless for the purpose that I had in mind. What I learned as far as my own interest is that the Callan-Symanzik analysis doesn't tell you *anything* about the long distance behavior of gauge theories! Within the next sort of day, it dawned on me that if it's no good for long distances, it *is* just what you need for short distances to tell you about scaling. All of a sudden I realized that I had something that was potentially interesting. I called up Sidney Coleman; I told him I was very excited. I told him why, and he said, 'Um hum. That's very interesting— except for one problem, which is that David Gross and a student of his had worked on the same calculation, and they said it comes out the other way.' I said, 'It's pages and pages'—I could do it now in two pages or one, but at that time it was pages and pages—and I said, 'I checked it, and I think I got it right.' He said, 'Those guys don't make mistakes.' " Politzer left soon thereafter with his wife for Maine, where they were staying in the house of a friend. "It rained a lot, and I thought it was important to check my thing, so I spent a lot of time checking the stuff. I got the same numbers, came back, and said, 'Sidney, I got the same numbers.' He said, 'Yes, I know. David and Frank found their mistake, and they've submitted a paper to *Physical Review Letters* because it's important.' " Politzer dashed off an article of his own, and the two proofs were published back to back in the issue of November 15, 1973.[49]

In retrospect, asymptotic freedom was clearly in the air. A Soviet physicist, I. B. Khripovich, had done the calculations around 1972 but had made a sign mistake, and thought gauge theories were not asymptotically free. Tony Zee, a visiting professor at Rockefeller, was also looking for negative coupling constants, but happened not to hit upon Yang-Mills theories.[50] Once the discovery was made, influential physicists like Gell-Mann and Weinberg were enthusiastic—Weinberg proposed that if the color force got stronger at greater distances, it trapped the quarks—and their excitement drove the realization through the scientific community.[51] More important, the two teams of Gross and Wilczek and Georgi and Politzer worked for months to obtain predictions of the effects of asymptotic freedom on scaling. They showed up as slight deviations from simple scaling that were duly found. (These deviations were due to the presence of colored gluons, which Bjorken had not considered when he did his figuring. Amazingly, scaling gave an impetus to quantum chromodynamics, as did the violation of scaling.)

Asymptotic freedom came like the opening of a curtain onto a previously hidden stage, revealing the strong interaction in its full dimensions. Many physicists had pictured elements of the scene—Gell-Mann came closest to encompassing it in its entirety—but none till then had fully grasped the flawless elegance of nature's conception. Whereas the electroweak theory was clumsy, quantum chromodynamics was pretty—a perfect, unbroken symmetry. Eventually, the excitement around it built like a tsunami, from an initial slight ripple into a tidal wave. The words *unbroken Yang-Mills SU(3) color with asymptotic freedom* rang through a hundred transcontinental wires; preprints of new work came to departments, were instantly photocopied a dozen times, and then a dozen times more. In such moments of collective ferment, a scientific sodality moves like a single, self-absorbed organism enraptured by the most pleasing and unexpected interior imagery. Emotional valences tip readily at these times, and theorists who previously held back from constituent models found themselves interpreting the evidence with a more kindly eye. At a stroke, the picture seemed both beautiful and real: Inside every proton are three quarks and numberless gluons, all their colors blending into white.

One of the oldest riddles known is the question, "Is there a smallest piece of matter?" Plato held the elemental things to be unbreakable geometrical shapes, whereas Aristotle stated that all substance was infinitely divisible. Numerous other suggestions surfaced in the intervening millennia. Made uncomfortable by the lack of progress, Immanuel Kant asseverated in the *Critique of Pure Reason* that the phrasing of the question makes it unanswerable. The human mind, Kant wrote, can pose certain questions about nature which have contradictory but perfectly logical answers. One is whether everything in the world is made of simple parts; both the affirmative and negative answer, or so Kant thought, can be proven, which means that the way we think about such subjects is inadequate.

One can speculate endlessly about whether there are particles that can be subdivided infinitely. Quantum chromodynamics does not pretend to answer the question. In the manner of science, however, it does provide a definite answer to what happens when you actually go out and try to do so with the basic components of our world, hadrons. Suppose you begin shooting electrons at a proton, trying to knock loose one of its constituent quarks. As the quark is kicked farther away from its partners, something strange occurs; the virtual gluons whirling between the quarks begin exchanging gluons among themselves. The greater the separation, the more intricate and powerful the web of interactions. Eventually, the energy needed to separate the quark still farther from the snarl of gluons becomes sufficiently great that a new quark-antiquark pair is created *ex nihilo* from the vacuum. The antiquark bonds to the quark separating from the proton to create a meson; the new quark meanwhile pops right back into the proton, leaving it with the same number of quarks as before.

"The whole process is rather hard to visualize," Glashow told us, laughing. "It's like a prison where you don't restrain the prisoners at all except to keep them in the jail. Inside, they can do what they bloody well please, but they simply can't get out. Of course, what happens if you try to pull a prisoner out of the jail is sort of curious in this quantum world, because as you tug on the prisoner, you get him out all right, but you produce a new prisoner-antiprisoner pair. The second prisoner will get stuck in the jail and you will end up with a prisoner and an antiprisoner." He paused for a moment, struck by the worry, common to physicists, that his audience may not be following him. "It's . . . ah, a bit counterintuitive."[52]

Quarks are not parts of protons and neutrons in the same way these particles are parts of atoms; they are not just another rung down on a ladder. According to SU(3), quarks and the gluons that bind them are perpetually unseen constituents of the final rung, ghosts in the machine, concealing their existence from the world in the act of comprising it.

□　□　□　□　□

The power and elegance of quantum chromodynamics—SU(3)—and what some physicists have called "quantum flavordynamics"—SU(2) × U(1)—quickly led physicists to splice them together as SU(3) × SU(2) × U(1), a standard model of elementary particle interactions. The standard model had a curious birth, for it was rushed together headlong after much confusion, and then celebrated as complete well in advance of experimental proof. Theorists intoxicated with the thought of containing all elementary particle interactions in a single coherent package regarded the show as almost over, whereas their colleagues on the machines, as is only proper, recognized only that a new fad had overtaken the pencil pushers. At Brookhaven, at CERN, at Fermilab and SLAC, physicists settled down to put SU(3) × SU(2) × U(1) to the rack of experiment.

17

Neutral Currents/Alternating Currents

IT IS IMPOSSIBLE TO LEARN ABOUT SCIENCE—OR ANYTHING ELSE, FOR THAT matter—without asking a lot of stupid questions. Every scientist has a humiliating memory of revealing ignorance by posing a particularly naïve question to a teacher or a senior colleague. Etched into the brain, the sarcastic response is savored, years later, for the lesson it imparted; being caught short may not be the most comfortable way to learn, but one seldom forgets the result.

While poring over the epochal alpha particle experiment by which Ernest Rutherford divined the existence of the nucleus, we had the notion that we would better understand the discovery if we repeated the experiment with a practicing scientist. We had an enticing image of ourselves watching as the physicist set up the radioactive source, the thin gold foil, and the scintillation screens. We envisioned scribbling notes blindly as the lights were turned out and the little telescope adjusted to count the flashes at each angle. Making a few quick calculations on a scrap of paper, our physicist would announce triumphantly that the evidence indicated that atoms have solid, massy, positively charged centers.

An obvious candidate for this signal honor was Samuel Devons, a former Cavendish physicist who had taught a course on the art of experiment at Barnard College in Manhattan. We broached the idea one day to him and had the embarrassing experience of hearing a kind man attempting not to laugh in our faces; we had obviously made his day. He answered simply enough, but a guffaw kept creeping into the edges of his voice. "In principle, the experiment is simple," he said. "In practice, it would be nearly impossible. First of all, there's the problem of working with radioactive materials. You'd have to find a strong radioactive source—do you want to wait a month to see a flash?—and you'd have to make a new source daily and you would need a certification by a health officer to ship highly radioactive materials around New York City. Do you think you'd get one here at Columbia or anywhere else? No way!

"Okay, so that's one problem, getting permits. Maybe you could," he said dubiously. We were beginning to feel exceedingly foolish, a sensation that Abdus Salam once told us opens the mind to the spikes of insight. "The main problem, though, is that experiment is a *craft*, like making an old violin. A violin isn't a very complicated-looking gadget. Suppose you went to a violin maker and said, 'Could you kindly help me make a Stradivarius? I'm interested in violin-making, and I'd like to see how it was done.' He'd smile at you just like I did. Because craft is a knowledge you have in your finger-tips, little tricks you learn from doing things, and they don't work and you do them again. You have little setbacks, and you think, how can I overcome them? And then you find a way. Every time your equipment changes you forget all the old techniques and have to learn new ones. And you have to know them, because when you're pushing your equipment to the limit it's bloody easy to get spurious results. You're scratching at the ground all the time, and you don't know what you've missed. Every experimenter has made terrible errors at one time or another, and knows of instances where friends have fallen on their faces because they got spurious results and published too early. And yet, you've *got* to push what you know to the limit. If you don't, someone else is going to do it first. And that's dreadful, being beaten. Everyone's got a closetful of discoveries they missed because they were too cautious or some other fellow was cleverer. There was a whole Austrian school working on the same things as Rutherford at about the same time, and nobody's heard of them today. Why not? Rutherford was just a little more daring and crafty."[1]

Looking back at experiments from the vantage of the present, discov-eries appear inevitable, drawing experimenters on like beacons, and errors or sidetracks en route seem like evidence of inattention or stupidity. Such subtle, false teleology is hard to purge from the history of science; it is hard not to say, for instance, that *of course* neutral currents were there, and *of course* SU(3) × SU(2) × U(1) was right, and it was only a matter of time before experimenters found out. Graduate students studying today's text-books find few hints of the long and tumultuous period of testing which the theory underwent. Although the results now appear in the *Review of Particle Properties* as a few lines summarizing an apparently orderly series of confir-mations, the trials of the standard model lasted throughout the 1970s and provoked considerable human conflict.

Experiments focused initially on the electroweak theory of Glashow, Salam, and Weinberg, because this part of the standard model was put together first and because it clearly pointed to the existence of new physical phenomena. When SU(2) × U(1) was first proposed, all known examples of the weak force involved some gain or loss of electric charge by one or more of the particles in the interaction, and hence a change in their identities. The presence of Z^0s in the theory, however, implied that there could also be *neutral* weak interactions, exchanges of Z^0s with no alteration of particle identities or charges. Such currents had never been observed, and their existence was routinely discounted; finding them would impressively con-firm the theory.[2] In two clear, urgent, and important calculations during the latter months of 1971, Weinberg carefully redid SU(2) × U(1) and noted that it predicted the ratio of neutral currents to charged currents—that is, weak interactions with Z^0s compared to those with W^+s or W^-s—to be between one-eighth and one-quarter. Such effects, Weinberg said, "are just on the verge of observability." Extant data gave a lower limit of about one-eighth with a hefty margin of error, which "neither confirmed nor refuted" the theory.[3]

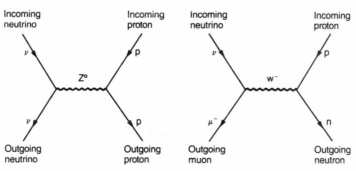

latter months of 1971, Weinberg carefully redid SU(2) × U(1) and noted that

Here was a clear target. According to the electroweak theory, between one-eighth and one-quarter of the time, neutrinos and antineutrinos interacting with protons would emit a Z rather than a W, and no particles would change identities. If SU(2) × U(1) had been renormalized a few years earlier, this prediction could not have been checked, because neutrinos were too hard to work with. By happenstance, exactly the device for such experiments had just become available—a bubble chamber, then the world's largest, which had been built in France over the course of seven years. First proposed in February 1964 by André Lagarrigue, Paul Musset, and A. Rousset of the École Polytechnique in Paris, the new chamber was specifically designed to examine neutrinos.[4] Almost twenty times bigger than any previous European chamber, it consisted of a tank of freon swaddled in a thousand tons of steel and copper. The steel served as a shield from errant particles; the copper consisted of a few kilometers of thick wire spooled around the iron and plugged into a six megawatt generator, which converted the entire assembly into a powerful electromagnet that bent the particle tracks inside. Louis Leprince-Ringuet, the old cosmic ray physicist, dubbed the project "Gargamelle," after the monstrous, big-bellied mother of Gargantua in Rabelais's *Gargantua and Pantagruel*. Gargamelle took six years to construct; the team that planned to use it swelled to more than fifty physicists from eight institutions across Europe.

As Gargamelle came into readiness, the renormalization of 't Hooft and Veltman burst onto the stage. Theorists at CERN learned of the work early from Veltman. In November 1971, Jacques Prentki received a preprint of the first of Weinberg's two papers from an Italian friend at MIT. Written across the front in large letters were the words "MOLTO IMPORTANTE!!!" (*Very important!!!*)[5] Six weeks later came the second paper. Prentki, Bruno Zumino (the theorist who had told Veltman of Weinberg's 1967 paper), and Mary Kay Gaillard talked about SU(2) × U(1) and neutral currents to a group of experimentalists in a room of CERN Building 17, the annex constructed to house Gargamelle.

CERN is a complex of long boxy buildings scattered higgledy-piggledy across a mile-long stretch of Swiss farmland. Experimenters are always knocking walls down or implanting new trailers to house facilities, and great cranes stand perpetual watch over the site. Gargamelle had been tucked into the side of one of the experimental halls on Rue Pauli, a few hundred yards southeast of the Proton Synchrotron, then CERN's largest accelerator. Around it were several floors of catwalks and cheap little metal offices, in one of which Gaillard, Prentki, and Zumino talked about electroweak models. They agreed with Weinberg that the most direct proof of the theory would be some evidence of the Z^0 and that the requisite evidence would come most likely from examining neutrino-electron interactions in a bubble chamber.

Such interactions would occur when one of the muon neutrinos in the beam bounced off an electron in a freon molecule; the only way this could happen is if the two had exchanged $Z°$s. The neutrino would continue on its merry invisible way, and the electron would, in the subsequent photograph, seem to suddenly erupt out of nowhere, spiraling outward in the magnetic field. (In a charged current, the neutrino changes its identity, becoming a positive or negative muon.) According to the theorists, the scanning team in the basement of Building 17 should be asked to look for the thin, circling tracks of such isolated electrons. The "signature" of the neutral current would be both the ejected electron and, more important, the *absence* of a muon.

Gargamelle's *raison d'être* was to study the general domain of neutrinos, rather than neutral currents in particular, which were dismissed in two short paragraphs of the 194-page proposal.[6] Five members of the Gargamelle team had set low limits on neutral currents as recently as a year before.[7] Despite the theorists' pleading, the experimenters were not impressed with the hand-waving about Ws and Zs. They refused to give high priority to neutral currents, although they did agree to ask the scanners to report any single electrons they saw. Such events would be easy to identify, and looking for them would not be a burden.

Easy to identify, perhaps, but rare. Rare enough, in fact, that Garga-melle project director Musset doubted whether one would be seen in hundreds of thousands of photographs. After spending half a decade immersed in the delicate task of building Gargamelle, the completion of the chamber had left Musset somewhat at loose ends. He decided to occupy his time by looking into neutral currents more closely; his willingness to do so, he told us later, was partly due to his isolation from physics, and his ignorance of the lengthy series of experiments that had failed to observe neutral currents. Musset thought that the team should look for hadronic events (interactions between neutrinos and protons or neutrons) which should occur about two thousand times more often. Unfortunately, hadronic neutral currents—the diagram above pictures one—are much messier than neutrino-electron neutral currents. The reason is that after the neutrino hits the proton, the proton flies off and interacts strongly with any nearby particle, producing a spray of secondary pions, lambdas, and so forth. In hadronic interactions, too, the signature of a neutral current is the absence of a muon. Experimenters would have to be certain that they could identify *every* particle in the melee, and that none of them were muons.

On the other hand, Musset thought that the past experimenters who had not seen neutral currents had probably killed the very signal they sought. They had been worried about the neutrons created by interactions between neutrinos and the hundreds of tons of metal shielding around their detector. These neutrons might slip into the chamber and smack into a proton, creating something that mimicked a neutral current. The initial neutron would be invisible, because, like a neutrino, it is uncharged, and bubble chambers cannot "see" neutral particles. Because no final muon would necessarily be produced, such an event could look very much like a neutral current. To pick out the phony neutral currents, past experimenters had reasoned that these secondary neutron interactions would on the whole have less energy than neutrino interactions, because some energy would be expended making the neutron, the way the second bounce of a tennis ball is always less high than the first. The experimenters had therefore adopted the general practice of throwing out all interactions below a certain energy level.

Musset believed, however, that whereas the escaping muon in charged currents can be photographed and its energy accounted for, the escaping neutrino in neutral currents cannot, because it is not seen. Nobody knew how much energy the neutrino carried off. If it took away even a small amount, the event would seem to have too little energy, and experimenters would consign it to the trashcan as the work of a neutron. Making rough calculations, Musset figured that the energy cutoff had effectively concealed all but 5 to 10 percent of the neutral currents. "It was essential to have the same criteria," he said, "to have a really good comparison, a fair comparison between neutral currents and charged currents."[8] He argued that calculating the energy of charged currents should be done by ignoring the muon, thus putting the two types of interaction on an equal footing. He was unable to convince his colleagues, who thought the idea was ridiculous.[9]

Musset decided to take a look for himself. Together with Antonino Pullia and B. Osculati of CERN and two visitors, Ugo Camerini and William Fry of Wisconsin, he went to the scanning rooms in the basement after the scanners had gone home. They turned on the lights, unspooled the film, and let the big black-and-white images play on the formica tables. The five men spent almost a year in the room, working until the early hours of the morning, making drawings and measurements of the tracks that curled soundlessly in the liquid, shooting out like sheaves of wheat held by invisible hands. Musset was helped through this dull task by his myopia, which allowed him to squint closer to the photograph than someone with normal sight. Some time at the beginning of 1972, he came across an event that took his breath away: an interaction well in the back of Gargamelle, but centered on the line of the beam. There was no muon. The event was energetic. It could not be caused by a neutron because on average neutrons could only penetrate two feet of freon without interacting and this event took place over three yards into the chamber. It could not be caused by a cosmic ray because the interaction was right on line with the beam. One picture was not enough to prove anything, but Musset showed it to collaboration head Lagarrigue as a way of exciting his interest.

Years before, in Paris, Musset was on a team that had been beaten to the discovery of the eta meson by physicists at Brookhaven. He had come to the conclusion that the Europeans were too hidebound, too obsessed with precision, and feared that by the time the CERN group understood everything to its satisfaction, it would again be scooped by a few Americans with less regard for niceties. Musset knew American experimenters at Fermilab were examining high energy neutrino interactions. If neutral currents existed, he wanted to find them first.

Designed by Robert Wilson, one of the original accelerator builders on Ernest Lawrence's team, Fermilab is a kingdom of cowboy electronics.[10] A lifelong maverick, Wilson had strong notions about aesthetics and a firm belief in frugality that produced an odd set of priorities in the laboratory's design. When designing Fermilab, Wilson brought with him a graphics specialist named Angela Gonzalez, who was given carte blanche on color— so long as the paint was cheap. The laboratory bought an entire community of tract houses to acquire the site; these became experimental buildings, and Gonzalez ordered them painted solid dark Rustoleum blue, solid dark Rustoleum green, or solid dark Rustoleum pink. In the middle of the campus, the laboratory headquarters rises fifteen stories in a sort of giant H, fat at the base, curving toward the top, an atrium in the middle climbing to the roof— a lone skyscraper in the prairie. The building was named Robert Rathbun Wilson Hall after Wilson retired in 1978. Cooling water for the accelerator flows nearby through canals adorned by small fountains; the water steams in cold weather, fogging the lower windows of Wilson Hall, and is the permanent home to a gaggle of Canadian geese that never migrates from the warm scientific pools.

Despite constraints of budget and bureaucracy, Fermilab was built Wilson's way, making it, in some sense, one of the largest monuments to a personality since Peter the Great laid the foundations for Saint Petersburg. Ten square miles of grassland cut and shaped and molded by a single individual, Fermilab is an utterly eccentric celebration of cerebration. Wilson was from Wyoming, a square-backed, bowlegged man with a hard sunbaked face and square silver-framed glasses. He liked cowboy boots, horses, and buffalo. He liked buffalo so much that eighty of them now live on the Fermilab range, mean-eyed shambling beasts that the physicists avoid when they can. He stood an enormous obelisklike sculpture of his own fashioning in front of the central building and surrounded it with a reflecting pool. He built three experimental halls—the meson, neutrino, and proton buildings—awarded each a distinctive roof shaped like one of the Platonic solids, and placed them to be seen to maximum advantage from the top floor of the laboratory headquarters.

Wilson liked things big and he liked things cheap, and he didn't mind if a few people got scars doing what he wanted. He gave Fermilab the most powerful accelerator in the world for next to nothing, which in physics terms meant about $250 million. He told physicists they could have cranes to move around their equipment or a roof above their heads, but not both. Some of the experimental buildings didn't have floors, because Wilson saw no reason why people couldn't wear boots and walk on dirt. Wilson left no room to tear out experiments; it was cheaper every now and then to knock down a wall.

The main ring at Fermilab is four miles around and marked by an earth berm that resembles, in its simplicity and perfection, the legacy of a Stone Age religion—a technological Stonehenge. Twenty feet beneath the surface is a tunnel with a pipe to carry the protons and about a thousand boxcar-shaped magnets to bend them around the circle. Figuring to save a quarter of a million dollars, Wilson ordered the magnets welded into the beam line without the expensive seals that would allow them to be removed if they failed. Moreover, the tunnel itself was just a tunnel, nothing like the climate-controlled rings at CERN. By 1971, when the accelerator was ready to be tested, Wilson's decisions about the tunnel and magnets had proven cata-

strophic. When the hot humid summer nights settled over the Illinois farmland, moisture seeped through small hairline cracks in the insulation of the
magnets and burned them out one by one. Nobody knew how many magnets
had cracks; the technicians just had to sit and wait for the next one to blow;
meanwhile, the first experimenters sat on their hands, their detectors dark,
their careers on hold, hoping for some beam. For almost a year, Wilson
watched powerlessly as his decrees caught up with him. Although the first
200 GeV protons made it through the ring in March 1972, not all of Wilson's
colleagues forgave him for the ordeal.

A passionate advocate of unfettered scientific research, Wilson frequently was pressed by Congress to explain why the taxpayer should spend
millions of dollars to fund an enormous, expensive gizmo whose sole use was
to let physicists chase subatomic particles. One exchange, between Wilson
and a skeptical Senator John Pastore of Rhode Island, has become legendary
in the research community. Pastore asked, "Is there anything connected
with the hopes of this accelerator that in any way involves the security of this
country?"

"No, sir," Wilson said. "I don't believe so."

"Nothing at all?"

"Nothing at all."

"It has *no* value in that respect?"

"It has only to do with the respect with which we regard one another,
the dignity of men, our love of culture. It has to do with, are we good
painters, good sculptors, great poets? . . . It has nothing to do directly with
defending our country except to make it worth defending."[11]

One of the teams sweating through Fermilab's rough early days was
situated in the neutrino building and composed of a baker's dozen physicists
from Harvard, the University of Pennsylvania, and the University of Wisconsin. The team was known as the HPW collaboration from the initials of its
component universities; the senior collaborators were David Cline (Wisconsin), Alfred K. Mann (Pennsylvania), and Carlo Rubbia (Harvard), and the
initial goal was to find the W. All three men were specialists in weak interactions; Cline and Mann had previously conducted searches for neutral currents with K mesons, and had not found them. Rubbia, who had the hunch
that weak interactions were ripe for a theoretical breakthrough, specialized
in building large and complicated detectors. The initial round of experiments
at Fermilab would be the first to probe a new high energy range, and
competition for time slots was, under Wilson's laissez-faire approach, not
always gentlemanly. Although the HPW team submitted the first formal
proposal to the lab in April of 1970, some members of the group were not
certain they were ever going to run.[12]

Cline, Mann, and Rubbia set up their detector at the end of a kilo-meter-long pile of dirt. After flying into a target, protons from the main ring produced a gush of pions, which decayed mostly into muons and muon neutrinos before slamming into the dirt. The dirt absorbed everything but the neutrinos, which poured into the detector, where the experimenters ardently hoped they would do something interesting. The HPW detector was designed as a long series of vertical slabs somewhat like a fifty-foot-tall Dagwood sandwich turned on its side. Incoming neutrinos first passed through four alternating layers of liquid scintillator (a fluid that flashes when charged particles pass through) and spark chambers (long flat boxes that make sparks when charged particles pass through). Directly behind this segment was the second half of the detector, a magnet consisting of four iron blocks, each four feet thick, interspersed with four more spark chambers. If the neutrinos interacted in the first half of the detector, the various scin-tillators and spark chambers could record the trajectories of the particles. Of these, all but muons would be stopped by the iron in the second half of the detector. The idea was to use the sparks created by muons in the second half of the apparatus to trigger the detector in the first to record the event. By setting the detector to fire only when muons passed through the iron, the collaboration could avoid scanning the hundreds of thousands of useless events that bubble-chamber physicists had to go through. Unfortunately, because the detector went off only when a muon passed through, the device was not capable even in principle of looking for neutral currents, which are characterized by the *absence* of a muon.

In addition to working on the HPW detector, Rubbia, a fast-talking, ambitious man with an apparently endless capacity for juggling widely sepa-rated projects, was running another experiment at Fermilab and one at CERN. After 't Hooft's work, Rubbia said later, Steven Weinberg "took a tremendous amount of time telling [me] how nice his theory was." "Brain-washed," as he put it, into looking for neutral currents, Rubbia insisted that the collaboration modify the detector to trigger even without a muon. He was supported by Larry Sulak, a young Harvard colleague with as much ambition as himself. Sulak assembled the new trigger.[13] Right away, in the small dribbles of the beam they could get before the magnets blew, they started seeing interactions that lighted up the first part of the detector, but not the second—events with no muon.

The HPW collaboration did its work in the Fermilab neutrino building, a boxy affair with corrugated metal walls. One end is a bubble-chamber assembly area with a twenty-sided geodesic roof whose panels were made from soft-drink cans. Somebody suggested to Wilson that you could cut off the ends of Coke cans and sandwich the resultant cylinders between sheets of colored plastic to produce a cute, stained-glass effect on the cheap. Wilson tried it, producing a weird particolored dome that leaked every time it rained. Wilson didn't mind. The floor was dirt, and because the building was sunk into the earth, the place was full of puddles anyway. Come spring, they dried up.

"It was pretty terrible," Sulak said later about his move to Fermilab. "First of all, you had absolutely minimal support. I mean, you were out in the boonies. Wilson didn't believe in any frills at all, where frills meant buildings or sump pumps or cranes. No cranes! He had an objection to cranes. He had an objection to trailers as opposed to his so-called por-tacamps; we bought at Harvard a super trailer, filled it up with all the stuff to bring out, and then we were forbidden to put it at the experiment because of aesthetic reasons. And had to stick it a quarter-mile away at a farmhouse. Then the watertable was high and the foundation of the building leaked, and the water would come in and it'd get filled with frogs and the frogs would rot and stink, they would die because they couldn't get out. It was just terrible." Sulak spent hours scooping out the corpses of unlucky amphibians from the detector building and wondering what the muonless events were. "But the major problem was waiting forever and ever to take beam and not being able to do it."[14]

Yet a third hunt for neutral currents began at about the same time. It is a measure of the size of contemporary high energy physics that the third search occurred *inside* the first, when Robert Palmer of Brookhaven took a sabbatical at CERN in the fall of 1971. He wanted to work on the Gargamelle experiment, but the software to analyze the scanning data was not yet ready. Realizing that he would have to leave Geneva by the time the computer programs were complete, Palmer and two other American visitors, William Fry and Ugo Camerini, began to wonder what they could do with the scan-ning data alone. When the scanners went over the first Gargamelle runs, they projected the film on long tables and went over the tracks with a device somewhat like the mouse used in some home computers. By centering the mechanism on each vertex of every track, the scanners created a rough computer "recording" of the event on a keypunch card. Camerini and Palmer realized that they could pore through the cards to search for neutral currents. All they had to do was to figure out what arrangement of holes on the cards indicated highly energetic events with no obvious muon candidate in the rear of the chamber. They found them. Palmer wrote: "I am not sure if Ugo and Fry believed the results at the time, but I was very excited and rushed about trying to interest the other members of the group. I remember having little success and suspected that the already published negative re-sults from the earlier work of the group had a lot to do with it." Incredibly,

Musset was making rough calculations at about the same time with about the same results, but the two team-mates did not run into each other because one man directly measured the photographs themselves while the other examined the computer cards made from them. In any case, both were rebuffed by their colleagues for the crudity of their analyses. As part of a large group, neither man could publish on his own. Palmer returned to Brookhaven in the beginning of 1972 and began telling his friends at Brookhaven that the laboratory should quickly finish its own neutrino experiments, for neutral currents existed and were waiting to be discovered.[15]

It all seems a long time ago now [Palmer wrote]. The originals of [my calculations] are a little yellow, and the angers and frustrations have faded too. Yet I have not altogether forgotten that it was I who first saw these neutral currents and it was I who one night (these moments always come late at night) made that plot and believed, alone in all the world, that God if he exists had decided to have a neutral current.[16]

Palmer's colleagues paid little attention to his claim, partly because of doubts about the accuracy of the measurements on which it was based and partly because a Brookhaven experimenter, Wonyong Lee, had reanalyzed some old neutrino experiments performed at the lab in the 1960s, and concluded, in May 1972, that "there is no evidence for the existence of neutral currents" in the events studied.[17] Moreover, theorists worried by the evident nonexistence of neutral currents had invented ways to save the basic features of $SU(2) \times U(1)$ by making the model a bit uglier or adding extra features.[18] One of the earliest and most ingenious end runs around the apparent absence of neutral currents was made by Glashow and Howard Georgi, then a postdoctoral fellow.[19] By positing the existence of two heavy leptons, they were able to contain the whole electroweak theory in an $SU(2)$ group, making do with just the Ws and the photon and excising the Z. Part of the theorist's sense of elegance comes from doing the most with the smallest number of constituents; the Georgi-Glashow model satisfied this predilection admirably, and many of the growing band of gauge theorists thought, well, maybe one didn't need neutral currents after all. . . .

Perhaps inspired by Musset's superb photograph of a neutral current candidate, Lagarrigue wrote to CERN Director-General Willibald Jentschke in April 1972, informing him that the search for neutral currents was now going to be one of the top priorities of the laboratory neutrino program.[20] Although neutral currents had moved into the forefront, there was no agreement about how best to search for them. Some members of the collaboration focused on the cleaner but much more rare electron-neutrino interactions, whereas Camerini, Fry, Musset, Pullia, and Osculati worked on the hadronic search—formalizing, in effect, an extant informal division of the team. By July 1972, the hadron subgroup had set up ground rules for classifying events.[21]

The HPW group got a bit of beam over the Thanksgiving and Christmas breaks of 1972. Physics lore has it that accelerators work best late at night and during holidays, when the hot shots in the control room who like to tinker with the beam are not around. The great detector in the neutrino building triggered for about one hundred and fifty events with no obvious muon. Recorded on big drums of computer tape, the data were taken first to Wisconsin and then to the fourth floor of Lyman Laboratory at Harvard, where Rubbia, Sulak, and a crew of undergraduates began to analyze it by a process similar to that underway at CERN. The computer singled out events having more than a minimum amount of energy; photographic prints of the pertinent tracks were then developed and blown up. The HPW collaborators built templates to measure the lines of sparks photographed in the detector. By the spring of 1973, they, too, had begun to suspect that the muonless events were for real. They were well aware of what the CERN group was finding. A Mercury of scientific rumor, Rubbia shuttled between his experiments in Cambridge and Geneva, bringing the message to each group that the other was hot on the trail.[22]

Around New Year's Day 1973, a research student named Franz Hasert, a Gargamelle collaborator from West Germany in the subgroup looking for electron-neutrino events, noticed that one of the scanners had discovered a curious interaction that apparently consisted of a muon and a photon, the latter subsequently becoming an electron and positron. Surprised by this odd interaction—three leptons coming from a neutrino!—he went back to the original photograph and realized immediately that the event had been misclassified. The picture actually showed an electron suddenly spiraling out of nowhere across the black background of the chamber, precisely what would occur in the event of a neutral current electron-neutrino interaction. Helmut Faissner, the senior Aachen physicist, was elated by this "picture-book example" of a neutral current. After scanning over 725,000 pictures, luck had allowed them to find the object of their search.

All that remained, Faissner said optimistically, was to check out the background. That took another half a year.[23]

Meanwhile, on January 31, 1973, Musset in the other CERN subgroup presented initial evidence for neutral currents at the annual New York meeting of the American Physical Society, then one of the largest and most important gatherings in the physics world.[24] Upon his return to Geneva, he urged Lagarrigue to help convince the collaboration; no begging was needed, because the discovery of the single electron at Aachen had dramatically changed the climate of opinion. Meanwhile, the hadronic team went through the events one by one, furiously arguing the criteria. They had two hundred and thirteen candidate events, each of which had to be listed and drawn and measured and studied and restudied.[25] To resolve the arguments and doubts, Musset took to carrying big stacks of eighteen-by-twenty-four-inch blowups around from lab to lab, logging thousands of miles as he criss-crossed the Continent.[26] Some physicists couldn't see the kinks; others worried about contamination from strange particles or cosmic rays; nobody wanted to make a false claim that would be subject to much attention and excitement. On the other hand, the Gargamelle group did not want to be beaten to the punch.[27]

Narrowing down the candidates was difficult, tedious, imprecise work, akin to the process whereby a mechanic gains the experience necessary to diagnose the weakness in a motor from the sound it makes. The coils made by electrons and positrons were easy enough to pluck out of the mass of lines in a photograph, but distinguishing between muons and pions was difficult. Both have almost the same mass and charge—one recalls cosmic ray physicists confusing them for a decade in the "ten-year joke"—and the tracks in bubble chamber photographs provide precious few clues to distinguish them. Pions usually end their lives with a small shower, and muons do not, but both sometimes simply stopped in the chamber, absorbed by a freon molecule. Yet distinguishing them was crucial: Mistaking a muon for a pion meant removing the final muon—producing a "neutral current" by an act of misidentification.

We once asked Musset to show us how he went about distinguishing muons and pions. His office was still in the old Gargamelle building, although the bubble chamber itself had been hauled off to a museum of science in Paris. Musset had a few minutes free, so he went across the hall, rummaged in a few old boxes, and emerged with half a dozen yellowing photographs of Gargamelle events. The images were grainy and sometimes washed out by the flare escaping from the edge of the light units. Together with the photograph was a drawing of the interaction and an evaluation sheet telling us that we were looking at Antineutrino Film Roll 592, Camera View 6, Photograph 125. Two tracks emerged from nothing, one short and stubby, the other long and branched. According to the remarks on the evaluation sheet, the event had been moved from the charged to the neutral category several times in 1972 and 1973, changing its identity in the back-and-forth of discussion. Musset's initials were on the analysis sheet.

From the size of the bubbles in the short track, Musset said, he could easily identify it as a heavy proton. "Then there is only one track [besides the proton coming out of the interaction], which is somewhat energetic because it's straight. There is a small kink, here—" his finger poked a point midway along the line "—that you can see, maybe." Kinks indicate where strongly interacting particles glanced off a nucleus. Musset said, "That's important on this particular event, because if you don't take this kink into account, then we have the wrong measurement of the momentum and energy. The energy looks to be less because the kink simulates bending, if you like. We had to measure only up to the kink."

Was that the reason the event had been reclassified?

Yes, Musset said. "This is a good example of an event which can be lost if you are not careful." Spotting the kink had first changed the identification of the particle from a muon to a pion, thus changing the interaction from "charged" to "neutral." Then measuring only up to the kink had boosted the figure for the total energy of the interaction, enabling the interaction to survive the energy cut. Seeing the kink, he said, involved a willingness to lie flat on the scanning table with an eye along the track; interpreting the photographs involved craft, educated judgment, and the kind of knowledge that resides in the fingertips. His myopia helped, too, for he could bring the photographs closer to his face.

At one level, we said, the whole edifice of theory boiled down to whether you saw a bend in a curve when you looked down a line. "Yes," Musset said. "And that was one of the difficulties, because I *never* convinced the scanning girls to put their eye on the table."

He put Photograph 125 atop an old khaki filing cabinet and invited us to look down the tracks. Neither of us could see the kink, though we strained our eyes considerably in the effort.

"It's there," Musset said, amused. He took off his glasses and flattened his cheek against the print. His hair spilled onto the table. "You just have to have the eye," he said, squinting nearsightedly.[28]

Robert Palmer heard Musset speak in New York. Frustrated by his inability to convince anyone on the Gargamelle team with his rough-and-ready calculations, he found himself in the bizarre position of having made a discovery—or thinking he had—from someone else's data. Figuring that the data had been made public by Musset, he tried to publish through the back door—scooping the CERN team on their own results—by writing a *theoretical* article, his first and only, with calculations taken from the Gargamelle data. Unfortunately, he sent it to the CERN journal, *Physics Letters*, where it arrived on May 24. There it sat for months, unpublished, frozen by a bureaucratic deity well disposed to the Gargamelle group.[29]

On April 12, 1973, the Gargamelle neutral current group, which had grown to include members from all the labs in the collaboration, met as an ensemble at CERN to go over every proposed instance of a neutral current. By the end of two days, they had reduced the total to 157 firm candidates.[30] Set down single-spaced on six sheets of paper, the events were listed by roll and frame number with summary comments:

570/253	OUT	1 GeV
570/399	*OK*	E badly known
605/033	OUT	possible mu
436/725	OUT	entering track
571/331	*(OK)*	Check energy (cosmic [ray])
491/263	OUT	error on frame number
443/108	*OK*	
558/316	*OK*	
558/646	*OK*	energy?
795/605	OUT	possible mu kink

The question remained whether they had really accounted for background; this was discussed in a second meeting the next month. To the dismay of some collaboration members, Musset jumped the gun at the end of May. While the team was still counting candidates, he told a French congress that the neutral currents "observed in the [Gargamelle] experiment are at the predicted level. . . . [O]ne may conclude that the current experiments cannot eliminate the Weinberg model."[31] Although the action was bold enough to alarm a few members of the team, Musset's style of presentation was sufficiently cautious that the announcement attracted little attention. Amazingly little, perhaps, in view of the consequences. Backed by Rousset, Musset had the go-ahead by mid-June to write a full paper broadcasting the discovery of neutral currents.

Throughout the spring, Sulak and Rubbia were preoccupied by a single worry, which was known as the problem of wide-angle muons. Although it filled almost the entire building, the HPW detector was long and thin, and it seemed quite possible that muons produced in charged current neutrino interactions could shoot out the sides before reaching the back half of the apparatus designed to detect them, making charged currents look like neutral ones. (Meanwhile, Mann and Cline were principally concerned with understanding the charged currents and the other research subjects the experiment had been built to study.) Years before, at a conference in Europe, Musset had argued with Rubbia about whether the detector would miss wide-angle muons, and Rubbia had said there would be no problem.[32] Quantifying that assertion involved setting up a "Monte Carlo," an elaborate computer simulation of the experiment, and seeing how many wide-angle muons should be produced. The trouble with such modeling, as the team well knew, is the immutable data processing law of GIGO—Garbage In, Garbage Out. If the assumptions they fed in were wrong, the predictions would be worthless. Similar questions beset the CERN group's simulations of neutron interactions. Both groups could therefore be making the same false claim. Nonetheless, Sulak felt good enough about the signal that by mid-June he, too, began to write a paper announcing the discovery of neutral currents.

Meanwhile, the two CERN subgroups prepared to publish. On July 3, the single electron event was submitted to *Physics Letters* from Aachen. The paper finished on a positive note: "We conclude that the probability that the single event [described in the paper] is due to non-neutral current background is less than 3%."[33] Still, the collaboration was uneasily aware that one picture rarely proves anything.

In Geneva, Musset and other teammates finished the rough draft of the hadron paper on July 4. Worrying about background continued until the last possible minute, with Musset bombarded by suggestions, arguments, counterarguments.[34] During a stopover in Europe, Rubbia wrote to Lagarrigue on July 17 that the HPW group had "approximately one hundred unambiguous [neutral current] events" and that they were in the final stages of writing a paper. Rubbia offered to ensure that the European team's work was mentioned in the American paper if the Europeans reciprocated. In a burst of chauvinism, Lagarrigue refused point-blank. He sent Rubbia a letter on July 18 informing him that by the time he opened it the Gargamelle collaboration would already have made a formal announcement.[35] The honors fell to Musset, who spoke in a large lecture hall on the CERN campus on July 19, 1973. The hadronic paper was submitted a week later.[36] For the first time, CERN had soundly beaten the Americans.

"I am not a theorist," Musset said to us. "So I am not able to understand all the beauty of the theories. But I can understand *facts*. It was to me so marvelous—to discover something which had been hidden for centuries."[37]

On July 25, the same day that the hadronic paper appeared in the office of *Physics Letters*, Rubbia received a notice from the district director of the Boston office of the U.S. Immigration and Naturalization Service, explaining that Rubbia's failure to fill out a routine visa extension form meant that he must depart the United States on or before July 29.[38] Notwithstanding his professorship at Harvard, Rubbia could not placate the federal bureaucracy;

he left on July 27, unable to return for three months. The ejection had fatal consequences, for it removed the chief link between Sulak, the junior collaboration member who had done most of the neutral current analysis, and Cline and Mann, the two senior members of the widely scattered group.

At first there was little problem. After amending the draft of the paper to suit comments from team members and others, Sulak took it personally to the offices of the *Physical Review Letters* on August 3. The preprint, distributed to every high energy physics laboratory in the world, announced a ratio of neutral currents to charged currents that was high but not inconsistent with $SU(2) \times U(1)$.[39]

As the news of neutral currents redounded through the physics world, Cline and Mann began to get cold feet. To begin with, Cline had done a half-dozen experiments with strange particles that showed that neutral currents did not exist. He could not understand how he could spend years not finding them in strange particles and then suddenly turn them up with neutrinos. Identifying real muonless events seemed so difficult that they were not at all sure of the analysis. Moreover, both distrusted Monte Carlo simulations.[40] Finally, another collaborator had just examined a second batch of data from Fermilab, taken that spring at a higher energy, and found that the total number of neutral current candidates was small enough to cut the ratio of neutral to charged currents in half. It was hard not to wonder if further work would make them completely disappear.[41]

"It's similar to the situation we are facing now," Cline said to us later. He was then working with Rubbia on an enormous experiment at CERN, smashing together protons and antiprotons to discover the W and Z. "We have some Z^0 decay results that are interesting, and are faced with a tough decision. Do we wait and run the experiment in the same way, or do we change the detector to improve our perception of the results? It's not easy. If you change something in the detector, you introduce new parameters. We did two things in that experiment. First, we modified the detector to pick up the wide-angle muons. Second, we got a more intense neutrino beam. These changed the parameters more than we were able to judge, and that led to what I have to call a disaster."[42]

The modification of the detector seemed harmless enough; Cline and Mann simply stuck in a foot-thick piece of iron before the last spark chamber in the first part of the detector. They hoped that this would block many strongly interacting particles, turning the spark chamber perforce into a muon detector. Although they would have liked to put in more iron, they were cramped by the small buildings that were part of Wilson's legacy. Still, because the spark chamber was only a foot behind the upstream end of the detector, rather than four feet, it would pick up more wide-angle muons.[43]

Although Rubbia was out of the country and thus unable to participate fully, he went along with the changes. However, he was confident enough to discuss the first results at a conference in Bonn at the end of August, and another at Aix-en-Provence two weeks later.[44] The two CERN papers appeared in the same issue of *Physics Letters* on September 3; Palmer's paper came out on September 17.

Even as news of the HPW finding swept Fermilab in early August, a group from Caltech headed by Barry Barish and Frank Sciulli began talking with Wilson about checking the result. Wilson had kept the Caltech group in competition with the HPW team since the start of the laboratory, giving first shot at the beam to the latter only after prolonged skirmishing. As the discovery became the subject of growing incredulity—*Physics Today*, the monthly trade magazine of the physics community, reported that "many experimenters are skeptical that either group has demonstrated the existence of neutral currents"—Barish et al. became, naturally enough, interested in confirming or killing the effect.[45] They were asked to work up a proposal in early September; on October 24, they delivered it. They were happy to state why their detector would not be plagued by wide-angle muons, they wrote, "But the burning question now is: *Is the effect real?*"[46]

Cline and Mann were asking themselves the same question. The first test run on the new detector took place on September 28, and it was quickly apparent that they were not picking up muonless tracks.[47] They were almost relieved when the referees at the *Physical Review Letters* bounced the HPW neutral currents paper, complaining that the wide-angle muon problem had not been handled satisfactorily.[48] Conferring via transcontinental phone lines, Cline, Mann, and Rubbia agreed to hold the paper by the simple expedient of not answering the referees' questions. (Sulak was not informed of what had transpired.) By the beginning of November, Mann had begun work on a "no neutral currents" paper that stated the group's "disagreement with recent observations made at CERN and with the predictions of the Weinberg model."[49]

Still in Europe, Rubbia did not actively participate in the revamped HPW experiment. A certain malicious glee nonetheless informed his encounters with the Gargamelle team, which was alarmed by the report that their overseas colleagues were not going to confirm their discovery. A powerful figure in his own right at CERN, Rubbia transmitted his views to Jentschke. The CERN director was appalled at the prospect that workers at his laboratory might have committed a grievous faux pas, which could have unhappy budgetary consequences on an international body zealously protective of its public image. He summoned Lagarrigue, Musset, Rousset, and the rest of the Gargamelle team to a meeting and grilled them about the validity of their experiment.

"Carlo came in from CERN," Sulak remembered. "I was at the mailbox [in Lyman Laboratory] with Karl Strauch [of Harvard] and Carlo came in, so proud of how much trouble the Gargamelle people were having because of their terrible results in the inquisition that he had arranged, having the Director General check them out and find out where they went wrong in their organization—" here Sulak does a sudden imitation of Rubbia's distinctive voice "—'seence the eh-vents rrreally wer-ren't there.'

"I said, '*What do you mean, they aren't there?* The data hasn't changed! What the hell's going on here?' "

The teams were sufficiently large and the experiments sufficiently long that one man, Bernard Aubert, participated in both at various stages of the tale. He spent the latter half of 1973 at Fermilab, from where, on November 5, he telephoned a French colleague, Jean-Pierre Vialle, at Orsay to inform him of the Fermilab group's apostasy. Musset happened to be in Vialle's office when the call came through. "Vialle told me, 'Aubert says they have started to see no neutral currents.' So I said, '*Please*, can I have the telephone?' And then I discussed a little bit with Aubert and immediately I found some problem with their experiment that to me prevented their answer from being conclusive. I always had this prejudice—or this opinion— that their experiment was not very good for that subject." On the next day, somewhat to Lagarrigue's dismay, Musset held a previously scheduled press conference at the French Physical Society and strongly laid claim to the discovery of neutral currents. The disbelief in the room, Musset recalled later, could have been cut with a knife. Another CERN experimenter, Jack Steinberger, had informally computed the background and explained away every single instance of the neutral currents. At the press conference, the head of the French Physical Society bluntly asked: "Are you sure this stuff's not junk?"[50]

On November 13, Cline, Mann, Rubbia, and D. D. Reeder of Wisconsin drafted a letter to Lagarrigue to the effect that the level of neutral currents had basically dropped to zero. Although the letter, on Wilson's advice, was never officially posted, Rubbia showed it around on his next jaunt to CERN: salt in the wound.[51]

Years later, we met Sulak at a hokey Mexican restaurant in a small Ohio town near one of his experiments. The hour was late, the waitresses dressed in cowboy outfits and anxious to go. A band thrashed away at some old songs by the Rolling Stones. Sulak had ordered something messy in a tortilla and a Dos Equis; a waitress hovered over his shoulder, asking every few minutes if she could clear the plate. To her dismay, we began to ask Sulak about the neutral currents experiment. He put his food down and answered in a burst. A big physics collaboration has all the tension associated with the cast and crew of a play; sometimes things can inexplicably go sour, dissent wrack the project, and mire the team in anger for a while.

"They wanted to try something," he said, "and they did something crude rather than doing the right job. There was no space to put anything in, so they put the fattest thing they could put in, and that was like a foot thick." That was the reason? we asked. Lack of space? "Yes. The building was too small, so they put in whatever they could. And everyone knows that a twelve-inch piece of iron doesn't shield hadrons for beans. You look in the book." He dug into his pocket, produced the 1984 Particle Properties Data Booklet, a pamphlet four inches high and two-and-a-half inches wide with over a hundred pages of tiny print. "You don't have to be smart. The book's been around for ages, and people know how to calculate the shielding power. You look up a standard table that everybody's had for years that has shielding power—wait'll I find the doggone thing. Here's iron, Fe." He squinted in the dim light. "You look in here, 'nuclear interaction length,' eighty-two centimeters—almost a yard." He threw the book on the table in disgust. "So if eighty-two centimeters is what you have to do to filter something out, and someone puts in something *this* thick"—spreading his hands a foot apart— "you can imagine that any hadron, any pion or whatever, that comes in there is going to make a shower and something has a good chance of coming out the back."

He took a bite, cooled down a bit. "It's very difficult, even today, to understand the propagation of hadronic showers through matter. The thing is very complicated, with lots of pions and neutrons and stuff that you don't really know how to model. To understand, for every pion going into a foot-thick block, how many come out the back, that is nontrivial. So you have to do all kinds of tests, which is what in the end [Richard] Imlay [of Wisconsin] and company finally did.

"But smart people don't put themselves into a situation where they have to understand something which is un-understandable. The reason the original magnet was four one-meter-thick pieces of iron was just to avoid this problem from Day One, and never have to calculate what happens in the middle of the iron because you never looked in the middle."

On November 29, Imlay completed a preliminary study of what is called "punch-through"—hadrons that pierced the iron shield. He combed the literature for examinations of the likelihood that pions would set off counters after going through given lengths of iron, found two measurements, and made from them a rough-and-ready calculation of the punch-through. The level seemed a lot higher than Mann and Cline had guessed. Imlay, Aubert, and T. Y. Ling then scanned thousands of frames of film to look for charged currents—events with muons in the rear half of the detector—that also had secondary tracks in the back. These secondary tracks would be due to hadrons that punched through the thin iron slab. Presumably, this would also have happened in the case of neutral currents, with the hadrons now fooling the detector into thinking they were muons, and making neutral currents appear as charged currents. Imlay, Aubert, and Ling soon saw that their first estimate of punch-through was, if anything, low.[52]

Cline was aware of Imlay's work but didn't think the punch-through would add up to much. On December 6, a week after Imlay's first memorandum, Cline gave a Friday afternoon wine-and-cheese talk at Fermilab about the progress of the experiment. As he wrote on the transparencies illustrating his talk, the collaboration's "very preliminary results!!!" showed that the level of neutral currents was "very likely too small to be consistent with [the] Weinberg model—also CERN data, if due to [the] Weinberg model. . . ." Neutral currents were "not confirmed by the present experiment" and the few surviving candidates simply needed "further study."[53]

In retrospect, the talk was a ghastly public-relations blunder. Playing by the rules of physics, an informal afternoon chat is not a release of experimental data. Nonetheless, it was treated as such by the hundreds of theorists and experimenters who wanted to know what was going on. The HPW team's apparent flip-flop spread with astonishing speed through the physics world even as people within the collaboration maintained that they *were* finding neutral currents. If they stuck to their original findings, the heat of publicity would treat that as a second reversal. The relentless curiosity of their colleagues thus would have translated the routine business of sifting through the data into a set of humiliating about-faces.

Cline's seminar began a week of heated discussion within the group. By December 13, Cline had changed his mind, admitting "the distinct possibility that a muonless signal of order ten percent is showing up in the data. At present I don't see how to make these effects go away." They decided to publish the original paper. It appeared in *Physical Review Letters* on April 8, 1974, six months after the original submission. Another CERN paper came out in *Nuclear Physics* on June 24. When the Caltech group announced that it, too, had observed neutral currents, their reality was at last accepted.[54]

Nobody ever won a Nobel for the discovery. Lagarrigue and Musset were logical choices as the heads of the CERN search, but Lagarrigue died suddenly in 1975 and Musset was killed in the summer of 1985 in a climbing accident near Mont Blanc. In addition, both the HPW and CERN groups had lost a measure of credibility. The lengthy brouhaha, which provided cliff-hanger entertainment to the physics community, was mocked as "the discovery of alternating neutral currents." But now that neutral currents existed in neutrino interactions, the most pressing question became: Why hadn't they been found in strange particle decays? Even the most presumptuous theorist was unlikely to argue that dozens of past experiments were incorrect. "Three or four, maybe," Glashow said. "But this was something like *twenty*. People were walking around wringing their hands and asking, 'What's going on?' So I put them out of their misery and *told* them. It was charm."[55]

In April 1974, Glashow spoke before a conference of meson specialists that was held at Northeastern University, in Boston. In a speech entitled "Charm: An Invention Awaits Discovery," he explained to the assembled experimenters that by virtue of the mechanism described in the Glashow-Iliopoulos-Maiani paper, the fourth quark prevented neutral currents from occurring in strange particles. He set down his reasoning on the blackboard, cheering up some in the crowd when he botched the mathematics the first time through. Undeterred, Glashow challenged the meson physicists in the room: They were rightfully the ones to discover charm, because the lightest charmed particles would be charmed mesons, he said; but because they weren't looking, charm would be found by "outlanders"—other sorts of experimenters, such as the neutrino teams at CERN and Fermilab. By the time of the next meson conference, Glashow said, "There are just three possibilities: One, charm is not found, and I eat my hat. Two, charm is found by [meson specialists], and we celebrate. Three, charm is found by outlanders, and *you* eat *your* hats."[56]

Two months later, Iliopoulos threw down the gauntlet before the Rochester conference in London. An excellent orator, Iliopoulos delivered a tub-thumping defense of $SU(3) \times SU(2) \times U(1)$. There are, he asserted, two fundamental classes of matter: quarks and leptons. There are four members of each class, shuffled neatly into two families. Electromagnetism, the weak force, and the strong force are covered by two gauge theories consistent with each other, the whole being described in a grand synthesis called $SU(3) \times SU(2) \times U(1)$.

From this, he said, "a very simple and beautiful picture emerges." With the acceptance that Yang-Mills gauge theories described all interactions, an easy step led to a startling insight. *$SU(3) \times SU(2) \times U(1)$ is nothing other than the broken remnant of a single unified gauge group that existed in the distant past.* Finding the unified theory now depended on plucking out the right group.

The synthesis demands that charm exist, Iliopoulos said, and he bet the assembled physicists bottles of fine wine that the fourth quark would be found before they met again. At stake, he said, was more than just another quark or a couple of models. At stake was the possibility of unification itself.[57]

18
Charm and Parity

ALTHOUGH GLASHOW AND ILIOPOULOS DIDN'T KNOW IT, THEIR BETS WERE safe. Evidence for charm had already been collected, and more would follow. A team of experimenters at Brookhaven was turning up traces of the fourth quark even as Glashow spoke, although they ultimately played only a small role in the drama that charm became. The first definitive traces were found months later by two different teams, one from MIT and one from Berkeley and Stanford. Yet when all was said and done, it took another year and a fourth sighting before charm was accepted. The finding of charm is now described by a few quick lines in undergraduate textbooks, a circumstance that has prompted Nicholas Samios to remark that such volumes should be printed with some combination of sweat and blood.[1]

Since the revelation of the second neutrino, Samios and Robert Palmer, like Musset and his colleagues at CERN, had planned an experiment to study neutrino interactions by shooting them into a big bubble chamber. At Brookhaven they constructed a seven-foot bubble chamber—much smaller than Gargamelle, but still impressively large—and filled it with liquid hydrogen. The nucleus of a hydrogen atom consists of only a single proton, and therefore when it was struck by a neutrino only two particles would be involved. By contrast, Gargamelle used freon, which is a complicated molecule with 159 neutrons and protons. Neutrinos hitting hydrogen nuclei, Samios said, "is not a messy collision. If you saw something happen, you could measure and calculate it to a gnat's ass."[2] Brookhaven approved the experiment in August 1973.

Shortly afterward, Glashow and Samios met at Brookhaven. Glashow, who was then a consultant for the laboratory, proposed to Samios that the new experiment would be a good way to look for charm. A speeding neutrino might interact weakly with a proton inside the chamber—that is, a W^+ would pass from one to the other. Upon absorbing the W^+, one of the down quarks inside the proton could change to a charmed quark, and for a fraction of a second—10^{-13} seconds, according to theory—a charmed baryon would come into existence. The charmed baryon would not survive long enough to be photographed. But when it decayed (and the charmed quark again changed its identity, becoming a strange quark), it would turn into a strange baryon—a lambda particle. The lambda, too, would eventually disintegrate into other particles, but it would hang around long enough to be detected.

As physicists had known for twenty years, strange particles are generally produced in pairs. (The reason is that they come from the production of a strange quark and a strange antiquark, which fly away from each other to produce two strange particles.) Glashow pointed out that if experimenters saw an exception to this rule—the creation of a single, solitary strange particle—it would be evidence for charm.

Physicists are always discerning the existence of things they can't see from things that they can, but Glashow's chain of reasoning here seems particularly oblique. He was assuming that the presence of a single strange particle meant that it had been produced by another particle (the charmed baryon) that did not live long enough to be seen, in whose interior was *another* type of particle (the charmed quark) that *inherently* could not be seen. Invisible pieces of invisible particles; Samios and Palmer were intrigued but skeptical.

The first neutrino interaction in the seven-foot bubble chamber happened in January 1974.[3] Except for occasional maintenance breaks, the AGS shot a billion neutrinos every two-and-a-half seconds into the bubble chamber until March 1975. Twenty-four hours a day, seven days a week, the three cameras atop the detector snapped pictures, capturing on seventy-millimeter film whatever they saw inside. By the end of the run, Samios's team had taken more than a half million slides. According to the theory, neutrino-proton interactions should be recorded on about two thousand of them, less than one percent.

The Brookhaven scanning team was highly regarded, largely because of the work of its supervisor, Milda Vitols. A Latvian emigré who had spent several years in displaced person camps before settling on Long Island, Vitols was working as a YMCA cashier when she spotted an ad in a local newspaper for a scanning job. She soon worked her way up to a supervisor, thanks to her unusual knack for explaining what the physicists were talking about to the other scanners.[4] She also had to encourage the scanners and keep their spirits up in the face of considerable pressure. Worried that the tedium of the job might make the scanners miss events, the physicists had them placed in separate booths so that they couldn't talk to each other while working. Moreover, the scientists would periodically have the scanners' work secretly redone, and assign them "efficiency numbers" that depended on how many events they had not spotted. "That's an important number," said Michael Murtagh, a physicist who worked closely with the scanners. "It goes into the published results. To have an idea of how many events you had, one of the factors is how many events the scanners missed on the film, and to calculate that you needed to know the efficiency of the scanners."[5] Scanners, however, found it humiliating.

Being electrically neutral, a lambda would not leave tracks, but its decay would leave two particles whose tracks would be visible. Vitols told the scanners to look for two rays appearing out of nowhere and shooting off in a V. (This would be the decay products of the lambda.) If the scanners saw a V, she said, they should look to see if the vertex pointed back to the vertex of another set of tracks. (This would be the locus of the initial collision between the neutrino and the proton.) If they found a second group of lines near the first, Vitols said, the scientists would be very excited, because the configuration would mean that their quarry, the lambda, had glided silently between these two points for the brief flicker of its existence.

On May 3, 1974, Helen LaSauce, an ex-switchboard operator and mother of three, advanced her scanning machine to frame 6,967 of roll 27, and sketched a lone strange particle—the artifact of charm. Five tracks led away from one point of impact, while close by a pair of tracks parted from a second vertex. Camera angles 1 and 2 were unclear, so she decided to sketch angle 3.[6] It was near the beginning of the experiment; she had no idea she was seeing anything special. She just started drawing, her mind on the easy listening music pouring from the transistor radio on the desk. "They never told me much about what I was doing," LaSauce said. "The only way I knew I had found something important was when *they* [the physicists] asked me to remeasure it for the umpteenth time."[7]

At the end of the week, LaSauce had filled up her sketchbook. She passed it to Vitols, who glanced through the book, looking for events she thought would interest the physicists. When she came to frame 6,967, she pointed it out to Samios.

"I remember being in the control room," said Palmer, "and Nick came rushing in, waving this sketch and saying, 'It fits charm!' Now, I think this was the first strange particle we'd seen. The probability that we'd find a *single* strange particle the first time around was just nothing. And I remember thinking, okay, Nick's being wild again. Clearly one of those other tracks is another strange particle. I was skeptical because it was a new chamber, and we weren't really sure how to identify with certainty all the tracks. But it *was* a beautiful photograph. Not one but two electrons stopped in the picture, and that's something that gives us the information to calculate the dynamics of the interaction very precisely. Still, it took us seven months of round-the-clock calculation to conclude that, to an infinitesimal degree of uncertainty, the first set of five tracks consisted of four muons and a pion, and the second pair of a proton and another pion. We had found a lambda—and charm.

"But we'd only found *one* of them. We couldn't turn up another. You don't want to say you've found a new state of matter on the basis of one event, so we kept going back and redoing the calculations."[8]

Once before, Samios and Palmer discovered a particle on the basis of a single photograph: the omega minus. Made in a new detector with a new beam, frame 6,967 was nowhere near as clear-cut a case for discovery. Proving that one and only one of the seven tracks was a strange particle took months of endless figuring, while Samios and Palmer kept begging for more beam time to find a second.[9]

While Samios and Palmer and their team pored over frame 6,967, two other groups of experimenters came up with more definitive evidence of charm. Whereas Samios had found a charmed baryon, these teams found mesons made of charmed quarks. One team, whose head was Samuel Chao Chung Ting of MIT, drew protons from the same Brookhaven accelerator that Samios and Palmer had used. The other, led by Burton Richter, was working on the other side of the country at a new SLAC accelerator. The MIT and Berkeley-Stanford teams were run utterly differently; their joint discovery is a study in contrasting styles—open against closed, precision standing opposed to luck, an army versus a commune—brought into irremediable conflict by nature's chariness with her secrets.

The Patton of experimental physics, Sam Ting is famous for driving himself and his collaborators to exhaustion. A tall, stooped man with large features, Ting has the trick of commanding a room by speaking so quietly that everyone in it has to shut up to hear him at all. His manner is mild, almost affectless; his pants are baggy, his feet slightly splayed; but the look is misleading, for Ting is one of the world's great control freaks—a man of fierce ambition and abiding determination who has ruthlessly pursued his vision of physics for a quarter century. His colleagues love to recount tales of Ting sleeping on the floor by his equipment, or being physically carried out of experimental halls by lab officials who have slated his experiment to be replaced by another. The stories are exaggerated, but their circulation is a measure of the man. Renowned both for sudden, towering rages and for prodigies of self-discipline, Ting has demonstrated a gift for inspiration, command, and bureaucratic maneuvering not usually associated with the pursuit of knowledge. Today, he is putting together the largest experiment ever done, a matter of some 400 Ph.D.'s from more than a dozen nations, which is scheduled to begin on January 1, 1989, the inauguration of CERN's new Large Electron Positron machine. (The vast LEP will itself be *sixteen miles* around.) Ting planned to employ satellites to coordinate the reams of expected data to be analyzed by groups from the Soviet Union, China, the United States, and the thirteen member nations of CERN. "There are two kinds of experimenters," he once told us. "The first kind do what theorists tell them to do. The second kind follow their own ideas. I am of the second kind. I am happy to eat Chinese dinners with theorists, but to spend your life doing what they tell you is a waste of time."[10]

Ting was born prematurely in 1936 at Ann Arbor, Michigan, the child of two Chinese university professors on a brief visit to the United States.[11] Despite his accidental citizenship in the United States, Ting did not return to the United States for another twenty years. Knowing little English, he arrived at the Detroit airport with a hundred dollars and the determination to work his way through college. He spent the next six years surviving on scholarships, learning English, earning degrees in mathematics and physics from the University of Michigan. After accepting a post at Columbia in 1965, Ting became interested in a series of experiments on quantum electrodynamics.

Although quantum electrodynamics had never failed an experimental test, many theorists thought the theory would have to be replaced after particle accelerators reached several billion electron volts.[12] The first important high-energy test of quantum electrodynamics had been done at the recently built Cambridge Electron Accelerator, or CEA. Sponsored by MIT and Harvard, the 6 GeV CEA was one of the last of the university laboratories, built at a time when national laboratories such as Brookhaven were taking over all high energy research.[13] In 1964, a group of experimenters led by Francis Pipkin of Harvard used the machine to make high-energy photons and smash them against a target to create electron-positron pairs, which were then directed into a new type of detector called a double-arm spectrometer. This device took the streams of electrons and positrons issuing from the target and passed them through a series of magnets which bent them back to a point, like two streams of traffic that momentarily separate around an island and remerge. A set of counters on each arm act like "traffic flow meters" to measure the passage of each electron, and to ensure each electron is paired up with a positron on the other side. Pipkin's aim was to test the prediction of how many pairs would be produced at particular angles. It was, he said later, "the first opportunity to test the theory in a domain where you might expect the thing to go screwy."[14] At high energies, the pair production did not seem at all as theory said it should be, and Pipkin announced as much at a series of conferences in the spring of 1965. Indeed, quantum electrodynamics seemed to have gone screwy.

"When we first announced the thing," Pipkin said, "people weren't particularly perturbed. But when people began to think about it, they were." Quantum electrodynamics, which had originated in the work of Dirac, Heisenberg, and Pauli, and had culminated in the work of Feynman, Schwinger, and Tomonaga, was *the* successful physical theory. It was one thing to speculate on its inadequacy, completely another to imagine physics without it. Faced with a growing clamor of dismay, Pipkin's team began working on their equipment in the summer of 1965 so that they could redo the experiment and publish a complete paper. On the evening of Independence Day, when most of the rest of Cambridge was down along the Charles River watching the fireworks, Pipkin dropped by the CEA experimental hall, which was on the Harvard campus, across the street and down a bit from Lyman Laboratory. Working late that night was another team that was filling a big new bubble chamber for the first time. Because new chambers usually leak, they tend to be filled in frustrating fits and starts, with delays to seal problem areas. Both experiments were going smoothly.

Pipkin returned to Harvard for his shift at four o'clock in the morning. As he approached the CEA, he was stopped by a police line and a circle of fire trucks around the building. There were flashing lights, crackling walkie-talkies, ambulances with gaping doors—all the trappings of a modern catastrophe. As he learned later, the bubble chamber had caused an explosion. The windows through which the photon beam was to enter had cracked, and the liquid hydrogen had gushed out to the floor. (Because the filling had gone unexpectedly smoothly, the technicians had neglected to put on the metal safety caps for the windows.) Liquid hydrogen is as explosive as dynamite; hot vacuum pumps throbbed nearby. The blast blew the concrete ceiling off the building; coming down, it shattered on the steel girders and rained concrete on the technicians below. One person was killed, five more were badly injured, and all the equipment in the hall was smashed and twisted by the fire and explosion. Pipkin spent weeks with the subsequent investigation. With two team members hospitalized, the experiment was over. Nevertheless, they published their results.[15]

For someone like Ting, who had just mastered quantum electrodynamics, these results were distressing indeed. "It was a very important experiment, and so I thought that I should do it over again," Ting said. "Quantum electrodynamics was the only theory at that time—even now—that's really very accurate, from cosmological distances to satellite communications to very small distances. And so I was very surprised for someone to say that it is wrong." Ting's first thought was to redo the experiment at the CEA, and on November 9, 1965, he drove up to Cambridge to visit Pipkin. That happened to be the day of the great Northeast blackout; the two men met for the first time in the dark. Ting wanted to know what Pipkin had done, what problems he had encountered, and what suggestions he had if a similar experiment were to be performed. Pipkin responded to all these questions in detail. Then Ting asked how he could go about doing the experiment on the CEA. On this subject Pipkin was less encouraging. "It was made very clear to me that a young person with no background had no chance to do an experiment there."[16] That decision rankled for years. A few years later, when Ting was appointed a professor at MIT, one of his first conditions for accepting the post was that he *not* have to work at the CEA.

Annoyed, Ting contacted lab officials at the Deutsches Elektronen-Synchrotron (DESY), a slightly more powerful "lookalike" of the CEA in Hamburg. The director of the laboratory, Willibald Jentschke, knew Ting from his CERN days, and was encouraging. The following March, Ting took a leave of absence from Columbia to do the experiment in Germany. Just before he left, Ting ran into Leon Lederman, then director of Columbia's Nevis laboratory. Lederman expressed skepticism about Ting's ability to carry off an experiment of a sort he had never done before, and bet Ting twenty dollars that it would take him at least three years.

When Ting arrived in Hamburg, he immediately put into play the knowledge he had learned from Pipkin. Like Pipkin, Ting used collisions between photons and nuclei to produce electron-positron pairs, which he then studied with a double-arm spectrometer. The experiment had all the hallmarks of subsequent Ting experiments: meticulously executed, tightly scheduled, and performed by a totally dedicated international collaboration whose members worked up to seventy-two hours straight and slept by the machine. He has used that electron-positron experiment as a template again and again, and worked with some members of that group for the next twenty years. "Electron-positron collisions are very simple systems," he said to us.[17] He spoke carefully, slowly, laconically; clipped sentences with a slight accent. "They are easy to set up and understand. When they collide, they annihilate each other, and for a brief moment all you have is light. Then the photon of light decays into other things." Primarily, electron-positron pairs. "There aren't lots of other particles floating around, so you have a good idea of what you're doing."

Ting and his group carried off the experiment with such speed and precision that it was completed in several months. Quantum electrodynamics was strikingly confirmed. The refutation of Pipkin's CEA experiment was dramatic enough that Ting decided, in September of 1966, to fly to a Rochester conference at Berkeley to proclaim it. But Jentschke, DESY's director, tried to get Ting to soft-pedal the contradiction, arguing that a flat-out attack on the CEA work might wreck the credibility of the laboratory as a whole. "DESY was built with much help from the CEA. And here we had a result which really would do a lot of damage to the CEA. The question was, should I just present my data alone, or should I present my data as a comparison with the CEA data? I mentioned to the director that as long as people have published their data—published it in *Physical Review Letters*—I can quote it, I can put it on the same graph as mine. He said no, that's too impolite. I said, well, it's *not* too impolite, because after all, I may be wrong, too. If somebody presents his data, and publishes it, other people can quote it—that's the *whole purpose* of publication! We had a *very* strong disagreement about that. We had the disagreement in front of the DESY cafeteria. There were a lot of people going in and out, and he and I were talking, and people were surrounding us, looking on. My comment was, 'If you publish a result, you take the consequences for it!' "

Ting decided to go ahead and compare his results and Pipkin's on the same transparency, which he drew on the plane on the way over. At the conference he listened patiently while a Cornell group, working at a much lower energy on the Cornell accelerator, reported results which appeared to support Pipkin's.[18] A few minutes afterward, Ting quietly but firmly said that the Harvard and Cornell experiments were wrong, and that quantum electrodynamics was still valid. The announcement was a shock: An unknown thirty-year-old from a German knock-off of the CEA had baldly claimed that not one, but two, outstanding American teams had stumbled badly. "Science," Ting has remarked with evident satisfaction, "is one of the few areas of human life in which the majority does not rule."[19]

Ting followed up this experiment with more work on high energy photons, and became particularly intrigued by the way they could occasionally and spontaneously turn into three vector mesons, the rho, phi, and omega, and then back again. (The omega meson is not the omega minus, a baryon, but a meson known as "little omega"; *vector* means "spin-one.") They were enormously difficult to observe; out of a hundred thousand rho decays, only *one* produced an electron-positron pair for his detector. Theorists identified them as quark-antiquark combinations; up-antiup (rho), down-antidown (omega), and strange-antistrange (phi). Because they have exactly the same quantum numbers as photons, Ting called them "heavy photons," and has continued to study them throughout his career. "Always the same thing, always a little more precisely."

At the beginning of 1970, Ting began to have the hunch that there were more heavy photons in addition to the rho, phi, and omega. "The world is not so simple that a photon just changes to three vector mesons, and not more." The most suitable place, Ting thought, would be the new accelerator at Fermilab. He drew up a proposal—Fermilab proposal 144—and one of his collaborators, Min Chen of MIT, went to Batavia to design the detector. Construction lagged, however, and Fermilab officials seemed to be dragging their heels about approving the Ting group's proposal. "I think my style was really very different from the style of Bob Wilson, who was then the director," Ting said. "I did a lot of work, designing a beam, and we never got this approved.

"Then I decided I should propose this experiment at CERN. So I showed up at CERN. And the management of CERN at that time said, 'This is crazy; we know there's no heavy photons. Besides, we have more important things to do.' So I was also very discouraged at the CERN PS [Proton Synchrotron]. And then I decided, maybe I should do it on the ISR, the Intersecting Storage Ring [another CERN accelerator]. So I submitted a proposal there. Well, the management of the ISR at CERN said, 'This is fine, but you are a completely American team—we need some European balance.' And so fundamentally I also didn't get it approved. I did not get rejected, but I did not get the thing approved." As the year 1970 wore on, Ting flew back and forth between Hamburg and Chicago trying to get his various proposals approved, with no luck. He became angry and despondent.

"From Hamburg to Chicago you spend a lot of time and effort to design a beam, and design lots of other things, and you get a flat 'no' from these people. And there was *no reason* for it. They had the beam, they had the money, they have the backup. And basically it was a conflict of style. I became discouraged because people were not using a physics reason to disapprove it. You see, if people say 'no' to you, and give you a physics reason, you can understand. You may not agree, but you can understand. If people say, 'You don't have enough Europeans on your team'—this is not a physics reason." On top of his depressed mental state, Ting began to develop an ulcer. On the advice of his physician, and lab director Jentschke, Ting took several months off from physics to relax.

When he returned, it was with a vengeance. His various proposals still unapproved, he decided to make one last try. At the beginning of December 1971, he called Chen from DESY at Hamburg to announce that he had decided to quit Fermilab and do the experiment at Brookhaven. Chen violently opposed the idea; he had just spent six months completing a design for a Fermilab detector, and was ready to build. A few days later, Chen received a letter from Ting instructing him to quit Fermilab immediately; in place of a closing salutation, Ting wrote, "I will not accept any other alternatives or suggestions."[20] Chen was back at MIT within the week. Between Christmas and New Year's, Ting and Chen got together to write a proposal for Brookhaven, submitting it on January 11, 1972.[21]

They planned to build an experiment on one of the five beam lines in Brookhaven's slow extraction experimental hall. The Alternating Gradient Synchrotron (AGS), the lab's biggest accelerator, can deliver its protons in two ways; It can spit them in one short bunch a few microseconds long, or pour them out in a long bundle of about a second's duration. An experiment is usually suited for either the one or the other, there are separate halls for each, and the AGS beam is usually in only one mode or the other. Samios and Palmer's bubble chamber experiment was in the fast extraction hall, because they wanted to make sure that any events that occurred did so during the bubble chamber's brief period of maximum sensitivity. Most detectors, on the other hand, work best with slow extraction, where the events are spread out so they can be counted more easily. Once delivered to a hall, the beam is subdivided into different beam lines so that many experiments can work at once.

Ting proposed to slam slowly extracted protons from the AGS into a target to produce a hail of particles, mostly protons, neutrons, and pions. These would be ignored; the team was interested only in new, heavier cousins of the rho, phi, and omega. If any were produced, some fraction of them would decay into electron-positron ($e^+ e^-$) pairs. Any sudden upsurge in the number of such pairs at higher energies would give away the presence of new heavy photons. As ever, a highly sophisticated double-arm spectrometer would count the pairs. The $e^+ e^-$ pairs would be produced in the horizontal plane and bent vertically, separating the measurements of angle and momentum, making them more precise and easier to check.

In May 1972, Brookhaven approved Experiment 598, as it was now called, awarding it a thousand hours of beam time.[22] Ting, Ulrich Becker, and Chen spent the next year constructing the detectors, magnets, and beam pipes, and in the fall of 1973 began installing them in the slow extraction hall of the AGS. Because Ting was examining the direct collision of the protons and the target, instead of watching the decay of secondary particles, the beam line was brought into the slow-extraction experimental hall, where it had to be covered with shielding. To Ting's dismay, he needed much more shielding than planned—ten thousand tons of concrete. The CEA had just been shut down due to budget cuts, and, profiting from Cambridge's misfortune, Ting's group was able to borrow all of its old shielding. In addition to the concrete, they used a hundred tons of lead and five tons of Borax soap, which contains boron, a neutron stopper. The team worked on the A beam line, next to Experiment 614, which was led by Mel Schwartz, Ting's former colleague at Columbia, who had since moved to Stanford. The two experiments would alternately extract protons from the A line.

Ting's detector was enormous in every way: in size, intricacy, sensitivity, and cost. Situated near the center of the slow-extraction hall, the detector—an eighty-foot V swaddled in great concrete blocks—rose in a giant-sized jumble to the ceiling. Fist-thick cables coiled from the apparatus to a small shelter against one of the dull green walls; here the experimenters lived for days on end.

In April 1974, Ting, Becker, and Chen asked for beam and started shaking down their equipment by looking at the phi meson. When the number of electron-positron pairs produced by the collision is graphed against the total energy, the phi is indicated by a hill whose summit, or maximum number of pairs, is at 1.02 GeV (alternatively, 1,020 MeV). To

Ting's mind, the detector had to measure the energy of the electron-positron pairs with great accuracy, because the energy was equivalent to the mass of the particle that produced them. Consequently, his detector was fantastically precise—needlessly so, he was told, for the vagaries of quantum physics smeared the mass value over a broad range. For this reason, particles are spoken of as having a wide or narrow "width," depending on the spread in their mass value; typical widths are on the order of several hundred million electron volts. Ting's detractors complained that making a detector able to detect gradations twenty times smaller than the natural width of the narrowest known particles was ludicrous. "What people worried about was the cost," Ting told us. "The cost was over five million dollars. That was more than ten years ago, when five million dollars still was a lot of money. Most of the theoretical physicists said, 'But there's no particle with a narrow width, and we know it. There's no particle there anywhere. You are just wasting your time.' So I did have a lot of hard times to get the money."

Brookhaven shut down for its summer break in May.[23] Near the end of July, the machine was switched back on and ready to deliver protons to the slow extraction hall. Experiment 598 ran for two weeks, taking data at 3.5 to 5.5 GeV. They found nothing. On August 7 the beam was switched over to Mel Schwartz's experiment, which used the beam for another two weeks.

On August 22, Ting got more beam, this time tuning his detector to between 2.0 and 4.0 GeV. The first few days were difficult; the AGS frequently malfunctioned, and on two successive evenings electrical storms put the whole accelerator out of commission for several hours. Nevertheless, the team ran until the last minute when the AGS was switched off for a maintenance break at four o'clock in the afternoon of Wednesday, September 4.

On every other experiment, a single analysis team dissects the results. Ting's fanatic care is such that he always has had *two* data analysis groups independently crunching the same numbers. Each team wrote their own computer programs to find the energies of each electron-positron pair and to plot the results on a graph. Bumps in the curves would indicate a new particle—a heavy photon. Captained by Becker and Chen, the two analysis squads started working on the data right after the AGS was turned off for a maintenance break on September 4.[24]

Sometime in the next day or two, Terry Rhoades of Brookhaven drove from the lab to Cambridge with the 2-4 GeV data tapes to analyze them with Becker. Each had written half of the computer program; Rhoades and Becker stitched the two parts together and ran them through on great piles of cards through the MIT computer. It was late Saturday night.

When the computer center sent the numbers back, Becker said, they stared at them in "intrinsic disbelief." The printout was a rough graph, with asterisks indicating the number of events at each energy, and the energy in gradations of .025 GeV (25 MeV).[25]

2.875	0
2.900	0
2.925	0
2.950	1***
2.975	0
3.000	3*********
3.025	1***
3.050	6*****************
3.075	17***
3.100	15***
3.125	5***************
3.150	1***
3.175	0
3.200	0
3.225	0

Nearly all the events were piled up at 3.1 GeV; instead of a hill, they were looking at a needle. No known subatomic object had such a narrow width. Something was obviously wrong with the computer program; if a routine is incorrectly written, the computer often reacts by shelving all the data into one category. Becker and Rhoades began to shout at each other, each sure that the other had screwed up one of the thousands of statements in the program. They argued until midnight, combing through the cards, neither man giving in. "Since we are both not particularly fond of making compromises," Becker said, "we went to the computer center and threw everybody else out." Brazenly yanking everyone else's cards, they rudely inserted their rechecked cards in front of the stack. This brought down the wrath of another experimenter, Wit Busza, who had been waiting for hours to use the machine; Busza shouted that he was going to call the lab director then and there to protest this behavior. Unperturbed, Becker and Rhoades waited for the cards to go through. Paper clattered from the printer: a needle at 3.1 GeV, the sharpest peak ever seen.[26]

At the same time, Chen began his analysis at Brookhaven. Chen's data were in three separate parts, and when he sat down to total the result by hand, he was astounded to see that most of them were right atop 3.1 GeV. Skeptical, he rechecked his analysis, and got the same result. "I ran into Lee Grodzins," Chen told us, "who was teaching the same course as I was at the time. I yelled at him, 'I just found a narrow resonance!' [Resonance is another term for a short-lived particle.] And he *laughed* at me. He said, 'Maybe the whole world is full of narrow resonances!' "[27]

Becker and Chen each called Ting that Sunday to tell him the news. He told them to come down to Brookhaven immediately and called a joint meeting of the analysis groups. They met Monday morning, September 9, in one of Brookhaven's small conference rooms. Afraid of having committed a disastrous error, both groups sat in silent expectation for what is today recalled as a very long time. Finally, someone in Chen's group said, "Our analysis looks peculiar—" and suddenly everyone was shouting at each other.

Afterward, Chen was overcome by excitement. Delighted, he raced into the central cafeteria, his arms full of cards and charts. Experiment 598, he proclaimed, had found a particle like nothing that had ever been seen—a spike. Ting was enraged. He had seen too many experiments get in trouble by prematurely releasing data. It was made known that he would be extremely angry with anyone who talked about 3.1 GeV outside the group.

Ting was unsure of what to do next. The experiment had ended its scheduled run, having logged only 470 hours of the 1,000 approved hours of beam time. To get more beam time, which would give him more data and allow him to test his results, he would have to wait. The AGS maintenance break was scheduled to continue until October 2, and when the machine was switched on again fast extraction experiments would run. After that, he would have to jockey for beam with the other slow extraction experiments, including 614, Schwartz's experiment. It would be at least six weeks before Ting could run again. He was in an exquisite dilemma: Experimenters must push their equipment to the limit, and announce their findings before others get there first. But they thereby court the risk of overestimating their equipment—a lesson Ting knew well, having established his reputation because of Pipkin's inability to recheck his results.

Against the advice of some group members, Ting decided to wait. In the meantime, he wrote a polite letter to lab officials asking for more time, and followed it up with a less polite phone call.[28]

"If I may make a bad example," Becker said, "suppose you go out with a net to catch a butterfly, and you catch an elephant in it instead; how would you move? The answer is, damned carefully, so the net doesn't break!"

Ting's decision was risky principally because SLAC had a new accelerator, known as SPEAR, that could discover the spike in a day—if the operators knew where it was. Whereas the $e^+ e^-$ pairs in Ting's experiment were by-products of collisions between protons and nuclei in a target, the Stanford Positron-Electron Accelerating Ring used electrons and positrons as the beam itself; Ting was painfully aware of the consequences entailed by this difference. Being the shrapnel from explosive impacts, the electrons and positrons in Experiment 598 shot out with a wide range of energies, some flying away violently, others dribbling out to the side. The detector could thus survey a broad energy spectrum, but it could not focus precisely on any given value because the electron-positron pairs came out more or less randomly. Only Ting's obsessive insistence on fine-tuning the detector to pick up every detail had allowed the group to find the spike at all. Indeed, an earlier experiment with a less precise detector had found only a puzzling "shoulder" in the 3–4 GeV energy range.[29]

SPEAR, on the other hand, circulated electrons and positrons around a ring for hours at a time, pushing them to ever-greater energies. The electrons traveled one direction about the circle, the positrons the other; they met in the middle and destroyed each other, producing minute, savage flashes of energy that turned into other particles. (Today we would say that the energy, which is in the form of photons, "promotes" to the status of real particles the virtual quarks and leptons in the vacuum.) Such machines were built because accelerator physicists realized that accelerators might be able to reach higher energies by letting sprays of particles slam into each other instead of into stationary targets. The difference is the difference between driving a car into a motionless tree and having a head-on collision with another vehicle; the head-on is much more violent.

Controlling the particles in their course around the ring was impossible unless they all traveled with the same momentum. By contrast, imagine trying to guide a mix of speeding race cars and lumbering vans around a narrow race track. Because there was little range in the momentum of the electrons and positrons, the energy—or, alternatively, the mass—of the produced particles was restricted to that narrow range. (The reason is the conservation of energy.) Thus, if each beam was given an energy of 1.5 GeV, for instance, the collision energy of the two beams was $1.5 + 1.5 = 3.0$ GeV, and only events of that specific energy could be seen. To switch from 3.0 to 3.2 GeV, physicists had to turn off the whole accelerator and reset the magnets and injection beam so that the beams each had an energy of 1.6 GeV. The result is that the new SLAC machine was a superb probe for any given energy, but useless for sweeping broad ranges. Because the designers of the machine had thought in terms of broad bumps, rather than spikes, the limitations of the machine became apparent mostly in retrospect.

Hence the MIT team's fears: If SPEAR was set to a collision energy of exactly 3.1 GeV, the machine would find the spike instantly. The machine operators need only hear the magic number 3.1, and X would mark the spot. If they sat at 3.2 GeV, they would miss it. During the debate over whether to announce the peak, Chen called an MIT colleague who was visiting Stanford and ever so casually asked what energy SPEAR was running. When he heard that it was skipping between 4.0 and 6.0 GeV, Chen breathed freely; SLAC was in the wrong terrain, and had missed it.

Completed in April 1972, SPEAR was the culmination of a decade-long struggle by Burton Richter of Stanford. Initial speculation about colliding beams centered on beams of protons; Richter, on the other hand, thought of electrons and positrons. In the late 1960s, the Italians built ADA, a tabletop model of an electron-positron ring that was so tiny that it was filled by shooting in a burst of electrons, then flipping the accelerator upside down and firing in the positron beam through the same port. The Italians then built a bigger version, as did Russian and French physicists. Throughout, Richter pushed fruitlessly to have an e^+ e^- machine in the United States. He was lucky: SPEAR eventually caught the fancy of John P. Abadessa, then the controller of the Atomic Energy Commission, who figured out a bit of bureaucratic financial legerdemain to smuggle the accelerator past Congress as a particularly expensive experiment. And, in fact, the ring was pasted onto the extant SLAC accelerator for a price tag just two million dollars higher than Ting's detector alone.[30]

After having spent much of his career struggling to build SPEAR, Richter was the obvious choice to take the first crack at using the ring. A plainspoken, easy-going man, he put together a group from Stanford and Berkeley that worked closely and informally on a preliminary sweep of the energy domain. "We wanted to look for new phenomena," Richter said. "And what we decided was to start at 2.4 GeV and move in steps of 200 MeV. So we went 2.4, 2.6, 2.8, 3.0, 3.2, and at that time we could only get up to about 4.8 or 5 GeV. Because we were not—nor was anyone else—thinking of a new kind of quark in a very narrow resonance." (Resonance, again, is another name for an unstable, strongly interacting particle.)

The SPEAR group was primarily examining a ratio known in the jargon as R, which is, roughly speaking, the number of hadrons divided by the number of muons produced in electron-positron annihilations at a given energy. Ordinarily, high-energy e^+ e^- collisions make mu-antimu pairs, but if the energy is equivalent to the mass of a hadron, showers of mesons and baryons can pop out. A sudden jump in R thus indicates the existence of a new particle. From the first, there were odd things about the results. As early as January 1974, a SPEAR team member named John Kadyk first noticed that the value of R for the run at 3.2 GeV was about a third higher than expected.[31] After the completion of the first series of runs, in June, another team member named Martin Breidenbach followed up on Kadyk's observation by taking data around that point, at 1.55, 1.6, and 1.65 GeV per beam— that is, 3.1, 3.2, and 3.3 GeV total energy, respectively—to check whether this was a machine error; he also took additional data in the area of another unusually high spot, around 4.0 GeV total energy. At a first glance, everything seemed normal. "A preliminary analysis of that data was done," Richter said, "and nothing very dramatic showed up. The group had been working twenty-four hours a day, seven days a week, for more than nine months, and we all collectively collapsed for the summer." SPEAR shut down in July for a three-month break. When the accelerator was switched on again, some members of Richter's team began detailed analysis of Kadyk's and Breidenbach's data, while others set about retuning the machine to run at a higher energy, 4.0 to 6.0 GeV, hoping to find something there. To Ting's relief, they spent October hopping around that energy region, far away from 3.1 GeV.

Back on the other coast, Ting's frustration increased when the AGS was restarted on October 2, for problems developed immediately with a newly developed experimental magnet in the fast extraction beam line, and the AGS had to shut down for a week to fix it.[32] Even when, on October 11, the machine was switched on and fast extraction experiments began, the new magnet continued to cause delays.

The wait made Ting rethink his decision to postpone the announcement of what he was now calling the *J* particle. He had the chance to make a dramatic announcement on October 17 and 18, when MIT hosted a festival to honor the retirement of Victor Weisskopf. A full-blown bash celebrating one of the university luminaries, the gala gave Weisskopf, an enthusiastic amateur musician, his chance to conduct the Boston Symphony. If Ting used the occasion to announce the existence of the narrowest resonance ever discovered in an energy region that had been thought previously to be well understood, the splash would be enormous. He didn't do it. He knew he was sitting on something very odd, and odd results are usually incomplete.

When Becker gave a previously scheduled seminar on Experiment 598 at MIT on October 22, he had to disguise the narrow peak, which he did by cleverly presenting the number of collisions over a sufficiently wide energy range to smear out the peak. Martin Deutsch, head of the MIT's nuclear physics laboratory, was not fooled. A kind of mentor for Ting, Deutsch had gone out of his way to arrange funds for Experiment 598, and had followed its progress carefully. He glanced at Becker's graph, and mentally spread out the data, differentiating what Becker had integrated. Taking Becker aside, Deutsch told him to publish immediately. Becker said he would, once Ting was ready, and asked Deutsch to help expedite publication once it came time. Fancy mathematical skills were not needed to figure out Ting's team had something, Deutsch said later. "All you had to do was to take one look at Min Chen's face."[33]

That same day, October 22, the AGS was finally switched back to the slow extraction experiments, and Ting got his chance to check out the narrow peak. "There were many, many tests, okay?" Ting told us. "The first is to change the thickness of the target, let's say by a factor of two, and see whether the counting comes off by a factor of two. If it's a scattering from the side of the magnet or something, it won't. Or change the magnet current by ten percent. A true peak better show up in the same place, and not move. Or change both magnets to positive polarity, and see what happens. Plug up the magnet to a smaller aperture, and see whether the rate changes. All kinds of tests. We were constantly going in and out of the experimental area, which is doubly sealed because of the radiation; and we drove the Brookhaven people completely crazy."[34]

Ever meticulous, Ting had laid in a reservoir of good will with the AGS control staff by constantly showing his appreciation of their round-the-clock care for the machine. Now they repaid his concern by whipping the AGS through its paces at top speed and with great care.[35] The narrow peak didn't disappear. Chen kept calling his friend at SLAC.

Newly returned from his first trip to China, Mel Schwartz of Stanford, the head of the group that shared Ting's beam, stopped at Brookhaven on October 22 to check the progress of his experiment.[36] Schwartz's assistant, Jayashree Toraskar, filled him in on the progress of Experiment 614, and then relayed to him some exciting news about Experiment 598 that she had heard from a member of Ting's team: There was a bump at 3.1 GeV.[37] Schwartz walked over to congratulate Ting.

The encounter was a classic clash of opposites. Husky, brash, an all-American go-getter who thought of frankness and openness as among the cardinal human virtues, Schwartz cheerfully asked Ting, a private, driven, careful man, about a secret he was not ready to announce and had energetically tried to keep to himself. Moreover, the result was a spike on the kind of accelerator that wasn't supposed to be able to detect them, and Schwartz was on his way to SLAC, where they could find it instantly if the word got out. In Schwartz's recollection, the following conversation ensued:

SCHWARTZ: *Sam, I hear you got a bump at 3.1.*
TING: *No, absolutely not. Not only do I not have a bump, it's absolutely flat.*
SCHWARTZ (insulted by Ting's caginess): *I'll make you a bet. Ten dollars you got a bump.*
TING: *Absolutely. I'll bet.* (They shake hands. Schwartz is furious. Ting goes to his trailer and hangs up a notice saying, "I owe M. Schwartz $10.")[38]

Long after the event, Schwartz explained to us the reason for his annoyance. "If somebody came up to me and said, 'Mel, I hear you got a bump,' I would say, 'Look, I got one, but I really wish you would not discuss it, because I don't honestly know for sure that I got anything yet. OK? Since you somehow or other found out in one way or another, let me tell you what it is. *This* is what I think I see, *this* is why I don't quite believe it, *these* are the tests I'm going to do. That's the way a rational human being reacts to somebody coming in there and saying, 'Mel, I hear you got a bump.' Now, you realize that Sam Ting, *if* he had leveled with me at that point, *if* he had shown me the thing, *if* he had talked to me in those terms, there wouldn't be *any issue* as to who discovered what. He would have got full credit. By doing what he did, he put himself in the position where he ended up having to *share* it. For what did I do? I flew back to SLAC that very same day, and I said to my guys, 'Hey, I just made ten bucks.' 'What do you mean, you just made ten bucks?' 'I just made a bet with Sam Ting that I'm guaranteed to win.' 'What's the bet?' 'Sam Ting tells me he hasn't got a bump. But I know he's got a bump at 3.1 because his people have told me he's got a bump.' "

Three days later, on October 25, Deutsch telephoned Chen to ask whether the results Becker had presented in the seminar had been confirmed by the recent data. Chen hedged. Growing angry, Deutsch began to bellow at Chen that he had better publicize whatever it was he was sitting on before SPEAR got to it. Loyal to his boss, Chen didn't say anything, but when he finished talking, he pressed Ting to publish.[39]

One member of the SPEAR analysis team was Gerson Goldhaber, a good-natured, able physicist whose office is in the Lawrence Berkeley Laboratory, high on the hills overlooking the San Francisco Bay. We drove up there to meet him one morning in February, when Goldhaber had just returned from a skiing trip; his features above his full beard were red from the winter glare. Crowded by books and journals, a self-portrait was hung near the ceiling; foot-thick stacks of papers blocked every chair. Taped to the side of his desk was a fortune cookie prophecy: "You will finally solve a difficult problem that will mean much to you." A bumper sticker on the wall proclaimed: "Physicists do it with charm." Two empty champagne bottles, souvenirs of charm, rested atop file cabinets. Ultimately, Goldhaber played a big role in directing SLAC to 3.1 GeV, and he did it for reasons that, in retrospect, are entirely incorrect.

"In October," he said, "Roy Schwitters [a member of the SPEAR group now at Harvard] decided that we needed to write a paper on this experiment, and he started to work on it. To write a paper you have to look at the data carefully. And then he noticed that there was a very funny point; at 3.1 GeV there was some inconsistency in the data." Although the second pass through the 3–4 GeV region had not picked up a bump, the pattern of decays was a little odd. Schwitters told Goldhaber about the funny pattern on October 22. "We wanted to publish this paper, and with this funny fluctuation you can't say that you've measured something." Goldhaber set out to find the reason for the odd behavior, and thought he saw something exciting. "It looked like there were K^0s [neutral K particles] in these events, more than one would have expected. My student, Scott Whitaker, who was also working on this data, looked at it, and he somehow thought he saw an excess of *charged* Ks in these events. And then we really got very excited because we knew about the charm hypothesis and we knew that charm should decay into strange particles! As it turned out, my observation of excess K^0s was partly statistical [due to random fluctuation] and partly some were misidentified. There was really no great excess in this data. As Schwitters put it, this was the red herring of the century. It got us thinking that we have to go back and look at this data."

Richter was out of town at that moment, giving a lecture series at Harvard. Goldhaber knew it was going to be difficult to convince him, for SPEAR had just been reset for the 4–6 GeV energy range. "And when you decide on a program," Goldhaber said, "you don't just then go and jump around and do other things. You have to retune the beam and all the magnets had to be reset. So this going back to 3.1 GeV was not completely trivial." Talking over the situation with other group members, he decided that the excess of K particles merited turning the machine back. Richter was due to return in the first two or three days of November; Goldhaber would push him then.[40]

By the end of October, Experiment 598 had fully confirmed the existence of the J, a particle with a mass of almost exactly 3.1 GeV. But something else now prevented Ting from announcing it—"He got greedy," in Chen's words.[41] If there was one unexpected bump, there might be more. Ting knew about the previous experiment that had found a gentle shoulder instead of the sharply peaking J. The shoulder seemed to be centered around 3.4 GeV; perhaps it indicated an even larger, sharper spike there.[42]

Ting was not interested in whether the new particle played in existing theoretical speculations. But some of his collaborators made clever guesses; on Halloween, for example, Min Chen casually wrote at the top of log book page 195: "Study trigger rate for charm particle."

Later that day, Ting inveigled the Brookhaven control room head "Woody" Glenn into postponing some scheduled tinkering with the AGS to give him more running time; he thanked them by bringing a large, carved-out pumpkin full of beer. On the following days he prodded them for increases in the beam intensity, and then made several changes in the targets, trying to see whether this would produce more bumps. Meanwhile, Becker, Chen, and nearly everyone else on the team was demanding that they announce the one they already had. "There is an old Chinese proverb," Chen said drily. "Something about a bird in the hand being worth more than two in the bush. Well, Ting wanted those birds, and more."

Goldhaber called Richter's office on Monday, November 4. For the rest of that week Richter, Goldhaber, Schwitters and other experimenters argued the merits of pushing forward or going back. Although Richter was sympathetic to Schwitters and Goldhaber, there were several obstacles in the way of setting back the energy. First of all, another experiment was being run at the time, which would have to be bumped. Second, Richter wasn't even sure that the machine *could* be switched back. So many improvements had been made in the machine—now called SPEAR II—to allow it to run at a higher energy that he was not sure it could be detuned. The final straw that convinced Richter to try it was Goldhaber's discovery that the K particle production was abnormally high.

Goldhaber said, "On Friday the 8th he called me up to tell me that he has figured out how we can run in that region, and we're going to try it that weekend."

The next morning, Saturday, November 9, Gerson Goldhaber called the SPEAR control room to see how the new run was going. A hastily built maze of metal shelving crammed with humming power sources, noisy computer printers, and beeping control gauges, the control room was in the center of the ring, which stood in a pool of concrete at one end of the two-mile SLAC accelerator. Richter was there—Richter was practically always there when the beam was on—and he told Goldhaber that they were going at 1.57 GeV per beam, 3.14 GeV total energy, and that R was unusually high. Goldhaber drove to SLAC that afternoon, where he discovered they had set the beam at 1.5 GeV, and nothing special was happening. He sat in front of a little computer screen and keyed the computer to show him reconstructions of events, tallying the leptons and hadrons. They were not too difficult to distinguish at a glance: Only two charged "prongs" come out of events that produce pairs of muons or electrons, whereas events that make hadrons create a barrage of particles. At 11:15 P.M., the energy was eased up to 1.56 GeV, and suddenly the number of hadron interactions jumped to twice what it had been. Richter and Goldhaber followed the interactions, with the beam being tuned ever more finely, until three that morning; convinced they had really found some sort of peak, Goldhaber retired to his motel room.

Goldhaber came back later Sunday morning to find they had tuned the beam still more finely, and that R had risen still higher. "By noon the cross section [the likelihood of producing hadrons] had gone up by a factor of seven. So I said, 'We have to go and write this up. This is just so, so unbelievable; we have to write a paper on this. This is such a fantastic effect.' And Burt agreed with me, and so I went off and started writing." For the moment, he called the particle SP(3105); SP for SPEAR, and 3105 for its precise location.

"I picked up a piece of computer output and started writing on it. In the meantime, by about one or two o'clock in the afternoon, we had got an extra factor of ten, so the increase was a factor of *seventy*. I had worked with hadronic resonances, and there you usually get a factor of 1 and ½ percent, 30 percent, 50 percent—no factors of seventy. I had never seen anything like that in my physics career!" In the meantime, someone had brought out a bottle of champagne, and celebration commenced with celerity. "There was a very funny scene," Goldhaber said, "where we went on taking data and standing around, but at the same time we started drinking champagne and having cookies."

Clad in jeans and sneakers, Richter sat at a table covered with computer output paper, a single champagne glass perched precariously atop one stack, counting hadrons and leptons. News spread across the campus, and gradually the control room became jammed. The phone calls to other laboratories began. A simmer of euphoria, the clean pleasure of discovery, coursed through the talk and the cigarette smoke.

Later, Goldhaber said to us, "we realized that nobody calls a particle by two letters. So I looked in my little book—" he hunted around his desk in vain for the Particle Properties Data Booklet "—oops. Today I haven't got it. You know what they look like? Okay. So I was looking through that to see if there was any letter that hasn't been used. While talking with George Trilling on the phone I came up with this—psi. I said, 'Here's a letter that hasn't been used, and it resembles SPEAR, sort of close. It isn't in use, and it's one that you could pronounce.' "

They decided to announce the discovery Monday morning.[43]

That same day, Sunday, November 10, Ting left Brookhaven to attend a meeting of the SLAC Program Advisory Committee, of which he was a member. The meeting was scheduled for the next morning, and Ting's trip had been planned for months. Just before his departure, he accidentally ran into Mel Schwartz, who asked him about the ten dollars he was owed. Irritated by Schwartz's constant pressure, Ting said he had nothing.

When Ting's flight landed in San Francisco, he was paged by Chen, calling from Brookhaven. Chen said he had completed a paper on the discovery and had inked in all the graphs, and made one last plea to submit it for publication. Ting said he understood how Chen felt, but told him to go to bed and catch up on some lost sleep.

At the hotel, Ting found another frantic message, this one from Deutsch, who had learned that SPEAR had found something at 3.1 GeV. Horrified, Ting called Brookhaven and said to prepare to announce right away; he would announce it at SLAC the next morning.

Back at the Brookhaven control room, Austin McGeary was the AGS engineer on the graveyard shift. His shift started out quietly; his log entries record the typical small difficulties of maintaining a large and complex machine:

While at Linac, Dely found a very small H_2O leak on the tygon hose on the Buncher in LEBT—It should be checked periodically. . . .

Lost a little over 2 hours starting shortly after midnight when a +15V P.S. in the linac pipe O.T. bucket crapped out. By bad luck we replaced the supply with another powermate 1521 which although checked OK on the bench would not regulate with even the load of 1 P.C. card. . . . This experience and many others of this type point up the real need for a spare locker. . . .

He was a bit surprised when two members of Ting's team asked him for the keys to the Xerox machines at three in the morning; they had to do some copying immediately, they said, and the machines had been switched off. McGeary hunted around in vain for the keys. A few moments later, the women were back to ask for the phone numbers of some scientists who lived out of town. For some hours they refused to tell him why they wanted to wake up everybody they knew.

0600 Exp. 598 has found a new particle!!! Ms's Wu and Schultz have been rushing around madly trying to make Xerox copies of some data and calling outside. They are obviously very excited. Computer printout for data taken since start of this running period shows a very prominent resonance at a mass of 3.1 GeV.[44]

A few hours later, on the other coast, Ting arrived early at Panofsky's office for the committee meeting, and Panofsky told him about SLAC's discovery. Richter arrived a few minutes later, and when he heard Ting's news, said, "Sam! It's the same thing! It has to be right!"[45]

Later that day, Ting relayed the news to the director of the particle physics laboratory in Frascati, Italy, where an electron-positron collider had spent seven years running up and down between 1 and 3 GeV. Within two days, chagrined physicists there had pushed up its energy to 3.105 GeV, becoming the third team to confirm the existence of the particle. The three papers by the three groups on the new particle arrived at the offices of *Physical Review Letters* within a week and were published back to back on December 2.[46] In the meantime, on November 21, SLAC had discovered what Ting had suspected was there but which he had been unable to find—a second spike, which Richter called the *psi prime.*[47]

The particle that Ting and Richter discovered is now called the "J/psi" by everyone except Ting and Richter, who refer to it as the *J* and the psi, respectively. (The two men won the Nobel Prize in 1976.) The J/psi electrified the physics world for many reasons, including its simultaneous discovery on two different types of machines. Unsurprisingly, the simultaneity has often led to speculation in the sometimes claustrophobic high energy physics community that somehow word about the peak spread from Brookhaven to SLAC and played a role in the decision by SLAC officials to turn back the machine. The most often cited possible channel for this information is Mel Schwartz, who learned of Ting's possible new finding in October and in his annoyance with Ting broadcast it to anyone who would listen. One member of Experiment 598 told us: "They were at 4.2 the week before, and they reported a rise in *R*. It was a new result! When you discover something new, you explore it. Why go backwards? You can *always* go backwards—next week, next month, next year. Why go backwards *then?*"

Richter's team members deny that they knew anything about Ting's result, and point to the logical series of steps by which they zeroed in on 3.1 GeV. One physicist, not on Richter's team, even angrily pointed out that Ting, in fact, had decided to announce his discovery only after hearing of SLAC's. "If you aren't ready to announce something, then you really aren't sure yet you have made a discovery," he said. Others at SLAC think that Ting would have been scooped entirely if he hadn't taken that trip to SLAC for the committee meeting.[48]

Although such arguments over priority are the private mania of physicists, they reverberate through any historical inquiry. The discovery of the J/psi was the work of two detectives, each on the trail of something, although not quite sure what it was. Each occasionally caught glimpses of the other; Richter learned of Becker's seminar at Harvard from Deutsch, members of Ting's group followed the course of SLAC; and both were immersed in the sea of rumor, speculation, and gossip that is the natural medium of physics. Still, both were astonished to find that they had converged on the same quarry at the same moment.

The argument over priority masks the difference in the two teams' contributions. Ting found the J on the AGS, a machine barely powerful enough to produce the particle; finding it was a virtuoso turn, a feat of unrepeatable expertise akin to watching a jazz musician squeeze out a beautiful solo from a battered and broken saxophone. But Ting could do little more than establish its existence. Indeed, his experiment ended where Richter's began. A superior tool for the task, SPEAR could—and did—hold the psi up to a clear light, examining it with the attentive care of a jeweler. For the next few months, the California group mapped and measured thousands of psis. Ting may have had the particle earlier, but our knowledge of its properties comes from Richter.

We asked Martin Deutsch whether he thought the discoveries were independent. "Oh, the discoveries were independent—absolutely. I'm convinced of that. But the physics—that's something else. Physics is *never* independent. How independent can you be in such a strongly interacting environment?"[49]

The announcement of the J/psi in November of 1974 is now known in the physics community as the November Revolution. Like many revolutions, its meaning was not clear at first; the unprecedented narrowness of the peak fostered a host of theoretical speculations. Some thought the new particle was the Higgs boson; others suggested that it was the long-awaited intermediate vector boson; still others hoped it might be a "colored" particle, evidence that quarks really did have integer charges and were not confined. Chen called Glashow on the morning of the eleventh. Glashow went to MIT and by lunchtime was sure that the J/psi was a meson formed of a charmed quark and an anticharmed quark, which he dubbed "charmonium."[50]

"Around six months after the psi discovery," Richter said, "it was very, very hard not to think that it was charm. Then there was a problem that left us all doubting." The problem was that if the J/psi truly were charmonium, then experimenters should turn up other mesons with charmed quarks. "We set out specifically to look for charmed mesons. It was still the same experiment—the same machine, the same detector, the same people. And we couldn't find the damned things. It was getting pretty disappointing. I was beginning to wonder about charm. Everyone was—except Shelly Glashow. Shelly is one of the most self-confident people I have ever met. He was quite sure at the time that it was the fault of the stupid experimenters. But we couldn't turn up anything. That's how matters stood when I left for CERN on sabbatical."[51]

The case for charm was only slightly bolstered when Samios and Palmer publicized their single bubble chamber photograph at a meeting in March 1975. They stuck their necks out for charm, but the Gargamelle group, which announced a tentative sighting of a charmed meson at the same congress, left the actual description of the event to a graduate student—a sure sign of unease.[52]

When Richter left on sabbatical that summer, various members of the SPEAR group continued the hunt for charmed mesons, and continued to find nothing. In April 1976, Goldhaber attended a conference in Madison, where he gave a talk about the psi. Another collaboration member reported on SPEAR's failure to turn up charmed mesons after an exhaustive search. Charm wasn't there—a message that displeased Glashow, who was in the audience. Following the conference, the two men flew to Chicago together in the same plane, an occasion Glashow seized upon to harangue Goldhaber for the entire trip. "No bets this time," Glashow said, "just imperatives. I insisted he take another look."[53]

"I told Shelly I would," Goldhaber recalled. "I would spend another month looking. Based on the reaction of the physicists I had spoken with, it would be a month well spent even if I didn't find anything. When I arrived back in Stanford, a new batch of data had just come in. I liberalized the criteria a bit, because they had been rather strict. Within three days, I found the mesons. I called Shelly, and he was elated. He knew it meant the Nobel."[54]

Goldhaber's data convinced even the most die-hard skeptics. In Paris, Iliopoulos received his bottles of wine. And at the next meson conference, the program director, Roy Weinstein (now at the University of Houston), decided to repay Glashow. Near the end of the conference, Weinstein reread Glashow's hat speech of two years before. Whereas Goldhaber, Richter, Samios, and Ting were not meson specialists and were thus outlanders, and whereas according to the terms of the bet everyone present had to eat their hats, as head of the honorable assembly he was duty-bound to hold them to it. He proceeded to pass out candied Mexican hats.

"They were extremely peppery," Weinstein said. "Not a pleasant candy. But everyone ate them with relish, and agreed that Shelly's prediction had been magnificent."[55]

□   □   □   □   □

When theorists first put forth their ideas, they frequently point out the experiments that would suffice to prove or disprove them. For a theory to be right, it must, after all, be capable of being wrong. Often, however, these crucial experiments are not the ones that finally sway the scientific community. Erroneous experimental results, theoretical blind alleys, and mistaken interpretations of each by the other can lead physics down a different path than what would seem to be the straight and logical route. A test that at one moment appears to be a mere exercise can suddenly loom as the arbiter of a theory's fate. So it was with the standard model at the beginning of 1977. Oddly enough, this decisive test was a reprise of the one that began to put $SU(3) \times SU(2) \times U(1)$ on its feet; it involved, once again, a prediction about neutral currents.[56]

The discovery of neutral currents showed that Z^0s could be exchanged between neutrinos and hadrons or neutrinos and electrons; to round out the picture, experimenters wanted to know whether Z^0s were exchanged between electrons and hadrons as well. At low energies the strength of the Z^0 effect would be minuscule compared to that of electromagnetism, and directly measuring such a small contribution was out of the question. But the weak force has a sign which talismanically identifies its presence: parity nonconservation. Experimenters considered experimental setups in which the interaction between electrons and hadrons would show an effect if parity was not conserved in the interaction, and no effect if it was conserved.

Most high-energy physicists wrote off the task as hopeless. But the task was taken up by atomic physicists, scientists who studied the interactions of

electrons and nuclei; sometimes regarded as bookkeepers by particle physicists, they would take a certain pleasure in carrying off a tabletop experiment that could do something a huge particle accelerator could not. The
general theory of "atomic parity violation" experiments was first described
by the French husband-and-wife team of Claude and Marie-Anne Bouchiat
in 1974.[57] Electrons surround the nucleus in many types of elaborate orbits
described by Schrödinger wave equations. In the case of extremely heavy
atoms, the charge on the nucleus is so strong that some of these electrons are
pulled close enough to come within the range of the weak force. According to
SU(2) × U(1), the electrons and the protons and neutrons in the nucleus
should exchange Z^0s during these close encounters. The Bouchiats calculated that the presence of the Z^0s could be detected by their effect on the
light emitted by the electrons.

The fundamental idea is this: All photons have a spin, which has a
direction that can be viewed as a combination of left- and right-handed.
Similarly, if someone tosses a ball, the ball spins as it flies through the air.
Using the direction of motion as a reference point, the actual spin can be
described as a combination of two different spins, left- or right-handed along
the direction of motion. A beam of light that is circularly polarized has all of
its photons spinning in one direction. Linearly polarized light is the special
case in which the spin of every photon is an equal mix of left or right circular
polarization. In some atomic parity experiments, linearly polarized light is
absorbed and then emitted by atomic electrons. Because the electron feels
the weak force and is surrounded by virtual Z^0s, it interacts differently with
the right-handed and left-handed components of the light wave. The outcome is that the linearly polarized light is emitted with its polarization in a
slightly different direction than it would have if the electron-photon interactions respected parity; the spin is tilted, so to speak. The angle of tilt is
proportional to the difference in absorption rates of right and left. This
emitted light in turn interferes with the incoming light, and what the experimenter actually measures is the shift produced by the interference between
these two.

One of those impressed by the Bouchiat's paper was Edward Fortson of
the University of Washington in Seattle, who had read Weinberg's 1967
paper when it first came out. "He [Weinberg] had a certain way of looking at
the weak interactions in analogy with electromagnetism," Fortson said. "I
didn't get much physics out of it, just the concept."[58] He had thought
periodically about testing Weinberg's concept for parity-violating effects for
several years, but had thought it hopeless. When someone showed him the
Bouchiats' paper in 1974, Fortson realized that he could do it.

According to the electroweak theory, the angle of the tilt in the polarization would be minute in the extreme—about 10^{-5} degrees, or roughly the

angle between the two sides of a nail viewed from five miles away. Nevertheless, Fortson felt that he could arrange his equipment to measure it. He would need a finely tunable, intense laser to provide a precise source of light. In addition, Fortson had to use an atom with a large, heavy nucleus to pull the electron close enough to the nucleus to have the weak force come into play. Only one heavy element gave off light that matched Fortson's laser: bismuth, a white, brittle metal, similar to lead in many respects and its next-door neighbor on the periodic table; in the early days of chemistry, the two elements were often confused. But using an element with eighty-three electrons was a mixed blessing, for it made the exact calculations of each orbit sufficiently complicated that it was not easy to know which approximations were justified and which would lead one astray. If one took the figuring seriously, however, $SU(2) \times U(1)$ predicted that the bismuth spectral line should have a shift of about -25×10^{-8} radians. (Radians are a scientific measure of angle: one degree is just over one-hundredth of a radian; and -25×10^{-8} radians is about 1.43×10^{-5} degrees, that is, a few ten thousandths of a degree, with the minus sign here indicating direction.) In the winter of 1975–76, Fortson's team began to perform their experiment and found a markedly lower tilt of around -8×10^{-8} radians. Fortson was interested—experimenters always like to set the theorists on their ears—but unsure whether he was seeing nothing but an error in his own calculations.[59]

Meanwhile, another team of atomic physicists led by Patrick Sandars of Oxford University had begun a similar experiment in England. Sandars received his doctorate at Oxford at the time of the discovery of parity nonconservation and had been fascinated by the subject ever since. The discovery of neutral currents had pricked his interest, and when he ran across the Bouchiats' paper he decided to try to detect a parity-violating neutral current. Like Fortson, he decided on bismuth, although he chose a different wavelength of light to measure, one for which the Weinberg-Salam model predicted a spectral shift that was also about -25×10^{-8} radians. Sandars began to get his first numbers in early 1976. They, too, were much less than predicted, about $+10 \times 10^{-8}$ radians.[60]

In August 1976, Sandars boarded a plane for the United States, where he planned to discuss atomic parity at a conference in Berkeley. He flew to Seattle, for he wanted to talk to Fortson about the latter's experiment. Fortson met him at Sea-Tac airport, and they exchanged numbers. Fortson's results were slightly to one side of zero, Sandars slightly to the other. At the time, what seemed most significant was that both were much less than predicted by the electroweak theory. Pleased, the two men continued together to California, where they announced their results in a spirit of exuberance.

In December the two groups published a joint letter in *Nature* in which they concluded that "the optical rotation in bismuth, if it exists, is smaller than the values . . . predicted by the Weinberg-Salam model."[61] The same issue of *Nature* carried a comment about the implications for unification by British physicist Frank Close, who concluded that "the clear blue sky of summer now has a cloud in it. We wait to see if it heralds a storm."[62] The storm began to break early the next year, when both Fortson and Sandars released updates on their experiments that confirmed their earlier numbers. The Oxford and Washington results appeared to dash the hopes not only of the standard model but of the entire approach to unification that had been growing ever since the renormalization of gauge field theories.

In such circumstances, advocates of unification had only two ways out—pull a fast trick or turn a deaf ear. Neither alternative was palatable. Some tried to incorporate clever ideas to make the $SU(2) \times U(1)$ conserve parity in atoms after all, but in so doing learned that the model was almost fiddle-free. "In an odd way," Weinberg said, "that made it *more* convincing. You really couldn't do anything much to it without dreadful consequences."[63] Nonetheless, some theorists tried, producing, among other things, a batch of "left-right" theories that restored parity symmetry to the weak interactions at the price of invoking the existence of something else.[64] Others hoped that the atomic parity experiments were wrong or that the atomic physicists were confused by the complexity of the bismuth atom. Certainly particle physicists were baffled; most were simply unable to evaluate the veracity of the experiments.

In March 1978 two Russians reported that they had observed optical rotation of the predicted amount in bismuth.[65] But the score was still two to one against Glashow-Weinberg-Salam, and physicists in the West had no way of evaluating the reliability of the Soviet experiment.[66] With the experimental and theoretical situation in chaos, rumors began to percolate that a SLAC version of the same neutral current experiment was in progress. Based on a logic and a set of assumptions that could be followed by particle physicists, the experiment, to their minds, was a proper experiment, done on a proper accelerator.

The idea for the SLAC experiment came from Charles Prescott, a short, precise man who got his Ph.D. at Caltech in 1966. He, too, had first become interested in neutral currents when he read Weinberg's 1967 paper. ("I didn't understand it, but I was aware of why it was interesting.") Prescott was intrigued by the way that it linked electromagnetism, which conserves parity, with the weak interaction, which does not.[67] In 1969 Prescott, then at the University of California at Santa Cruz, approached Richard Taylor, head of one of the SLAC experimental groups, with an idea for testing the general implication of the Weinberg paper that parity-violating neutral currents existed between electrons and hadrons.

The mutual respect between Taylor and Prescott is masked at first glance by the contrast in their personal styles. Taylor is burly, generously proportioned, expansive, a broad and happy man who puts his feet on the desk, scattering papers, and bids his sentences adieu with a great wave of the hands; Prescott speaks softly, in a neat monotone, hands folded in his lap, a tie closely knotted about his neck, a trim, reserved, and thoughtful player in the complex game of physics. In the parity experiment, Taylor had the role of the go-getter, the worrier about precision, the jack-of-all-trades; Prescott brought inspiration and calculation. (But as in any good collaboration the roles often switched.) Taylor had been incredulous—"We said, '*Shee-it*! how you going to do *that*, Charlie?' "—the first time Prescott approached him with the idea. His participation gained, Taylor kept riding Prescott and the other members of the group to make the experiment ever more sensitive. Any effect would be little, and they had better not miss it.

Scattering particles from each other is, in a sense, the basic experiment of twentieth-century physics. In Prescott's first proposal, he hoped to demonstrate that the recoiling protons were polarized as a result of parity-violating collisions involving Z^0s. Gradually he became convinced that the difficulties in measuring this small polarization were insuperable.

At about this time a Yale group led by Vernon Hughes proposed experiments at SLAC with polarized electrons and polarized protons. (Polarized electrons, like polarized photons, have all their spins pointing in one direction.) Hearing of this, Prescott realized that the scattering of polarized electrons from unpolarized targets would test parity violation. The experiment was simple in essence, knottily difficult in specific execution. In SU(2) × U(1), electrons can scatter from protons in two ways, by an exchange of photons or by an exchange of Z^0s. Parity is conserved in the former, violated by the latter. Prescott's experiment would be performed with the spins of the electrons polarized either in the direction of motion or against it. These two configurations are related by parity reflection. If parity were perfectly conserved, the number of electrons scattered from the protons in a target would be exactly the same in the one direction as in the other; if parity were not conserved, the number would be slightly different—very slightly different.[68]

Prescott wanted to employ the Yale team's equipment to make polarized electrons, which he would measure after scattering with the same apparatus that had been used by the scaling experiments. A SLAC-Yale collaboration was formed; Prescott left Santa Cruz to join Taylor's group at SLAC.

The experiment was approved in February 1972, but events of the following year forced Prescott and Taylor to rethink things from the ground up. "It was about this time," Taylor said, "that 't Hooft showed that the Weinberg stuff was renormalizable. That increased the respectability of Weinberg by an enormous factor, and got every bugger and his brother proposing parity violation experiments. We were ahead, but we weren't alone any more." The discovery of neutral currents that next year caused theorists to scrutinize the implications of the electroweak theory in substantial detail. According to the new bout of calculations, the effect Prescott and Taylor were looking for should have a relative magnitude of about one part in five thousand, which was so small that to see it the detector would have to count a hundred times more electrons on every pass than had been planned. The experiment therefore had to be revamped; the Yale source of polarized electrons was replaced. The house expert in solid state physics, E. L. Garwin, teamed up with laser specialist D. C. Sinclair to design and build a new high intensity source of polarized electrons. A second detector was constructed from pieces of the old one. Both had to be more precise than any built before, and both had to be built quickly if the experiment wanted to play a role in the outcome of $SU(2) \times U(1)$.

For some two years, the group sweated to design tests to ensure that any observed variation in scattering was due solely to the polarization of the electrons. A second big effort went into measuring properties of the beam—position, angle, intensity, and so on—to be certain they did not change when the polarization was switched. The remodeled experiment was approved in 1975.

"There was tremendous pressure at every point in the experiment," Taylor said. "To build that source, learn how to run it, and make it work was an awesome thing to have done on such a short time scale. And then there was all of the electronics that went with the measurement itself. I mean, building electronics in such a way that you could measure this thing—you know, it's very small! It's like standing on top of Hoover [Tower, overlooking Stanford University] and watching ten thousand people streaming into Stanford every day and trying to decide whether there was one more or one less today than there was yesterday. And the ten thousand has to be that *squared* to determine that statistically. So you're dealing with a *hundred million* counts, all right? And we had to deal of course with much *more* than that. We needed to detect *ten billion* scattered electrons in order to see what was happening!"

In March 1978, a decade after Prescott's first proposal, the experiment began. Within a few days, it was clear that they had an effect on the order predicted by $SU(2) \times U(1)$. When they changed the direction of the beam polarization, the counting rate changed as well. "You'd run for awhile and be above the line [the value for unpolarized electrons], then you'd change polarization and it dropped below." Taylor held one hand above an imaginary line, and then plunged it below. "I remember that in just three days you went from wondering whether the experiment was going to work to being pretty sure that you knew that there was parity violation. That, I mean, *that's* why you do this business. That feeling of knowing something before anybody else. It's, ah, it's why you're here."

During the final runs a powerful new check was designed, of which Taylor was particularly proud. Because the accelerator beam was bent by magnets to reach the experimental target, the direction of polarization at the target depended on the beam energy. The bending of the electrons caused the polarization to sweep around in direction as the energy changed. "When the energy was changed by twenty percent, the polarization would go from pointing *that* way"— Taylor pointed one way with his hands —"to pointing *that* way. This would reverse the effect that we were supposed to see in the scattering. But it would *only* reverse it if it were due to polarization."

After the experimental runs were over, Prescott and Taylor insisted that the collaboration spend three months rechecking the data before they told anybody what the numbers were. Taylor said, "We were really worried about credibility. We didn't want to come out with a preliminary number and then change it in three months. I mean, we wanted to give one number as the result of the experiment and have that number sit there. We wanted to do the analysis before we talked. So we had to give Charlie three months."

He was asked whether in retrospect the group had worried too much about nailing down the last decimal places. Had they been too precise? The question made Taylor's jaw pop open in stupefaction. A long silence ensued as he groped for an articulable response. Finally, he exploded, "You *never* know the answer to that question! I mean, because you *don't* go ahead and do the—I mean, we *didn't* do a sloppy experiment to find out whether it was going to make trouble [for the atomic experiments]! You do the *very best* you can!" We wondered if any part of the group's concern for precision had originated in the vicissitudes of the numbers from the atomic parity experiments. He said, "Those experiments looked okay. On the other hand, the theory that went with those experiments was not in very good shape. So the fact that they were getting lower answers—it wasn't clear whether that was because they didn't know their bloody wave functions or because there was no effect. Now, some theorists were saying that there is no parity violation. They must have had little to do in those days, because many of them went off and all those left-right theories just *grew*—I mean, there were *acres* of them, like spring flowers." (Later, when the theories were no longer needed, some people referred to the "SLAC massacre" of the left-right theories.)

The three months were over on June 12, 1978. In his quiet, careful way, Prescott stood before a packed auditorium hall at SLAC and took the crowd through the polarized electron experiment step by step, methodically ticking off the checks, the backups, the alternative procedures, the theoretical figuring, the experimental run-throughs. All doubts vanished; after years of uncertainty, the standard model had been proven by an experiment distinguished for its novel approach and delicate execution.

Years after Prescott's exposition, Taylor still savored the memory. "Nobody, I mean *nobody*, could see any way in which the answer could be anything other than the answer that Charlie had put on the board. Every time you looked for a flaw, it was clear that we had covered it from one side or another, and the answer was just boxed in. And you couldn't—it was just *there*." Taylor sighed. "It was quite a moving thing. I mean, Charlie had spent ten years of his life and of many of the rest of us—the life of the group, which was my responsibility, had been poured into the experiment for at least three years at full strength. Now it was working."

At the end of Prescott's talk, there was applause, as there is at the end of many lectures. But witnesses recall this show of hands as different: subdued, a soft pattern that went on and on, white and powerful and slow, until it filled the room like a fall of water. They were applauding the experiment, the craft and dedication of its executors, but the sound was more than that: It was a recognition of accomplishment; the physicists were quietly saluting themselves—the long, elegaic salute given to the end of an age. When the clapping died down, Prescott asked for questions. There were none.[69]

VI
Unification

19
The Beginning of Time

"AFTER THAT EXPERIMENT BY PRESCOTT AND TAYLOR," FRANK WILCZEK SAID
with admiration, "it was no longer a question of changing SU(3) × SU(2) ×
U(1), but of explaining it."[1] In hand were theories of the strong, weak, and
electromagnetic interactions—the three elementary particle forces—each
written in the same language, and all shown to be in accord with experiment.
No infinities or anomalies troubled the waters; experiments poked lights into
corners, lifted up carpets, and found nothing amiss, no unexpected cracks—
everything in apple pie order. Save for the embarrassing absence of the W
and Z, which were too heavy to be produced in extant accelerators, SU(3) ×
SU(2) × U(1) seemed by 1980 to be marvelously robust. (Three years later,
Carlo Rubbia's team found the W and Z exactly where predicted, after a long
and expensive search with a new accelerator at CERN.)[2] A new orthodoxy, in
Bjorken's phrase, the standard model was suitable for rapid inscription into
textbooks.[3]

Few things seem as dull to theorists as solved problems, although the
definition of "solved" varies from one physicist to another. Even as experimentalists spent the early 1970s struggling with charm and neutral currents,
some theoreticians began to relegate SU(3) × SU(2) × U(1) to the status of
an old, and therefore uninteresting, idea. They started to look for a reason
that the trio of elementary particle interactions possessed such consonant
structures. An obvious hope was that the similarity indicated some deeper,
unifying principle, as yet undiscovered, from which these theories sprang. In
the usual practice of contemporary physics, by the time that experimental
verification of the standard model had progressed to the point where such
speculation became the rational and logical step to take, less methodical
physicists had already taken it, however haltingly, for other reasons.

Fittingly, the first attack on unification was made by the generation of
physicists that rose to prominence in the 1950s, Glashow, Salam, and Weinberg among them. They did not realize the full implications of initiating such
a search, for it continues to the present day and is still transforming the
practice of physics itself. Indeed, in the view of some worried scientists, the
current full-tilt push toward unification has more chance of derailing physics
than accomplishing its goal.

In the spring of 1984, we met Abdus Salam at midday in a New York City hotel; he had promised to talk of unification—at length and carefully, in a quiet room where he could think. Salam was in town for the annual convention of the American Association for the Advancement of Science, a huge affair that seemed to fill the midtown area with laboratory dwellers on holiday. Scores of seminars on every possible subject progressed simultaneously, followed by bewildered members of the press and an amazing number of ordinary citizens with strong convictions about science. Salam stood out in the bustling Hilton lobby, a zone of calm in a billowing brown suit. He was surrounded by representatives of the Pakistani community in New York. It developed that an elaborate luncheon had been arranged for him. Unfortunately, the arrangers had neglected to inform the guest of honor. As a compromise, we were invited along, riding in a limousine at breakneck speed to a Pakistani restaurant near Gramercy Park.

A room had been set aside and specially decorated for the occasion. The banquet table was reserved for Salam and the men, while the veiled women sat to the side. Cooks had been working since the day before to prepare the many kinds of rice, beautifully colored by aromatic spices, that lay molded in bowls across the table. Course after course appeared from the kitchen; the finale was a great heap of soft candies. Salam was given no time to eat; besieged by books to sign, hands to shake, babies to kiss, and youths eager for career advice, he spent the meal administering to each entreaty with unflappable reserve. He was asked to pose for photographs with every man in the room. Batteries of cameras at the ready, three reporters from three different Pakistani papers came to interview Salam, sequestering him for half an hour of popping flashes and whirring tape recorders. "This happens all the time," Salam said afterward, in his hotel room. "It's extraordinary, the number of Pakistanis who have arrived in different places. It's practically impossible to get any physics done." He propped himself sleepily on the bed. "One always eats too much at these affairs," he said.

The 1971 renormalization of Yang-Mills theories came as a godsend to Salam, whose reputation was then at a low ebb. He had been identified, somewhat unfairly, with a program to mix the eightfold way and spin into a symmetry called, variously, $SU(6)$, $U(6,6)$, and $SU(12)$. Great hopes were invested in this program; when it was thoroughly debunked, Salam said later, he was demolished.[4] A field theorist who had lost the respect of even the few other field theorists, he was elated at the sudden and unexpected surge of interest in the old $SU(2) \times U(1)$ model he had developed with John Ward. We asked if he had been inspired by 't Hooft's work on gauge fields to look again into using them for unification, as he had once tried in the early 1960s.

"No." Flatly. "I think 't Hooft's work was very nice, but one took it for granted. Even if it had not come, we wouldn't have stopped. If I think that an idea has some sense, then I always act on the premise that the technical difficulties have *already been solved*."[5]

A consequence of such an attitude is that Salam moves ahead of the wave, skipping from idea to idea, embedding half a dozen notions in omnibus papers whose erratic quality often infuriates the more formally inclined. In Isaiah Berlin's term, he is a fox, rather than a hedgehog, whose thought is "centrifugal rather than centripetal"; turbulent, multileveled, and sometimes self-contradictory; seeking to answer many questions in many ways at the same time.[6] Emblematic of his restless and fertile progress is his development, with a colleague, Jogesh Pati—in the summer of 1972, before neutral currents, before charm, before the parity experiments—of the first steps beyond SU(3) × SU(2) × U(1).

That summer, as the Gargamelle group held its first formal debates over neutral currents, Salam was joined at his Trieste institute by Pati. Longtime acquaintances, Pati and Salam had first met in the late 1950s, when Pati was finishing his doctorate at the University of Maryland. Pati was born in Orissa, on the Bay of Bengal, and came, in 1957, to the United States, where he has remained ever since, except for the odd year teaching in Trieste or India. Upon his return from a year in New Delhi, he quite naturally stopped in Trieste for a few summer months by the Adriatic Sea.

At the time, a good proportion of the theoretical energy in particle physics went into examining SU(2) × U(1) and the apparent absence of weak neutral currents; more would soon be used to erect quantum chromodynamics. Like Salam, Pati felt (as he said afterward) "that the heart of the matter [lay] somewhere else. Even if SU(2) × U(1) was eventually borne out—at least at low energies—fully, by experiment, it seemed to us, in seventy-two, that there was a lot of arbitrariness in the choice of gauge interactions. . . ." In contrast to most theorists, they were dissatisfied with the theory. Although it accounted compellingly for weak interactions it did not shed light on other, perhaps more fundamental, issues. They did not want to ask questions about the details of weak, strong, and electromagnetic forces. They wanted to know why there were three interactions, and not one or two. Why, they asked, is electric charge quantized? Why do quarks and leptons both have spins of one-half if they are so different? Why, in fact, does nature use quarks and leptons, and not just one or the other? The first answer they came up with was that quarks and leptons were somehow the same thing.[7]

"One of us said, 'Why don't we assume that?' " Salam recalled. "And the other said, 'I was just thinking the same thing, but do you really believe it's true?' And we examined it further, tabulating all possibilities; and then for a number of days we were just dazed by the audacity of it. This was 1972, remember." He shook his head in amusement. "So early, summer seventy-two! The audacity of it! That quarks and leptons are not two distinct types of matter!"

Paradoxically, Salam's unfashionable predilection for quarks of integral charge was of decisive help. "I think there is something unaesthetic in putting particles together which have one-third charge and charge one in the same multiplet. Wouldn't you agree? If you tried to teach a child that these were the same particles, he would say, 'What the hell? You have charge 1, 0, −1, and then you suddenly have −⅓, ⅔, −⅓, ⅔—what's going on?' It sounds very ugly. The formalism does allow you to do this, and we said so in our paper. But if you have all integer charges, then you are in a better—well, you can even teach it to a child."

Early enthusiasts for color, Salam and Pati had also dreamed up something quite like quantum chromodynamics, but without asymptotic freedom and with integral quarks. They also believed in charm. By putting all these ideas into a box and shaking them, they came up with a theory that bears all the defects of great originality. At once a marvel of prescience and a jury-rigged tower of speculation, the Pati-Salam model is brilliant in general architecture, but is stuck together with shoe polish and string around the cracks. They realized quickly that the strong interactions change the color of quarks, whereas weak interactions change their type—their "flavor," in the jargon. To match the four known leptons—electron, muon, and two types of neutrino—they brought in charm, the fourth quark. And to put quarks and leptons together, they baldly asserted that leptons were nothing but a fourth quark color, lilac, split off from the rest by some cosmic symmetry-breaking mechanism invoking a set of no less than fourteen Higgs bosons. The full array fit into a four-by-four matrix for color and flavor, or $SU(4) \times SU(4)$. They called the theory "electronuclear unification."

	red (r)	blue (b)	green (g)	lilac (l)	
up (u)	u_r	u_b	u_g	u_l	(electron neutrino)
down (d)	d_r	d_b	d_g	d_l	(electron)
strange (s)	s_r	s_b	s_g	s_l	(muon)
charm (c)	c_r	c_b	c_g	c_l	(muon neutrino)

Although both men wanted to explore their pet notion further, the work was interrupted by Pati's return to Maryland and a trip by Salam to China. In Beijing, Salam was asked by Benjamin Lee to speak at the September 1972 conference for the opening of Fermilab. His schedule, as ever, tightly packed, Salam could not make it to the United States in time for the session on weak interactions, but he did send ahead a rough outline of the unification theory. James Bjorken briefly described it, without comment; the idea was obscured in the blizzard of theoretical excitement about gauge theories.[8]

Pati and Salam worked more on the idea during the fall, but it was difficult to make headway while on different continents. A disagreement arose over how to handle a trouble they came across in the course of working through their model. As a general rule, particles in the same family can change into each other. Putting quarks and leptons together thus tends to imply that a quark could become a lepton, and vice versa. If a quark could turn into a lepton, then one of the quarks inside, say, a proton, might suddenly turn into an electron or neutrino. The proton would then cease to exist. In other words, unifying quarks and leptons was tantamount to predicting that the basic building block of atoms was unstable. This meant that all ordinary matter, everywhere, would eventually disintegrate. Indeed, the two men had to worry about why any atoms were left.

"It was obvious the thing would happen," Salam said. He did not recall any particular moment when the realization had sunk in. "But we shouted at each other over whether we should put it into this paper or not. In fact, Pati was more impatient, and I was more cautious—in this respect. I said we should work it out properly before writing it up." Eventually the two men compromised by mentioning it in a whisper. They came up with four unified models, singled out the simplest, $SU(4) \times SU(4)$, and discussed proton decay in the appendix for the other three. They admitted that their proton lifetime was "ridiculously low," but thought that proton decay could be avoided in $SU(4) \times SU(4)$—incorrectly, as they later realized the next summer at Trieste when they worked out the model. Bad news: Proton decay was inevitable, and the lifetime must be in excess of 10^{28} years.[9]

A first paper was sent off to *Physical Review Letters*, which gave it a polite brushoff. Saying that the paper was anything but urgent, they suggested that a longer article be published at leisure in the *Physical Review* proper. Annoyed, Salam telephoned Samuel Goudsmit, an old friend and the chief editor of the whole *Physical Review* complex, who overruled the referee and printed the paper quickly.[10] (A longer article did appear later in the *Physical Review*.)

Pleased, Salam flew to Paris in August 1973, where he took a train to a meeting in the south. In his Nobel address, he evoked the scene:

I still remember Paul Matthews and I getting off the train at Aix-en-Provence for the 1973 European conference and foolishly deciding to walk with our rather heavy luggage to the student hostel where we were billeted. A car drove from behind us, stopped, and the driver leaned out. This was Musset whom I did not know well personally then. He peered out of the window and said: "Are you Salam?" I said "Yes." He said: "Get into the car. I have news for you. We have found neutral currents."[11]

Neutral currents dominated the Aix-en-Provence conference. Having seen the first HPW preprint, Weinberg said with cautious enthusiasm, "[T]here is now at last the shadow of a suspicion that something like an SU(2) × U(1) model . . . may not be so far from the truth."[12] T. D. Lee asked Salam for his thoughts on the momentous discovery. To the exasperation of the experimenters who had spent more than a year worrying over the interpretation of bubble chamber photographs, Salam blithely ignored neutral currents— they were a settled issue for him—and extolled the virtues of looking for proton decay. "Nobody took it very seriously," he said. "I think it was so outrageous that . . ." He made a gesture of shrugging indifferently.

That December Salam again boosted proton decay at a conference at the University of California at Irvine. In the audience were Feynman, Gell-Mann, and Glashow. There, too, the reaction was unenthusiastic, except for Frederick Reines, one of the experimenters who first tested for the presence of the neutrino. Concurrently with his continuing research into neutrino behavior, Reines had measured the proton lifetime by looking in his detectors to see if any of their protons had decayed. He arranged for Salam, who was flying to meet Pati in Maryland afterward, to be a dinner guest of William Wallenmeyer, the director for high-energy physics of the U.S. Department of Energy. Salam was to take Pati along, and the two were instructed to sell Wallenmeyer on the idea that the DOE should fund a proton decay experiment. "I still remember that we paid for the dinner out of our own pockets!" Salam said. "So the whole thing just drifted in the hole. Nobody took it very seriously."[13]

At the Irvine conference, Salam also learned that he and Pati were not alone in trying to unify the three particle forces. Sheldon Glashow and Howard Georgi had come up with a unified theory that not only put quarks and leptons in the same family but described all the elementary particle forces with the same coupling constant to boot.

A third-year postdoctoral fellow at Harvard, Georgi began collaborating with Glashow almost as soon as they met. Glashow thought their talents were perfectly complementary. Sitting in his office he would spin out one idea after another, and snarl in mock fury as Georgi shot them down. "He comes at you in the morning with ten ideas he's had since the day before," Georgi told us. "You have to tell him what's wrong with them, which means you have to have this kind of automatic computer in your brain to spit back responses. But if you can't figure out what's wrong with an idea right away, then you go to work on it."[14]

Although to outsiders the collaboration seemed anything but harmonious, Glashow and Georgi enjoyed it immensely. In 1973 and 1974 alone they wrote four important papers together, one of which set out the first grand unified theory.

When SU(2) × U(1) came together, Gell-Mann has recalled, he was fond of needling his colleagues by informing them that it was not a fully unified theory. "It's a *mixing*," he said. "That's why you have a mixing angle, because the electromagnetic and weak forces are not really unified. A truly unified theory would show how both the weak and the electromagnetic forces are different aspects of the same interaction."[15] A real unification theory, in other words, would not stick together two groups like $SU(2) \times U(1)$, but embed both in a single, larger structure.

The notion that the weak and electromagnetic forces could be *directly* unified occurred to Glashow on the day in October 1973 when he first heard about $SU(3)$ of color, the gauge theory of quantum chromodynamics. He saw immediately the rough outline of the theory that he was looking for. He needed to find a group large enough to contain both $SU(3)$ and $SU(2) \times U(1)$ as constituents. Such a group would permit him to put quarks and leptons into one family and would allow for a unified description of the strong, weak, and electromagnetic forces. The obvious difficulty was that these forces didn't *look* as if they could be unified, inasmuch as the strong force is a hundred times more powerful than electromagnetism, and both are enormously stronger than the weak interaction. Somehow the differing strengths would have to be reconciled.

The relative strength of a force between two objects depends in part on such variables as their masses and the distance between them. Above all, it depends on the coupling constant, which describes the force's absolute strength. To produce a truly unified theory of the strong, weak, and electromagnetic forces, one would have not only to put the quarks and leptons together, but find some way to describe all three elementary particle forces by means of a single coupling constant. Glashow laid all this out for Georgi, who is an expert model builder, and the two sat down to work out a theory.

Georgi knew of the two papers by Pati and Salam, and so was prepared for the idea of quarks and leptons in one family, and proton decay. "The introductory remarks in one of them is really extremely prescient," Georgi said, "and more or less right on. They didn't quite understand or connect the U(1), but they understood almost everything, and then the rest of the paper is not really much fun because it's an explicit model that doesn't do the things that they clearly wanted it to." Georgi laughed. "You see, the reason that they didn't get, and perhaps don't deserve, more credit for all of this was that they were at the time extraordinarily confused about the nature of the strong interactions. At the time, Abdus's primary concern was the Quark Liberation Front. He hated quark confinement and would go around with these buttons that said 'Free the Quarks.' However, if you go back and look at the group theoretical structure, you'd see that it can be translated into something more modern. You don't have to do this awful thing to the quarks. And then it really does look like the precursor of grand unification."[16]

They set out to find a fully unified model; some idea of the naturalness of the step may be indicated by the fact that it took them less than twenty-four hours. (By contrast, Einstein spent half a lifetime on his unified field theories.) It was a day that Georgi remembers well. True to form, he and Glashow spent the afternoon arguing furiously about how to proceed. Unable to hammer out a solution, each left Lyman Hall in some distress. Later that evening Georgi sat down at his desk at home to work on the problem further.

"I first tried constructing something called the SO(10) model, because I happened to have experience building that kind of model," Georgi said. "It's a group in ten dimensions. The model worked—everything fit neatly into it. I was very excited, and I sat down and had a glass of Scotch and thought about it for a while.

"Then I realized this SO(10) group had an SU(5) subgroup. So I tried to build a model based on SU(5) that had the same sort of properties. That model turned out to be very easy, too. So I got even more excited, and had another Scotch, and thought about it some more.

"And then I realized this made the proton, the basic building block of the atom, unstable." Georgi, like Salam and Pati before him, realized that he could not avoid proton decay, although for a different physical reason.[17] If the proton was unstable and would eventually fall apart, so would all atoms—and thus all matter. If the model on his scratch pad was true, then the Universe would ultimately disintegrate. He said, "At that point I became very depressed and went to bed."

The next morning he arrived in Glashow's office with "some good news and some bad news." The good news was SU(5), of course, and the bad news was proton decay. His colleague didn't take the bad news hard. "It wasn't shattering," Glashow said. "I mean, we know the sun will burn out in a few billion years. This is *known*. It's a *fact*. Spaceship earth and all that—poof! That matter falls apart a long, long time afterward is scarcely an upsetting idea. It's bad enough as it is."[18]

Glashow *was* troubled by the question of why the protons in the Universe had not already decayed, given SU(5)'s implication that they could have. The two men raced upstairs to the Harvard physics library to look up what Reines had found to be the minimum lifetime of the proton. The figure was 10^{27} years, trillions of times the present age of the cosmos.[19] Georgi and Glashow sat down to figure out how to make the theory predict that protons would hold out that long.

"My immediate reaction was that we had to make the virtual particles that bring about proton decay very heavy," Glashow said. "That meant they wouldn't appear very often. This we could do thanks to Steve, who had taught us in his 1967 paper how to give the intermediate vector particles a big mass. Only in our case they had to have a *much* bigger mass—a thousand trillion times bigger than anything that had ever been seen." In fact, the proton-decay particle was later calculated to have a mass of almost a billionth of a gram—nearly heavy enough to be weighed on a scale.

Glashow loved the idea of monstrous subatomic particles. "It was another way for him to throw rocks at the establishment," Georgi said. "Nobody we knew had ever even *talked* about elementary particles that heavy. Not even within twelve orders of magnitude of that."

Georgi and Glashow sent SU(5) to *Physical Review Letters* in January of 1974. They had decided not to be shy about what they were doing, and had given the paper the most imposing name they could think of: "Unity of All Elementary-Particle Forces."[20] Only three and a half pages long, the paper linked nearly all of the significant discoveries in quantum field theory that had been made in the past quarter-century. The mathematical techniques it employed came from group theory and Yang and Mills; articles by Gell-Mann, Glashow, Higgs, 't Hooft, Salam, Schwinger, Weinberg, and others were cited. Quarks, symmetry breaking, and gauge fields played a role. All were staples of the theoretical larder; Georgi and Glashow were simply the first to combine them into a meal. Historically speaking, the paper is a culmination of fifty years of physics—even if SU(5) is wrong, as it probably is, for the argument gathers into itself everything that came before.

The paper opens with a fanfare as brassy as that in any recent scientific work:

We present a series of hypotheses and speculations leading inescapably to the conclusion that SU(5) is the gauge group of the world—that all elementary particle forces (strong, weak, and electromagnetic) are different manifestations of the same fundamental interaction involving a single coupling strength, the fine-structure constant. Our hypotheses may be wrong and our speculations idle, but the uniqueness and simplicity of our scheme are reasons enough that it be taken seriously.

They then proceed to lay out the ground rules, acknowledging their sources. (The emphasis is Georgi and Glashow's.)

Our starting point is the assumption that weak and electromagnetic forces are mediated by the vector bosons of a gauge-invariant theory with spontaneous symmetry breaking. *A model describing the interactions of leptons using the gauge group SU(2) × U(1) was first proposed by Glashow, and was improved by Weinberg and Salam, who incorporated spontaneous symmetry breaking.*

In the next paragraph they introduce what was then little more than a favorite hypothesis of Glashow's: charm.

To include hadrons in the theory, we must use the Glashow-Iliopoulos-Maiani (GIM) mechanism and introduce a fourth quark [c] carrying charm. . . . The next step is to include strong interactions. We assume that strong interactions are mediated by an octet of neutral vector gauge gluons *associated with local color SU(3) symmetry.* . . .

They now had the entire standard model, which was still years away from being demonstrated in experiment.

Thus, we see how attractive it is for strong, weak, and electromagnetic interactions to spring from a gauge theory based on the group $\mathbf{F} = SU(3) \times SU(2) \times U(1)$. *Alas, this theory is defective in one important respect: It does not truly unify weak and electromagnetic interactions. The* $SU(2) \times U(1)$ *gauge couplings describe two interactions with two independent coupling constants; a true unification would involve only one.*

Next Georgi and Glashow hunt through the list of groups that Cartan had compiled decades before, coming up with three that could provide the basis for a truly unified electroweak theory. These three have already been used in models made by Weinberg and, later, Georgi and Glashow. None is reconcilable with $SU(3)$. This allows the writers to conclude, with some pleasure, "We see we cannot unify weak and electromagnetic interactions independently of strong interactions." They then declare themselves forced to consider the "outrageous possibility" that a single group could account for all three. Methodically they set down all of the groups of sufficient size. As it happens, there are nine. One by one, these are eliminated, except the last, which was $SU(5)$—the model that Georgi had created over shots of Scotch a few months before.

The last half page of the paper was devoted to characterizing $SU(5)$ and its implications, which are multifarious. To start, within this grand unified theory, as within the Pati-Salam model, quarks and leptons are kith and kin. Particles within a family can decay into one another, and thus Georgi and Glashow had to postulate the existence of a brand-new, never-before-seen force that would allow for just that possibility. They named this force the *superweak*, and suggested that it is mediated by a set of extremely heavy vector bosons, which Glashow nicknamed "ponderons" or "vector basketballs." These particles weigh too much to be emitted often, but when a quark does emit such a particle, that quark turns into a lepton. If the quark is part of a proton, the proton will fall apart. The pieces of the proton will, in turn, decay into electrons and positrons, which will eventually meet each other to form photons—light. This light will ultimately traverse an utterly empty plenum. Because protons are a nonrenewable resource, the Universe will end as it began, with light. Eventually, the light will dissipate, and the cosmos will settle into a long darkness. *Obeat lux.*

In their paper Georgi and Glashow, afraid of the reaction that this implication would provoke, refrained from mentioning proton decay until the end:

Finally, we come to a discussion of superweak interactions and SU(3)-colored superheavy vector bosons. In addition to mediating such bizarre interactions as $K^0 \rightarrow \mu^+ e^-$ they make the proton unstable.

"Unity of All Elementary-Particle Forces" appeared in February 1974 and initially met with little favor. Most theorists didn't notice it in the excitement provoked by $SU(3) \times SU(2) \times U(1)$. Even the select public that did read it found proton decay hard to swallow. Bjorken, for instance, told us he had been increasingly excited and convinced by the paper until he came to the very end. "It is a very tightly constructed and convincing paper," he said. "Then, all of a sudden, they hit you with proton decay out of the blue. 'Oh, by the way, the proton decays.' I lost my enthusiasm and decided I didn't like the paper. Later, after I'd thought about it, I also became skeptical of the difference in scale between the tremendously heavy particle they had to assume and all the other particles we know about. I'm still not convinced, by the way."[21]

"And *that* was the reaction of the *good* physicists," Georgi said. "To most people, it was just another group. You know, 'First it was $SU(2)$ and then $SU(3)$—' " he shifted into a sudden, surprising imitation of the Dumb Skeptic " '—now it's $SU(5)$ and soon it'll be $SU(18)$.' But we, too, didn't know how seriously to take it. Remember, this was before neutral currents, before charm, before any of the components had been found which would make people want to take the theory seriously."[22]

In addition, the $SU(5)$ paper did not provide any clear reason for supposing that the coupling constants could be made equal; this was provided a few months later, in a clever paper that Georgi wrote with Weinberg and another Harvard postdoc, Helen Quinn. "Hierarchy of Interactions in Unified Gauge Theories" was published in *Physical Review Letters* in August 1974. In it the authors used the renormalization group to examine the behavior of the coupling constants of the strong, weak, and electromagnetic interactions. Pointing out that the coupling constants are not true constants at all, but vary with the energy of the interactions, they climbed hand-over-hand up the scale of energy, rising fifteen orders of magnitude. The strength of electromagnetism and the weak interactions increases, while the strong force grows feebler. At 10^{14} GeV and higher, the coupling constants converge. Where the coupling constants merge, the three forces reveal their unity.

"That paper also was sparked by the excitement about quantum chromodynamics," Weinberg said. "We had Politzer at Harvard, and of course we also knew what Gross and Wilczek were doing, and at a certain point it just occurred to Georgi, Quinn, and me that if the strong interaction is getting weaker when you go to high energy [which is equivalent to small distances], then at *very* high energy it might be comparable to the electroweak." The

paper was very much in Weinberg's mode of operation; although it spoke of
SU(5), it was applicable to an entire class of models. "I find it hard to believe
in specific models. Those results, to a certain accuracy—ten percent accuracy—are valid whatever the grand unified theory is, within a broad
range."[23]

Now, 10^{14} GeV is a lot of energy for one particle to have. It corresponds
to a temperature of about 10^{28} degrees Fahrenheit—one so extreme that it
had not been reached in the Universe since the first fractions of an instant
after the Big Bang. Thus, whereas Georgi and Glashow had demonstrated
that uniting the strong, weak, and electromagnetic forces is theoretically
possible, Georgi, Quinn, and Weinberg showed that these forces had actually been united only at the beginning of time, after which they had separated like curds and whey. Whereas Georgi and Glashow could only confess
that the proton decays "with an unknown and adjustable rate," Georgi,
Quinn, and Weinberg gave a strong prediction of the average proton lifespan:
10^{32}—ten quadrillion quadrillion—years. Beyond the specific number, however, the Georgi-Quinn-Weinberg paper demonstrated to physicists the
compelling need to unite particle physics, the study of the smallest objects in
existence, with cosmology, the study of the Universe as a whole.

Cosmology began over a century ago, it can be said, with the Czech
mathematician Christian Doppler's prediction of the eponymous Doppler
effect in 1842.[24] Three years later, the existence of the effect was demonstrated by one Christoph Baillot, a Dutch scientist who persuaded a band of
trumpeters to play on an open railway car as it passed a group of musicians
with perfect pitch. The speed of the car changed the frequency of the
trumpet blast, an effect familiar to anyone who has heard the change in the
pitch of an ambulance siren as it zooms by. It turns out that light waves, too,
can manifest this effect, a fact of little consequence until Edwin Hubble
discovered in the 1920s that there were millions of galaxies in the cosmos,
and that the Doppler shift in the light they gave off indicated that they were
flying from us at great speed. It looked to Hubble as if the whole Universe
consisted of the speeding fragments of a monstrous explosion.

Attracted by the novelty of the idea, some astronomers leaped on it;
others were leery of postulating a definite beginning to the Creation, for fear
of invoking the question, "Well, what came before *that?*" The fortunes of the
theory of the Big Bang, as the initial explosion was soon dubbed, waxed and
waned until the early 1960s, when two Bell Laboratories researchers, Arno
Penzias and Robert W. Wilson, gave themselves the problem of ridding a big
radio dish of a strange type of interference. One of the first satellite trackers,
the big dish in New Jersey was designed to pick up signals from Telstar.
Penzias and Wilson at first guessed that the interference was due to a terres-

trial source, possibly the family of pigeons nesting on the antenna. After scraping off what Penzias obliquely referred to as a "white dielectric substance," the interference remained, no matter where they pointed the dish. Years later they realized they were picking up the residual radiation from the Big Bang, which floods space like the morning heat from a campfire the night before. (Wilson and Penzias shared the 1978 Nobel Prize for their discovery.)[25]

The combination of the stellar Doppler shift and the cosmic background convinced cosmologists of the reality of the Big Bang, but few of them had then a strong idea of where to proceed. In the 1970s, the advent of unified theories persuaded many that there was a great deal of physics that happened in the first microseconds of creation. (They tended to assert their hypotheses loudly, despite the paucity of empirical data; the saying runs that cosmologists are often wrong, but never in doubt.) Although the standard model was constructed to describe the behavior of particles at ordinary energy levels, it contained within it, much to most theorists' surprise, plausible grand unified theories with a vastly longer reach; united with cosmology, elementary particle physics can present a picture of the history and evolution of the Universe even as it seeks to explain the ties among ordinary matter and forces.

On the far side of the Big Bang is a mystery so profound that physicists lack the words even to think about it. Those willing to go out on a limb guess that whatever might have been before the Big Bang was, like a vacuum, unstable. Just as there is a tiny chance that virtual particles will pop into existence in the midst of subatomic space, so there may have been a tiny chance that the nothingness would suddenly be convulsed by the presence of a something.

This something was an inconceivably small, inconceivably violent explosion, hotter than the hottest supernova and smaller than the smallest quark, which contained the stuff of everything we see around us. The Universe consisted of only one type of particle—maybe only one particle—that interacted with itself in that tiny, terrifying space. Detonating outward, it may have doubled in size every 10^{-35} seconds or so, taking but an instant to reach literally cosmic proportions.

Almost no time passed between the birth of the Universe and the birth of gravity. By 10^{-43} seconds after the beginning the plenum was already cooler, though hardly hospitable: every bit of matter was crushed with brutal force into every other bit, within a space smaller than an atomic nucleus. But the cosmos was cool enough, nonetheless, to allow the symmetry to break, and to let gravity crystallize out of the unity the way snowflakes suddenly drop out of clouds. Gravity is thought to have its own virtual particle (the graviton), and so the heavens now had two types of particles (carriers of forces and carriers of mass), although the distinction wasn't yet as clear as it is in the Universe today.

At 10^{-35} seconds the strong force, too, fell out of the grand unified force. Less time had passed since the Big Bang than it now takes for a photon to zip past a proton, and yet the heavens were beginning to split. Somewhere here, too, the single type of mass-carrying particle became two—leptons and quarks—as another symmetry broke, never to be complete again. The Universe was the size of a bowling ball, and 10^{60} times denser than the densest atomic nucleus, but it was getting colder and thinner rapidly.

One ten-billionth of a second after the Big Bang, the firmament reached the Weinberg-Salam-Glashow transition point, and the tardy weak and electromagnetic forces broke away. All four interactions were now present, as well as the three known families of quarks and leptons. The basic components of the world we know had been formed.

"Let me draw the whole picture for you," Glashow said to us once as he scratched a line in white chalk from one edge of his office blackboard to the other. "This represents the Universe from the beginning to the end of time." After thinking a moment, he drew a second line in purple chalk; it was just a bit shorter. "We live in the fortunate era—the era in which there is matter. Matter first appeared 10^{-38} seconds or so after the Big Bang, and will all disappear maybe 10^{40} seconds from now." He hunted around in the tray below the board, finally happening upon a piece of brown chalk, which he used to plot a line considerably shorter than the first two. "Within the fortunate era of matter there is a somewhat shorter period in which atomic nuclei exist, because the Universe is cool enough to permit them to form. And within that"—a red line now, barely a foot long—"an even tinier domain in which there are atoms." He made a dot near one end of the red line. "Then you have a brief ten billion years or so when things are palatable on earth.

"After that, eventually it all winds down. We won't be bothered when, say, ten percent of the protons go. If we're still around, we may not even notice. But when ninety-nine percent go, you won't have enough left in one place to make a person, and it will be unpleasant. After this point there's a long period of decline, and a very boring period it is, too. This is not a view we could have gotten without grand unified theories. We may not like it, but we have no choice."[26]

Is the picture true? At the end of the 1970s, the standard model was well enough established that experimenters and theorists were ready to look at something that pushed unification further. Unfortunately, one of the predictions of SU(5) is that there are no further elementary particles to be found—not until the grand unification energy of 10^{15} GeV is reached. Such energies cannot be reached by any accelerator conceivable today. "It would take an accelerator ten light-years long," Glashow said, "which, given current budgetary constraints, is unlikely." Grand unification in general and SU(5) in particular did have a second major testable prediction: proton decay, a question Glashow phrased as, "Are diamonds forever?"[27]

□ □ □ □ □

Sunrise never seemed to happen at Kolar. The long slow foredawn, shadowed and almost purple, hung in the air with a last promise of coolness for an hour or more—then was abruptly gone. Harshly the Indian sun went up, and the dust rose to meet it. The miners flinched from the heat as they waited for the koepis to rattle them beneath the surface. Straggling workers joined the line, brushing aside goats, chickens, and somnolent cattle. Great steel drums the size of a house turned with a clank, the metal cable smeared with oil. Bunched with the miners were physicists and technicians, distinguishable by the clipboards in their hands. The small elevator arrived; the crowd slowly moved toward it. Seven thousand five hundred feet below, the oldest running proton decay experiment began another day.

Cosmic ray physics has long been an Indian specialty, because India's high mountains and deep mines enabled its physicists to do important research with inexpensive instruments. Since the 1950s, cosmic ray studies have been carried out at the Kolar Gold Fields, an immense and almost exhausted mine in south central India. There, in 1964, M. G. K. Menon, the assistant director of the Tata Institute of Fundamental Research in Bombay, participated on the Anglo-Indian-Japanese team that detected the first cosmic ray neutrino; twenty years later, he was the senior physicist in the proton decay experiment.

In 1966 Menon succeeded the physicist Homi Bhabha as director of the Institute when the latter died in a plane crash. (India's premier research center, the Tata Institute was founded in 1945 by the trust of a steel magnate, Sir Dorab Tata.)[28] A precise, soft-spoken, patriotic man who has devoted much his life to fostering Indian science and technology, Menon also became his nation's delegate to the United Nations Advisory Committee on Science and Technology. He immediately hit it off with the Pakistani representative, Abdus Salam, and when the two met at United Nations conferences in Geneva or New York they would retire to a corner of the assembly rooms during the more long-winded speeches and chat about physics.

At a meeting in the summer of 1973, Salam told Menon about proton decay. By hooking detectors to a box of heavy material, Menon could scrutinize the box to see if its protons decayed. Because proton decay was expected to be quite rare, the detector should be kept far away from background cosmic radiation that might confuse the issue. A deep mine like Kolar would be ideal. Menon replied that testing the lifetime Salam and Pati had predicted—about 10^{29} years—would require at least a hundred-ton detector, large even by accelerator physicists' standards and huge for a cosmic ray physicist. Nonetheless, Menon decided to give it a shot. In 1978, Menon, now a member of the Indian Planning Commission, V. S. Narasimham, and Badanaval Sreekanton, Menon's successor at the Tata, received permission to go ahead. They had decided to collaborate with scientists from the Japanese universities of Osaka City and Tokyo.

The entrance to the Gifford shaft of the Kolar gold mine is in a large, hangarlike structure criss-crossed by steel girders and the cables from the koepi to the hoist. The morning we visited, monkeys played and chattered in the struts. An amiable security guard escorted us down the shaft; as the lightless rattling box hurtled to the seventieth level, more than six thousand feet below our feet, we shared a moment of claustrophobic alarm. At the bottom was a small bright shrine to the Hindu gods. Compressed air from above blew down the tunnels, keeping the temperature to a bearable eighty degrees. "The temperature is ninety-one at the hottest," Narasimham said. "It's bearable, for Indians used to the climate. But it gives problems to our electronics."

We walked a hundred yards to a second shaft, descended another few hundred feet, then followed the tracks of the bandie cars down a tunnel. A cardboard sign hung from the ceiling: FIRST EVER COSMIC RAY NEU-TRINO INTERACTION RECORDED HERE IN APRIL 1965. Nara-simham had been at the mine then. A slightly built man with saturnine features, long thinning hair, and somewhat protruding eyes, he had spent years of his life waiting for cosmic rays inside Kolar. Two bandies full of chopped stone were wheeled past us by wiry sweaty men with cloths around their heads. The rock above pressed down like a heavy cloak, smothering. Just before the track rounded a corner was a niche blocked off by a chain-link fence.

PROTON STABILITY EXPERIMENT OF
TATA INSTITUTE OF FUNDAMENTAL RESEARCH, BOMBAY
AND
OSAKA CITY AND TOKYO UNIVERSITIES, JAPAN
Site: 80th Level Heathcote Shaft, KFG.
2300 meters depth.

Fifteen feet behind the gate was a wall of long rectangular pipes, each eighteen feet long and five inches square, stacked Lincoln Log–fashion thirty-four layers high. The finished product was an iron grid of rusted pipe ends, each with a dusty valve and a painted number; a passage to the right led to the data-taking room, where the computers were kept much cooler than the people who attended them. Purchased from Japanese construction firms, the pipes were altered into inexpensive proportional tubes—a kind of Geiger counter. Engineering students had threaded a single wire down the length of each pipe, sealed it full of argon, and linked the array together to a high voltage line. The central wires and the walls surrounding them acquire opposite charges. If a charged particle passes through, a spark is created. Particle interactions are traced and measured by counting the lines of triggered tubes.

A second, larger detector was near completion. Perspiring freely, we walked down the dimly lit passage; the rock above had grown heavier with the passage of time. We were led into a cavern, roofed with I-beams, eighty feet long and thirty-five-feet wide, excavated at the risk of a hundred lives. Punctured by long supportive bolts and coated with a yard-thick layer of protective concrete, the walls seemed to tremble with the immensity of the planet above them. Dominating the space was a rough iron cube, twenty feet on a side, that rose like a monolith in the crepuscular light. A corona of garish sparks from welding equipment danced across its crown. Ringed by scaffolding, the three-hundred-ton detector was tended by a dozen quickly moving men. Stripped to the waist, workers clambered over its face like ants. Blasts of compressed air thundered into the space, and Narasimham had to shout above them. In the roar, the dust, the splash of water, and the shriek of machinery, he yelled calmly in our faces, his shirt flapping in the hot ventilating breeze. "Nobody knows when the proton decays, but everybody seems to think it must," he said. "People used to think it was immortal and hardly anyone looked. Now everyone is looking, and they are sure it decays."[29]

Others who looked, with whom the Kolar group was in keen competition, included experimenters working in a salt mine near Cleveland, Ohio; in the Mont Blanc tunnel on the French–Italian border; in the Silver King mine near Park City, Utah; and the Kamioka lead and zinc mine near Takayama in western Japan. All took boxes of iron or tanks of water—reservoirs of protons, in either case—to secluded areas and watched to see if any decayed. Time passed but convincing evidence of proton decay did not materialize, and the accepted minimum lifetime of the proton kept rising until it was well above the value predicted by SU(5).

But despite this apparent failure to clinch unification, the genie was out of the bottle. The discipline of physics had an apparently incurable case of unification fever. The result was a constant outpouring of publications, conferences, and journals devoted to unification. Although the failure to find proton decay would dissipate the immediate enthusiasm, unification no longer seemed like an impossibly distant goal. It seemed like something that was only one breakthrough away, and every physicist on Earth, experimenter and theoretician alike, wanted to be there when it happened.

20
The End of Physics

THE LUCASIAN PROFESSORSHIP OF MATHEMATICS OCCUPIES A CURIOUS POSI-
tion at Cambridge University, for it has traditionally been occupied by a
great theoretical physicist. Newton and Dirac were both Lucasian Professors,
and its current occupant is Stephen Hawking, easily the most well-known
cosmologist in the world. Unfortunately, Hawking is famous less for his
considerable accomplishments than for his long struggle with a degenerative
neural disease that has kept him in a wheelchair for many years, impeded his
ability to speak and write, and may eventually kill him. On April 29, 1980,
Hawking assumed the chair with an inaugural lecture entitled, "Is the End
in Sight for Theoretical Physics?" The answer, he said, is probably yes. A
student read the speech for him:

*In this lecture [Hawking's text began] I want to discuss the possibility that the
goal of theoretical physics might be achieved in the not-too-distant future, say, by
the end of the century. By this I mean that we might have a complete, consistent,
and unified theory of the physical interactions which would describe all possible
observations. Of course one has to be very cautious about making such predictions: We have thought that we were on the brink of the final synthesis at least
twice before. At the beginning of the century it was believed that everything could
be understood in terms of [classical] mechanics. All that was needed was to
measure a certain number of coefficients of elasticity, viscosity, conductivity, etc.
This hope was shattered by the discovery of atomic structure and quantum
mechanics. Again, in the late 1920s Max Born told a group of scientists visiting
Göttingen that "physics, as we know it, will be over in six months." This was
shortly after the discovery by Paul Dirac . . . of the Dirac equation, which
governs the behavior of the electron. It was expected that a similar equation
would govern the proton, the only other supposedly elementary particle known at
that time. However, the discovery of the neutron and of nuclear forces disappointed these hopes. We now know in fact that neither the proton nor the neutron
is elementary but that they are made up of smaller particles. Nevertheless, we
have made a lot of progress in recent years and, as I shall describe, there are
some grounds for cautious optimism that we may see a complete theory within the
lifetime of some of those present here.[1]*

Hawking has been but one among many physicists who have hoped
that unification is at hand, although few have been willing to state their hopes
so boldly. Today, unification is the research topic for an entire new generation
of theorists. The subject has split into branches and subdivisions, and there
are unification conferences and unification journals and even unification text-

books. It is the rare theorist who has not tried his hand at putting things together.

Physics has always progressed by drawing together seemingly disparate phenomena into one framework. Newton's recognition that the force that made apples fall was identical to the force that kept the earth in its orbit was a unification, as was the approximately contemporaneous realization that lightning, static electricity, and Saint Elmo's fire were all manifestations of one phenomenon: electricity. In the nineteenth century, Maxwell synthesized electricity, magnetism, and optics into his theory of electromagnetism.

These successes in turn elicited more grandiose but less happily conceived unification schemes. As Freeman Dyson has pointed out, "The ground of physics is littered with the corpses of unified theories."[2] The early nineteenth-century chemist Claude Berthollet spent years trying to demonstrate that gravity was a sort of chemical attraction. Michael Faraday, Maxwell's great predecessor, anticipated Einstein by eighty years when he tried to establish a relation between electromagnetism and gravity in 1850. The beginning of gauge theory was a mistaken stab at unity by Hermann Weyl. Einstein's long failure with unified field theories is well known; in retrospect, a principal stumbling block was his refusal to treat the forces in atomic nuclei as fundamental, a folly that reached its height when he published a paper unsuccessfully attempting to prove "that the elementary formations [i.e., particles] that make up the atom are held together by gravitational forces."[3] Schrödinger, too, spent the last years of his life obsessively pursuing the chimera of a unified field theory. In the 1930s, Yukawa tried to unify the strong and weak forces, but the eventual success of his work owed nothing to unification. Heisenberg launched, two decades later, a unified field theory that started as a collaboration with Pauli. When Pauli withdrew, Heisenberg pressed on. To Pauli's fury, Heisenberg claimed during a radio broadcast in February 1958 that a unified Heisenberg-Pauli theory was imminent, and only a few small technicalities remained to be worked out. Rumors swept the press. Pauli responded by mailing his friends a letter consisting of a blank rectangle, drawn in pencil, with the caption: "This is to show the world I can paint like Titian. Only technical details are missing."[4]

Mindful of the recent record of failure, unification aficionadoes today practice their craft with a measure of irony, for they are as embarrassed by the loose speculation required as they are entranced by the sweetness of the problem. Almost every practicing physicist has been approached by eager cranks with unified theories, sweaty would-be Einsteins with equations of the world that sew up all loose ends; now look at the theoretical legions invading the territory of the nuts. (The presence of the crackpots is in its own way a measure of the attractiveness of the idea.) A complete unified theory would mean the end of physics. Which is not to say that physics equations would vanish, physics experiments would stop, and all physicists would be out of a job. Many millions of loose ends would need tying up; connections would still need to be drawn; effects, to be understood; applications, to be devised and exploited. Science would continue, but all of the fundamental questions that physics can pose would have been answered, and our knowledge of force and matter would henceforth change only in particulars and not in outline.

In the *Critique of Pure Reason*, Immanuel Kant argued that some aspects of nature will remain forever unknown to us, because our minds must impose a structure on our sensations for us to have any experience of the world at all. As a consequence, we are led inevitably to make certain suppositions about nature that are, in actuality, by-products of the organizing activity of our own brains. Such presuppositions—Kant showed their number, and said they apply to more than science—are unprovable in theory but indispensable in practice. He called them "regulative ideas," and listed the unity of nature as a cardinal example. Because of it, the structure of science itself draws physicists toward unification. Given that the standard model contained the answers to all ordinary physical questions, it is then little wonder that an orgy of unifying came after its completion, as theorists followed the impulse built into the science, and no surprise that the sudden explosion of unified theories in the 1970s brought explosive and diverse reactions.

"To my mind," Gerard 't Hooft said, "the most successful programs of unification always came about when there was some urgent question to be answered. In most cases, the urgent question was that something seemed to be wrong in the present understanding, and the correct answer then turned out to be that the only way to get it right was to say that this effect and that effect cancel, and the only way to make them cancel was to put them into one big theory together and show that they were the same force. Then you get something like unification. That's the way I view, say, the unification of the gravitational forces with special relativity which gave general relativity. That was an enormously far-reaching theory, but it arose because there was an obvious question to be answered: How do you reconcile the notion that the effect of two bodies on each other only depends on their masses, nothing else, and on the other hand keep relativistic invariance for the whole set of equations? If you try to answer it, you run into all sorts of difficulties unless you assume, precisely as Einstein did, curved space-time and all that. There was no other solution. So it *had* to be right. That's the way, to me, you should derive a theory. What Einstein did later was try to put electromagnetism together with gravity. However, this time, he didn't have a good motive. All he wanted was unification. That is, he didn't have as a motive that something would be wrong with physics if he didn't do it. And that's why I think he did not succeed."

We asked if this applied to the current unification efforts.

"I'd make the same objection," 't Hooft said. He had been working on putting forces and particles together for a decade, and felt that in many respects he did not have that much to show for it. "There is no obvious physical need. There is an aesthetic need, but not one of purely mathematical logic. And my conviction is that as long as that need is not there, it's unlikely that it will work in this simple way. It may work, so people should continue trying it, but it may well be just like the fate that struck Einstein— that although it looks from an aesthetic point of view to be an obvious thing to try, nature is more subtle than this. The trick is not to try to put things together which do not really belong together, but rather to search for places in the world where there are discrepancies, where different ideas are clashing that ought to be described in one and only one way."

He was asked what kind of discrepancy he had in mind.

"Well, a very important discrepancy I'm interested in, like many other people, is quantum gravity. Because we still don't have a good way of reconciling gravitation with quantum mechanics. There still isn't. I'm sure that whenever somebody finds a way to do that, he'll solve millions of problems in ordinary physics." We remarked the long-standing difficulties in the theory of quantum gravity. "Well, there simply *isn't* a theory. A theory is completely lacking. People claim that they have ideas of theories abut it—Stephen Hawking is doing a lot on it. He has made, you know, some brilliant contributions. But the most fundamental theory, Lagrangian or Hamiltonian, or a proper description of Hilbert space, is missing. And so we just switch on gravity, which is a pain in the neck."

The difference between the domains of gravitation and the elementary particle forces is the difference between the proton and the plenum. The queen of the interactions, gravitation is easily ignored when looking into the microscope of a particle accelerator; at any other time, its gentle, firm, omnipresent pull dominates the firmament. Gravitation takes place on great, epic terrain—the bald ridges and smooth valleys of space-time itself— whereas quantum theory deals in billionths of millimeters, microscopic ferment, entities beneath visibility and beyond visualization. Big/small, classical/quantum, geometry/algebra, the sole obvious point of contact of these most opposite of physical constructs is that their arena of play is the same Universe.

As the standard model and the various grand unified theories drew together the elementary particle forces, the absence of gravitation became more and more conspicuous. In a sense, the very elegance of general relativity has impeded the urgent task of putting gravity into quantum terms.

Too successful to offer much scope for theoretical tinkering, Einstein's monument stood, a seamless wall of mathematics, in proud isolation from the rest of physics. If quantum mechanics and relativity were uneasy spouses in quantum field theory, how much more troubled would be a quantum theory of gravity itself.

In any quantum treatment of gravitation, the force is transmitted by a massless spin-two particle called the graviton. The graviton gives rise to infinities intractable enough to make veteran renormalizers conclude that straight quantum gravity simply is not finite. "Something is missing there," 't Hooft said. "We are all trying very hard to make it all work, but it turns out to be conceptually extremely difficult. One of the big difficulties is that we realize that some of our well-known concepts have to be abandoned, because nature isn't going to be as simple as it appears now. However, we cannot abandon everything at once, because then there is nothing left to work on."[5]

The problem faced by 't Hooft and his coevals is unprecedented in the history of physics. They believe that matter and energy were one just for an instant, at the dawn of time, in the ravening fire of the Big Bang. Thus the phenomena they seek to describe existed only at unimaginable energies that can never be reproduced in the laboratory. Experiment consequently seems almost helpless. Theorists have spent the two decades since SU(5) trying to bootstrap themselves to unification without benefit of data, working out the right theory by pure mathematical ingenuity and physical intuition. They court the risk of divorcing theory entirely from experiment, and turning physics, the prototype of an empirical science, into what Georgi has called "recreational mathematical theology."

Notwithstanding such fears, a dizzying variety of unification theories sprang up in the 1970s and 1980s. Each one bolder than the next but all unproven, they went by various overlapping names—quantum gravity, supersymmetry, Kaluza-Klein, supergravity—the list of permutations was as long as the list of theorists working on the subject. Of all available tries at unification, what is called "superstring theory" is the most remarkable because its predictions are more than usually bizarre, because its history is more than usually chaotic, because it is apparently renormalizable, because it has a tenacious body of adherents, and because, when asked about it, Murray Gell-Mann told us flatly that he believed that some version of string theory some day would be the theory of the whole world. "It's a fantastic thing," he said. "It's a candidate. It's *the* candidate."[6]

Superstring theory arose from the unexpected recent marriage of two wild ideas: (1) the Universe has extra, hidden dimensions; (2) subatomic particles ultimately are not little points but little strings. The idea that the Universe had hidden dimensions was first proposed by the German physicist

Theodor Kaluza in 1919.[7] Inspired by Einstein's four dimensional theory of relativity as well as the Unified Field Theory, Kaluza tried to incorporate electromagnetism into a *five* dimensional form of general relativity. Kaluza's work was brought into accord with quantum mechanics by a thirty-two-year-old Swedish physicist named Oskar Klein; in Klein's version, the fifth dimension was hidden, curled up into a minute circle and playing no real role in our world.[8]

For a while, theorists found the work of Kaluza and Klein exciting, but it didn't seem to go anywhere, and the idea languished for nearly half a century. Then, in the 1970s, it was revived in a strikingly different context, the multidimensional "string theories." Based on work by Gabriele Veneziano, an Israeli, and elaborated by Yoichiro Nambu and a dozen other theorists, the string model at first dealt just with the strong interactions.[9] Its adherents regarded hadrons as little one-dimensional strings rather than points. Mesons were strings with a quark at one end and an antiquark at the other; when the meson forcefully struck another particle, the string snapped, producing two new strings. Isolating a single quark was thus as impossible as creating a piece of string with just one end. The visualizability of the theory broke down for baryons, which had to be imagined as strings with *three* ends. Although the mathematical properties of the string model were fascinating and elegant, its equations seemed to contain a horrific panoply of ghosts, infinities, anomalies, unobserved spin-two particles, and impossible particles that travel faster than light. Many of these could be removed by artful equation-juggling, but only at the price of assuming that space-time has more than the usual number of dimensions—twenty-six, in fact. (*Twenty-six dimensions*!? The physicists who discovered this didn't even *try* to explain what on earth it could mean.)[10] In 1974, two theorists at Caltech, John Schwarz and the late Joël Scherk, who had worked on an alternative string model with only ten dimensions, realized that the unwanted spin-two particle might be the quantum of gravitation.[11] At a stroke, what had been a troubled theory of the strong force was converted into an excellent candidate for a unification theory.[12]

Because gravitation, unlike the strong force, is a manifestation of the structure of space-time, extra dimensions are not necessarily disastrous. Schwarz and Scherk could use a Kaluza-Klein-like device to ensure that the extra dimensions are perpetually hidden from view, squashed into tiny, unvisualizable balls at each point in space. Moreover, string theories naturally could be extended to "supersymmetry," a method of classifying together particles of force and particles of mass. (For this reason, the theory bears the name of *super*string theory.)[13] The ideas of Schwarz and Scherk were sufficiently off-beat that they attracted little interest until 1984 and 1985, when Schwarz and a colleague proved that superstrings were not only completely

free of ghosts and anomalies—that is, they are mathematically consistent—
but that they are consistent for just two versions of the theory.[14] These
immediately became candidates for a Theory of Everything.[15] Physicists
found the thought of deriving the Universe from the requirements of consistency alone to be irresistible, enchanting, marvelous; unlike Candide, who
lived in the best of all possible worlds, we might live in the *only* possible
world.[16] Theorists have descended upon superstrings in droves, despite its
penchant for predictions that even physicists consider bizarre, such as the
existence of "shadow matter" in the Universe, matter invisible to us, that can
only be detected by gravitational effects and nothing else. Although there is
as yet not a scrap of experimental evidence for superstring theory, it is
completely renormalizable and does not appear to be in conflict with anything we know so far—no mean feat for a physical theory nowadays.

At the very least, superstring theory is a textbook example of a theoretical bandwagon, of how a clever mathematical conceit can suddenly become *démodé*, dominate discussion and conference proceedings for months
and even years, ultimately withering for lack of contact with experiment. At
the very most, it is, as Gell-Mann put it, "the theory of everything—gravity,
weak, strong and electromagnetic interactions plus a lot of other things all
together—a completely unified theory of nature." If Gell-Mann is right, the
books of future historians of science may well treat the construction of the
standard model as a lengthy parenthetical interlude between the first ink-
lings of superstring theory after the First World War and its successful application to nature sixty years later.

Howard Georgi once remarked that there is little need for string theories and other unified theories because they only apply to phenomena like
the Big Bang that can never be approximated by experiment. He advocated
what are called "effective field theories," the suggestion that at different
energy realms different field theories are applicable. Just as it would be
foolish for engineers who build bridges and design cars to use quantum
mechanics, it is nonsensical for particle physicists in an $SU(3) \times SU(2) \times U(1)$ world to try to go much past $SU(5)$. The phenomena beyond grand
unification are so ephemeral, so distant in time, or so heavy that they play no
role even in the subatomic domain probed by the largest particle accelerators. Despite his status as a godfather to the unification movement, he
professed to find most unification theory unappetizing. "The most interesting question at the moment is what exactly breaks $SU(2) \times U(1)$," he said.
"We still don't know what's giving mass to the W and the Z. We just know
that symmetry is broken. It's an absolutely open question whether it's a
Higgs or a dynamical mechanism or something that we haven't thought of. I
regard that as the only question that I can see at the moment that is both

obviously fundamental and obviously physics. Unification is clearly fundamental, but it may not be physics if you can't see any of the effects."

In 1984, Steven Weinberg came to Harvard to give a lecture series on
string theories. Georgi greeted him by writing a limerick on the blackboard
before Weinberg's first talk.

> *Steve Weinberg, returning from Texas*
> *Brings dimensions galore to perplex us.*
> *But the extra ones all*
> *Are rolled up in a ball*
> *So tiny it never affects us.*

"And," Georgi said, "it bothers me a little that it never affects us."[17] In his
view, unification theories in general and string theories in particular may
inherently be concerned with the hows and whys of phenomena seen only
during the unreachable holocaust of the first instants of creation. If reaching
the energy scale of grand unification requires an accelerator whose length is
measured in light-years, reaching the energy of full unification could only be
done in a machine the size of the galaxy. Because such machines are absent
from any version of the future, Georgi has argued that despite their formal
elegance, mathematical rigor, and beautiful complexity, unification theories
may ultimately be no better than attempts to calibrate the end of the world
by examining permutations of the number 666.

Contemptuous of idle philosophizing, practicing physicists tend to be
uninterested in the metaphysical overtones of their craft. They define the
end of physics operationally, as the day when no government will pay to test
further a future unified theory, and resist speculating about why physicists
keep trying to put such theories together. When we asked Glashow one day
why he had immediately jumped to the idea of a larger, unified gauge group,
he responded by reading a passage written in 1927 by one of his Harvard
predecessors, Percy Bridgman.

*Whatever may be one's opinion as to the simplicity of either the laws or the
material structures of Nature, there can be no question that the possessors of such
a conviction have a real advantage in the race for physical discovery. Doubtless,
there are many simple connections still to be discovered, and he who has a strong
conviction of the existence of these simple convictions is much more likely to find
them than he who is not at all sure that they are there.*[18]

We asked why unification was the necessary outcome of physics.

"It's *not* necessary," Glashow said. "All I can say is that if you have the
faith, you have an advantage. Physicists in the past who have looked for
simplifying, unifying assumptions have done well. Better than physicists
who haven't. But there's no *reason* that things get simpler. They could
become more and more chaotic and more and more complicated. They may,
at some point. But so far things are getting simpler. I can't say simple, but

simpler."[19] Nonetheless, the difference between unification as a long-range goal and unification *now* was important to Glashow. Supersymmetry, supergravity—none of it was to his taste. String theory, he told us in 1985, "is sociologically interesting as an example of a theoretical bandwagon, and not much else." In 1995, his opinion had not changed.

He is not alone. Julian Schwinger, for one, told us that unification was a "fad," a "grand illusion" that is not "a theory in the usual sense but an aesthetic and emotional glow about how things would work if only we could compute them." He dismissed the current push toward unification as simple theoretical hubris. "It's nothing more than another symptom of the urge that afflicts every generation of physicist—the itch to have all the fundamental questions answered in their own lifetimes."[20]

Across the city at Caltech, Richard Feynman grimaced with annoyance when we brought up the subject of tying together the three theories of the standard model. He let it be known that he didn't like the group terminology and that he had doubts about the ambition. "There isn't any theory today that has SU(3), SU(2) x U(1)—whatever the hell it is—that has any experimental check," Feynman snapped. Biting off the words, he quickly listed several serious difficulties with existing unification theories. His voice boomed sarcasm, crowding his big office. "Now, these guys are trying to put it all together. They're *trying* to. But they haven't." He was asked if he felt there'd been any progress toward unification. "*No*," he said. "For the following reason." He stopped, frowned. "Wait a second. It's a crazy question! Because we now know that in Einstein's time he was nowhere near unification. So to say that we're nowhere closer than that time, that's ridiculous. We're certainly closer. We know more. And if there's a finite amount to be known, we obviously must be closer to having the knowledge, okay? I don't know how to make this into a sensible question. It always looks like you're close to unification. We're always trying to put stuff together, okay? The thing that's different between the present time and the time of Einstein is the enormous amount of new phenomena that Einstein knew nothing about." Irritation crept into his voice. He literally twisted with agitation; the discussion was veering into the philosophical. "Electricity and gravity in the 1920s—I'm talking about the 1920s rather than the 1930s—looked close to being unified. The Schrödinger equation, even when it came, was still a differential equation like gravity. So you could say, oh, they're just some sort of differential fields or equations of the world, and it looks like you might be able to unify them someday. But in the meantime, a whole lot of new phenomena came. *It's a dumb question.* Cancel everything I said." He slammed his hand down on the desk. "We know more than we did then. That's true." He abruptly stood up, muttered goodbye, and stalked out the door without another word.[21]

Astonished, we watched him walk with long, urgent strides down the long corridor, drumming his knuckles on the walls as he passed. He turned his head as he went, and glared back in our general direction. Graduate students dodged out of the way as Feynman careened down the hall. "It's goddamned useless to talk about these things!" he shouted back at us. Doors opened along the hallway; heads craned out. "It's a complete waste of time! The history of these things is nonsense!" Feynman paused before turning the corner, and took in a lungful of air. "You're trying to make something difficult and complicated out of something that's simple and beautiful," he said, loudly, vanishing around the corner.

In the corridor, a respectful moment of silence. Murray Gell-Mann poked his head out of his office. "I see you've met Dick," he said mildly.

"You know," Steven Weinberg said, "their wasn't that much of an intellectual discontinuity from $SU(3) \times SU(2) \times U(1)$ to grand unification." We had just ordered lunch in the Harvard Faculty Club. Around us were open jackets and open wine bottles, loosened ties and clattering silver: the furniture of academic meals. Weinberg talked about going up fifteen orders of magnitude in his paper with Georgi and Quinn, the giddy audacity of cranking through the numbers across such an enormous range. The fire alarm rang. For a moment or two, everyone in the room looked about with the polite incomprehension customarily awarded to signals of disaster. Waiters shooed out the crowd. Weinberg brushed off his trousers and sat on the steps of a nearby building. We asked what he meant by the lack of intellectual discontinuity in grand unification. "I'm not saying this in any critical spirit. What you were really talking about was a new symmetry structure imposed on the good old dynamics of quantum field theory. But with strings, you really have a new dynamics. It's still within the framework of quantum mechanics. But that's almost the only thing that has remained. String theories *look* like field theories over an enormous range of energies, up to the fundamental scale, which is somewhere in the neighborhood of 10^{16}, 10^{17}, 10^{18} GeV. But if you really get up to the fundamental scale, then they stop looking like field theories altogether. They really are a new kind of dynamics."

Mutterings of false alarm; people started to return to the faculty club, although the fire alarm was still ringing because no one could figure out how to turn it off. As we filed in, we asked Weinberg whether strings change our understanding of the birth of the Universe. Despite the hubbub and the jostling, he spoke readily and concisely. "It doesn't really answer any questions, because if you let the clock run backward and imagine what the Universe looks like as you go to earlier and earlier times, you still see a singular state." That is, properties like the energy density shoot up to infinity as you go back in time, and at the beginning is a white-hot point unexplained by current physics.

On another occasion, Weinberg remarked, "I'd say the period from the mid-sixties to the mid-seventies was enormously exciting, progressive, the best time we've had in physics since the late forties. Unfortunately, since then experiment and theory have gotten out of touch with each other. It's not really the fault of anybody, it's just the logic of the way the subject has developed. It's been the most frustrating decade, really, just awful!—in the sense that the thing that the brightest theorists are doing does not directly bear on any experiment that's about to be done or can be done in the foreseeable future. Supergravity, Kaluza-Klein, grand unification, all of that stuff, with a few little exceptions, can't be tested experimentally. And where it can, it hasn't been terribly impressive. Look what's happened with supergravity. The people who've been working on it for the past ten years are enormously bright. Some of them seem brighter than anyone I knew in my early years. They have elaborated these theories of supersymmetry, supergravity, and superstrings in a way that I think is unprecedented in the history of science for a theory that has *no experimental support whatsoever.*

"I've done it myself, I'm not badmouthing them. I think it was the right thing to do, because, as I say, you do what you can, and this was the best that could be done." He was discouraged by the implicit ironies: So many good theorists with so many good ideas who think they're so close to unification— only to find that proving the theories is utterly impossible with any foreseeable technology. "We just can't go on doing physics like this without support from experiment," he said. "The experimentalists do great things—discover the W and Z—and God bless them, it's wonderful. But the theory has moved to the point where these experiments are not helping. I hope that with the next generation of accelerators, we'll get out of this morass."[22]

Weinberg spoke in 1985. Today, a decade later, his hope is unrealized; to a daunting extent, particle physics is exactly where it was when he was interrupted by the faculty club fire alarm. With huge effort, experimenters have filled in a few missing pieces from the standard model. Notable among them is the top quark, its discovery finally announced at Fermilab in March 1995, after many false starts and early intimations. (The top turned out to be amazingly heavy—its mass is about that of a gold atom.) But neither that discovery nor any other provided decisive indications of how to move beyond the standard model. Indeed, some of the experiments have closed down possibilities. Studies of the Z° at CERN and Stanford in 1989 strongly suggested that three and only three families of elementary particles exist; the standard model's list of quarks and leptons seems exhaustive. (No new quark families; no more neutrinos.) Dashing hopes, nobody has turned up any of the welter of mirror particles predicted by supersymmetry. (No squarks, as the supersymmetric mirror quarks are called; no sneutrinos.) Experiments to determine whether neutrinos might have mass have been inconclusive. The Higgs has never been observed. And proton decay—which probably must exist if unified theories are to be constructed—has not been confirmed, though a second generation of experiments is under way.

With the standard model, high-energy physicists seem to have become victims of their own success. They have explained everything in reach yet are unsatisfied by the answers; worse, they have no way to go beyond them. To cite just one pressing question, they want to know why mass is scattered in such apparently random fashion. Why is the muon 200 times heavier than the electron? Why does the top quark have a mass almost *40,000* times greater than that of the up quark? Why not 4,000 times, or 400? If the Higgs field provides all particles with mass, why does it provide them with *those* particular masses? And what is the mass of the Higgs itself? For that matter, why do the coupling constants have their values? Surely nature has not chosen such a sloppy picture for its final statement. At higher and higher energies, physicists say, the standard model *must* break down somewhere. But where? How? Why? Despite thorough inspection of every corner in the subatomic realm, nothing so far has shown a road ahead. The prospect is of endless tantalization and stretches of tedium.

Even more dismaying, no projects now on the drawing board seem guaranteed to turn up something new. In the past, physicists have always forged ahead by building bigger accelerators, which have allowed them to explore terrain at ever higher energies. Faced with the dilemma of the standard model, their initial reaction was to propose the largest pure research project ever attempted: the Superconducting Supercollider. This vast accelerator—a fifty-mile ring in the drylands outside Waxahachie, Texas—was intended to smash together protons at 20 trillion electron volts, an energy theorists believed sufficient to find the Higgs boson, thus pinning down one of the greatest unknowns in the standard model. Yet the U.S. Congress stunned scientists the world over by killing the project in the fall of 1993. Years of effort and billions of dollars had already been spent; bulldozers had already excavated ten miles of tunnel. But as price estimates rose from $4.4 billion to $11 billion and doubts arose about the management of the project, the government balked. Its decision marked the end of the postwar era of partnership among science, industry, and government—the environment in which the standard model had been developed.

The only current project of similar scope is at CERN, which plans to spend $2 billion to expand its Large Electron Positron ring, opened in 1989, into a bigger ring called the Large Hadron Collider (though financing the machine has been the subject of contention among member states). The LHC, as it has been dubbed, is slated to reach 10 TeV in 2004 and 14 TeV in 2008— less powerful than the Superconducting Supercollider would have been but at least with some chance of existing. If completed, the LHC may well be the last large particle accelerator ever built. To be sure, a few more ambitious projects are on the drawing board: a muon collider, backed by scientists at Brookhaven, and a Next Linear Collider, which would shoot bundles of electrons from both ends of a long, straight accelerator, having them collide in the middle. The idea, which has champions at Stanford and KEK in Japan, would avoid, at least in theory, some of the complications necessary to bend electrons around a circle. But such projects are bound to be enormously expensive; even if funded, they are at least a decade away. And in an era of austerity there is no guarantee that any of them will be approved.

Recoiling from this dark prospect, some high-energy physicists wonder if their discipline has indeed come to an end. Perhaps this generation may see the end of physics, as Hawking suggested, not because it has accomplished its intellectual mission but because it has exhausted the interest and resources of the governments that have supported it. Other physicists worry that the vast teams of Ph.D.s needed to run the projected accelerators will be so unpleasant to work in that the best and the brightest will shun the field. Time and again, the lions of the last generation of physics, now leaders of huge experimental groups, told us that, were they young and beginning their careers, they would think twice about working in such a huge group.

In response, some physicists have attempted to devise clever, quick, inexpensive ways of peering beyond the standard model—conjuring up high-energy phenomena with low-energy equipment. One way is to hunt for rare events: processes that just maybe could occur, decays that might not be forbidden, particles that could be spotted by the lucky. All over the world, these experiments are now tucked into the corners of laboratories, small and untended apparata which hope to snare occurrences rare enough to have been overlooked in earlier experiments. Proton decay is the archetype—a phenomenon whose mere existence would be enough to shake the community. (No need to spot it twice. Just one gold-plated event and unification is in business!) By and large, though, these experiments are risky, as the failure to establish proton decay demonstrates; worse, they cannot rule out any subject of investigation, for the failure to find a phenomenon may only mean that the background was prohibitively high or the equipment insufficiently sensitive.

Other physicists have chosen a different route: high precision. Like art scholars who learn new insights about old masters from microscopic inspection of brushstrokes, the physicists hope to measure familiar parts of nature with such incredible accuracy as to shed new meaning on the whole. Many such projects are under construction, but we were drawn to one with both a fascinating chance to peek into the future and great historical resonance—the (g-2) experiment at Brookhaven National Laboratory.

□   □   □   □   □

Workmen unbolted the wagon wheel from the big silver ring, and the ceiling crane lifted it away. Dangling from the hook, the spokes silhouetted against the ceiling, the assembly bobbed like a spider drifting on its line. Underneath, the ring rested on a system of metal shims that workers would use to nudge the ring into its proper position; several of them were already measuring its placement and tightening bolts. A team of engineers looked on from a computer room on a sort of mezzanine, a room whose floor was torn up in preparation for installation of computer cables. Satisfied that the morning's work was nearly complete, Gerry Bunce ordered pizzas from Alfredo's, a local joint celebrated at Brookhaven for its quick deliveries to the laboratory bench. He had to twist around pieces of apparatus that seemed to fill every available inch of floor space. Things were cramped, he said. "As you can see"—waving his hand toward the metallic curve of the outer magnet, nearly scraping the walls—"the experiment just barely squeaks inside."

The building, like much else in the experiment, was a matter of new wine in old bottles. It had once housed the bubble chamber that Nick Samios and company used to discover the omega minus, a big step en route to the construction of the standard model. Now, three decades later, Bunce and his colleagues were seeking evidence of forms of matter beyond the standard model, by precisely measuring the effects of the swarm of virtual particles that envelops the muon. Because a spinning muon continually emits and reabsorbs virtual particles, which are always themselves spinning, a magnetic field "sees" a muon as having a slightly different spin than it otherwise would have. The muon's spin direction in a magnetic field is thus fractionally different from what it would be if it were not releasing and taking in virtual particles. Measuring this deviation is also a reworking of the past. A kissing cousin to the Lamb shift, it is known formally as the anomalous magnetic moment and informally as $(g\text{-}2)$. (The name, one recalls, came from the realization that the magnetic moment of an electron in a world without virtual particles would be exactly 2; the real magnetic moment, g, is slightly larger than 2, with the difference being $g\text{-}2$.) In 1947 John Nafe and Edward Nelson measured $(g\text{-}2)$ for the electron, an experiment that spurred Julian Schwinger to renormalize quantum electrodynamics, thus proving the validity of the theory. Today, the Brookhaven team is hoping that measuring $(g\text{-}2)$ for the muon would help them go beyond theory.

Schwinger worked out the value of $(g\text{-}2)$ for the electron. That $(g\text{-}2)$ is now the most precisely calculated number in the history of science. In the early 1980s, Toichiro Kinoshita, a theorist at Cornell, spent hundreds of hours on the world's biggest computers to evaluate the mind-bogglingly complex equations necessary to predict the contributions of every known type of particle. Eventually he arrived at a value for $(g\text{-}2)$ of the electron—.001159652460, plus or minus a few trillionths.

Meanwhile, he was working on $(g\text{-}2)$ for the muon. In addition to calculating the interactions when muons emitted and reabsorbed virtual photons, he had to take into account what happened when the photons split into virtual electrons and positrons, and what happened when the virtual electrons and positrons interacted with the magnetic field. Maddeningly, the virtual photons could also split into virtual quarks and antiquarks, so the contributions of the strong force had to be included. And through the weak force the muon could emit and reabsorb a virtual Z^o, emit and reabsorb a W and a virtual neutrino, emit a virtual neutrino, turning into an electron in the process while also emitting another neutrino—the possibilities were endless. Kinoshita found himself trucking with equations that had 20,000 terms; he gobbled up supercomputer time in three continents. Eventually he accounted for every known type of particle. His final prediction for the muon $(g\text{-}2)$—.00116591688, with the usual small margin of error.

That number—.00116591688—was the target at which the experimenters at Brookhaven were shooting. If their work was accurate enough, any deviation would signal something new. Frustratingly, though, even if Bunce and colleagues detect the presence of new forms of matter, they will be unable to pin down their identities; the experimenters would be like sea explorers forced to turn back with land visible on the horizon. Nevertheless, they would have made a vital contribution to the geography of the subatomic world.

"We're seeking a factor of twenty improvement in precision over the previous $(g\text{-}2)$ measurement [made at CERN in 1979]," Bunce said. "That's not going to be easy—it was a great experiment! But it's a fantastic opportunity to see if there's something new out there." His glance came to rest at the empty pipe protruding from the wall, future conduit for the experimental muons. "This was the last magnet ring to be installed," he said. "But much remains to be done. Tomorrow, we start making measurements on the magnets. Then we have to hook up the refrigeration system and connect up the beam pipe that will lead the particles from the accelerator into the—"

Bunce stopped; the pizza had arrived. As the delivery driver waited, Bunce moved about the floor, gathering a collection to pay for the food. He escorted the driver to the frigid parking lot.

As it often does, the Long Island weather had taken a blustery turn. The wind had picked up, shaking the scrub oak and pitch pine. While the sun continued to stream confidently and unfalteringly through patches of blue, white clouds covered most of the sky, and thick black columns indicated the presence of snowstorms on other parts of the island. Anything could happen.

The pizza van rumbled off. Bunce stood outside, good hand jammed into his jacket pocket. It was possible to imagine that an air of stillness and expectation hung over the scene. Eventually, smiling, he turned back to work. Inside the recycled building, amid reused pieces of equipment in a half-completed room, we saw a dozen or so members of the $(g\text{-}2)$ experiment gathered around boxes of pizza. Food steaming in one hand, writing implements in the other, they were honing their strategies to tease the secrets of matter from a measurement a decimal place or two more precise than ever before. "Hey!" Bunce said, laughing. "Save me some!"[23]

Afterword
A Note on People and History

> For those who participated [in the development of quantum mechanics], it was a time of creation; there was terror as well as exaltation in their new insight. It will probably not be recorded very completely as history. As history, its recreation would call for an art as high as the story of Oedipus or the story of Cromwell, yet in a realm of action so remote from our common experience that it is unlikely to be known to any poet or any historian.
>
> —J. ROBERT OPPENHEIMER, 1953

FOR ANYONE WRITING THE HISTORY OF PHYSICS, OPPENHEIMER'S WORDS STAND as a warning—and a challenge. Like the story of quantum mechanics, the story of the standard model is set in a territory so remote from common experience that it is almost impossible for outlanders to picture. Indeed, many physicists believe that the quantum land they are exploring can only be navigated with the symbolism of advanced mathematics; plain English is not up to the task. We could not accept this argument, not least because it implicitly shuts out most people from one of humankind's grandest intellectual achievements.

As Oppenheimer said, the historian of science is confronted by a forbidding array of abstract concepts and technical language. Yet historians of music have never been daunted by the impossibility of describing the abstract realms of harmony and timbre to the musically unschooled. To convey nuances with absolute precision, they might have to resort to musical notation; but they regularly enable lay readers to glean something of the musical issues involved without filling the page with sharps and semiquavers. To be sure, historians of music can assume their readers' familiarity with the music itself—an assumption that cannot be shared by the historian of science. Nonetheless, we were too reluctant to forego the advantages of ordinary speech not to make the attempt.

Mathematics has its own beauty and power, as its users will attest. And, as physicists say, most concepts in their field are clearly defined and explained only in mathematical terms. But integrals and derivatives cannot create a novelistically rich portrait of a Rutherford or a Stueckelberg, evoke their joy and frustration as the quest twists and turns, allow a step back from the personal to evoke the overarching sweep of the story, or celebrate the soaring ambition of the collective effort. These can only be accomplished with the resources of everyday language. Because we wanted to do all of these things, we tried to strike a balance, describing just enough of the detail to depict the science accurately while avoiding the slide into technical digressions.

No balance is perfect. This edition of *The Second Creation* provided us with a chance to correct errors, reshape the narrative, and answer complaints— some of them, anyway. After reading the first edition of this book, for instance, a few particle physicists complained that certain technical points were incompletely developed. Because, as we noted at the beginning of this book, our ambition was to write about people, not protons, we have for the most part ignored these criticisms. Here and there, though, we discovered that we had strayed into outright error: a misspelled name, a mistaken identification of nationality, and, in one unfortunate circumstance, the insertion of the word "relativistic" into a sentence in which it does not belong. We are particularly embarrassed to admit that we are among the legions who swallowed Leon Lederman's whopper that the shielding around his Nobel-winning two-neutrino experiment was stripped from the U.S.S. *Missouri*. (The *Missouri* was still intact.) At least we are not among those who also believed his claim that during the Cuban missile crisis he feared the U.S. Navy would take back the shielding for his experiment, which was completed in 1961, a year before the incident. In this edition, we have rectified these mistakes; we thank the people who drew our attention to them, including the kind man from Shreveport, Louisiana, who apparently pored through the entire text for typos. We are especially grateful in this regard to Yasuo Shizume, the able physicist who was our Japanese translator. Naturally, we also took this occasion to update the book where necessary. As a result, this edition of *The Second Creation* is thoroughly corrected and revised. But the chance to go through it again has only convinced us anew that our original impulse was sound—to tell the story from the point of view of the individuals who put together the standard model, the science being given as needed.

Character matters in such an approach, because in science, as in other human endeavors, the deeds of its practitioners are intimately linked to their personal histories and proclivities. True, when the historian's camera lens is trained on large-scale or long-duration events—changes in systems of land tenure or the conception of the Western family—the character of individuals can be irrelevant. But when the subject is a closely-knit community of constantly interacting people, personality assumes a major role. Character becomes important to history.

So baldly put, this seems mundane—a truth as old as Thucydides, who found that the events of the Peloponnesian War would be senseless without reference to the characters of its key participants. It would be unthinkable, for instance, to write a history of the Impressionist movement, or the spread of

Freudian therapy, or of the adoption of the U.S. Constitution, without describing the major protagonists and how they interacted to shape the events. The standard model, we believed, could be no exception.

To our astonishment, we learned over the years that certain sociologists and historians thought otherwise. At one scientific conference, for instance, we were drawn aside and courteously upbraided by a leading sociologist of science. Citing a phrase in our description of Sheldon Glashow—we depicted him as "surrounded by an oracular veil of cigar smoke"—the sociologist asked us what those words contributed to the subject matter. He then gave us his answer: Nothing. Indeed, he argued, our "local color," as he called it, tended to reinforce an erroneous picture of science as an enterprise that uncovers eternal truths about nature. Instead, he said, scientific experiments embody interpretations that scientists make about nature, interpretations that can and do change, and what we think of as scientific laws are really the expression of a social consensus among scientists—the laws are "constructed," in the jargon. Emphasizing personality, in this sociologist's view, places scientists on a pedestal. Surrounded by his oracular veil of cigar smoke, Glashow becomes a pundit, a sage, a priestlike dispenser of immutable truths.

The same objection might be raised for other details in *The Second Creation*: the candied Mexican hats consumed after Glashow won a bet with meson physicists, the handkerchief that Emmy Noether kept stuffed in her blouse and waved to illustrate a point, the flash of Julian Schwinger's florid silk tie as he vanished from students' sight after teaching a class, and the comfortable slippers which Abdus Salam kept underneath his desk in Trieste. What is each one of these points but another ultimately mystifying oracular veil?

By contrast, we regarded such "gratuitous details" (to use Irving Howe's ironic term) as an integral part of the story we were telling, for they reveal character and how it works in shaping a given situation. The oracular veil of smoke evinced the theatrical presence that explains part of the influence that Glashow—one of the more distinctive and influential characters among postwar theoretical physicists—has had on his peers. Cigars loom large in small rooms; they can seem to confer a prophetic presence on their handler, which we underscored with the word "halo." That prophetic presence, in turn, suggests why Glashow's highly speculative notions, such as charm, were given much attention and no small credence far in advance of any empirical data—and why, in the November Revolution, the use of charm to explain the J/psi was so rapidly proposed. For us, the cigar smoke served as a "gratuitous detail" to draw the reader's attention to the kind of presence that this physicist has had among his colleagues, and thus to illuminate one aspect of the collective social process that produced the standard model.

Likewise for the other details. The bet helped reveal the gamelike quality theoretical physics has for many of its practitioners and the way that theorists and experimenters jocularly thumb their noses at each other. That male students and colleagues found Noether's handkerchief ridiculous and unfeminine illuminates the sexist environment in which she had to work, which in turn affected her career. The flash of the tie was emblematic of Schwinger's obsessive secrecy, which led to his spending years developing an inaccessible notation for his hermetically sealed work, which in turn had much to do with how little of it was eventually incorporated into standard physics despite the immense achievement it represented. And the slippers were mute testimony of the lonely efforts of a person from the Third World to make a home in an unfamiliar environment. It would be possible to spell out each of these points in the form of an explicit sociological study—on theatricality, jokes and gambling in science, sexism, idiosyncracy, mentoring, the anxieties of Third-World participants in the international scientific community. But we were entrusting ourselves and our readers to the resources of narrative, among which are the use of arresting detail to crystallize character, situations, and events vividly and economically.

A few other historians and sociologists objected to our use of personal detail, but to make just the opposite criticism. One reviewer in a scientific journal remarked that this book's "colorful journalistic style" made it "difficult to put down." To our surprise, this was not meant entirely as a compliment. Denouncing the fact that the book had in many places elided technical niceties, the reviewer, a professional historian of science with a degree in physics, remarked with a near-audible sigh that the reader of our book "will have to glean what he can of the intellectual structure [of the standard model], while getting a sense of the excitement of the chase, the conflict of personality, and the rest of the human drama." Because that was exactly our intention, we found the critical tone of the remark baffling. It implied that genuine history of science is entirely a matter of chronological rendering of technical achievements, with everything else an unnecessary distraction. We disagree.

As readers of *The Second Creation* will have noted, we eagerly embraced the chance to interview our subjects, and the interviews figure abundantly in our narrative. Historians often have a low regard for oral interviews, deriding reliance on them as "journalistic," a term of scorn in the trade. To be sure, oral interviews can be a treacherous source of information and must be used carefully. But we believe that interviews are essential in obtaining a picture of the workings of contemporary science. People in general rely ever less on the letters, memos, and so on that are the traditional fodder of historians, but the decline may be especially steep in the case of scientists, who used electronic mail for decades before the rest of the world heard about the information superhighway.

Even where written scientific texts are available, such documents, as the biologist Peter Medawar has noted, disguise the historical process of creation in ways that only oral interviews can elucidate. We once learned of an experiment, for instance, in which two team members disagreed sufficiently with the others that they parted from their colleagues. Anger was felt on both sides, and the quarrel shaped the final stages of the experiment. In keeping with the standard practice of science, which is to pretend that personal differences have no place, the dispute was carefully kept off the printed record. Indeed, the paper reporting the experiment thanked the two mutineers "for their continuous interest and support." Oral interviews, for all their flaws, probably would be the only way to reconstruct this kind of event—a personal interaction, shaped by individual character, that played a small role in the development of science.

To pass over the role of character in particle physics would contribute to the impression that science, unlike other kinds of human activity, is conducted in a pure, rarefied environment by automatons without history or motive. Thinking, even scientific thinking, always belongs to the world of appearances, of concrete historical environments; there is a constant byplay between scientists and their situation and between scientists and their subject, each constraining and shaping the other. To paraphrase Marx, human beings make science but not any way they please. Sociologists may focus on the human beings making the science, whereas scientist–historians may be more impressed by their inability to make it any way they please. We hoped our depiction would balance both aspects of the human activity called science, showing readers that physicists are, like the rest of us, social creatures; we hoped, too, to portray how their activity is shaped by its play with nature.

Far from using descriptions of people as a sweetener, we wanted to employ them to answer Oppenheimer's challenge. The terrain of high-energy physics *is* forbiddingly strange. But its inhabitants, the tribe of physicists, become recognizable, even familiar, as human beings, though in their professional work they speak a technical language. Who better to accompany us as guides through the magnificent cliffs and valleys of the standard model than these individuals themselves? And in whose voice better to hear the terror and exaltation that accompanied the action that unfolded there?

—June 1995

Notes

ABBREVIATIONS FOR WORKS FREQUENTLY CITED

AdP: *Annalen der Physik*

AHES: *Archives for the History of the Exact Sciences*

AHQP: Archives for the History of Quantum Physics, American Institute of Physics, New York City

AIP: American Institute of Physics

BPP: Brown, L. M., and Hoddeson, L., eds., *The Birth of Particle Physics*. New York: Cambridge University Press, 1983.

CI: Berthelot, A., et al., eds., *Colloque International sur l'Histoire de la Physique des Particules*. In *Journal de Physique* (coll. C-8), 43, no. 12 (supp.), December 1982.

CQ: Pickering, A., *Constructing Quarks: A Sociological History of Particle Physics*. Chicago: University of Chicago Press, 1984.

CR: *Comptes rendus des Séances de l'Académie des Sciences*

DSB: Gillispie, C. C., et al., eds., *Dictionary of Scientific Biography*. New York: Scribners, 1970.

HDQ: Mehra, J., and Rechenberg, H., *The Historical Development of Quantum Theory*. New York: Springer-Verlag, 1982.

HPA: *Helvetical Physica Acta*

HSPS: *Historical Studies of the Physical Sciences*

JETP: *Soviet Physics. Journal of Experimental and Theoretical Physics* (Translation)

LET: Hermann, A., Meyenn, K. V., and Weisskopf, V. F., *Wolfgang Pauli, Wissenschaftlicher Briefwechsel mit Bohr, Einstein, Heisenberg, u.a.* New York: Springer-Verlag, 1979 (vol. 1), 1985 (vol. 2), quotation translated by the authors where necessary.

NC: *Nuovo Cimento*

NP: *Nuclear Physics*

PL: *Physics Letters*

PM: *London, Dublin, and Glasgow Philosophical Magazine and Journal of Science*

PR: *Physical Review*

PRL: *Physical Review Letters*

PRpts: *Physics Reports*

PRSA: *Proceedings of the Royal Society* (London), section A

PT: *Physics Today*

PTP: *Progress in Theoretical Physics*

PZ: *Physikalische Zeitschrift*

RMP: *Reviews of Modern Physics*

SIL: Pais, A., *"Subtle Is the Lord . . .": The Science and Life of Albert Einstein*. New York: Oxford University Press, 1982.

TP: Kevles, D. J., *The Physicists*. New York: Knopf, 1978.

ZfP: *Zeitschrift für Physik*

CHAPTER 1

1. Interview, Gerry Bunce, Brookhaven, 26 January 1995; E821 Collaboration, "Design Report, BNL AGS E821," 1 March 1995; Farley, F.J.M., and Picasso, E., "The Muon g-2 Experiments," in Kinoshita, T., *Quantum Electrodynamics*, Singapore: World Scientific, 1990.

2. Cited in Guillemin, V., *The Story of Quantum Mechanics* (New York: Scribners, 1968): 19.

3. Quoted in Badash, L., *Isis* 63: no. 1 (January 1972): 52. The "decimals" is given as the opinion of an "eminent physicist." See also, Galison, P., in Graham, L., et al., eds., *Functions and Uses of Disciplinary Histories*, Vol. VII (Amsterdam: D. Reidel, 1983): 35.

4. This story has been recounted in many places, including Romer, A., *The Restless Atom.* Rev. ed. (New York: Dover, 1982), pp. 15–29; and Shamos, M., *Great Experiments in Physics* (New York: Dryden, 1959): 213ff. Becquerel's memoirs are in *Memoires de l'Académie des Sciences* 46 (1903): 1. Translations of Becquerel's original papers are in Romer, A., ed., *The Discovery of Radioactivity and Transmutation* (New York: Dover, 1964).

5. Cited in Glasser, O., *Wilhelm Conrad Röntgen und die Geschichte der Röntgenstrahlen* (Berlin: Springer–Verlag, 1959): 26.

6. *CR* 122: 150, séance of 20 January 1896.

7. Becquerel, H., *CR* 122: 420, séance of 24 February 1896.

8. Becquerel, H., *CR* 122: 502, séance of 2 March 1896.

9. Becquerel's memoirs note scornfully that one had only to reread the publications of one G. Le Bon "to realize that at the moment he did them, the author didn't have the slightest understanding about the phenomenon of radioactivity" (trans. by authors), Becquerel, *Memoires, op. cit.*, p. 6.

10. *Ibid.*, p. 3.

11. Standard books on Thomson include Strutt, J. W., Lord Rayleigh, *The Life of J. J. Thomson, O. M.* (New York: Cambridge University Press, 1943); Thomson, J. J., *Recollections and Reflections* (London: G. Bell and Sons, 1936); and the odd, amusing Crowther, J. G., *British Scientists of the 20th Century* (New York: Routledge and Kegan Paul, 1952).

12. See, for example, Moralee, D., et al., eds., *A Hundred Years of Cambridge Physics* (2d ed.), Cambridge University Physics Society, 1980.

13. According to Pais, A., *RMP* 49, no. 4 (October 1977): 925, the same announcement was made three months earlier by the Prussian physicist Johann Weichert, who played no role in subsequent events. Walter Kaufman of Berlin also found *elm.*

14. Thomson, J. J., *PM*, series 5, 44, no. 269 (1897): 293. (For a good general history, see Anderson, D. L., *The Discovery of the Electron* [Princeton: Van Nostrand, 1964]). The title of this journal is worthy of attention, as it gives some insight into the origins of physics, and its relation to philosophy. Once upon a time, physics and philosophy were allied disciplines, and a physicist was equally as likely to be referred to as a "natural philosopher," someone who sought truth by rational argument in the natural instead of the human world. This alliance began to break up at the end of the last century; we now place them in separate sections of the curriculum and suppose that different talents are required to pursue each. In current practice, physicists often use *philosophy* to mean "poor physics." Some relics of the old terminology have persisted into this century; Einstein, for instance, taught in the Philosophical Faculty at the University of Zürich. Another is the persistence of the word *philosophical* in the title of this journal. In Rutherford's time, articles treating most branches of science could be found in the *Philosophical Magazine.* Today, with the same title, it is exclusively a journal of solid state physics.

15. The influential German chemist Wilhelm Ostwald went further, arguing that "In fact, energy is the unique real material in the world, and matter is not a carrier, but a manifestation of it" (*Zeitschrift für physikalische Chemie* 9 [1892]: 771, trans. by authors).

16. Thomson, *Recollections, op. cit.*, p. 341.

17. Becquerel, H., *CR* 130: 106, séance of 26 March 1900.

18. Cited in Eve, A. S., *Rutherford* (New York: Macmillan, 1939): 15. Rutherford's original letters to his mother and fiancée seem to have disappeared. Eve is the standard biography, but see also: Badash, L., ed., *Rutherford and Boltwood: Letters on Radioactivity* (New Haven: Yale University Press, 1969); Birks, J. B., ed., *Rutherford at Manchester* (New York: W. A. Benjamin, 1963); Oliphant, M., *Rutherford: Recollections of the Cambridge Days* (New York: Elsevier, 1972); and Wilson, D., *Rutherford: Simple Genius* (Cambridge, Massachusetts: MIT Press, 1983). For the discovery of the nucleus, we have greatly relied on the excellent Heilbron, J. L., *AHES* 4 (1967): 247.

19. Quoted in Wilson, *Rutherford, op. cit.*, p. 90.

20. Oliphant, *Rutherford, op. cit.*, p. 29.

21. Rutherford, E., *PM*, series 5, 47, no. 284 (January 1899): 116.

22. Eve, *Rutherford, op. cit.*, pp. 54–55.

23. The figure is from Champlin, ed., *The Young Folks Cyclopedia* (New York: Holt, 1916). Typical of the awe with which radium was regarded is the 1921 edition of *The American Educator* encyclopedia (Chicago: Ralph Durham), which unhesitatingly defines radium as "the most valuable and possibly the most wonderful substance in the world. . . . In proportion to its weight radium is a hundred times more valuable than diamonds, being worth $3,200,000 an ounce."

24. Rutherford, E., *PM*, series 6, 5, no. 49 (February 1903): 177; *Nature* 79, no. 2036 (5 November 1908): 12.

25. Becquerel, H., *CR* 136: 1517, séance of 22 June 1903; *CR* 141: 485, séance of 11 September 1905.

26. This conclusion was too much for Becquerel as well. Because a particle's momentum is the product of its mass and its velocity, Becquerel argued that it was not the velocity that was increasing but the *mass*; the alpha particle somehow must be collecting other bits of matter in flight, like a snowball rolling downhill.

27. Rutherford, E., *PM*, series 6, 11, no. 61 (January 1906): 166.

28. *Ibid.*, p. 174.

29. Cf. Eve, *Rutherford, op. cit.*, p. 183.

30. Letter, W. H. Bragg to E. Rutherford, 1 October 1908, quoted in Heilbron, *op. cit.*, p. 262; see Wilson, *Rutherford, op. cit.*, p. 286.

31. Marsden, E., in Birks, J. B., ed., *Rutherford at Manchester, op. cit.*, p. 1.

32. Geiger, H., and Marsden, E., *PRSA* 82, no. 557 (31 July 1909): 495.

33. Probably under Rutherford's guidance, Geiger and Marsden had written, "If the high velocity and mass of the α particle be taken into account, it seems surprising that some of the α particles, as the experiment shows, can be turned within a layer of 6×10^{-5} cm. of gold through an angle of 90°, and even more." To produce the same effect with a magnet, they noted, an "enormous field . . . would be required" (*Ibid.*, p. 498).

34. Crowther, J. A., *PRSA* 84, no. 570 (15 September 1910): 226.

35. On 9 February 1911, for instance, Rutherford wrote to his friend William Henry Bragg, another expert in radioactivity: "I have looked into Crowther's scattering paper carefully and the more I examine it the more I marvel at the way he made it fit (or thought he made it fit) JJ's theory. . . . I believe it was only by the use of imagination and failure to grasp where the theory was inapplicable that led him to give numbers showing such an apparent agreement." Bragg was delighted, writing back three days later: "Your opinion of Crowther's paper agrees with mine; do you know, I think it is quite an immoral paper, very nearly dishonest, though I am sure the dishonesty was not intentional." With cheerful unfairness, Bragg continued, "I must say, I have often found myself wrong when I have felt inclined to condemn utterly. But I think the censure is just, this time. . . . [Crowther's article] has the worst fault an experimental paper can have, because it unctuously brings round a lot of facts to suit a theory backed by a great name, and it is so jolly cock–sure. . . ." Quoted in Heilbron, *op. cit.*, pp. 294-5.

36. Rutherford, E., *Proceedings of the Manchester Literary and Philosophical Society*, series 4, 55, no. 1 (March 1912): 18.

37. The term *nucleus* came into use quickly. It may have been suggested by Bohr (Interview, G. Hevesy, *AHQP* [25 May 1962]: 2). Rutherford first used it six months later in *PM*, series 6, 24, no. 142, (October 1912): 453.

38. A Japanese physicist, Hantaro Nagaoka, had previously proposed a somewhat similar model (*Proceedings of the Tokyo Physico–Mathematical Society*, series 2, no. 2 [February 1904]: 92; *PM*, series 6, 7, no. 42, [May 1904]: 445). But G. A. Schott pointed out that the model envisioned by Nagaoka was unstable (*PM*, series 6, 8, no. 45 [September 1904]: 384); after some argument, Nagaoka accepted. For a discussion, see Yagi, E., *Japanese Studies in the History of Science* 3 (1964): 29.

39. Rutherford, E., *Radioactive Substances and their Radiations* (Cambridge, England: Cambridge University Press, 1913).

40. Many histories have attributed the problems of the Bohr atom to radiative instability, but mechanical instability was the more fundamental issue, scientifically and historically speaking. See Heilbron, J. L., and Kuhn, T. S., *HSPS* 1 (1969): 211.

CHAPTER 2

1. Interview, J. Franck, *AHQP* (7 December 1962): 11. The frustrated collaborator was Dirk Coster.

2. The following account is based on the articles in Rozental, S., ed., *Niels Bohr* (New York: Wiley, 1967), particularly Rosenfeld, L., and Rüdinger, E., p. 40; Heilbron, J. L., and Kuhn, T. S., *HSPS* 1 (1969): 211; Bohr's letters in Rosenfeld, L., ed., *Niels Bohr Collected Works* (New York: Elsevier, 1972); autobiographical writings in Bohr, N., *Essays 1958–1962 on Atomic Physics and Human Knowledge* (New York: Wiley, 1963); interviews in the *AHQP*; and the sources cited below.

3. Interview, Niels Bohr, *AHQP* (1 November 1962): 1.

4. Letter, Niels Bohr to Margrethe Nørlund, quoted in Rosenfeld and Rüdinger, *op. cit.*, p. 40. Heilbron and Kuhn (*op. cit.*, p. 223), give the date of the letter as 26 September 1911.

5. Interview, Niels Bohr, *AHQP* (1 November 1962): 6. In the interests of readability, we have switched the first two and last two sentences and slightly changed the punctuation in the transcript.

6. Letter, Niels Bohr to Harald Bohr, 23 October 1911, in Rosenfeld, *op. cit.*, p. 529. Our view here differs from Heilbron and Kuhn, *op. cit.*, who argue that Bohr enjoyed himself at Cambridge. In her *AHQP* interview, Margrethe Bohr discusses her husband's discouragement (Interview, Margrethe Bohr, *AHQP* (30 January 1963): 2–3. Wheeler, J., in Steuwer, R., ed., *Nuclear Physics in Retrospect: Proceedings of a Symposium on the 1930s* (Minneapolis: University of Minnesota Press, 1977): 236ff, also describes Bohr's unhappiness.

7. Bohr, N., *Essays, op. cit.*, p. 31.

8. Planck, M., *AdP* 4, no. 3 (1 March 1901): 553.

9. Planck, M., *Scientific Autobiography and Other Papers*, trans. by Gaynor, F. (New York: Philosophical Library, 1949). His hostility to the atomic theory is described on pp. 32–33. Also useful is the introduction by M. von Laue, which quotes Planck's teacher's advice on p. 8. Much of the following material is covered in Planck, M., *Physikalische Abhandlungen and Vorträge*, (Braunschweig: Vieweg, 1958), especially vol. 3, which has historical articles. The same story is given in Meissner, W., *Science* 113, no. 2926 (26 January 1951): 75. A good popular account is in Segré, E., *From Atoms to Quarks* (San Francisco: W. H. Freeman, 1980): 61ff.

10. In his fine biography of Einstein, the physicist Abraham Pais writes that Planck's mathematical reasoning was unjustifiable "by any stretch of the classical imagination" (*SIL*, p. 371). Kuhn, T. S., *Blackbody Theory and the Quantum Discontinuity, 1894–1912* (New York: Oxford University Press, 1978), chap. 4, argues the opposite.

11. Hermann, A., *Frühgeschichte der Quantentheorie 1899–1913* (Baden: Mosbach, 1969): 32. Planck said, "I had to obtain a positive result, under any circumstances and at whatever cost."

12. A very small number, h is equal to 6.6262×10^{-27} erg–sec.

13. Interview, J. Franck, *AHQP* (7 September 1962): 6.

14. Einstein, E., *AdP* 17, no. 1 (9 June 1905): 132. Einstein hypothesized that freely moving electromagnetic radiation of some types behaves "as though it consisted of distinct independent energy quanta of magnitude $[h\nu]$." (He used other, then-current symbols instead of $h\nu$.) He noted that if light "behaves like a discontinuous medium consisting of energy quanta of magnitude $[h\nu]$, it is reasonable to inquire if the laws of emission and transformation of light are so constituted as though the light were composed of these same energy quanta." (p. 143).

15. Quoted in *SIL*, 382 and 357, respectively.

16. The invention of the term *photon* is in Lewis, G. N., *Nature* 118, no. 2981 (18 December 1926): 874.

17. Letter, E. Rutherford to W. H. Bragg, 20 December 1911, quoted in Eve, A. S., *Rutherford* (New York: Macmillan, 1939): 208.

18. Interview, Niels Bohr, *AHQP* (31 October 1962): 4; (1 November 1962): 13. Interestingly, as shown in Heilbron and Kuhn, *op. cit.*, p. 246, Bohr's decisive rejection of Thomson's model, which took place at this time, was to some extent based on a mathematical error.

19. Letter, Niels Bohr to Harald Bohr, 19 June 1912, Rosenfeld, *op. cit.*, p. 559.

20. Letter, Niels Bohr to Margrethe Nørlund, quoted in Rosenfeld and Rüdinger in Rozental, *op. cit.*, p. 49. There dated "the beginning of July."

21. Most of this memorandum is reprinted in L. Rosenfeld's introduction to Bohr, N., *On the Constitution of Atoms and Molecules* (Copenhagen: Niels Bohr Institute, 1963).

22. In his note to Rutherford, Bohr admitted that he wouldn't even try to give "a mechanical foundation (as it seems hopeless)" to this hypothesis. He added, "This seems to be nothing else than what was to be expected, as it seems to be rigorously proved that [the usual set of physical assumptions] is not able to explain the experimental facts. . . ." (*Ibid.*, pp. xxvi–xxvii).

23. Interview, Niels Bohr, *AHQP*, no. 2 (1 November 1962): 13.

24. During this time, Bohr also worked on the absorption of alpha particles, work which helped stimulate his more well-known studies of the atom. At the honeymoon's end, Bohr showed his bride the pleasures of smoggy Manchester when he took the alpha-ray paper to Rutherford.

25. Balmer, J., *AdP*, series 3, 25, no. 1 (15 April 1895): 80.

26. Rosenfeld in Bohr, *Constitution, op. cit.*, p. xxxix.

27. Hamiltonians were invented by the eponymous William Rowan Hamilton, a nine-

teenth–century Irish mathematical prodigy who is said to have mastered a dozen languages by the time he was nine. Appointed to the astronomy faculty of Trinity College, Dublin, while still a twenty–two–year–old student at the school, Hamilton married unhappily, struggled with alcoholism, and wrote reams of bad poetry, of which he was extremely proud. He was a poor astronomer, maintaining the observatory only with the reluctant assistance of three of his sisters. Shortly after his appointment, in 1827, he recast the science of optics into a more general form, inventing the function that now bears his name. Soon after, in 1833, he used the methods of his optics to reformulate Newtonian mechanics as a whole.

28. Letter, Niels Bohr to Ernest Rutherford, 6 March 1913, quoted in Eve, A. S., *Rutherford* (New York: Macmillan, 1939): 220–21. Bohr wanted Rutherford's opinion, but also lacked standing to submit it himself. Until the 1930s, young scientists had their papers "communicated" to journals by their seniors.

29. Letter, Ernest Rutherford to Niels Bohr, 20 March 1913, quoted partially, *ibid.*, pp. 220–21, and Rosenfeld and Rüdinger, *op. cit.*, p. 54.

30. Bohr, N., PM, series 6, 26; no. 151 (July 1913): 1; no. 153, (September 1913): 476; no. 155, (November 1913): 857.

31. Millikan, R., in *Nobel Lectures 1922–1941* (New York: Elsevier, 1965): 54.

32. Radon, in 1902.

33. Rutherford, E., *PM*, 37, no. 222 (June 1919): 536. The story goes that Rutherford was reprimanded for missing a war meeting and that he responded, "I have been engaged in experiments which suggest that the atom can be artificially disintegrated. If it is true, it is of far greater significance than a war." The statement is sufficiently prescient as to make one skeptical of its veracity. (Jungk, R., *Brighter Than A Thousand Suns*, trans. by J. Cleugh [New York: Harcourt Brace Jovanovich, 1958]: 1).

34. Speech before 1920 Cardiff meeting of British Association for the Advancement of Science in Rutherford's *Collected Papers*, vol. 2 (New York: Interscience, 1964).

35. Curie, I., CR, 193, séance of 21 December 1931, p. 1412; Joliot, F., *ibid.*, p. 1415; Curie, I., and Joliot, F., CR, 194, séance of 11 January 1932, p. 273.

36. Chadwick, J., *Nature* 129, no. 3252 (27 February 1932): 312. See also, Chadwick, J., in *Proceedings of the Tenth International Congress of the History of Science*, Ithaca, N.Y., 28 August–2 September 1982 (Paris: Hermann, 1964): 159.

37. Quoted in Wilson, D., *Rutherford: Simple Genius* (Cambridge, Massachusetts: MIT Press, 1983): 391. Punctuation and capitalization slightly altered for readability.

38. Interview, Samuel Devons, his office, 23 May 1984. See also, Devons, S., *PT* 24, no. 12 (December 1971): 38.

CHAPTER 3

1. Interview, Howard M. Georgi III, his office, Harvard University, 14 June 1983.

2. Georgi, H., and Glashow, S., *PRL*, 12, no. 8 (25 February 1974): 438.

3. Interview, Howard M. Georgi III, restaurant in New York, 29 January 1985.

4. Interview, Howard M. Georgi III, his office, Harvard University, 14 June 1983.

5. The first use of the term *quantum mechanics* seems to have occurred around 1919 (Interview, P. A. M. Dirac, AHQP [10 May 1963]: 6–9). Max Born claimed the first usage occurred in Born, M., ZfP 26, no. 6 (20 August 1924): 379 (Born, M., *My Life: Recollections of a Nobel Laureate* [New York: Scribners, 1975]: 215).

6. The literature about Einstein and relativity is vast, and cannot be more than alluded to here. Two authoritative biographies are *SIL* and Hoffman, B., and Dukas, H., *Albert Einstein, Creator and Rebel* (New York: Viking, 1972). See also Schilpp, P., ed., *Albert Einstein: Philosopher–Scientist* (New York: Tudor, 1949). Many popular explanations of relativity exist; one of Einstein's own is *Relativity*, trans. by Lawson, R. W. (New York: Crown, 1961).

7. See SIL. Special relativity, which Einstein developed in 1905, is the branch most germane to particle physics.

8. Lorentz, H., *Versuch einer Theorie der Electrischen und Optischen Erscheinungen in Bewegten Körpern*, reprinted in his Collected Papers, *op cit.*, vol. 5, 1.

9. See the discussion in *SIL*, chap. 6.

10. Thomson, J. J., *PRSA* 96: no. 678 (15 December 1919): 311.

11. *Times* (London), 6 November 1919, p. 12.

12. *New York Times*, 9 November 1919, p. 1.

13. *Ibid.*, 10 November 1919, p. 17. There were articles on Einstein daily for the next week.

14. Quoted in interview, Walter Heitler, *AHQP* (18 March 1963): 4. As Einstein complained, "In the principle of relativity, everything is so clear. But in quantum theory it is

horrible. What a mess it is in!" (In interview, James Franck, *AHQP* (9 July 1962): 12). Franck was unsure of the date, but the substance of Einstein's remark suggests that it must have been made after 1915.

15. Strictly speaking, the charge *e* of an electron is 4.8×10^{-10} absolute electrostatic units; the charge of the proton is the same.

16. More technically, these quantum numbers are *n*, *1*, and *m*, which are, respectively, the number of nodes in the electron wave equation (which is directly related to the allowed energies): the angular momentum; and one of its three space components, the Z-axis projection of the angular momentum.

17. Pauli, W., *ZfP* 31: nos. 5/6 (19 February 1925): 373. See also letter, Wolfgang Pauli to Alfred Landé, 24 November 1924, *LET*. At the time, it was thought these quantum numbers might be properties of the nucleus, rather than the electrons.

18. This account is based on the *AHQP* interviews with both men; Goudsmit, S. A., *PT* 29: no. 6 (June 1976): 40; Uhlenbeck, G., *ibid.*, p. 46; van der Waerden, B. L., in *Theoretical Physics in the Twentieth Century* (New York: Interscience Publishing, 1960), 199; and letters among van der Waerden, Goudsmit, and Uhlenbeck on *AHQP* microfilm 66, section 6.

19. Uhlenbeck, *op. cit.*, p. 43.

20. Interview, George Uhlenbeck, *AHQP* (31 March 1962): 11.

21. Interview, George Uhlenbeck, his office, Rockefeller University, 10 May 1984.

22. Kronig, R. d. L., in *Theoretical Physics, op. cit.*, 21; Van der Waerden, *op. cit.*, pp. 211–12.

23. Uhlenbeck, *op. cit.*, p. 43.

24. Recently Heisenberg has become a controversial figure as historians debate his role in the German atomic bomb project. Because this controversy has little to do with his scientific work, we do not treat it here. Sources consulted for this book include Hermann, A., *Heisenberg* (Hamburg: Rowolt, 1976). Heisenberg's own autobiographical and philosophical writings include *Physics and Beyond: Encounters and Conversations* (New York: Harper & Row, 1971); *Physics and Philosophy* (New York: Harper & Row, 1962); *Philosophical Problems of Quantum Physics*, trans. by Hayes, F. C. (Woodbridge, Connecticut: Ox Bow Press, 1979); "Theory, Criticism, and a Philosophy," in International Center for Theoretical Physics, ed., *From A Life of Physics*, Supplement to the IAEA Bulletin (Vienna: International Atomic Energy Agency, n.d.); the lengthy romanticized *AHQP* interviews; and *HDQ*.

25. Heisenberg, *Physics and Beyond, op. cit.*, pp. 8–9.

26. Interview, Werner Heisenberg, *AHQP* (30 November 1962): 3.

27. Heisenberg, *Physics and Beyond, op. cit.*, pp. 8–9.

28. Born, M., *My Life, op. cit.*, pp. 212–13.

29. See Paulsen, F., *The German Universities*, trans. by Perry, E. D. (New York: Macmillan, 1895) for an institutional description.

30. Heisenberg, "Theory," *op. cit.*, pp. 34–35.

31. Letter, Heisenberg to Pauli, 9 July 1925, *LET*.

32. This account follows Heisenberg's own account of events. But see Beller, M., *AHES* 33, no. 4, winter 1985, p. 337. For a more skeptical view, see Mackinnon, E., *HSPS* 8 (1976): 137.

33. Heisenberg, W., *Physics and Beyond, op. cit.*, p. 60.

34. *Ibid.*, p. 61 (both quotes).

35. The term *commute* was introduced a year later by a number of physicists, including Paul Dirac (see Dirac, P. A. M., in *History of Twentieth Century Physics* [New York: Academic Press, p. 129]). The term first appears in print in Born, M., Heisenberg, W., and Jordan, P., *ZfP* 35, no. 8/9 (4 February 1926): 557.

36. Heisenberg, W., *ZfP* 33, no. 12 (18 September 1925): 879. Heisenberg concludes, "Whereas in classical theory [*ab*] is always equal to [*ba*], this is not necessarily the case in quantum theory" (p. 884). The bracketed letters are used in lieu of some notation otherwise not used in this account. Mackinnon, *op. cit.*, pp. 178–81, notes that Heisenberg's mathematical manipulations were not entirely correct.

37. Interview, Werner Heisenberg, *AHQP* (22 February 1963).

38. It is not clear whether Heisenberg mentioned his new work in the public lecture, which was on aspects of line spectra. Cf., interview, Werner Heisenberg, *AHQP* (5 July 1963): 1–2.

39. A fine popular exposition of matrices can be found in Kramer, E., *The Nature and Growth of Modern Mathematics*, 2d ed. (New York: Princeton University Press, 1982).

40. *HDQ*, vol. 3, 43–44; see also Born, *My Life, op. cit.*, pp. 218–20.

41. This breakdown was relatively minor; Born had a more complete breakdown in

1928-29. The recounting of these developments in Born, *My Life, op. cit.*, p. 218ff (from which the quotation is taken), should be used with care. More reliable chronologies if nothing else are provided in the massive, exasperating *HDQ*, vol. 3, especially chap. 2.

42. Born, M., and Jordan, P., *ZfP* 34: no. 11/12 (28 November 1925): 858, received on 27 September.

43. Interview, Werner Heisenberg, *AHQP* (27 February 1963): 22.

44. Letter, Heisenberg to Pauli, 23 October 1925, *LET*.

45. Born, Heisenberg, and Jordan, *ZfP* 35, no. 819, 4 February 1926, p. 557. The article was received on 16 November 1925.

CHAPTER 4

1. Letter, Werner Heisenberg to Wolfgang Pauli, 16 November 1925, *LET*.

2. This is a necessarily truncated summary of a complex series of arguments carefully put in Beller, M., *The Genesis of Interpretations of Quantum Physics, 1925–1927* (Ph.D. dissertation, University of Maryland, 1983). It would be difficult to overstate the extent to which we have been assisted by Dr. Beller's thesis, portions of which were published, in somewhat different form, in *Isis* 74, no. 274 (December 1983): 469 and *AHES*, 33, no. 4 (Winter 1985): 337. Other useful secondary sources are MacKinnon, E., *HSPS* 8 (1977): 137; MacKinnon, E., in Suppes, P., ed., *Studies in the Foundations of Quantum Mechanics*, PSA, 1980; van der Waerden, B. L., in Mehra, J., ed., *The Physicist's Conception of Nature* (Boston: D. Reidel, 1973): 276; Hendry, J., *The Creation of Quantum Mechanics and the Bohr-Pauli Dialogue* (Dordrecht: D. Reidel, 1984), and the sources cited below.

3. Pauli, W., *ZfP* 36, no. 5 (27 March 1926): 336, received on 17 January.

4. For Pauli's description of his solution, see *Naturwissenschaften* 18, no. 26 (27 June 1930): 602. Heisenberg's delight is in his letter to Pauli, 3 November 1925, *LET*, vol. 1.

5. Dirac, P. A. M., *PRSA* 110, no. 755 (1 March 1926): 561. A little later, almost identical reasoning was published in Wentzel, G., *ZfP* 37, nos. 1/2 (22 May 1926): 80.

6. Strictly speaking, they could not calculate the intensity of the hydrogen spectrum lines, and this information was necessary to go on to more complicated systems.

7. A good biography of Louis de Broglie is *Louis de Broglie, physicien et penseur* (Paris: Albin Michel, 1953). See also Segrè, E., *From X-Rays to Quarks* (San Francisco: W. H. Freeman, 1980): 149–53; the thesis itself is de Broglie, L., *Recherche sur la théorie des quanta*, University of Paris, defended 25 November 1924 (Paris: Masson et Cie, 1924).

8. Schrödinger, E., *AdP*, series 4, 79, no. 4 (13 March 1926): 361. Curiously, this initial formulation of Schrödinger's work was nonrelativistic, that is, it did not take into account the Einsteinian effects of the very rapid motions of the electron. Schrödinger had initially worked on a relativistic model, but had gotten what he thought were wrong answers; in fact, the apparent discrepancy was caused because he did not know about the spin of the electron. Thus, he published a nonrelativistic paper.

9. The first instance of this term appears to be in a letter from Schrödinger to Albert Einstein, 28 April 1926, quoted in the "Schrödinger" entry in the *DSB*.

10. Bernstein, J., *Science Observed* (New York: Basic Books, 1982): 149.

11. Schrödinger, E., *AdP*, series 4, 79, no. 6 (6 April 1926): 489.

12. Typical positive reactions are quoted in Jammer, M., *The Conceptual Development of Quantum Physics* (New York: McGraw-Hill, 1966): 271.

13. Letter, Max Born to Erwin Schrödinger, 6 November 1926, *AHQP*.

14. Schrödinger, E., *AdP*, series 4, 79, no. 4, *op. cit.*, p. 375. English versions are in Schrödinger, E., *Collected Papers on Wave Mechanics*, trans. by Shearer, J. F., and Deans, W. M. (London: Blackie & Son, 1928). We use these translations, but amend awkward diction. The original papers are photographically reprinted in Schrödinger, E., *Die Wellenmechanik* (Stuttgart: Ernst Battenberg Verlag, 1963).

15. *Ibid.*, p. 506 (both quotes trans. by authors; emphasis in original).

16. *Ibid.*, p. 514.

17. Heisenberg's letter to Born is described in Born, M., *My Life: Recollections of a Nobel Laureate* (New York: Scribners, 1975): 233; the letter to Pauli was on 8 June 1926 (*LET*, vol. 1). "And now for an unofficial remark about physics. The more I think about the physical side of Schrödinger's theory, the more loathsome I find it. . . . When Schrödinger writes about the visualizability of his theory, I find it crap [*Mist*]."

18. Heisenberg, W., *Physics and Beyond: Encounters and Conversations* (New York: Harper & Row, 1971): 72; Mehra, J., *The Birth of Quantum Mechanics* (Geneva: CERN service d'information scientifique, 1976): 39.

19. Letter, Wolfgang Pauli to Pascual Jordan, 12 Apr. 1926, *LET*, vol. 1. There is an English translation in van der Waerden, B. L., *op cit.*, p. 282ff. As van der Waerden makes clear, the letter also contains what is now called the Klein-Gordon equation.

20. Schrödinger, E., *AdP*, series 4, 79: no. 8 (4 May 1926): 734. Two weeks after Schrödinger submitted his paper, an American physicist, Carl Eckart, of the California Institute of Technology, independently sent in a third equivalence proof. (Eckart, C., *Proceedings of the National Academy of Sciences* 12, no. 7 [15 July 1926]: 473).

21. See, for example, his letter to Lorentz, 6 June 1926, in Przibaum, K., ed., *Letters on Wave Mechanics*, trans. by Klein, M. J. (New York: Philosophical Library, 1967): especially 64–65. For the attitude toward older physicists, see *HDQ*, vol. 3, p. 167.

22. Schrödinger, *AdP*, *op cit.*, p. 736.

23. Letter, Erwin Schrödinger to Hendrik Lorentz, 6 June 1926, in Przibram, *Letters, op cit.*, pp. 63–65. Schrödinger writes, "But what I had not clearly recognized yet at that time [a few weeks before] was and is that the rules set up for this purpose by Born, Jordan, and Heisenberg are actually false if one applies them to generalized coordinates. . . ." Heisenberg later confessed that throughout his career he had used "rather dirty mathematics," but said that this forced him "always to think of the experimental situation . . . [and] somehow you get closer to reality than by looking for the rigorous methods." (Heisenberg, W., in International Center for Theoretical Physics, ed., *From a Life in Physics*, Supplement to the IAEA Bulletin [Vienna: International Atomic Energy Agency, n.d.]: 39).

24. Van Vleck, J., in Price, Chissick, and Ravensdale, eds., *Wave Mechanics: The First Fifty Years* (New York: Wiley, 1973): 26. Van Vleck dryly remarks, "The last days of the old quantum theory were the golden age of empiricism, where physicists often obtained correct answers by appropriate doctoring of formulas based on questionable theory, and some of this empiricism survived in the early days of quantum mechanics" (p. 31). Heisenberg, Born, and Jordan had developed their methods for a simple, two-dimensional model, then substituted three-dimensional terms in the equations, an erroneous procedure that Schrödinger, at least, had spotted.

25. Schrödinger, E., *AdP*, series 4, 80: no. 13 (13 July 1926): 437. The steady drumbeat of Schrödinger's papers—one a month for five months—must have seemed particularly alarming to Heisenberg, who spent the same time struggling with his increasingly unwieldy matrices.

26. Heisenberg, *Physics and Beyond*, *op cit.*, p. 73.

27. Schrödinger, E., *AdP* 83, no. 15 (9 August 1927): 956.

28. Heisenberg, W., in Rozental, S., ed., *Niels Bohr* (New York: Wiley, 1967): p. 103. We have slightly amended the unlovely translation. Another version of the same story can be found in Heisenberg, *Physics and Beyond*, *op cit.*, pp. 73–76.

29. Interview, I. I. Rabi, his apartment, New York City, 9 May 1984. See also, *TP*, pp. 213–14. (All quotes from the same interview.)

30. See, for instance Born, *My Life*, *op cit.*, p. 226, and his *AHQP* interview.

31. Born, M., *ZfP* 38, no. 11/12 (14 September 1926): 803.

32. Letter, Pauli to Heisenberg, 19 October 1926, *LET*.

33. *Ibid.*

34. Letter, Heisenberg to Pauli, 28 October 1926, *ibid.*

35. There is amazingly little biographical writing about Jordan. We have relied on his *AHQP* interview.

36. Jordan, P., *ZfP* 30, no. 4/5 (29 December 1924): 297; Einstein, A., *ZfP* 31, no. 10 (21 March 1925): 784.

37. Jordan, P., and Kronig, R. d. L., *Nature* 120: (3 December 1927): 807.

38. This point is developed at length in Jammer, *Conceptual Development*, *op cit.*, p. 331. As Beller has pointed out, however, Jammer's conclusions are outdated.

39. Jordan, P., *Naturwissenschaften* 15, no. 5 (4 February 1927): 105, especially 108. See also, Beller, *AHES*, *op. cit.*

40. Letter, Heisenberg to Pauli, 5 February 1927, *LET*.

41. This is a reconstruction based on the several somewhat contradictory accounts Heisenberg wrote of his train of thought. See, for instance, interviews, Heisenberg, W., *AHQP* (22 February 1963): 26 (5 July 1963): 10, and (11 February 1963); Heisenberg, *Physics and Beyond*, *op cit.*, pp. 77–78; and Heisenberg in Rozental, *Niels Bohr, op cit.*, p. 105ff. The inconsistencies in these reports are due in part to the difficulty of conveying in ordinary language the creative jumble of a mind at work. A good synthetic account is in Robertson, P., *The Early Years, The Niels Bohr Institute 1921–1930* (Copenhagen: Akademisk Forlag, 1979): 117–23.

42. Letter, Heisenberg to Pauli, 23 February 1927, *LET*. We have slightly amended the letter by using modern notation.

43. Interview, Heisenberg, W., *AHQP* (25 February 1963): 17.

44. This point has been brought out very clearly by Beller, *Genesis, op cit.*, p. 245ff. The sources here are those in note 41.

45. Interview, Heisenberg, W., *AHQP* (25 February 1963): 17.

46. Heisenberg, W., *ZfP* 43: no. 3/4 (31 May 1927): 172.

47. Letter, Erwin Schrödinger to Max Planck, 4 July 1927, reprinted in Przibaum, *Letters, op cit.*, pp. 17–18.

48. Beller, *Isis, op cit.*, pp. 490–91. The synthesis of Heisenberg's and Schrödinger's approaches was acknowledged at the fifth Solvay Conference on Physics in October 1927, when many physicists viewed quantum mechanics as essentially complete (*Proceedings of the Fifth Solvay Conference* [Paris: Gauthiers-Villars, 1928]). Einstein, however, violently objected; in a now famous series of discussions each day at breakfast he proposed a hypothetical experiment designed to "trick" a particle in some Rube Goldberg–like manner into revealing its exact position and momentum at the same time. Bohr, Pauli, and Heisenberg anxiously tried to answer Einstein's question at the table; by dinner, they had succeeded, and Einstein retired to dream up another experiment. When the conference closed, Einstein's every objection had been satisfied, except for his gut feeling that something was wrong with quantum mechanics. See *SIL*, pp. 444–57, for an account of Einstein's opposition.

49. Wigner, E. P., in Wigner, E. P., ed., *Symmetries and Reflections* (Bloomington, Indiana: Indiana University Press, 1967): 172.

50. Well-known exemplars of this genre are Zukav, G., *The Dancing Wu Li Masters* (New York: William Morrow, 1979); and Capra, F., *The Tao of Physics* (New York: Random House, 1975).

51. Zukav, *op cit.*, p. 53.

52. Interviews, Robert Serber, his office, Columbia University, 1 and 8 April 1985.

53. Bridgman, P. W., *Harper's Bazaar*, March 1929, p. 451.

CHAPTER 5

1. Schweber, S., "Some Chapters for a History of Quantum Field Theory: 1938–1952," 1983 Les Houches lectures, unpublished ms., p. 19. We are grateful to Dr. Schweber for allowing us to see a draft of his excellent survey. This account is based primarily on Dirac's *AHQP* interviews and autobiographical writings, and *HDQ*, vol. 4.

2. Interview, P. A. M. Dirac, *AHQP* (4 January 1962): 5–6.

3. Interview, P. A. M. Dirac, *AHQP* (7 May 1963): 15.

4. The story is told that Peter Kapitza, a Russian physicist who passed the 1920s at the Cavendish, told Dirac to read *Crime and Punishment*, and later asked about his reaction. "It is nice," was the diplomatic reply, "but in one of the chapters the author made a mistake. He describes the Sun as rising twice on the same day." Cited in the anecdotal Gamow, G., *Thirty Years That Shook Physics* (New York: Doubleday, 1966): 121–22.

5. Quoted in *HDQ*, vol. 4: 11.

6. Dirac, P. A. M., *BPP*, p. 46.

7. Interview, Samuel Devons, his office, Barnard College, 23 May 1984. See also, interview, H. B. G. Casimir, *AHQP* (5 July 1963): 8.

8. Interview, I. I. Rabi, his apartment, New York, 9 May 1984.

9. Interview, H. B. G. Casimir, *AHQP* (5 July 1963): 8.

10. Heisenberg, W., *ZfP* 33, no. 12 (18 September 1925): 880.

11. Dirac, P. A. M., *The Development of Quantum Theory* (Oppenheimer Memorial Lecture; New York: Gordon and Breach, 1971): 23–24.

12. Dirac, P. A. M., in *Rendiconti della Scuola Internazionale di Fisica "Enrico Fermi,"* 57th Course (Rome: Università degli Studi, 1974): 134.

13. Letter, Heisenberg to Dirac, 20 November 1925, quoted in *HDQ*, vol. 4, pp. 159–60.

14. Dirac. P. A. M., in C. Weiner, ed., *History of Twentieth Century Physics* (New York: Academic Press, 1977): 125.

15. *Ibid.*, p. 128.

16. Schrödinger, E., *AdP*, series 4, 79, no. 4 (13 March 1926): 361.

17. Dirac, P. A. M., *BPP*, p. 44.

18. Interview, P. A. M. Dirac, *AHQP* (10 May 1963): 28.

19. Gamow, *Thirty Years, op cit.*, chap. 6; Dirac's *AHQP* interviews; interview, Werner Heisenberg, *AHQP* (27 February 1963): 17.

20. Williams, L. P., *The Origins of Field Theory* (Lanham, Maryland: University Press of

America, 1980); Maxwell, J. C., "On Physical Lines of Force, III," in Niven, W. D., ed., *The Scientific Papers of James Clerk Maxwell* (New York: Dover, 1966): 489.

21. Einstein, A., *AdP* 17: no. 1 (June 1905): 132.

22. As far back as 1910, when Paul Debye tried it, physicists had thought of trying to quantize the field, but they had not possessed the necessary quantum mechanical techniques (Debye, P., *AdP*, series *4*, 33, no. 16 [20 December 1910]: 1427).

23. He also derived the creation and annihilation operators first described in Einstein, E., *Verhandlungen der Deutschen Physikalischen Gesellschaft*, series *2*, 18, no. 13/14 (30 July 1916): 318.

24. Dirac, P. A. M., *PRSA* 114, no. 747 (1 March 1927): 243. (This and other important articles on quantum electrodynamics are reprinted in Schwinger, J., *Selected Papers on Quantum Electrodynamics* [New York: Dover, 1958].)

25. *Ibid.*, 243.

26. This was pointed out in similar language by Weisskopf, V. F., *BPP*, pp. 56–57.

27. Dirac, P. A. M., *PRSA* 114, no. 747 (1 April 1927): 243 (both quotes).

28. Schweber (*op. cit.*, p. 36) makes this point with elegance and care.

29. It is worth nothing that the concept of empty space was itself relatively new. Until the turn of the century, the Universe was supposed to be permeated by a substance known as the ether, intangible stuff which was thought to be the very fabric of the cosmos. When Dirac banished the vacuum, he got rid of a concept some older physicists had just learned to accept.

30. Dirac, P. A. M., *BPP*, pp. 50–52.

31. Klein, O., and Gordon, W., *ZfP* 45, no. 11–12 (18 November 1927): 751. Strictly speaking, Klein and Gordon recovered relativistic invariance for the Schrödinger equation, but changed the wave equation to an equation of motion.

32. Dirac, P. A. M., *BPP*, p. 51.

33. Dirac, P. A. M., *PRSA* 117, no. 766 (1 February 1928): 610; Dirac, P. A. M., *PRSA* 118: no. 779 (1 March 1928): 351.

34. The Klein-Gordon equation also had two energy values, which Dirac knew full well (see *ibid.*, p. 612).

35. Cf. Weisskopf, V. F., *BPP*, p. 63.

36. Møller, C., *AdP* 14, no. 5 (15 August 1932): 531; Klein, O., and Nishina, Y., *ZfP* 52, no. 11–12 (9 January 1929): 853; and Dirac's two electron papers.

37. Interview, Sir Rudolf Peierls, *AHQP* (18 June 1963): 5–6; confirmed in interview, Sir Rudolf Peierls, his office, Oxford University, 9 October 1984.

38. See *TP*, chap. 14.

39. The first major contribution by an American to quantum theory came in 1926, when Carl Eckart, of the California Institute of Technology, demonstrated the equivalence of Schrödinger's wave mechanics and Heisenberg's matrices. See chap. 4, note 20, above.

40. Interview, I. I. Rabi, his apartment, New York City, 9 May 1984.

41. Heisenberg, W., and Pauli, W., *ZfP* 56: no. 1–2 (8 July 1929): 1; *ibid.*, 59, no. 3–4 (2 January 1930): 168.

42. Rosenfeld, L., *ZfP* 76, no. 11–12 (12 July 1932): 729.

43. Dirac, P. A. M., *PRSA* 126, no. 801 (1 January 1930): 360; P. A. M. Dirac, *BPP*, p. 50.

44. Halpern L., and Thirring, W., *The Elements of the New Quantum Mechanics*, trans. H. Brose (London: Methuen, 1930–31): 101.

45. Weisskopf, V. F., *BPP*, p. 63.

46. Dirac, P. A. M., in *ibid.*, p. 51.

47. *Ibid.*

48. Dirac, P. A. M., *PRSA* 126 (*op cit.*): 360.

49. Weyl, H., *The Theory of Groups and Quantum Mechanics*, trans. by Robertson, H. P. (New York: Dutton, 1931, reprinted New York: Dover, 1930): 234; see also Oppenheimer, J. R. *PR* 35, no. 5 (1 March 1930): 562.

50. He did. Dirac, P. A. M., *PRSA* 133, no. 1 (1 September 1931): 60.

51. Interview, P. A. M. Dirac, *AHQP* (14 May 1963): 30. We have reversed the order of the two sentences.

52. Wilson, C. T. R., Nobel Prize lecture, 12 December 1927, in *Nobel Lectures in Physics 1922–1941.* (New York: Elsevier, 1965): 194.

53. Wilson came to his realizations over a period of years. The publications are listed in Blackett, P. M. S., "Charles Thomson Rees Wilson 1869–1959," in *Biographical Memoirs of the Royal Society*, vol. 6, 1960, p. 294ff.

54. The critical factor is the expansion ratio, that is, the volume of the chamber before and after the pistol is pulled out. With dust in the air, clouds form at a very small expansion ratio; without dust, the ratio rises to at least 1.25 (cf. Janossy, L., *Cosmic Rays* [Oxford: Clarendon Press, 1948]: 56–58).

55. Anderson, C. D., with Anderson, H. L., *BPP*, p. 136. Anderson discovered that adding ethyl alcohol to the vapor made the tracks brighter. The first person to use cloud chambers to study cosmic rays was the Russian scientist Dmitry Skobeltzyn, in the late 1920s. See Skobeltzyn, D., *BPP*, p. 111.

56. Interview, G. D. Rochester, his office, Durham University, 8 October 1984.

57. Anderson, C. D., *American Journal of Physics* 29, no. 12 (December 1961): 825.

58. Anderson, C. D., *PR* 43, no. 6 (15 March 1933): 491.

59. Anderson, C. D., *Science* 76, (9 September 1932): 238.

60. Hanson, N. R., *The Concept of the Positron* (Cambridge, England: Cambridge University Press, 1963): 139, note 2.

61. Blackett, P. M. S., and Occhialini, G. P. S., *PRSA* 139, no. 839 (3 March 1933): 699; Dirac, *Development, op cit.*, pp. 59–60.

62. Oppenheimer, J. R., and Plesset, M. S., *PR* 44, no. 1 (1 July 1933): 53; Fermi, E., and Uhlenbeck, G., *PR* 44: no. 6 (15 September 1933): 510.

63. Kevles, D. J., *The Physics Teacher* 10, no. 4 (April 1972): 175, provides some discussion of the excitement of the time.

67. Rutherford, E., comment after discussion, in *Structure et Propriétés des Noyaux Atomiques*, Rapports et Discussions du Septième Conseil de Physique tenu à Bruxelles du 22 au 29 Oct. 1933 (Paris: Gauthiers-Villars, 1934): 177–78 (trans. by authors).

CHAPTER 6

1. Letter, Pauli to Dirac, 17 February 1928, *LET*. Italics in original; "What do you think about this?" (Was meinen Sie dazu?) was a favorite Pauli phrase. These notions are discussed in a somewhat different fashion in Weinberg, S., *Daedalus* 106 (Spring 1977): 17, especially pp. 24–30.

2. We are leaving out factors of pi and the like.

3. Thomson, J. J., *PM*, series 5, 11, no. 68 (April 1881): 229.

4. See *SIL*, p. 157.

5. Lorentz, H., *Collected Papers*, Vol. 5. (The Hague: Nijhoff, 1937): 127. See also, Lorentz, H., *The Theory of the Electron* (Leiden: Teubner, 1916). This discussion is presented in considerably more technical form in Pais, A., *Developments in the Theory of the Electron* (Princeton: Institute for Advanced Study, 1947); and Pais, A., in Salam, S., and Wigner, E. P., eds., *Aspects of Quantum Theory* (New York: Cambridge University Press, 1972): 79.

6. Interview, Sir Rudolf Peierls, his office, Oxford University, 9 October 1984.

7. In a letter to Oskar Klein on 16 March 1929, Pauli still wanted to "discuss with you this perpetually obscure question: can the self-energy be eliminated by a simple permutation of the factors [in the equations] . . ."? *LET*.

8. The colleague was Victor Weisskopf; cf., Pauli's entry in *DSB*. We are indebted to Prof. Abdus Salam for the story about Pauli's assistant, Ralph Kronig. There is no full biography of Pauli.

9. Pauli, W., *Encyklopädie der mathematischen Wissenschaften*, vol. 5, part 2 (Leipzig: Teubner Verlag, 1921).

10. Pauli, W., *ZfP* 31, no. 10 (21 March 1925): 765.

11. Cited in Darrigol, O., "Les débuts de la théorie quantique du champs (1925–1948)," Thèse pour le Doctorat de Troisième Cycle, Université de Paris-Panthéon-Sorbonne, unpublished, 1982, (catalog no. I 8211–4): 6.

12. The literature on Oppenheimer is both large and curiously incomplete; his physics contributions have received scant attention compared to his military work and his trial. See, for example, Rabi, I. I., Serber, R., Weisskopf, V. F., Pais, A., and Seaborg, G., *Oppenheimer* (New York: Scribners, 1967); Oppenheimer, J. R., *Letters and Recollections*, Smith, A. K., and Weiner, C., eds. (Cambridge, Massachusetts: Harvard University Press, 1980); and the strongly criticized dual biography, Davis, N. P., *Lawrence and Oppenheimer* (New York: Simon and Schuster, 1968).

13. According to *HDQ* 1A (Preface, p. xxv), Oppenheimer knew Sanskrit only through English translations.

14. Letter, P. Ehrenfest to Pauli, 26 November 1928, *LET*. See also, Born, *My Life* (New York: Scribners, 1975): 210ff.

15. Letter, Pauli to P. Ehrenfest, 15 February 1929, *LET*.

16. Interview, George Uhlenbeck, his office, Rockefeller University, 10 May 1984.

17. It was thought at first that the paper might be signed by Heisenberg, Pauli, and Oppenheimer; it would be a sequel to the Heisenberg–Pauli formulation of quantum electrodynamics. Circumstances seem to have intervened, and although Oppenheimer used the ideas of the other two men, he wrote and published the paper himself. Interview, J. Robert Oppenheimer, *AHQP* (20 November 1963): 22. (We are grateful to Dr. Abraham Pais for drawing this to our attention.) The interaction was not entirely ignored—only what physicists call the "off-shell" part of it (see chap. 16).

18. Oppenheimer, J. R., *PR* 35, no. 5 (1 March 1930): 461. Concurrent calculations were made by Ivar Waller, another young theorist influenced by Pauli. Waller, however, explored the self-energy of an electron floating through space, not bound in a hydrogen atom (Waller, I., *ZfP* 62, no. 9/10 [15 June 1930]: 673).

19. Conversation, Abraham Pais, a cafeteria in Rockefeller University, 31 October 1984. Others have said that Oppenheimer may have been too influenced by the fatalistic world view of the Hindu religion (Rabi, I. I., in Rabi, et al., *Oppenheimer, op cit.*, p. 7; Elsasser, W. M., *Memoirs of a Physicist in the Atomic Age* [New York: Science History Publications, 1978]: 52–53).

20. The names for the orbitals come from the old spectroscopists' approximate description of the lines they produce. "Sharp," "principal," "diffuse," and "fundamental" have become S, P, D, and F. See chapter 7.

21. Oppenheimer, J. R., and Furry, W., *PR* 45, no. 4 (15 February 1934): 245. "A simple change in notation" is, of course, hindsight.

22. *Ibid.*, p. 253, italics added.

23. *Ibid.*, p. 260.

24. *Ibid.*, p. 260–61.

25. Dirac, P. A. M., in *Structure et propriétés des noyaux atomiques, Conseil de Physique Solvay, VII, rapports et discussions* (Paris: Gauthiers–Villars, 1934): 203. The Solvay Conference was held in late October, just as Oppenheimer and Furry were working on their paper.

26. The quotation is from Dirac's original English-language manuscript, "Theory of the Positron," Churchill College Archives, Section 2, File 8, p. 4.

27. *Ibid.*, p. 6.

28. Letter, Pauli to Heisenberg, 14 June 1934. Pauli said Dirac's "subtraction physics" filled him with "disgust." See also, letter, Pauli to Dirac, 1 May 1933 (both *LET*).

29. Weisskopf, V. F., *Physics in the Twentieth Century* (Cambridge, Massachusetts: MIT Press, 1972): 10–12.

30. Weisskopf, V. F., *ZfP* 89, no. 1/2 (15 May 1934): 27–39.

31. Weisskopf, V. F., *ZfP* 90, no. 11/12, (17 September 1934): 817–18.

32. Interview, Victor F. Weisskopf, his office, CERN, Geneva, 24 September 1984.

33. Interview, Werner Heisenberg, *AHQP* (12 July 1963): 11.

34. Heisenberg, W., *ZfP* 90, no. 3/4 (10 August 1934): 211.

35. They must also be gauge invariant, as will be discussed in chapter 10. One of the first to point this out directly was Peierls, R., *PRSA* 146, no. 857 (1 September 1934): 420.

36. This was first noticed by Mott, N. F., *Proceedings of the Cambridge Philosophical Society* 27, no. 2 (April 1931): 255 (read 26 January 1931); Sommerfeld, A., *AdP*, series 5, 11, no. 3 (29 September 1931): 257; and Bethe, H., and Heitler, W., *PRSA* 146, no. 856 (1 August 1934): 83.

37. Letter, Pauli to Ralph Kronig, 20 November 1935, *LET*.

38. Interview, Victor F. Weisskopf, his office, CERN, Geneva, 24 September 1984. Dr. Weisskopf has kindly allowed us to put together here a number of statements he made at different points during our conversation.

39. Interview, Robert Serber, his office, Columbia, 6 May 1985.

40. The National Research Council was established during World War I as an attempt by the American Academy of Sciences to help in the war effort. It was to promote scientific research that would benefit and protect "the national security and welfare." In 1919, the NRC received money to begin the first national program of postdoctoral fellowships in U.S. history. (The history of this program is recounted in the excellent book *TP*, especially pp. 110–13 and 149–51.)

41. Interviews with Robert Serber, Edward Uehling, and Wendell Furry; Franken, P., in ter Haar, D., and Scully, M., eds., *Willis E. Lamb, Jr.: A Festschrift on the Occasion of his 65th Birthday* (New York: North-Holland Publishing, 1978): VII; Rabi, I.I., in *Oppenheimer, op cit.,* p. 6; Lamb, W., *BPP*, p. 311, especially pp. 313–14.

42. Serber, R., *PR* 48, no. 1 (1 July 1935): 49; Uehling, E. A., *PR* 48, no. 1 (1 July 1935): 55. The Caltech spectroscopic results are described in the next chapter.

43. Uehling, E. A., *ibid.,* 61.

44. Interview, Robert Serber, his office, Columbia, 26 November 1984.

45. Weisskopf, V. F., *Det Kongelige Danske Videnskabernes Selskab, Mathematisk-fysiske Meddelelser* 14, no. 6 (1936). Serber and Uehling had considered only unchanging fields; Weisskopf treated varying external fields as well.

46. Serber, R., *PR* 49, no. 7 (1 April 1936): 545.

47. Interview, Robert Serber, his office, Columbia, 6 May 1985.

48. Bloch, F., and Nordsieck, A., *PR* 52, no. 2 (July 1937): 54.

49. Cf., *BPP*, p. 212, 270; Serber interviews cited above.

50. Serber, R., *PR* 49 (*op. cit.*): 546.

51. Interview, Robert Serber, a restaurant in Manhattan, 22 November 1983 (all quotes).

52. Weisskopf, V. F., in *Physics in the Twentieth Century* (Cambridge, Massachusetts: MIT Press, 1972): 13.

53. Heisenberg, W., *AdP*, series 5, 32, no. 1/2 (8 April 1938): 20–33.

54. This attempt is abundantly documented in *LET*, vol. 2.

55. Dirac, P. A. M., *PRSA* 167, no. 929 (5 August 1938): 148 (here he also had to assume that signals could travel faster than light within the electron); Dirac, P. A. M., *Annales de l'Institut Henri Poincaré* 9: no. 2 (1939): 13; P. A. M. Dirac, *Communications of the Dublin Institute for Advanced Studies*, series A, vol. 1, 1943 (see p. 35).

56. Sakata, S., and Hara, O., *PTP* (Kyoto) 2, no. 1 (January–February 1947): 30; Pais, A., *Verhandlungen Koninklijke Akademie van Wetenschappen te Amsterdam*, 19, 1947.

57. Weisskopf, V. F., *PR* 55, no. 7 (1 April 1939): 678 (minutes of the American Physical Society meeting in New York, 23–25 February 1939); Weisskopf, V. F., *PR* 56, no. 1 (1 July 1939): 72.

58. Interview, Victor Weisskopf, his office, CERN, 24 September 1984.

CHAPTER 7

1. Telephone interview, Markus Fierz, 21 September 1984. For the hydrogen atom, alpha is the velocity of the electron in the ground state divided by the speed of light; it also determines the position of every other energy level in the atom, making it of crucial importance to spectroscopists.

2. For example, at the conclusion of the paper in which he formulated both the full version of quantum electrodynamics and the basic subtraction method, Heisenberg asserted that "a contradiction-free union of the conditions of quantum theory with the corresponding predictions of field theory is only possible in a [theory] that provides a particular value for Sommerfeld's constant $e^2/\hbar c$." (Heisenberg, W., *ZfP* 90, no. 3/4 [10 August 1934]: 231, trans. by authors.) See also, letters, Heisenberg to Pauli, 8 June 1934 and 25 March 1935, *LET*; Born, M., *Proceedings of the Indian Academy of Sciences*, 2A (1935): 533.

3. We are grateful to Richard Learner for providing us with this information.

4. The story is recounted in Chandrasekhar, S., *Eddington: The Most Distinguished Astrophysicist of His Time* (New York: Columbia University Press, 1985). Spoofs of Eddington abounded in the literature of the period. See, for example, Beck, G., Bethe, H., and Riezler, W., *Naturwissenschaften* 19 (1931): 39, and Born, M., *Experiment and Theory in Physics.* (Cambridge: Cambridge University Press, 1944): 37. The latter connects 137 to the Book of Revelations.

5. A major proponent of the importance of studying the physical constants was Raymond T. Birge (1887–1980), who went to Berkeley in 1918 and became chairman of its department in 1932.

6. Spedding, F. H., Shane, C. D., and Grace, N. S., *PR* 44, no. 1 (1 July 1933): 58.

7. Houston, W. M., and Hsieh, Y. M., *Bulletin of the American Physical Society* 8, no. 6, (24 November 1933): 5; Houston, W. V., and Hsieh, Y. M., *PR* 45, no. 2 (15 January 1934): 130 (meetings of Stanford American Physical Society meeting, 15–16 December 1933). Quotation from *PR*. Their long paper (see below) was received September 16, although not published until five months later. Houston's name is pronounced "Hoo-stun," not "Hew-stun."

8. Kemble, E. C., and Present, R. D., *PR* 44, no. 2 (15 December 1933): 1031.

9. Kent, N. A., Taylor, L. B., and Pearson, N. H., *PR* 30, no. 3 (September 1927): 266.

10. Telephone interviews, Robley Williams, 20–21 November 1984.

11. Williams, R. C., and Gibbs, R. C., *PR* 44, no. 5 (15 August 1933): 325; also *PR* 44, no. 12 (15 December 1933).

12. Confusingly, this line is called the alpha line for entirely unrelated reasons.

13. Gibbs, R. C., and Williams, R. C., *PR* 45, no. 3 (1 February 1934): 221.

14. Houston, W. V., and Hsieh, Y. M., *PR* 45, no. 4 (15 February 1934): 263.

15. *Ibid.*, p. 263.

16. *Ibid.*, p. 272.

17. Telephone interview, Edward A. Uehling, 20 July 1984.

18. Uehling, E. A., *PR* 48, no. 1 (1 July 1935): 55, especially p. 61.

19. Williams, R. C., and Gibbs, R. C., *PR* 45, no. 7 (1 April 1934): 475.

20. Spedding, F. H., Shane, C. D., and Grace, N. S., *PR* 47, no. 1 (1 January 1935): 38. The quote below is from p. 38; see also pp. 43–44.

21. Telephone interviews, Robley Williams, 20–21 November 1984.

22. Williams, R. C., and Gibbs, R. C., *PR* 49, no. 5 (1 March 1936): 416 (minutes of Saint Louis American Physical Society meeting, 31 December 1935–2 January 1936).

23. Houston, W. V., *PR* 51, no. 6 (15 March 1937): 446. The distinguished European experimenter, Hans Kopfermann, had also weighed in on the side of the discrepancy–finders (Kopfermann, H., *Naturwissenschaften* 22, no. 14 [6 April 1934]: 218). His initial results fed the desire of Pauli and Weisskopf to keep at the self-energy problem. Then Kopfermann gave the problem to his student, Maria Heyden, who found no discrepancy, perhaps because of the poor quality photographic materials she used (Heyden, M., *ZfP* 106, nos. 7/8 [3 August 1937]: 499).

24. The atomic velocity causes a change in frequency known as a Doppler effect, after Johann Christian Doppler (1803–53), the Austrian physicist and astronomer who first predicted its existence in 1842.

25. The apparatus is sufficiently clever to warrant description. The hydrogen is kept in a 20-centimeter U–shaped tube (a Wood's tube, after R. F. Wood, who invented it in 1928) plunged into a closed tank of liquid air. Physicists connect a 2000–volt electrical source to terminals sealed into the tube. This excitation causes the hydrogen to radiate—that is, to emit light. The light goes into the interferometer, whose heart consists of two unbelievably flat glass plates about the size of a pocket watch, clamped face to face in a small, doughnut–shaped frame. The two inside faces of the plates are coated with a ghost–thin film of mirror silvering and are perhaps a centimeter apart. The light from the tube of hydrogen passes through the first plate, hits the mirror side of the second, bounces back, hits the mirror side of the first, bounces forward, and continues back and forth as many as thirty times before dissipating. At each reflection, however, a small amount of light makes its way through the silvering and passes through the forward lens. The thirty or so successively reflected beams emerge out of the interferometer out of alignment with each other, crests and waves ever so slightly out of step. By coating the interferometer just right and adjusting the spacing, an experimenter can "tune" the equipment so that in some directions the various crests reinforce each other. The reinforced crests can be focused to form a series of glowing, photographable concentric rings.

26. Bethe, H., in *Handbuch der Physik*, 2d ed., H. Geiger and K. Scheel, eds., vol. 24, pt. 1 (Berlin: Julius Springer, 1933): 273, especially sections 10, 43, and 44.

27. The Stark effect and the Zeeman effect, as it happens. Franck was just passing through. Bethe joined the Cornell faculty, and was in more contact with the experimenters (Letter, Hans Bethe to authors, 21 October 1985).

28. Telephone interview, Hans Bethe, 4 December 1984.

29. Telephone interview, Robley Williams, 20 November 1984.

30. Williams, R. C., *PR* 54, no. 8 (15 October 1938): 558.

31. Telephone interview, Robley Williams, 21 November 1984.

32. Pasternack, S., *PR* 54, no. 12 (15 December 1938): 1113. The quote below is from p. 1113. Houston was still working on the problem, as can be seen in the work of his student C. F. Robinson, *PR* 55, no. 4 (15 February 1939): 423 (minutes of the Los Angeles American Physical Society meeting on 19 December 1938). The presentation seems not to have been elaborated into a formal paper, however.

33. Telephone interview, Robley Williams, 20 November 1984. Pasternack died in 1982.

34. Telephone interview, J. W. Drinkwater, 19 October 1984.

35. Drinkwater, J. W., Richardson, O., and Williams, W. E., *PRSA* 174, no. 957 (1 February 1940): 164.

36. *Ibid.*, p. 184.

37. *Ibid.*, p. 187.

38. Interview, Dick Learner, his office, Imperial College, University of London, London, 5 October 1984.

39. Biographical information from Lamb, W. E., *BPP*, p. 311; Lamb, W. E., in *A Festschrift for I. I. Rabi*, Transactions of the New York Academy of Sciences, series 2, 38 (4 November 1977): 82; Lamb, W. E., in Marshak, R., ed., *Perspectives in Modern Physics* (New York: Wiley, 1966): 261.

40. Telephone interview, Willis Lamb, 9 August 1984.

41. The proposal is reprinted in Lamb, *Festschrift, op cit.*

42. The technical description can be found in Lamb, W. E., *PR* 79, no. 4 (15 August 1950): TK, and Trigg, G. L., *Landmark Experiments in Twentieth Century Physics* (New York: Crane, Russak, 1975): 99.

43. Telephone interview, J. Bruce French, 11 July 1984; interview, Victor Weisskopf, his office, CERN, 24 September 1984.

44. We are here relying heavily on the account in Schweber, S., "Some Chapters for a History of Quantum Field Theory: 1938–1952," 1983 Les Houches Lectures, unpublished ms.

45. See, for example, letter, J. R. Oppenheimer to H. A. Kramers, 14 April 1947, National Archives, Oppenheimer Collection, Case File, Box 44, Kramers file. Oppenheimer writes that he was "planning to go to this strange meeting on Eastern Long Island in early June, though I am as much in the dark as you as to what it is all about."

46. Letter, V. F. Weisskopf to K. K. Darrow, 18 February 1947, cited in Schweber, *op cit.*, p. 131.

47. Weisskopf, V. F., "Foundations of Quantum Mechanics, Outline of Topics for Discussion," unpublished ms., April (?) 1947, National Archives, Oppenheimer Collection, Case File, Box 72, File "Theor. Physics Conf.—Corres.—Shelter Isl. '47."

48. Kramers, H., *NC* 15, (1938): 108; see also, Kramers, H., *Nederlandsch Tijdschrift voor Natuurkunde* 11 (1944): 134, and his Shelter Island outline (same as note 47).

49. Interview, Julian Schwinger, a restaurant in Los Angeles, 4 March 1983.

50. Interview, Robert Serber, his office, Columbia, 18 July 1984.

51. Telephone interview, Hans Bethe, 8 August 1984.

52. Interview, Richard Feynman, *AHQP*, 1966, p. 454.

53. Letters, J. Robert Oppenheimer to Alfred N. Richards, 1 December 1947; Alfred N. Richards to J. Robert Oppenheimer, 2 December 1947, both in Library of Congress, Manuscript Division, Oppenheimer Collection, General Case File, Box 72, folder labeled "Theor. Physics Conf.—Corres.—Poconos 1948."

54. Interview, I. I. Rabi, his apartment, New York City, 9 May 1984.

55. Nafe, J. E., Nelson, E. B., and Rabi I. I., *PR* 71, no. 12 (15 June 1947): 914; see also Kusch, P., and Foley, H. M., *PR* 72, no. 12 (15 December 1947): 1256.

56. The reconstruction is based on interviews and the surprisingly good account by Steven White in the *Herald Tribune*, 3 June 1947.

57. Interview, Julian Schwinger, a restaurant in Los Angeles, 4 March 1983.

58. See chap. 9 and Marshak, R., *BPP*, p. 381.

59. Schwinger, J., *BPP*, p. 332.

60. Interview, Hans Bethe, *AHQP* (9 May 1966): 168.

61. Telephone interview, Hans Bethe, 8 August 1984. Bethe was directly inspired by Kramers, whose idea of comparing the self-energies of a free electron and a $^2S_{1/2}$ electron was simpler than that of Weisskopf and French, who compared the $^2S_{1/2}$ and $^2P_{1/2}$ states (Letter, Hans Bethe to authors, 21 October 1985).

62. Oppenheimer Collection, National Archives, General Case File, Box 20, Bethe folder. After further numerical calculations by Bethe's students, the paper was submitted two weeks later.

63. Lamb, W. E., *BPP*, p. 324. Weisskopf's chagrin was touched with exasperation; he believed Bethe had used but not acknowledged the ideas Weisskopf had developed (Interview, Victor Weisskopf, his office, CERN, 24 September 1984). Indeed, Bethe's rough draft in the Oppenheimer ms. collection does not mention Weisskopf.

64. Telephone interview, Willis Lamb, 9 August 1984.

CHAPTER 8

1. Telephone interview, Lloyd Motz, 10 December 1984; biographical information also from Schwinger, J., *BPP*, p. 329.

2. Interview, I. I. Rabi, his apartment, New York City, 9 May 1984.

3. Interview, Julian Schwinger, a restaurant in Los Angeles, 4 March 1983.

4. Interview, George Uhlenbeck, his office, Rockefeller, 10 May 1984.

5. Telephone interview, Lloyd Motz, 10 December 1984.

6. Letters, Victor Weisskopf to J. Robert Oppenheimer, 22 and 29 August and 22 and 14 December 1948, Library of Congress, Manuscript Division, Oppenheimer Papers, General Case File, Box 72, folder labeled "Theor. Physics Conf.—Corres.—Poconos 1948"; letter, Victor Weisskopf to authors, 5 August 1985; telephone interview, J. Bruce French, 11 July 1984; Weisskopf, V., *Physics in the Twentieth Century* (Cambridge, MIT Press, 1972): 17–18.

7. Schwinger, J., *PR* 73, no. 4 (15 February 1948): 416 (received on 30 December 1947); French, J. B., and Weisskopf, V., *PR* 75, no. 8 (15 April 1949): 1240; Kroll, N., and Lamb, W., *PR* 75, no. 3 (1 February 1949): 388.

8. The New York City American Physical Society meeting was held at Columbia on 29–31 January 1948.

9. Interview, Julian Schwinger, a restaurant in Los Angeles, 4 May 1983.

10. Dyson, F., *Disturbing the Universe* (New York: Harper & Row, 1979): 55.

11. Letter, Sin-itiro Tomonaga to J. Robert Oppenheimer, 2 April 1948, Library of Congress, Manuscript Division, Oppenheimer papers, General Case File, Box 73, Tomonaga folder. See also, Tomonaga, S., *PTP* 1, no. 2 (August–September 1946): 27 (a Japanese version appeared in 1943).

12. Takabayasi, T., *BPP*, p. 280.

13. Night letter, J. Robert Oppenheimer to Sin-itiro Tomonaga, 13 April 1948, found in Tomonaga folder (note 11, above); Oppenheimer's letter is in General Case File, Box 72, folder labeled "Theor. Physics. Conf.—Corres.—Pocono—1948."

14. Letter, Sin-itiro Tomonaga to J. Robert Oppenheimer, 14 May 1948; *Physical Review* note of receipt, 1 June 1948, both in Box 73, Tomonaga folder (see above).

15. Schwinger, J., *PR* 73, no. 4 (15 February 1949): 416.

16. Weisskopf, V. F., *BPP*, p. 76.

17. Pauli's reaction can be seen in the letters to Oppenheimer of 6 January 1948, 22 February 1949, and (28?) February 1949 in the Oppenheimer Collection, Library of Congress, Manuscript Division, General Case File, Box 56, Pauli file. For Dirac's attitude, see, for example, Dirac, P. A. M., *BPP*, p. 39.

18. Feynman's high school recollections can be found in Feynman, R. P., *"Surely You're Joking, Mr. Feynman!"* (New York: W. W. Norton, 1985): 15–30.

19. Feynman, R. P., in Badash, L., *Reminiscences of Los Alamos* (Dordrecht, Holland: D. Reidel, 1980): 105. Feynman, *"Surely," op. cit.*, pp. 107–55, has a somewhat similar account.

20. Feynman, R. P., in "The Pleasure of Finding Things Out," *Nova* broadcast, WGBH, Boston, 25 January 1983.

21. Introduced to quantum physics in Dirac, P. A. M., *Physikalische Zeitschrift der Sowjetunion* 3, no. 1 (January 1933): 64.

22. Interview, Robert Serber, his office, 22 October 1984; interview, John Wheeler, Columbia University Faculty Club, 14 December 1983; interview, Richard Feynman, *AHQP*, vol. VII, pp. 450–86.

23. Interview, Richard Feynman, his office, Caltech, 22 February 1985 (all direct quotes).

24. Dyson, F. J., *PR* 75, no. 3 (1 February 1949): 486.

25. Interview, Julian Schwinger, a restaurant in Los Angeles, 4 March 1983; interview, Victor Weisskopf, his office, CERN, 24 September 1984.

26. Cited in Chandrasekhar, S., *Eddington, the Most Distinguished Astrophysicist of His Time* (New York: Cambridge University Press, 1985): 3.

27. Weisskopf, V. F., *BPP*, p. 78; confirmed during interview cited above.

28. His father's name was Stückelberg, with an umlaut; in the United States, he changed his name to Stueckelberg for convenience.

29. Interviews, Valentine Telegdi, CERN cafeteria, 24 and 25 September 1984; interview, Charles Ruegg, his office, Geneva Institute of Physics, 21 September 1984; interview, John Iliopoulos, his office, Rockefeller University, 8 May 1984; telephone interview, Jean Rivier, 2 October 1984; interview, André Petermann, his home, Geneva, 10 February 1984.

30. Stueckelberg, E. C. G., *HPA* 11, no. 3 (1938): 221, especially 242–43; pts. II, III (1938): 298. On p. 317, incidentally, he introduces the concept of baryon conservation, the violation of which is currently a subject of considerable theoretical interest. See also *AdP* 21 (1934): 367.

31. The classical approach is found in Stueckelberg, E. C. G., and Patry, J. F. C., *HPA* 13

(1940): 167; Stueckelberg, E. C. G., *HPA* 14 (1941): 51.
    32. Stueckelberg, E. C. G., *HPA* 28 (1945): 21; Stueckelberg, *HPA* 19 (1946): 241.
    33. Rivier, D., *HPA* 22, no. 3 (1949): 265.
    34. Stueckelberg, E. C. G., and Petermann, A., *HPA* 24 (1951): 317; Stueckelberg, E. C. G., and Petermann, A., *HPA* 26 (1953): 499.
    35. Interview, E. C. G. Stueckelberg, his home, Geneva, 10 February 1984.

CHAPTER 9
    1. Wulf, T., *PZ* 11, no. 18 (15 September 1910): 812. The paper gives the time, place, and temperature of the observations.
    2. Becquerel, H., *CR* 122: 559, séance of 9 March 1896.
    3. Elster, J., and Geitel, H., *PZ* 2, no. 38 (22 June 1901): 560. Elster and Geitel, two physicists at Wolfenbüttel, near Brunswick, Germany, were inseparable colleagues known as the Castor and Pollux of physics. Their collaboration was so extensive that the story is told that a man who looked like Geitel was once greeted, "Herr Elster!" He said, "I'm not Elster, I'm Geitel. And I'm not Geitel, either." Elster was the prototypical absentminded professor: Upon being informed he had put a stamp of greater denomination than necessary on a letter, he crossed it out and put a correct stamp on. Because he had a large and a small dog, he cut a big hole and a little hole in his apartment door. And so on.
    4. Wilson, C. T. R., *Proceedings of the Cambridge Philosophical Society* 11 (read 26 November 1900): 32.
    5. McLenna, J. C., and Burton, E. F., *PR* 16, no. 3 (March 1903): 184.
    6. Rutherford, E., and Cooke, H. L., *PR* 16, no. 3 (March 1903): 183.
    7. Piel, C., *Der Mathematische und Naturwissenschaftliche Unterricht* 1, no. 3 (February 1949): 105.
    8. Simpson, G. C., and Wright, C. S., *PRSA* 85, no. 577 (10 May 1911): 175; Simpson, G. C., *Meteorology. British Antarctic Expedition 1910-1913*, vol. 1 (Calcutta: Thacker, Spink, 1919).
    9. See, for example, Wulf, T., *PZ* 10, no. 5 (1 March 1909): 152; Kurz, K., *PZ* 10, no. 22 (10 November 1909): 834. For a contrary view, however, see Richardson, O., *Nature* 73, no. 1904 (26 April 1906): 607, and sources cited therein.
    10. Wulf, T., *PZ* 11, no. 18 (15 September 1910): 811.
    11. Hess entry, *DSB*.
    12. Hess, V., *Thought* 15, no. 57 (June 1940): 229.
    13. Gockel, A., *PZ* 12, no. 14 (15 July 1911): 595.
    14. Hess, V. F., *PZ* 12, nos. 22/23 (15 November 1911): 998.
    15. The photograph is reproduced in Kraus, J., *Our Cosmic Universe* (Powell, Ohio: Cygnus-Quasar Books, 1980): 196-200.
    16. Hess, V., *PZ* 13, nos. 21/22 (1 November 1912): 1090.
    17. King, L. V., *PM*, series 6, 23, no. 134 (February 1912): 248.
    18. Kolhörster, W., *Verhandlungen der Deutschen Physikalischen Gesellschaft* 16, no. 14 (30 July 1914): 719.
    19. For an account of the interactions between these groups, see Cassidy, D., *HSPS* 12, no. 1 (Winter 1981): 1.
    20. Darrow, K. K., *Bell System Technical Journal* 11, no. 1 (January 1932): 148.
    21. There are many sources of information about Millikan. See, for example, Millikan, R. A., *The Autobiography of Robert A. Millikan* (New York: Prentice-Hall, 1950); Kevles, D. J., *Scientific American* 240, no. 1 (January 1979): 142; *TP*, pp. 88-89, 231-42, and *passim*. Millikan's own writing should be enjoyed, but read with suspicion.
    22. The first was Albert Michelson, for his experimental proof of the nonexistence of the ether.
    23. Millikan, R., *Science* 57, no. 1483 (1 June 1923): 630.
    24. Cited in *TP*, p. 180.
    25. Millikan, R. A., and Bowen, I. S., *PR* 22, no. 2 (August 1923): 198; subsequently rewritten and reinterpreted in *PR* 27, no. 4 (April 1926): 360.
    26. Millikan, R. A., and Otis, R. M., *PR*, series 2, 23, no. 6: 778; subsequently rewritten and reinterpreted in *PR* 27, no. 6 (June 1926): 645.
    27. Hess, V. F., *PZ* 27, no. 12 (15 June 1926): 405; Kolhörster, W., *Naturwissenschaften* 15, no. 5 (4 February 1927): 126.
    28. Millikan, R. A., *Science* 62, no. 1612 (20 November 1925): 445; a revised version of this

paper is Millikan, R. A., *Proceedings of the National Academy of Sciences* 12, no. 1 (January 1926): 48. Millikan's full account is Millikan, R. A., and Cameron, G. H., *PR*, series 2, 28, no. 5 (November 1926): 851. It is interesting to compare Millikan's original description of his first conclusions with those given a year later.

29. Quoted in *TP*, p. 179.

30. Among the angry papers written in the dispute: Hess, V. F., *PZ* 27, no. 6 (15 March 1926): 126; Bergwitz, K., Hess, V. F., Kolhörster, W., and Schweidler, E., *PZ* 29, no. 19 (1 October 1928): 705; Millikan, R. A., *Nature* 126, no. 3166 (5 July 1930): 14. See also *TP*, p. 179.

31. Lemon, H. B., *Cosmic Rays Thus Far* (New York: Norton, 1936): 56–57.

32. Bothe entry, *DSB*.

33. Bothe, W., and Kolhörster, W., *ZfP* 56, nos. 11/12 (16 August 1929): 751.

34. Quoted in Kevles, *Scientific American, op. cit.*, p. 145.

35. Millikan, R. A., *Annual Report of the Smithsonian Institution*, 1931, p. 270, especially 282–83.

36. The results are summarized in Störmer, F., in *Zeitschrift für Astrophysik* 1, no. 4 (1930): 237.

37. Clay, J., *Verh. Koninklijke Akademie van Wetenschappen te Amsterdam*, 30, no. 9/10 (1927): 711.

38. Millikan, R.A., and Cameron, G.H., *Nature*, 121 no. 3036 (supp.) (7 January 1928): 20.

39. Bothe, W., and Kolhörster, W., *Sitzungsberichte der Preussischen Akademie der Wissenschaften zu Berlin* 24 (1930): 450. The real reason for their failure to discover the latitude effect is that cosmic rays are sufficiently powerful to be affected by the earth's magnetic field only near the equator, where it is strongest. The calculations were carried out successfully in Rossi, B., *NC* 8, no. 3 (March 1931): 85.

40. Hess, *Thought, op. cit.*, p. 234.

41. Jaffe, B., *Outposts of Science* (New York: Simon and Schuster, 1935): 399.

42. The letter is reproduced in Auguste Piccard, *A 16000 Metri*, trans. by D. S. Suardo (Milan: Mondadori, 1933) facing p. 300.

43. Jaffe, *op. cit.*, p. 400.

44. Millikan, R. A., *Cosmic Rays* (New York: Macmillan, 1939): 58.

45. Quoted in *New York Times*, 15 September 1932, p. 23.

46. Neher, H. V., *BPP*, p. 127.

47. *New York Times*, 31 December 1932, p. 1; 1 January 1933, p. 16; 5 February 1933, p. 81; see also the reworked American Association for the Advancement of Science speech in Millikan, R. A., *PR* 43, no. 8 (15 April 1933): 661.

48. Auger, P., and Skobeltzyn, D., *CR* 189, séance of 1 July 1929, p. 55; Blackett, P. M. S., and Occhialini, G. P. S., *PRSA* 139, no. 839 (3 March 1933): 699.

49. Interview, Pierre Auger, his apartment, Paris, 3 October 1984.

50. Rossi, B.; Auger, P., and Leprince-Ringuet, L., both in *International Conference on Physics, London 1934* (Cambridge: University Press, 1935): 233, 195.

51. Anderson, C. D., *BPP*, p. 146.

52. Bethe, H., and Heitler, W., *PRSA* 146, no. 856 (1 August 1934): 83.

53. W. Heitler, *The Quantum Theory of Radiation*, 1st ed. (Oxford: Oxford University Press, 1936 [Preface dated November 1935]). Heitler's remark was typical; see also Oppenheimer, J. R., *PR* 47, no. 1 (1 January 1935): 44, especially p. 47.

54. Anderson, C., *BPP*, p. 147.

55. Anderson, C. D., and Neddermeyer, S. H., *PR* 50, no. 4 (15 August 1936): 263.

56. Yukawa, H., *Tabibito*, trans. by L. Brown and R. Yoshida. (World Scientific, 1982): 79–80.

57. *Ibid.*, p. 119.

58. Schwinger, J., *BPP*, pp. 354–55.

59. Yukawa, *op. cit.*, p. 174.

60. Heisenberg, W., *Zfp* 77, no. 1/2 (19 July 1932): 1; 78, no. 3/4 (21 September 1932): 156; 80, no. 9/10 (16 February 1933): 587.

61. Yukawa, H., *Tabibito, op. cit.*, p. 195.

62. Yukawa, *ibid.*, pp. 194–195.

63. Fermi, E., *NC* 11, no. 1 (January 1934): 1.

64. Letter, Wolfgang Pauli to Lise Meitner, Hans Geiger, et al., 4 December 1930, *LET*; the hypothesis first printed in a comment by Pauli in *Structure et Propriétés des Noyaux Atomiques*,

Rapports et discussions du Septième Conseil de Physique, Institut International de Physique (Paris: Gauthier-Villars, 1934): 324.

65. Rasetti, F., comment in Fermi, E., *Collected Papers* (Chicago: University of Chicago Press, 1962): 538.

66. Tamm, I., *Nature* 133, no. 3374 (30 June 1934): 981; Iwanenko, *ibid.*, 981.

67. Yukawa, *Tabibito, op. cit.*, p. 201. The argument below is simpler than Yukawa's original thoughts.

68. Yukawa, H., *Proceedings of the Physico-Mathematical Society of Japan* 17, no. 2 (February 1935): 48.

69. *Ibid.*, p. 48.

70. *Ibid.*, p. 53.

71. Brown, L., Konuma, M., and Maki, Z., *Particle Physics in Japan, 1930–1950*, Research Institute for Fundamental Physics preprint 408, vol. 2 (September 1980): 31.

72. Tanikawa, Y., in Yukawa, H., *Scientific Works* (Tokyo: Iwanami Shoten, 1979): xvi.

73. Yukawa, *Tabibito, op. cit.*, p. 203.

74. Hayakawa, S., *BPP*, p. 88; the work culminated in Yukawa, H., and Sakata, S., *Proceedings of the Physico-Mathematical Society of Japan* 19, no. 12 (December 1937): 1084; sections III–V, published with various collaborators in the same journal the next year.

75. Brown, Konuma, and Maki, vol. 2, *op. cit.*, p. 29.

76. Tomonaga, S., et al., in Brown, Konuma, and Maki, *op. cit.*, pp. 14–15.

77. Neddermeyer, S., and Anderson, C. D., *PR* 51, no. 10 (15 May 1937): 884–86. The ms. was received March 30, 1937.

78. Street and Stevenson reported their results to the American Physical Society meeting on 29 April 1937 (an abstract appeared in *PR* 51, no. 11 [1 June 1937]: 1005; the information that it was delivered on April 29 is from the final note to the Anderson and Neddermeyer paper). They sent a two-page Letter to the Editor to the *Physical Review* on 6 October (Street, J. C., and Stevenson, E. C., *PR* 52, no. 9 (1 November 1937): 1003. Later it was discovered that experimenters had been photographing mesons for years without knowing their significance. See, for instance, Kunze, P., *ZfP* 83, no. 1–2 (6 June 1933): 1.

79. Oppenheimer, J. R., and Serber, R., *PR* 51, no. 12 (15 June 1937): 1113.

80. Stueckelberg, E. C. G., *PR* 52, no. 1 (1 July 1937): 41.

81. Stueckelberg, E. C. G., *Nature* 137, no. 3477 (20 June 1936): 1032; telephone interview, Jean Rivier, 2 October 1984; interviews, Valentine Telegdi, CERN cafeteria, 24 and 25 September 1984.

82. Anderson, C. D., *BPP*, p. 148. The letter itself is in Anderson, C. D., and Neddermeyer, S., *Nature* 142, no. 3602 (12 November 1938): 878; see also, Kemmer, N., *PTP* (Supp. 35 ext.), November 1965, 605.

83. *E.g.*, Millikan, R. A., *Electrons (+ and −), Protons, Photons, Neutrons, Mesotrons, and Cosmic Rays*, rev. ed. (Chicago: University of Chicago Press, 1947): 508ff.

84. Compton, A., *RMP* 11, no. 3 (July–October 1939): 122.

85. Euler, H., and Heisenberg, W., *Ergebnisse der Exakten Naturwissenschaften* 17 (1938): 1.

86. Oppenheimer, J.R., *PT* 19, no. 11 (November 1966): 58.

87. Interview, Bruno Rossi, his office, MIT, 1 November 1984.

88. Amaldi, E., *Scientia* 114, no. 1 (1979): 51.

89. Piccioni, O., *BPP*, p. 225. We have relied greatly on this and Conversi, M., *ibid.*, p. 242. We are grateful for interviews with Marcello Conversi (his office, University of Rome, 16 February 1984) and Oreste Piccioni (Fermilab, 3 May 1985).

90. Comment by Piccioni, O., *BPP*, p. 272.

91. Piccioni, *BPP*, p. 226.

92. Interview, Oreste Piccioni, Fermilab, 3 May 1985.

93. Conversi, M., and Piccioni, O., *NC* 2, no. 1 (1 April 1944): 40. After the war, the two men learned that Rossi had made an earlier determination by a different method (Rossi, B., and Nereson, N., *PR* 62, nos. 9–10 [1 and 15 November 1942]: 417; Rossi, B., Hilberry, H., and Hoag, J., *PR*, 56, 8 [15 October 1939]: 837–38). A group in occupied France was also working on the meson lifetime, *i.e.*, R. Chaminade, André Fréon, and Roland Maze, *CR* 218, no. 10 (6 March 1944): 402.

94. Tomonaga, S., and Araki, G. *PR* 58, no. 1 (1 July 1940): 90. As Piccioni put it, "Because at that time previous work had well established that 55% of the mesotrons were positive and 45%

negative, we planned to show that only 55% of the stopped particles decayed." (Piccioni, *BPP*, p. 229.)

95. Conversi, M., Pancini, E., and Piccioni, O., *PR* 71, no. 3 (1 February 1947): 209.

96. The full explanation was given in Fermi, E., Teller, E., and Weisskopf, V., *PR* 71, no. 5 (1 March 1947): 314.

97. Rochester, G. D., *CI*, p. 169.

98. Interview, George Rochester, his office, Durham University, 8 October 1984.

99. There is considerable literature on the Japanese "Manhattan Project," much of which is discussed in Phillip S. Hughes, *Social Studies of Science* 10, no. 3 (August 1980): 345 and Wilcox, R.K., *Japan's Secret War* (New York: William Morrow, 1985). The material shortages were exacerbated when a German U-boat bound for Japan with two tons of uranium was sunk by Allied vessels. See for example, Shapley, D., *Science*, 13 January 1978, p. 152.

100. Interview, Michiji Konuma, History of the Weak Interaction Conference, Wingspread, Racine, Wisconsin, 30 May 1984.

101. Brown, Konuma, and Maki, *op. cit.*, p. 32.

102. Much of this story is recounted in *BPP*. esp. p. 285.

103. Nishina, Y., Sekido, Y., Miyazaki, Y., and Masuda, T., *PR* 59, no. 4 (15 February 1941): 401; Brown, Konuma, and Maki, vol. 1, *op. cit.*, p. 55.

104. The club was first called *Meiso-kai* ("illusion-meeting"), then renamed *Meson-kai*.

105. Sakata, S., and Inóue, I., *Proceedings of the Physico–Mathematical Society of Japan* 16(1942): 232.

106. U. S. Strategic Bombing Survey, Records, Section 2, Japanese records, 13 b.(4), USSBS report from Yoshio Nishina, National Archives.

107. Sakata, S. and Inóue, I., *PTP* 1, no. 1 (1946): 143 (trans. of ref. 105).

108. Marshak, R., *BPP*, p. 385.

109. Weisskopf, V. F., *PR* 72, no. 6 (15 September 1947): 510.

110. Lattes, C. M. G., Muirhead, H., Occhialini, G. P. S., and Powell, C. F., *Nature* 159, no. 4047 (24 May 1947): 694.

111. Marshak, R. E., and Bethe, H. A., *PR* 72, no. 6 (15 September 1947): 506. Interestingly, a similar hypothesis had been made by Christian Møller, summarizing work by himself and Abraham Pais at an English conference (Møller, C., in *Fundamental Particles and Low Temperature Physics*, vol. 1, Cavendish Laboratory, Cambridge, 22–27 July 1946 [London: Taylor & Franci, 1947]: 184). The paper also postulates something like mu–electron universality.

112. Neither Rabi nor anyone else can recall where this now–famous comment was first made, but Rabi thinks that it was at an American Physical Society meeting in New York City.

113. Rochester, G. D. and Butler, C. C., *Nature* 160, no. 4077 (20 December 1947): 855.

114. Marshak, R., *BPP*, p. 376.

115. This and the following quotation are from a lecture by G. D. Rochester at the History of the Weak Interaction conference, Wingspread, Racine, Wisconsin, 31 May 1984.

116. Interview, G. D. Rochester, his office, Durham, 8 October 1984.

117. Leprince-Ringuet, L., *Les rayons cosmiques: les mesotrons* (Paris, Editions Albin Michel, 1945): 137–38, trans. by authors.

118. The history of the Jungfraujoch laboratory is recounted in Debrunner, H., ed., *50 Jahre Hochalpine Forschungsstation Jungfraujoch* (Bern: Wirtschaftsbulletin no. 23 of the Kantonalbank of Bern, October 1981).

119. The description of the Jungfraujoch is from our visit to the site in February 1984.

120. Interview, Antonino Zichichi, his office, CERN, 3 February 1984.

121. Cf. Marshak, R., *Meson Physics* (New York: McGraw-Hill, 1952): 359.

122. The term was coined in Wigner, E., *PR* 51, no. 2 (15 January 1937): 106.

123. A slightly more technical description is to say that isotopic spin is a vector orientated by the presence of an abstract charge space in a way analogous to the magnetic field's orientation of the ordinary spin. Heisenberg thus proposed that different projections in this abstract charge space of the Z axis of this imaginary isotopic spin would be a mathematically convenient way of representing the charge states within a family of particles. Just as the different components of ordinary spin are separated in energy by the magnetic field, causing the fine structure in spectra, so the various components of isotopic spin in a particle family are separated in mass by the effects of the electromagnetic force. This is the origin, by the way, of the slight mass differences between the proton and neutron.

124. The papers are, respectively, Tuve, M. A., Heydenberg, N., and Hafstad, L. R., *PR* 50, no. 9 (1 November 1936): 806; Breit, G., Condon, E. U., and Present, R. D., *ibid.*, p. 825; Cassen, B., and Condon, E. U., *ibid.*, p. 846.

125. Kemmer, N., *Proceedings of the Cambridge Philosophical Society* 34 (1938): 354.

126. Oppenheimer, J. R., and Serber, R., *PR* 53, no. 8 (15 April 1938): 636. They credited the idea to conversations with Gregory Breit, and used it to construct a theory for nuclear transitions that is beyond the focus of this work. Experimental demonstration of isotopic spin conservation in the realm of particle physics first occurred when Enrico Fermi and his collaborators showed it held true for pion–nucleon interactions (Fermi, E., "Pion Scattering in Hydrogen," Third Rochester Conference, 1952).

127. Pais, A., *Physica* 19, no. 9 (July 1953): 869.

128. Telephone interview, Abraham Pais, 5 February 1985.

129. We are here skipping over an important step for the sake of narrative clarity. The Japanese and, later, Pais further hypothesized something called "associated production," which is to say that the V particles could be created only in pairs. At the time, the evidence for associated production was not very good: By the spring of 1951, when the Japanese started their work, less than 150 V particles had been observed, and some of the pictures were poor (Butler, C. C., *Progress in Cosmic Ray Physics* [Boston: North–Holland, 1952] p. 60). Four Japanese articles on associated production were published back to back in *PTP*, a journal already known as a graveyard of good ideas. These were duly interred (Nambu, Y., Nishijima, K., and Yamaguchi, Y., *PTP* 6, no. 4 [July–August 1951]: 615; 619; Aizu, K., and Kinoshita, T., *ibid.*, p. 630; Miyazawa, H., *ibid.*, p. 631; Oneda, S., *ibid.*, p. 633; Pais, A., *PR*, 86, 5 [1 June 1952]: 672).

130. Pais, *Physics, op. cit.*, pp. 869–70.

131. Telephone interview, Murray Gell–Mann, 5 February 1985; see also Gell–Mann, M., *CI*, p. 395.

132. A word about the origins of the names. Hadron is credited to L. Okun; lepton from Møller, *op. cit.*, p. 184; baryon from Pais, A., *Proceedings of the International Conference of Theoretical Physics* (Kyoto and Tokyo, September 1953): 157.

133. Peaslee, D. C., *PR* 86, no. 1 (1 April 1952): 127.

134. Gell-Mann, M., "On the Classification of Particles," typescript, his files.

135. Nishijima, K., *PTP* 9, no. 4 (April 1953): 414 presents similar but much less clearly formulated work. Implicit but unexplained here is the notion of baryon conservation, which will be treated later.

136. Telephone interview, Abraham Pais, 5 February 1985.

137. The first publication of the rules was Amaldi, E., et al., *PT* 6, no. 12 (December 1953): 24. The identical text was published in *NC*, *CR*, *Nature*, and *Naturwissenschaft* in early 1954, and the proceedings of *Congrès International sur le Rayonnement Cosmique*, Bagnères de Bigorre, July 1953.

138. Leprince–Ringuet, L., "Discours de clôture," in proceedings of the *Congrès International sur le Rayonnement Cosmique*, Bagnères de Bigorre, July 1953, typescript, pp. 289–90.

CHAPTER 10

1. Interview, Sheldon Glashow, his office, Harvard, 2 December 1982.

2. The description of the ceremony is based on interviews cited below with the three winners and personal visits to Stockholm.

3. Weinberg, S., *RMP* 52, no. 3 (July 1980): 515.

4. Weinberg, S., *PR* 118, no. 3 (1 May 1960): 838.

5. See, for instance, the discussion of gauge invariance in Wigner, E., "Invariance in Physical Theory," *Proceedings of the American Philosophical Society* 93, no. 7 (December 1949): 521, especially 524–26. It is worth noting that a striking exception to this statement is the study of crystal formation.

6. According to *SIL* (p. 140), the terms *Galilean invariance* and *Galilean transformations* were coined in 1909.

7. For more, see *ibid.*, pp. 121–26 and 140–44.

8. Weinberg, S., *Gravitation and Cosmology* (New York: Wiley, 1972).

9. Interview, Steven Weinberg, his office, University of Texas, 28 November 1984.

10. Literature on Emmy Noether includes Weyl, H., *Scripta Mathematica* 3, no. 3 (July 1935): 201, which is the printed version of Weyl's Memorial Address on Noether; Brewer, J. W., and Smith, M. K., eds., *Emmy Noether, a Tribute to Her Life and Work* (New York: Marcel Dekker, 1981); Dick, A., *Emmy Noether, 1882–1935* (Basel: Birkhäuser Verlag, 1970 [Beihefte zur Zeitschrift Elemente der Mathematik no. 13]); Kramer, E., *The Nature and Growth of Modern Mathematics* (Princeton: Princeton University Press, 1982): 656–79; and her *DSB* entry.

11. Kimberling, C., in Brewer, and Smith, *op. cit.*, pp. 10–12.

12. Quoted in Reid, C., *Hilbert* (New York: Springer Verlag, 1970): 127.

13. See, for instance, Mehra, J., *Einstein, Hilbert, and the Theory of Gravitation* (Boston: D. Reidel, 1974).

14. Quoted in Weyl, *Scripta Mathematica*, *op. cit.*, p. 207.

15. Quoted in Kimberling, *op. cit.*, p. 18 (trans. by authors).

16. It is said that following the memorial address Weyl sent an obituary notice to the *New York Times*, upon which an editor commented, "Who is Weyl—have Einstein wrote something, as he is the mathematician recognized by the world" (Kimberling, *op. cit.*, p. 52). Einstein did so, and his letter to the *Times* was printed on May 4, 1935. He wrote, "In the judgment of the most competent living mathematicians, Fräulein Noether was the most significant creative mathematical genius since the higher education of women began."

17. Weyl, *Scripta Mathematica*, *op. cit.*, p. 219.

18. The photograph is reproduced in Brewer and Smith, *op. cit.*, between pp. 17 and 18.

19. Noether, E., "Invarianten beliebiger Differentialausdrücke" and "Invariante Variationsprobleme," *Nachrichten von der Gesellschaft der Wissenschaften zu Göttingen*, 1918, pp. 37–44 and 235–57.

20. Strictly speaking, Noether's theorem applies only to continuous symmetries, such as those for space and time, and not discrete symmetries, such as the similarity between an object and its mirror image.

21. The argument is most fully developed in Weyl, H., *Raum, Zeit, und Materie*, 3rd ed. (Berlin: Springer Verlag, 1920).

22. The German terms are *Masstab invarianz* and *Eich invarianz*, respectively.

23. Weyl's most full description of gauge invariance was Weyl, H., *The Theory of Groups and Quantum Mechanics* (London: Methuen, 1931).

24. Weyl, H., *Mathematische Zeitschrift* 2, nos. 3–4 (30 October 1918): 384; Weyl, H., *AdP*, series 4, 59, no. 10 (20 June 1919): 101. We have not seen a somewhat earlier paper in the *Sitzungsberichte der Koniglichen Preussen Akademie der Wissenschaften* (1918, p. 465).

25. The history of this incident is covered in more detail in *SIL*, p. 341; Yang, C. N., *Annals of the New York Academy of Sciences* 294, no. 1 (8 November 1977): 86. Einstein was exactly right. Indeed, if you take a clock around a room, its wave function phase *does* change (Yang, C.N., *Proceedings of the International Symposium on the Foundations of Quantum Mechanics* [Physical Society of Japan, 1984]: 5).

26. London, F., *Naturwissenschaften* 15, no. 8 (15 February 1927): 187; London, F., *ZfP* 42, nos. 5–6 (14 April 1927): 375. Gauge invariance was independently rediscovered in Klein, O., *ZfP* 37, no. 12 (10 July 1926): 895; Fock, V., *ZfP* 39, nos. 2–3 (2 October 1926): 226. See also Gordon, W., *ZfP* 40 (1927): 119.

27. Weyl, H., *ZfP* 56, nos. 5–6 (19 July 1929): 330.

28. Among the exceptions to this generalization is Peierls, R., *PRSA* 146, no. 856 (1 August 1934): 420.

29. Yang's recollections of Fermi and his graduate school days can be found in Segrè, E., ed., *The Collected Papers of Enrico Fermi*, vol. 2 (Chicago: University of Chicago Press, 1965): 673. It is reprinted in Yang, C. N., *Selected Papers, 1945–1980* (San Francisco: W. H. Freeman, 1983): 305.

30. Telephone interview, Chen Ning Yang, 3 April 1983.

31. Telephone interview, Robert Mills, 7 April 1983. According to Mills's modest assessment, his main contribution was to suggest means of handling the self-interactions of the gauge field quanta.

32. Yang, C. N., and Mills, R. L., *PR* 96, no. 1 (1 October 1954): 191–92, 195. We have slightly amended the notation by ignoring the difference between the terms for the field and four-vectors. Moreover, we have replaced "$\pm e$" by "± 1" for reader convenience.

33. We are grateful to Dr. Abraham Pais for allowing us to see some of the manuscript of his *Inward Bound* (Oxford University Press, 1986), which documents this point.

34. Yang, C. N., *Selected Papers 1945–1980, With Commentary* (San Francisco: W. H. Freeman, 1983): 20.

35. Telephone interview, Chen Ning Yang, 3 April 1983.

CHAPTER 11

1. Goldhaber, M., and Goldhaber, G., *PR* 73, no. 12 (15 June 1948): 1472.

2. Chadwick, J., *Verhandlungen der Deutschen Physikalische Gesellschaft* 16 (1914): 383.

3. For example, Ellis, C. D., *PRSA* 99, no. 698 (1 June 1921): 261; Ellis, C. D., *PRSA* 101, no. 708 (1 April 1922): 1.

4. Ellis, C. D., and Wooster, W. A., *PRSA* 117, no. 776 (1 December 1927): 109.

5. Meitner, L., and Orthmann, W., *ZfP* 60, no. 3–4 (14 February 1930): 143.

6. Among the places Bohr urged this are, Bohr, N., *Journal of the Chemical Society* (1932): 382–88; Bohr, N., in *Convegno di Fisica Nucleare della Reale Accademia d'Italia* (Rome: Reale Academia, 1932): 119.

7. Letter, Wolfgang Pauli to Hans Geiger, Lise Meitner, et al., 4 December 1930, *LET*.

8. Interview, V. F. Weisskopf, *AHQP*, 10 July 1963, p. 12.

9. See citations below.

10. We owe this metaphor to Steven Weinberg, who attributes it to George Gamow.

11. Interview, Maurice Goldhaber, his office, Brookhaven National Laboratory, 21 December 1983.

12. This scene has been described by two of the three listeners: Amaldi, E., *CI*, 261ff; Segrè, E., *Enrico Fermi, Physicist* (Chicago: University of Chicago Press, 1970): 72.

13. Rasetti, F., in Fermi, E., *Collected Papers*, vol. 1 (Italy 1921–38) (Chicago: University of Chicago Press, 1962): 539.

14. The first note is E. Fermi, *Ricerca Scientifica* no. 4 (December 1933): 491. The follow-up papers are *NC* 11, no. 1 (January 1934): 1; and *ZfP* 88, no. 3/4 (19 March 1934): 161.

15. Letter, Ernest Rutherford to Enrico Fermi, 23 April 1934, quoted in Fermi, *Collected Papers*, vol. 1, *op cit.*, p. 641.

16. Fermi, *NC*, *op. cit.*, sec. 9.

17. Gamov, G., and Teller, E., *PR* 49, no. 12 (15 June 1936): 895.

18. Fermi, *NC*, *op. cit.*, sec. 10. The curve actually used by later experimenters, however, was more complicated. It was first described in Kurie, N. D., Richardson, J. R., and Paxton, H. C., *PR* 49, no. 5 (1 March 1936): 368.

19. Ellis, C. D., and Henderson, W. J., *PRSA* 146, no. 856 (1 August 1934): 213ff (phosphorus); Alichanow, A. I., Alichanian, A. I., and Dzelepow, B. S., *ZfP* 93, no. 5/6 (19 January 1935): 350.

20. Konopinski, E. J., and Uhlenbeck, G. E., *PR* 48, no. 1 (1 July 1935): 7; Part 2, *ibid.*, p. 60, no. 4 (15 August 1941): 308.

21. A remarkably straightforward review article of developments until 1943 is Konopinski, E. J., *RMP* 15, no. 4 (October 1943): 209. Particularly illustrating the erroneous results produced by thick sources is Tyler, A. W., *PR* 56, no. 2 (15 July 1939): 125.

22. Wu, C. S., and Albert, R. D., PR 75, no. 2 (15 January 1949): 315. See also Wu, *RMP* 22, no. 4 (October 1950): 386.

23. The attempts are reviewed in Crane, H. R., *RMP* 20, no. 1 (January 1948): 278.

24. Bethe, H., and Peierls, R., *Nature* 133, no. 3366 (5 May 1934): 689.

25. See, for instance, Dancoff, S., *Bulletin of the Atomic Scientists* 8, no. 5, June 1952, p. 139.

26. Reines, F., *CI*, p. 238.

27. Letter, Enrico Fermi to Frederick Reines, 8 October 1952, quoted in *ibid.*, p. 241.

28. Reines, F., and Cowan, C. L., Jr., *PR* 92, no. 3 (1 November 1953): 830.

29. The telegram is reproduced in Reines, *CI*, *op. cit.*, p. 249. The discovery is reported in Cowan, Reines, et al., *Science* 124, no. 3212, (20 July 1956): 103.

30. Danby, G., et al., *PRL* 9, no. 1 (1 July 1961): 36.

31. Reines, F., Sobel, H., and Gurr, H. *PRL* 37, no. 6 (9 August 1976): 315.

32. Pontecorvo, B., *PR* 72, no. 3 (August 1947): 246.

33. Lee, T. D., Rosenbluth, M., and Yang, C. N., *PR* 75, no. 5 (1 March 1949); 905; Klein, O., *Nature* 161, no. 4101 (5 June 1948): 897; Puppi, G., *NC* 5, no. 6 (1 December 1948): 587; *ibid.*, 6, no. 3 (3 May 1949): 194; Tiomno, J., and Wheeler, J. A., *RMP* 21, no. 1 (January 1949): 153.

34. Dalitz, R. H., *PR* 94, no. 4 (15 May 1954): 1046. The history of parity is evoked in Yang, C.N., *CI*, p. 439.

35. Discussion comment, Martin Block, Weak Interaction Conference, Wingspread, Wisconsin, 30 May 1984. Separately confirmed in an interview the same day.

36. Yang, C. N., in Ballam, et al., eds., *High Energy Nuclear Physics*, Proceedings of the Sixth Annual Rochester Conference, April 3–7, 1956 (New York: Interscience, 1956): VIII–1.

37. *Ibid.*, VIII–22.

38. *Ibid.*, VIII–27; a misprint has been corrected. It is worth noting that Martin Block recalls Yang's reaction as being considerably more negative (Interview, Martin Block, Weak Interaction Conference, Wingspread, Wisconsin, 30 May 1984).

39. Quoted in Gardner, M., *The Ambidextrous Universe* (New York: Basic Books, 1964): 240.

40. This account is primarily based on Lee, T. D., in Zichichi, A., ed., *Elementary Processes at High Energy*, Proceedings of the 1970 Majorana School (New York: Academic Press, 1971): 830; Lee, T. D., "Broken Parity," in Lee, T. D., *Collected Works*, forthcoming; Yang, C. N., in Yang, C. N., *Selected Papers*, 1945–1980 with commentary (San Francisco: W. H. Freeman, 1983): 26–31; Wu, C. S., *Adventures in Experimental Physics*, Vol. Gamma, p. 101; Franklin, A., *Studies in the History and Philosophy of Science* 10, no. 3 (Fall 1979); 201.

41. Siegbahn, K., ed., *Beta- and Gamma-Ray Spectroscopy* (Amsterdam: North-Holland, 1955).

42. Lee, T. D., and Yang, C. N., *PR* 104, no. 1 (1 October 1956): 254–55. The order of the quotes has been changed.

43. Dyson, F., *Scientific American* 199 (September 1958): 74.

44. Interview, Gerald Feinberg, his office, Columbia University, 1 March 1985.

45. Wu, C. S., Ambler, E., Hayward, R. W., Hoppes, D. D., and Hudson, R. P., *PR* 105, no. 4 (15 February 1957): 1413.

46. Garwin, R. L., Lederman, L. M., Weinrich, M., *PR* 105, no. 4 (15 February 1957): 1415. See also, Weinrich, M., Ph.D. thesis, Columbia University, 1958, issued as Nevis Report NEVIS–56 (February 1958); Garwin, R., *Adventures in Experimental Physics*, Vol. Gamma, p. 124.

47. Friedman, J. I. and Telegdi, V. L., *PR* 106, no. 5 (1 March 1957): 1290.

48. Letter, Wolfgang Pauli to Victor Weisskopf, 17 January 1957, quoted in Yang, *Selected Papers, op. cit.*, p. 30.

49. Letter, Wolfgang Pauli to Victor Weisskopf, 27 January 1957, in Kronig, R. d. L., and Weisskopf, V. F., eds., *Collected Scientific Papers of Wolfgang Pauli*, vol. 1 (New York: Wiley-Interscience, 1964): xvii–xviii.

50. Yang, C. N., *Selected Papers. op. cit.*, p. 35; Bernstein, J., *New Yorker*, 12 May 1962, p. 58–59.

51. In the volume of *Adventures in Experimental Physics* devoted to the discovery of parity nonconservation, Goudsmit replied to the criticism, offering reasons to support his insistence that the original be rewritten (*op. cit.*, p. 137). All review papers on the subject, however, have treated all three experiments on an equal basis.

52. The history is covered in Franklin, op. cit.; Cox, R., *Adventures in Experimental Physics*, vol. Gamma, 1973, p. 145; and Grodzins, L., *ibid.*, p. 154. Physicists who raised the general question of parity before Yang and Lee include Purcell, E. M. and Ramsey, N. F., *PR* 78, no. 66 (15 June 1950): 807; Wick, G. C., Wightman, A. D., and Wigner, E. P., *PR* 88, no. 1 (1 October 1952): 101.

53. Interview, Martin Block, Weak Interaction Conference, Wingspread, Wisconsin, 30 May 1984.

54. E.g., letter, Victor Weisskopf to J. R. Oppenheimer, 3 February 1964, in Oppenheimer Papers, Library of Congress, General Case File, Box 20, Weisskopf file; and the historical accounts by Lee and Yang cited above.

55. Lee, T. D., in "High Energy Nuclear Physics," *Proceedings of the Seventh Annual Rochester Conference*, 15–19 April 1957, VII–1, VII–7.

56. The neutron beta decay experiment was Robson, J. M., *PR* 100, no. 3 (1 November 1955): 933; the neon experiment was Maxson, D. R., Allen, J. S., and Jentschke, W. K., *PR* 97, no. 1 (1 January 1955): 109; the helium was Rustad, B. M., and Ruby, S. L., *PR* 89, no. 4 (15 February 1953): 880; *ibid.*, 97, no. 4 (15 February 1955): 991.

57. Wu, C. S., in *Proceedings of the Seventh Annual Rochester Conference, op. cit.*, VII–20.

58. Sudarshan, E.C.G. and Marshak, R. E., "Origin of the Universal V-A Theory," Virginia Tech preprint, VPI-HEP 84/8, pp. 7–10.

59. Sudarshan, E. C. G. and Marshak, R. E., *Proceedings of the Padua Conference on Mesons and Recently Discovered Particles* (1957): V–14.

60. Telephone interview, Robert Marshak, 14 January 1985.

61. Gell-Mann, M., Caltech preprint CALT–68–1214, pp. 15–16.

62. Feynman, R. P., *"Surely You're Joking, Mr. Feynman!"* (New York: W. W. Norton, 1985): 250–51. The paper was Feynman, R. P., and Gell-Mann, M., *PR* 109, no. 1 (January 1958): 193. See also the later Sakurai, J. J., *NC* 7, no. 5 (1 March 1958); and Sudarshan, E. C. G., and Marshak, R. E., *PR* 109, no. 5 (1 March 1958): 1860.

63. See, for instance, the experiments referred to in Sudarshan and Marshak, *ibid.*, pp. 1860–62.

64. Goldhaber, M., Grodzins, L., and Sunyar, A., *PR* 109, no. 3 (1 February 1958): 1015.

65. Telephone interview, Murray Gell-Mann, 18 May 1984. He was not alone. Cf., Lee, T. D., in Zichichi, *Elementary Processes, op. cit.*, pp. 837-38.

CHAPTER 12

1. Schwinger's impression on his students—Conversation, Herb Goldstein, 3 November 1983; interview, I. I. Rabi, his apartment, New York City, 9 May 1984; interview, Sheldon Glashow, his office, Harvard University, 2 December 1982.

2. Martin, P. C., *Physica*, 96A, nos. 1/2 (April 1979, Schwinger *Festschrift* issue): 70.

3. Interview, Julian Schwinger, a restaurant in Los Angeles, 4 March 1983.

4. Quoted in Bernstein, J., *New Yorker*, 12 May 1962, p. 82.

5. Schwinger, J., *Annals of Physics* 2, no. 5 (November 1957). See also, Schwinger, J., "Théorie des Particules Elementaires," mimeograph, June 1957 Paris lectures, Glashow's files.

6. The scientist asked for anonymity.

7. Schwinger's predecessors apparently include the Japanese physicists Shoichi Sakata and Takeshi Inoue in 1942 and Mituo Taketani, Seitaro Nakamura, and two others after the war (according to Brown, L., et al., in Brown, Konuma, and Maki, eds., *Particle Physics in Japan, 1930–1950*, vol. 1 (Kyoto: Research Institute for Fundamental Physics, September 1980): 58–60.

8. Schwinger, *Annals of Physics, op. cit.*, p. 425.

9. Flato, M., Fronsdal, C., and Milton, K. A., eds., *Selected Papers (1937–1976) of Julian Schwinger* (Boston: D. Reidel, 1979): 82.

10. His attendance is recorded in "High Energy Nuclear Physics," *Proceedings of the Seventh Annual Rochester Conference*, 15–19 April 1957 (New York: Interscience, 1957).

11. Interview, Sheldon Glashow, his office, Harvard University, 2 December 1982. The discussion here has been slightly condensed with Professor Glashow's kind permission.

12. Biographical information and all quotes from interview, Sheldon Glashow, his office, Harvard University, 2 December 1982.

13. The description of Bronx Science is based on telephone interviews with Charles Hellman, 13 February 1983; Herman Gewirtz, 13 February 1983; and Daniel Greenberger, 16 February 1983, in addition to the interviews cited below.

14. Interview, Gerald Feinberg, a restaurant in New York, 27 January 1983 (this and subsequent quotations).

15. We are grateful to the librarians of the Bronx High School of Science for showing us the yearbooks and course descriptions.

16. Interview, Sheldon Glashow, his office, Harvard University, 2 December 1982.

17. Telephone interview, Sheldon Glashow, 8 April 1983.

18. Telephone interview, Silvan Schweber, 2 January 1984.

19. Telephone interview, Sheldon Glashow, 14 February 1983.

20. Letter, Sheldon Glashow to Gary Feinberg, 16 November 1954, Feinberg's files.

21. Telephone interview, Sheldon Glashow, 8 April 1983.

22. Interview, Sheldon Glashow, his office, Harvard University, 2 December 1982.

23. Interview, Sheldon Glashow, his office, Harvard University, 25 April 1983. The thesis quote is from Glashow, S. L., "The Vector Meson in Elementary Particle Decays," Harvard University Ph.D. Thesis, July 1958 (typescript), p. 75.

24. Glashow, S., *NP* 10, no. 1 (1 January 1959): 107.

25. Interview, Sheldon Glashow, his office, Harvard University, 2 December 1982.

26. Interview, Abdus Salam, his office, International Center for Theoretical Physics (*ICTP*), Trieste, 23 February 1984.

27. Salam, A., *NP* 18, no. 4 (September 1960): 681; Salam, A., and Komar, A., *NP* 21, no. 4 (December 1960): 624; Salam, A., *PR* 127, no. 1 (1 July 1962): 331.

28. Interview, Sheldon Glashow, his office, Harvard University, 2 December 1982.

29. Interview, Murray Gell-Mann, aboard a plane, 3 March 1983.

30. Glashow, S., *NP* 22, no. 4 (February 1961): 579.

31. *Ibid.*, pp. 579, 583, 584.

32. After mixing with the photon, Glashow's Z_s is known as the B, which is today's Z^0.

33. Glashow used (*NP* 22, op. cit., 585) $1/\sin\theta$, whereas Weinberg used $\sin\theta$.

34. Gell-Mann, M., in *Proceedings of the 1960 Annual International Conference on High Energy Physics at Rochester*, 25 August–1 September 1960 (New York: Interscience, 1960): 510.

35. Telephone interviews, Murray Gell-Mann, 18 May 1984 and 13 July 1983.

36. Glashow, *NP* 22, op. cit., p. 586.

37. Glashow, S., and Gell–Mann, M., *Annals of Physics* 15, no. 3 (September 1961): 437.

38. This and the following quote from interview, Murray Gell–Mann, aboard a plane, 3 March 1983.

39. Interview, Abdus Salam, a hotel near Fermilab, 3 May 1985.

CHAPTER 13

1. Salam's childhood is described in an interview with the *Illustrated Weekly of India*, 1 February 1981, p. 10. See also Salam, S., *Ideals and Realities* (Singapore: World Scientific, 1984).

2. Interview, Abdus Salam, an office at CERN, 19 September 1983.

3. Interview, Abdus Salam, his office, ICTP, 23 February 1984.

4. Salam, A., *PR* 82, no. 2 (15 April 1951): 217; Matthews, P. T., and Salam, A., *RMP* 23, no. 4 (October 1951): 311.

5. Interview, Abdus Salam, *Illustrated Weekly of India*, 1 February 1981, p. 12.

6. Shaw, R., "The Problem of Particle Types and Other Contributions to the Theory of Elementary Particles," Cambridge University Ph.D. thesis, unpublished, 1954.

7. Salam, A., *NC* 5, no. 1 (1 January 1957): 299. Salam showed that, just as the photon's masslessness is a manifestation of gauge symmetry, parity violation is a manifestation of chiral symmetry.

8. Letter, Wolfgang Pauli to Abdus Salam, 11 March 1957, courtesy of Professor Salam.

9. Interview, Abdus Salam, his office, ICTP, 23 February 1984.

10. Discussion comment, Abdus Salam, Weak Interaction Conference, Wingspread, Wisconsin, 30 May 1983. The manuscript is Salam, A., "On Fermi Interactions," unpublished ms., courtesy Professor Salam.

11. Salam, A., and Ward, J. C., *NC* 9, no. 4 (16 February 1959).

12. Interview, Abdus Salam, his office, ICTP, 23 February 1984. The paper is Salam, A., and Ward, J. C., *PL* 13, no. 2 (15 November 1964): 168. Of some interest is Salam and Ward, *NC* 19, no. 1 (1 January 1961): 165.

13. Interview, Abdus Salam, New York Hilton, 27 May 1984.

14. Salam, A., "On Fermi Interactions," unpublished ms., courtesy Professor Salam.

15. Interview, Abdus Salam, an office at CERN, 19 September 1984.

16. Superconductivity was first discovered by Heike Kamerlingh Onnes, a Dutch physicist who was fascinated by the behavior of matter at low temperatures, and the electrical resistance of metals. Onnes was the first to liquefy many gases, such as helium, which he then used as a coolant to study low–temperature electrical resistance. In 1911, Onnes was surprised to discover that somewhere near absolute zero the resistance of mercury abruptly dropped so low that he could not measure it—the first evidence of the phenomenon later called superconductivity. In 1913 he was awarded the Nobel Prize for his low temperature studies. Onnes was fanatic about precision; he once said that every physics laboratory should have the motto: "Door meten tot weten" ("Through measuring to knowing"). Onnes entry, *DSB*.

17. Bardeen, J., Cooper, L. N., and Schrieffer, J. R., *PR* 106, no. 1 (1 April 1956) 162; *Ibid.*, 108, no. 5 (1 December 1957): 1175. Equivalent work was done independently by Bogoliubov, N., *JETP* 34, no. 1 (July 1958): 41, and following articles.

18. Among the articles which contributed to this realization are Anderson, P. W., *PR* 112, no. 6 (15 December 1958): 1900; Nambu, Y., *PR* 117, no. 3 (1 February 1960): 648. Nambu's paper contains a more complete list of references in note 3.

19. Telephone interview, Jeffrey Goldstone, 23 January 1984.

20. Goldstone, J., *NC* 19, no. 1 (1 January 1961): 154.

21. Interview, Jeffrey Goldstone, an office at Columbia, 26 November 1984.

22. Interview, Steven Weinberg, his office, University of Texas, 28 November 1984; Nambu, Y., and Jona-Lasinio, G., *PR* 122, no. 1 (1 April 1961): 345. See also, Nambu, Y., *PRL* 4, no. 7 (1 April 1960): 380. Here we give rather short shrift to Nambu; his important work will be discussed further in the following section on strong interactions.

23. Interview, Steven Weinberg, outside Harvard University, 7 May 1985.

24. Interview, Abdus Salam, his office, ICTP, 23 February 1984.

25. Interview, Jeffrey Goldstone, an office at Columbia, 26 November 1984.

26. Goldstone, J., Salam, A., and Weinberg, S., *PR* 127, no. 3 (1 August 1962): 965–70. This paper related the Goldstone boson to relativistic effects, a subject to be discussed below.

27. Weinberg, S., *RMP* 52, no. 3 (July 1980): 516.

28. Interview, Sheldon Glashow, his office, Harvard University, 10 November 1982.

29. Schwinger, J., *PR* 125, no. 1 (1 January 1962): 397.

30. Anderson, P. W., *PR* 130, no. 1 (1 April 1963): 439.

31. *Ibid.*, 441, 442.

32. Telephone interview, Philip Anderson, 28 January 1985.

33. Klein, A. and Lee, B. W., *PRL* 12, no. 10 (9 March 1964): 266–68; Gilbert, W., *PRL* 12, no. 25 (22 June 1964): 713–14. The article is the last Gilbert—who was Salam's graduate student at Cambridge University in England—wrote before switching to biology; he was honored for his biology work with a Nobel Prize in 1980.

34. Priority should be given to the Belgians, if given to anyone (Englert, F., and Brout, R., *PRL* 13, no. 9 [31 August 1964]: 321). They did not show the "eating" mechanisms explicitly, however.

35. Interview, Jeffrey Goldstone, an office at Columbia, 26 November 1984; letter, R. Brout to the authors, 22 October 1984; Higgs, P., "SBGT and all that," Weak Interaction Conference, Racine, Wisconsin, 30 May 1984. Higgs' papers are: Higgs, P. W., *PRL* 13, no. 16 (19 October 1964): 508: Higgs, P. W., *PR* 145, no. 4 (27 May 1966): 1156–63. Brout's quote below is from letter, R. Brout to authors, 13 September 1985.

36. Interview, Steven Weinberg, outside Harvard, 7 May 1985.

37. Interview, Steven Weinberg, his office, University of Texas, 28 November 1984.

38. Telephone interview, Steven Weinberg, 30 September 1985.

39. Weinberg, S., *PRL* 19, no. 21 (20 November 1967): 1264.

40. *Ibid.*, p. 1266. We have replaced "meson" by "boson" and dropped subscripts.

41. Guralnick, G. S., Hagen, C. R., and Kibble, T. W. B., *PRL* 13, no. 20 (16 November 1964): 585; Kibble, T. W. B., *Oxford Conference on Elementary Particles, 19–25 October 1965* (Oxford: Rutherford High Energy Laboratory): 19; Kibble, T. W. B., *PR* 155, no. 5 (25 March 1967): 1554.

42. Interview, Abdus Salam, his office, ICTP, 23 February 1984.

43. Salam, A., in "Elementary Particle Theory," *Proceedings of the Eighth Nobel Symposium, 19–25 May 1968* (New York: John Wiley, 1968): 367.

44. Gell–Mann's conference summary does not even mention the idea and mentions Salam only once in passing (Gell–Mann, M., in *Proceedings of the Eighth Nobel Symposium, op. cit.*, p. 395).

45. Interview, Thomas Kibble, his office, Imperial College, 5 October 1984.

46. Coleman, S., *Science*, 206, 14 December 1979, p. 1290; Koester, D., Sullivan, D., and White, D., *Social Studies of Science* 12, no. 1 (February 1982): 73. Weinberg worked a little on trying to renormalize $SU(2) \times U(1)$, but got nowhere (Stuller, R.L., "Are Symmetry Broken Models of Weak Interactions Renormalizable?" Ph.D. thesis, MIT, February 1971, typescript).

47. Weinberg, S., *Coral Gables Conference on Fundamental Interactions at High Energy* (3rd) (New York: Gordon and Breach, 1971): 182.

48. Interview, Steven Weinberg, faculty club, University of Texas, 28 November 1984.

CHAPTER 14

1. We have been greatly assisted in writing chapters 14–18 by the history of particle physics by Pickering, A., *Constructing Quarks* (Chicago: University of Chicago Press, 1984). For information about early English laboratories, see Sviedras, R., *HSPS* 7 (1976): 405–36. See also *ibid.*, 2 (1970): 127–45.

2. Glashow, *RMP* 52, no. 3, p. 1319.

3. Letter, Leland Hayworth to T. H. Johnson, 9 September 1953, courtesy Brookhaven National Laboratory. Samios is paraphrasing part of p. 1.

4. Wideröe, R., *Wissenschaftliche Zeitschrift der Friedrich-Schiller-Universität Jena* 13, no. 4 (Winter 1964): 431; Wideröe, R., *Archiv für Elektrodynamik* 21, no. 4 (1928): 400; Wideröe, R., "Some Memories and Dreams from the Childhood of Particle Accelerators," unpublished ms., January 1983, AIP Archives, 2. He subsequently built a linear machine.

5. Lawrence, E. O., and Edlefson, N. E., *Science* 72 (10 October 1930): 376. Livingston, M., *PT* 12, no. 10 (October 1959): 18; for a description of Lawrence, see Childs, H., *An American Genius: The Life of Ernest Lawrence* (New York: E. P. Dutton, 1968); Davis, N. P., *Lawrence and Oppenheimer* (New York: Simon and Schuster 1968).

6. Rutherford's attitude from interview, Norman Feather, *AHQP*, 25 February 1971, p. 17. Results of accelerator, Cockroft, J., and Walton, E. T. S., *PRSA* 136, no. 830 (1 June 1932): 619; ibid. 137, no. 831 (1 July 1932): 229.

7. Livingston, *op. cit.*, p. 21.

8. The precise definition is the energy acquired when an electron is accelerated through a

potential difference of one volt. A kilocalorie, which is the unit typically used, is roughly equivalent to 10^{23} electron volts.

9. Wilson, R. R. in Marshak, R., ed., *Perspectives in Modern Physics* (New York: Wiley, 1966); McMillan, E. M., *PT* 12, no. 10 (October 1959): 24; a vivid picture of this era is in Hilts, P., *Scientific Temperaments* (New York: Simon and Schuster, 1982): 46–55.

10. Interestingly, Lawrence's original plan for the 100 MeV machine would not have worked because he didn't take into account the relativistic effects of accelerating particles to such high energies.

11. Some of this process is described in Needell, A. A., *HSPS* 14, no. 1 (Winter 1983): 99.

12. Talk by Robert Seidel at the International Symposium on Particle Physics in the 1950s, Fermilab, 3 May 1985 (proceedings forthcoming).

13. Roos, M., *RMP* 35, no. 1, (April 1963): 314; the errors were pointed out by Zweig, G., in Isgur, N., *Baryon 1980*, Proceedings of Fourth International Conference on Baryon Resonances (Toronto: University of Toronto, 1980): 454–55.

14. Barkas, W. H., and Rosenfeld, A. H., Lawrence Berkeley Laboratory publication UCRL–8030, 1958; Particle Data Group, *RMP* 56, no. 2 (April 1984).

15. Stueckelberg, E. C. G., *HPA* 11 (1938), 317.

16. Fermi, E., and Yang, C. N., *PR* 76, no. 12 (15 December 1949): 1739. It is worth recalling that the antiproton had not yet been discovered, although its existence was widely assumed.

17. Examples are Goldhaber, M., *PR* 92, no. 5 (December 1953): 1279: *ibid.*, 101, no. 1 (1 January 1956): 433; Miyazawa, H., *PTP* 6, no. 4 (July–August 1951): 631.

18. Engels, F., *Dialectics of Nature*, trans. C. Dutt (New York: International Publishers, 1979). Engels wrote another book on natural science which was published during his lifetime: *Herr Eugen Dühring's Revolution in Science* (now known as *Anti-Dühring*) (New York: International Publishers, 1939). V. I. Lenin also wrote an influential work on natural science, *Materialism and Empirio Criticism* (New York: International Publishers, 1927). For a typical contemporary exposition of Marxist natural philosophy and its connection to Marxist social philosophy see Sheptulin, A. P., *Marxist–Leninist Philosophy* (Moscow: Progress Publishers, 1978).

19. Sakata's recollections are in two essays in Hara, O., et al., eds., *Shoichi Sakata: Scientific Works* (Tokyo: Horei Printing, 1977). Quotations from pp. 393 and 370, respectively.

20. Sakata, S., *PTP* 16, no. 6 (December 1956): 686.

21. Ikeda, M., Ogawa, S., and Ohnuki, Y., *PTP* 22, no. 5 (November 1959): 715; Yamaguchi, Y., *PTP* (supp.) 11 (1959): 1; *ibid.*, p. 37.

22. Pevsner, A., et al., *PRL* 7, no. 11 (1 December 1961): 421. As a sign of the lack of favor met by Sakata's school, the paper contains no mention of the prediction.

23. Sakata, S., in Hara et al., *op. cit.*, pp. 360–61. The paper has an epigram from *Dialectics of Nature*: "Natural scientists may adopt whatever attitude they please, they are still under the domination of philosophy."

24. Gell–Mann, M., Caltech report CALT–68–1214, n.d., p. 3.

25. The story is described in an interview with Gell–Mann by Schultz, R., *Omni*, May 1985, p. 54.

26. Interview, Murray Gell–Mann, aboard a plane, 3 March 1983.

27. Gell–Mann, M., *PR* 106, no. 6 (15 June 1957): 1296. Also important here is Sakurai, J. J., *Annals of Physics* 11, no. 1 (September 1960): 1.

28. Interview, Murray Gell–Mann, his office, Caltech, 21 February 1985.

29. Gell–Mann, M., Caltech report CTSL–20 (unpublished), 15 March 1961, p. 25.

30. Gell–Mann, Caltech report CALT–68–1214, n.d., pp. 22–23.

31. Interview, Murray Gell–Mann, his office, Caltech, 21 February 1985. The last three sentences of the quotation come from a telephone interview on the same subject on 24 April 1985.

32. More exactly, the first two generators are the third axis of isotopic spin, which indicates the particle's electric charge, and "hypercharge," which is the strangeness plus the baryon number. The remaining six are all associated with special vector bosons that change the values of isotopic spin and hypercharge. SU(3) is a "special" unitary group because two of the members are identical; an ordinary unitary group would have different members. A simple and clear exposition of the work can be found in Chew, G. F., Gell–Mann, M., and Rosenfeld, A. H., *Scientific American* 210, no. 2 (February 1964): 74.

33. Gell–Mann, M., Caltech report CTSL–20, *op. cit.*; reprinted in Gell–Mann, M., and Ne'eman, Y., eds., *The Eightfold Way* (New York: W. A. Benjamin, 1964). Originally written as a

forty-six–page report in January 1961, the eightfold way was first published formally as a three–page section at the end of a much longer paper in *Physical Review*: Gell–Mann, M., *PR* 125, no. 3 (1 February 1962): 1067.

34. Ne'eman, Y., *NP* 26, no. 2. (July 1961): 222; interview, Yuval Ne'eman, Fermilab cafeteria, 3 May 1985 (all subsequent quotes); letter, Yuval Ne'eman to authors, 6 August 1985; interview, Abdus Salam, ICTP, 23 February 1984. Some of this story is available in Ne'eman, Y., *Proceedings of the Israel Academy of Sciences and Humanities*, Section of Sciences, no. 21 (Jerusalem: 1983), and Ne'eman, Y., in *Symmetries in Physics, First International Symposium on the History of Scientific Ideas*, Universitat Autonomia de Barcelona, Catalonia, Spain, September 1983, forthcoming.

35. When Gell–Mann first put together SU(3), he visited Berkeley, where some experimenter friends horrified him by presenting him with a coffee cup that had SIGMA-LAMBDA-ODD written around its circumference. They had found evidence that the sigma and lambda did not have the same parity, and hence were not in the same multiplet. Gell–Mann sweated for months before deciding the rumors must be wrong. Eventually he was proven right. (Tripp, R. D., Watson, M. B., and Ferro-Luzzi, M., *PRL* 8, no. 4 [15 February 1962]: 175.)

36. Prentki, J., ed., *Proceedings of the 1962 International Conference on High Energy Physics at CERN* (Geneva: CERN, 1962): 795–805.

37. Within SU(3) systematics, the pyramid-shaped decimet was not the only possible place for the deltas, xi-stars and sigma-stars. Another scheme, with twenty-seven particles, started with an upper layer of strangeness plus-one particles, above the deltas. Such particles could be produced by scattering K + or K⁰ mesons (with one positive unit of strangeness) on protons or neutrons. The husband-wife team of Gerson and Sulamith Goldhaber, working at Berkeley, had just reported negative results from such an experiment. For Ne'eman, this was conclusive. It pointed to the decimet as the unique SU(3) fitting solution. (Letter, Yuval Ne'eman to authors, *op. cit.*)

38. This story compiled from interviews: Gerson Goldhaber, a room at Fermilab, 2 May 1985; Yuval Ne'eman, a hotel near Fermilab, 3 May 1985; Murray Gell–Mann, telephone interview, 5 February 1985. Gell–Mann's prediction is Gell–Mann, M., comment in Prentki, *op. cit.*, p. 805.

39. Interview, Nicholas Samios, his office, Brookhaven, 12 February 1985 (this and following quotes).

40. Interview, Robert Palmer, an office at Brookhaven, 18 May 1983.

41. Gaston, J., *Originality and Competition in Science* (Chicago: University of Chicago Press, 1973): 83–88.

42. Ne'eman, *Symmetries, op. cit.*, p. 42. Later, when the particle was found, Samios forgot to call Ne'eman. He later sent some photographs and a note reading, "Please excuse the oversight, but you knew it existed before we did!"

43. Barnes, V. E., et al., *PRL* 12, no. 8 (24 February 1964): 204; Fowler, W. B. and Samios, N. P., *Scientific American* 211, no. 4 (October 1964): 36.

CHAPTER 15

1. Telephone interview, Robert Serber, 4 June 1983.

2. Interview, Murray Gell–Mann, aboard a plane, 3 March 1983 (all quotes in this section).

3. Gell–Mann, M., *PL* 8, no. 3 (1 February 1964): 214.

4. Well, actually, not so easy. When Gell–Mann assigned strangeness values to particles back in 1953, he knew nothing about strange quarks. As a consequence, he awarded strangeness +1 to mesons with strange *antiquarks*, which means that a particle with strangeness −1 has a strange quark inside it. The principle of one unit of strangeness per strange quark is correct; only the signs are reversed.

5. Telephone interview, Sheldon Glashow, 12 May 1983.

6. Gell–Mann, M., Caltech report CALT–68–1214, p. 30 (unpublished).

7. Interview, Jacques Prentki, his office, CERN, 7 February 1984.

8. Interview, Yuval Ne'eman, a room at Fermilab, 3 May 1985.

9. Ne'eman, *Proceedings of the Israel Academy of Sciences and Humanities*, Sections of Sciences, no. 21 (Jerusalem: 1983): 7–8.

10. Goldberg, H., and Ne'eman, Y., *NC* 27, no. 1 (1 January 1963): 1.

11. Letter, Victor Weisskopf to J. R. Oppenheimer, 13 February 1964, in Oppenheimer Collection, Manuscript Division, Library of Congress, General Case File, Box 77, Weisskopf folder; Schwinger, J., *PR* 135B, no. 3 (10 August 1964): 817 (received 23 March).

12. Zweig, G., CERN preprint 8182/TH401, 17 January 1964 (unpublished); Zweig, G., CERN preprint 8419/TH412, 21 February 1964 (unpublished); some of it was eventually published as Zweig, G., in Zichichi, A., ed., *Symmetries in Elementary Particle Physics*, Proceedings of the "Ettore Majorana" International School of Physics, Erice, Italy, August 1964 (New York: Academic Press): 192.

13. Gell-Mann, M., Caltech report CALT-68-1214, p. 33.

14. Zweig, G., in Isgur, N., ed., *Baryon '80: Proceedings of the Fourth International Conference on Baryon Resonances*, 14-16 July 1980 (Toronto: University of Toronto, 1981): 457 (both quotes).

15. The stimuli were the $V - A$ theory of the weak interactions, which Gell-Mann and Feynman showed was compatible with an intermediate vector boson, and a demonstration by Nambu that the Hofstadter nucleon structure experiments entailed the existence of a spin–one boson (Nambu, Y., *PR* 106, no. 6 [15 June 1957]: 1366).

16. Feinberg, G., *PR* 110, no. 6 (15 June 1958): 1482. Japanese physicists had first raised the possibility of two types of neutrinos immediately after the war (cf., Brown, Konuma, and Maki, *Particle Physics in Japan* [Kyoto: Research Institute for Fundamental Physics] mimeograph, vol. 1, p. 59). Many Western physicists came up with the same idea independently in the next decade; including Feynman and Gell-Mann at the December 1957 meeting at Stanford of the American Physical Society. For some references, see footnotes 4 and 5 in Danby, G., et al., *PRL* 9, no. 1 (1 July 1962): 95. Feinberg's article was by far the most influential.

17. The opinion is held so commonly that it would be unkind to cite specific sources.

18. Interview, Mel Schwartz, his home, 19 February 1985 (all quotes); Schwartz, Mel, *PRL* 4, no. 6 (15 March 1960): 306; Lee, T. D., and Yang, C. N., *ibid.*, p. 307. The Italian–Soviet physicist Bruno Pontecorvo presented a slightly different, more impractical version of the idea in Pontecorvo, B., *Soviet Physics–JETP* 37, no. 6 (June 1960): 1751.

19. Lederman, L., Schwartz, M., and Gaillard, J.–M., in *Proceedings of the International Conference on Instrumentation*, Berkeley 1960 (New York: Wiley-Interscience, 1960): sec. V.1d. This paper and several others are included in Schwartz, M., *Adventures in Experimental Physics*, Vol. Alpha (1972): 82.

20. Lederman, L., "Neutrino Physics," Brookhaven Lecture Series, no. 23, BNL 787(T–300), (9 January 1963): 5–6; story confirmed in interview, Leon Lederman, his office, Fermilab, 14 February 1984.

21. Telephone interview, Murray Gell-Mann, 3 June 1983; interview, Sheldon Glashow, his office, Harvard University, 7 May 1985. See, for example, Glashow, S. L., and Coleman, S., *PRL* 6, no. 8 (15 April 1961): 423. This paper and others are reprinted in Gell-Mann, M., and Ne'eman, Y., *The Eightfold Way* (New York: W. A. Benjamin, 1964).

22. Telephone interview, James Bjorken, 10 May 1983; other fourth quark papers include Hara, Y., *PR* 134B, no. 3 (11 May 1964): 701; and Amati, D., Bacry, J. Nuyts, and J. Prentki, *NC* 34, no. 6 (16 December 1964): 1732, which contains a list of further references in notes 4 and 5.

23. Bjørken, B. J., and Glashow, S. L., *PL* 11, no. 3 (1 August 1964): 255.

24. Glashow, S., *Scientific American*, 233, no. 4 (October 1975): 47. The charmed quark suppresses a second–order process, something more extensively discussed in the following chapter.

25. Interview, Sheldon Glashow, his office, 2 December 1982.

26. Interview, Steven Weinberg, outside Harvard, 7 May 1985.

27. Weinberg, S., *PR*, sec. B., 138, no. 4 (24 May 1965): 990.

28. The chain of papers that created SU(6) began with Gursey, F., and Radicati, L., *PRL* 13, no. 5 (3 August 1964): 173; Pais, A., *PRL* 13, 5 (3 August 1964): 175; and Gursey, F., Pais, A., and Radicati, L., *PRL* 13, no. 8 (24 August 1964): 299. The defects of the model are well summarized in the introduction to Dyson, F., ed., *Symmetry Groups in Nuclear and Particle Physics: A Lecture Note and Reprint Volume* (New York: W. A. Benjamin, 1966).

29. Gell-Mann, M., *Physics* 1, no. 1 (July–August 1964): 63. *CQ*, p. 122, note 63, cites data that Regge pole work and quark ideas constituted about 75 percent of the theoretical papers in *Physics Letters* from 1967 to 1969.

30. Regge, T., *NC* 14, no. 5 (1 December 1959): 951. See also the exhaustive review in Collins, P. D. B., *PRpts* 1, no. 2 (January 1971): 103, and Chew, G., *The Analytic S–Matrix: A Basis for Nuclear Democracy* (New York: W. A. Benjamin, 1966). These ideas were not wrong; indeed, on the mass shell, where quarks are not seen, field theory does manifest itself as nuclear democracy and Regge poles. But the language of quarks was ultimately seen as having explanatory power over a greater range.

31. Biographical material from interviews, Yoichiro Nambu, Weak Interaction Conference, Racine, Wisconsin, 31 May 1984; his office, University of Chicago, 13 February 1985 (all quotes from latter).

32. Nambu, Y., *PR* 117, no. 3 (1 February 1960): 648. References to other workers are in notes 3 and 4. Curiously, one reason for the scorn given to the Heisenberg-Pauli unification (see chap. 20) was that it relied upon a "degenerate vacuum" similar to that posited by Nambu a few years later.

33. Greenberg, O. W., *PRL* 13, no. 20 (16 November 1964): 598. Another way out involved trying to award quarks radial quantum numbers, like electrons in an atom.

34. Nambu, Y., in Perlmutter, A., Kursunağlu, B., and Sakmar, I., eds., *Symmetry Principles at High Energy*, Second Coral Gables Conference, 20-22 January 1965 (San Francisco: W. H. Freeman, 1965): 274. Similar ideas were presented somewhat earlier by Bacry, H., Nuyts, and van Hove, L., *PL* 9, no. 3 (15 April 1964): 279.

35. Han, M. Y., Syracuse University preprint NYO-3399-21/1206—SU-21 (unpublished); Han, M. Y. and Nambu, Y., *PR* 139B, no. 4 (23 August 1965): 1006.

36. The term *color* was first suggested by Lichtenberg, D. B. (*Unitary Symmetry and Elementary Particles*, first ed. [New York: Academic Press, 1970]). The colors were first called red, white, and blue, then changed to red, blue, and green, apparently in the mistaken belief that these are the primary colors (see the second 1978 edition of Lichtenberg, *op. cit.*, p. 221).

37. Quote above from Nambu, Y., in de-Shalit, A., Feshbach, H., and Van Hove, L., eds., *Preludes in Theoretical Physics* (New York: Wiley, 1966): 133. The article arose when, quite unexpectedly, Nambu received a letter inviting him to contribute to a *Festschrift*, a book of papers in honor of the sixtieth birthday of Victor Weisskopf, then the head of CERN. Although Nambu had not been a student of Weisskopf and did not know him well, he was happy to be asked. The paper, "A Systematics of Hadrons in Subnuclear Physics," contains all but two of the basic statements about the strong interaction that later became gospel. The exceptions are the Yang–Mills formulation, which Nambu almost but not quite embraced, and the fractional quark charge, which had few strong adherents except Murray Gell–Mann. Because he avoided fractional charge, Nambu's color was not confined. Gell–Mann, too, had been invited to contribute to the *Festschrift*, but his article never arrived. "I was terribly embarrassed about not contributing," he said. "Of all the people who should have contributed to that, I was certainly the—I was someone who *should* have contributed to that, being one of his students. I was asked, of course I was asked, but I have such trouble writing things. I was worried and confused about something, as usual, and I didn't produce anything in time for publication. So I never read it." The coincidence still rankles, and he returned to it a number of times in different conversations. "If I'd seen Nambu's paper, I probably would have pieced everything together. But who knows? Maybe I wouldn't have. But I was so ashamed. I never write articles on time. I always send them in months or years late. We all just sat there, stewing around, and it was so *needless*." (Interviews, Murray Gell–Mann, 3 March 1983, 22 May 1984, 5 February 1985, 21 February 1985.)

38. Interview, Howard M. Georgi III, his office, Harvard University, 14 June 1983. The discussion has been slightly condensed.

39. Interview, James Bjorken, his office, Fermilab, 13 February 1985 (all quotes).

40. Telephone interview, Richard Taylor, 26 February 1985 (all quotes).

41. Panofsky, W., in Prentki, J., and Steinberger, J., eds., *Proceedings of the Fourteenth International Conference on High–Energy Physics*, Vienna, 28 August–5 September 1968 (Geneva: CERN): 23 (quote below, p. 37).

42. E.g., Feynman, R., *Acta Physica Polonica* 24, no. 6 (December 1963): 697; interview, Richard Feynman, his office, Caltech, 22 February 1985.

43. Telephone interview, Henry Kendall, 26 February 1985.

44. Bjorken, J., and Paschos, E. A., *PR* 185, no. 5 (25 September 1969): 1975. Feynman himself did not publish anything on the model until 1972 (Feynman, R., *Photon–Hadron Interactions* [Reading, Massachusetts: W. A. Benjamin, 1972]).

45. Letter, Richard Feynman to authors, 18 September 1985.

46. Interview, Murray Gell–Mann, his office, Caltech, 21 February 1985.

47. *CQ*, pp. 140–52, carefully summarizes the experiments.

48. Interview, Steven Weinberg, outside Harvard, 7 May 1985.

CHAPTER 16

1. Biographical information and all subsequent quotes from interview, J. Iliopoulos, an office at Rockefeller University, 8 May 1984.

2. Interview, Claude Bouchiat, his office, École Normale Supérieure, 4 October 1984.

3. Interview, Jacques Prentki, his office, CERN, 7 February 1984.

4. See, for example, Feinberg, G., and Pais, A., *PR* 131, no. 6 (15 September 1963): 2724. The classification system was introduced in Lee, T. D., *NC*, section *A*, 59, no. 4 (21 February 1969): 579.

5. Bouchiat, C., Iliopoulos, J., and Prentki, J., *NC*, section *A*, 56, no. 4 (21 August 1968): 1150; Iliopoulos, J., *NC*, section *A*, 62, no. 1 (1 July 1969): 209.

6. Cabibbo, N., *PRL* 10, no. 12 (15 June 1963): 531. More exactly, Cabibbo said that the quark coupled in the weak interaction to the up quark is $d_c = \alpha d + \beta s = d \cos \Theta + s \sin \Theta$, where d and s are the quarks in the strong interaction and $\alpha^2 + \beta^2 = 1$. Historically, Cabibbo reintroduced a parameter into the eightfold way that Gell-Mann and Lévy (among others) had invented in 1959.

7. Cabibbo, N., and Maiani, L., *PL*, section *B*, 28, no. 2 (11 November 1968): 131; Cabibbo, N., and Maiani, L., *PR*, section *D*, 1, no. 2 (15 January 1970): 707.

8. Interview, Luciano Maiani, his office, University of Rome, 17 February 1984 (all quotes).

9. Interview, Sheldon Glashow, his office, Harvard, 25 April 1983.

10. This had been realized before in different ways by, for example, Feinberg and Pais, *op. cit.*, p. 2728. Many unobserved processes disappeared when the quark model was considered, but not, as Glashow, Iliopoulos, and Maiani realized, this one.

11. Glashow, S., Iliopoulos, J., and Maiani, L., *PR*, section *D*, 2, no. 7 (1 October 1970): 1285; quotes from p. 1290.

12. The anecdotes above are based on the interviews with Glashow, Iliopoulos, and Maiani cited above, and interview, Samuel Ting, an office at CERN, 9 February 1984.

13. Glashow, S. L., and Iliopoulos, J., *PR*, section *D*, 3, no. 4 (15 February 1971): 1043.

14. Glashow, S. L., in *Meeting on Renormalization Theory*, 14-18 June 1971 (Marseille: Centre de Physique Théorique, C.N.R.S., 1971): 159.

15. Interview, Sheldon Glashow, his office, Harvard, 2 December 1982.

16. Interview, Martinus Veltman, an office at Columbia, 12 December 1983 (all quotes).

17. Schwinger, J., *PR* 3, no. 6 (15 September 1959): 296. But see also, Gotô, T., and Imamura, T., *PTP* 14, no. 4 (October 1955): 396.

18. Veltman, M., *PRL* 17, no. 10 (5 September 1966): 553.

19. Bell, J. S., *NC*, section *A*, 50, no. 1 (1 July 1967): 129-30; see also, Bell, J. S., in *Rendiconti della Scuola Internazionale di Fisica "E. Fermi,"* 41st Course (Rome: Università degli Studi, 1966).

20. This is a truncated summary of a longer process. Using current algebra, S. L. Adler and W. I. Weisberger independently discovered a relation between the weak coupling constant and pion–nucleon interactions. Although the Adler–Weisberger relation fit the data tolerably well, one ran into Schwinger terms in its derivation. Veltman removed them by imposing gauge invariance. Bell realized that Veltman's conditions implied that current algebra was really a consequence of Ward identities, a set of equations that describe many of the properties of quantum field theory. Puzzling over Bell's paper, Veltman learned, during a casual encounter with Feynman, that the fields, in the form restated by Bell, were Yang–Mills fields. Feynman however insisted that they were applicable to the strong interactions. (Interview, Martinus Veltman, an office at Columbia, 13 December 1983.)

21. Interview, John Bell, his office, CERN, 6 February 1984. Bell told us his paper had the modest aim of trying to understand current algebra, and that Veltman had found more in the paper than its author had intended. "The illumination to Tini later was a much bigger thing— that gauge invariance is more than a technical device. He underestimated the intelligibility of the paper because he was looking for more than it had, and afterwards, when he found it in his own head, he exaggerated its depth."

22. Veltman, M., *NP*, section *B*, 7 (1968): 637.

23. Interview, David Politzer, his office, Caltech, 21 February 1985.

24. Fadeev, L. D., and Popov, V. N., *PL*, section *B*, 25, no. 1 (24 July 1967): 29. For a good review of the Soviet work, see Veltman, M., in Rollnick, H., and Pfeil, W., eds., *International Symposium on Electron and Photon Interactions at High Energies*, Bonn 1973 (London: North Holland, 1974): 429 and especially appendix.

25. Veltman, M., *NP*, section *B*, 21, no. 1 (1 August 1970): 288.

26. Letter, Gerard 't Hooft to authors, 30 July 1985; subsequent quotes from interview, Gerard 't Hooft, his office, University of Utrecht, 26 September 1984.

27. 't Hooft, G., *NP*, section *B*, 33, no. 1 (1 October 1971): 173.

28. Bell, J. S., and Jackiw, R., *NC*, section *A*, 60, no. 1 (1 March 1969): 47; Adler, S. L., *PR* 177, no. 5 (25 January 1969): 2426.

29. Cited in *CQ*, p. 178.

30. Interview, David Politzer, his office, Caltech, 21 February 1985 (all subsequent quotes).

31. 't Hooft, G., *NP*, section *B*, 35, no. 1 (1 December 1971): 167.

32. Interview, Martinus Veltman, an office at Columbia, 13 December 1983.

33. Lee, B. W., *PR*, section *D*, 5, no. 4 (15 February 1972): 823; Lee, B. W., and Zinn–Justin, J., *PR*, section *D*, 5, no. 12 (15 June 1972): 3121; *ibid.*, p. 3137; *ibid.*, p. 3155.

34. Interview, Steven Weinberg, his office, University of Texas, 28 November 1984.

35. Adler, S. L., and Bardeen, W. A., *PR* 182, no. 5 (25 June 1965): 1517. The full history of the anomaly is quite interesting, and we no more than touch on it here. It was shown to spoil the renormalizability of the model unless colored quarks were introduced (Bouchiat, C., Iliopoulos, J., and Meyer, Ph., *PL*, sec. *B*, 33, no. 7 [3 April 1972]: 519).

36. Adler, S. L., in Devons, S., ed., *High–Energy Physics and Nuclear Structure*. Proceedings of the Third International Conference on High Energy Physics and Nuclear Structure, New York City, September 1969 (New York: Plenum Press, 1970): 654.

37. Bardeen, W. A., Fritzsch, H., and Gell–Mann, M., in *Proceedings of the Topical Meeting on Conformal Invariance in Hadron Physics* (Frascati, May 1972); Bardeen, W. A., Fritzsch, H., and Gell–Mann, M., in Gatto, R., *Scale and Conformal Symmetry in Hadron Physics* (New York: Wiley, 1973): 139. An ancillary point: Color without colored hadrons is mathematically equivalent to parastatistical quarks without parahadrons. But parastatistics with real parahadrons is not like color with colored hadrons, which is why Gell–Mann et al. did not treat the two ideas similarly. Gell–Mann quote from telephone interview, Murray Gell–Mann, 21 February 1985.

38. Telephone interview, Murray Gell–Mann, 28 May 1985.

39. Fritzsch, H., and Gell–Mann, M., in Jackson, J. D., and Roberts, A., *Proceedings of the XVI International Conference on High Energy Physics*, Fermilab, September 1972 (Batavia, Illinois: National Accelerator Laboratory): 135. See also Gell–Mann's rapporteur talk on p. 333.

40. In their 1972 paper, Fritzsch and Gell–Mann proposed the scheme only tentatively, but it worked so well that the following year they joined with a Swiss physicist in proposing it explicitly: Fritzsch, H., Gell–Mann, M., and Leutwyler, H., *PL*, section *B*, 47, no. 4 (26 November 1973): 365.

41. Telephone interview, Murray Gell–Mann, 28 May 1985.

42. Symanzik, K., *Communications in Mathematical Physics* 18, no. 3 (1970): 227; Callan, C. G., Jr., *PR*, section *D*, 2, no. 8 (15 October 1970): 1541.

43. More exactly, the renormalization group describes such scale transformations in terms of changes of an effective coupling constant in the underlying field theory. (Stueckelberg, E. C. G., and Petermann, A., *HPA* 26, p. 499; Gell–Mann, M., and Low, F., *PR* 95, no. 5 (1 September 1954): 1300.

44. Weinberg, S., in Guth, A. H., Huang, K., and Jaffe, R. L., eds., *Asymptotic Realms of Physics* (Cambridge, Massachusetts: MIT Press, 1985): 1.

45. Wilson, K. G., *PR*, section *B*, 4, no. 9 (1 November 1971): 3174; *ibid.*, p. 3184; Wilson, K. G., *PR*, section *D*, 3, no. 8 (15 April 1971): 1818.

46. Interview, David Gross, an office at Brookhaven, 2 April 1985 (all quotes).

47. Interview, Frank Wilczek, his office, University of California at Santa Barbara, 22 February 1985 (all quotes).

48. Letter, Gerard 't Hooft to authors, 30 July 1985; letter, John Iliopoulos to authors, 22 August 1985.

49. Gross, D., and Wilczek, F., *PRL* 30, no. 26 (25 June 1973): 1343; Politzer, H. D., *PRL* 30, no. 26 (25 June 1973): 1346. See also, Gross, D. J., and Wilczek, F., *PR*, section *D*, 8, no. 10 (15 November 1973): 3633; *ibid.*, 9, no. 4 (15 February 1974): 980; Politzer, H. D., *PRpts C*, 14, no. 4 (1974): 129.

50. Telephone interview, Tony Zee, 5 October 1985; Zee, T., *PR*, sec. *D*, 7, no. 12 (15 June 1973): 3630. See also, Zee, T., *Thy Fearful Symmetry* (New York: Macmillan, 1986).

51. Weinberg, S., *PRL*, 31, no. 7 (13 August 1973): 494; Gross and Wilczek, *op. cit.*, 3633.

52. Interview, Sheldon Glashow, his office, Harvard, 2 December 1983.

CHAPTER 17

1. Interview, Samuel Devons, Barnard, 4 January 1984.

2. See, as a typical example, the discussion in Bernstein, J., *Elementary Particles and Their*

Currents (San Francisco: W. H. Freeman, 1968), p. 43. We have been greatly aided in constructing this account by the writings of Peter Galison and Andy Pickering cited below.

3. Weinberg, S., *PRL* 27, no. 24 (13 December 1971): 1688; Weinberg, S., *PR*, section *D*, 5, no. 6 (15 March 1972): 1415, quotes from p. 1415. The predictions involved interactions with one pion, but approximately the same held true for other reactions.

4. Lagarrigue, A., Musset, P., and Rousset, A., "Projet de Chambre à Bulles à Liquides Lourdes de 17m^3," typescript, 10 February 1964; Allard, J. F., et al., "Proposition de construction d'une grande chambre a bulles liquides lourdes destinées à functioner auprés du synchrotron à protons du CERN," typescript, 1 January 1965, both from Musset's files; other particulars provided by letter, Paul Musset to authors, 12 July 1984.

5. Interview, Jacques Prentki, his office, CERN, 7 February 1984.

6. On the other hand, Lagarrigue, at least, was aware of the puzzle posed by their nonexistence, saying in a speech about the uses of large bubble chambers: "The lack of neutral currents is one of the major points [to be studied] of the weak–interaction physics with neutrinos" (Lagarrigue, A., in Puppi, A., ed., *Old and New Problems in Elementary Particles* [New York: Academic Press, 1968]: 148).

7. Cundy, D. C., et al., *PL*, section *B*, 31, no. 7 (30 March 1970): 478. The upper limit, expressed as a ratio of neutral to charged currents, was $0.12 \pm .06$, the limit cited by Weinberg. See also, Gargamelle collaboration, "Proposal for a Neutrino Experiment in Gargamelle," CERN–TCC/70—12, 16 March 1970, which mentions the use of Gargamelle "to reduce this limit to 0.05" or even 0.03 through various tests that, in fact, were never performed (p. 6).

8. Interview, Paul Musset, his office, CERN, 2 February 1984.

9. This point is expanded upon in some detail in Pickering, A., "Making Meaning, or Editing and Epistemology: Three Accounts of the Discovery of the Weak Neutral Current," a talk before the Joint Seminar for the History and Philosophy of 20th Century Science, 26 October 1984, typescript, p. 15. It is interesting to note that one of Donald Perkins's graduate students, E. C. M. Young, had conducted a study of the earlier round of neutrino experiments at CERN with results that could, with hindsight, be interpreted as supporting the existence of neutral currents (Pickering, A., *Studies in History and Philosophy of Science* 15, no. 2 [June 1984]: note 34). However, Young claimed "reasonably good agreement" between the expected and observed background events (Young, E. C. M., "High Energy Neutrino Interactions," CERN Yellow Report 67–12, 21 April 1967, p. 56).

10. The following description of Fermilab's construction is based on visits to the laboratory in February and May 1985; interviews with Dick Carrigan, James Bjorken, Leon Lederman, Carlo Rubbia, Larry Sulak, and Samuel C. C. Ting; and Hilts, P. J., *Scientific Temperaments* (New York: Simon & Schuster, 1982): 17–99.

11. Quoted in Hilts, *op. cit.*, pp. 98–99.

12. Interview, Larry Sulak, an office at New York University, 28 November 1983. Harvard–Pennsylvania–Wisconsin Collaboration, "NAL Neutrino Proposal," Proposal 1, Fermilab Archives, Proposal Shelf, 15 April 1970; addendum, July 1970. Hereinafter the group will be referred to as "HPW Collaboration."

13. Interview, Carlo Rubbia, his office, CERN, 11 February 1984.

14. Interview, Larry Sulak, a restaurant in Ohio, 5 June 1985.

15. Palmer, B., "Very Preliminary Results of Neutral Current Search in the Neutrino–Freon Experiment," handwritten ms. in Dr. Palmer's files, May 1972; notes to talk at Brookhaven, handwritten ms. in Dr. Palmer's files, July 1972.

16. Letter, Robert Palmer to Peter Galison, 8 June 1983, Palmer's files.

17. Lee, W., *PL*, section *B*, 40, no. 3 (10 July 1972): 423. See also, Lee, B. W., *PL*, section *B*, 40, no. 3 (10 July 1972): 420.

18. Prentki, J., and Zumino, B., *NP*, section *B*, 47, no. 1 (September 1972): 99; Lee, B. W., *PR*, section *D*, 6, no. 4 (15 August 1972): 1188.

19. Glashow, S. L., and Georgi, H., *PRL* 28, no. 22 (29 May 1972): 1494.

20. Letter, André Lagarrigue to Willibald Jentschke, 12 April 1972, quoted in Galison, P., *RMP* 55, no. 2 (April 1983): 484. However, Lagarrigue's letter is pessimistic on whether the question of their existence could be resolved on CERN's low energy accelerators (Letter, Paul Musset to authors, 12 July 1984).

21. Baltay, C., et al., CERN technical memorandum TC–L/PA/UC/WF–/fv, 14 July 1972; *ibid.*, TC–L/WFF/ju, 14 July 1972. (All CERN memoranda from Musset's files unless otherwise specified.)

22. Galison, *op. cit.*, pp. 494–95; interview, Larry Sulak, a restaurant in Ohio, 5 June 1985; interview, Carlo Rubbia, his office, CERN, 11 February 1984.

23. Galison, *op. cit.*, pp. 486–87; we have rendered Faissner's "Bilderbuch–example" as "picture–book example."

24. Musset, P., *Bulletin of the American Physical Society* 18, no. 1 (winter 1973): 73.

25. Musset, P., CERN technical memorandum TC–L/BC/PM/fv, 19 March 1973; Musset, addendum, CERN technical memorandum TC–L/BC/PM/fv, 26 March 1973.

26. Interview, Paul Musset, his office, CERN, 2 February 1984.

27. Musset recalls meeting Rubbia at a French Physical Society conference in Dijon in early 1973. "I explained what kind of events we had, and that we were beginning to see a signal. I did not give any final number because it was too early. Rubbia said, 'Okay, we can do the same thing with the Fermilab apparatus.' So I knew that they certainly were working on that. But I had the attitude not to try to be informed about what their result was, because really what we had to understand was ours." (Interview, Paul Musset, his office, CERN, 2 February 1984).

28. Interview, Paul Musset, CERN bubble chamber photo storage room, 6 February 1984.

29. Letter, Robert Palmer to P. Galison, *op. cit.* Palmer also tried to get Brookhaven to do their own experiment; Brookhaven National Laboratory Memorandum to R. R. Rau from R. B. Palmer and N. P. Samios, 21 May 1973, Palmer's files. The publication is Palmer, R. B., *PL*, section *B*, 46, no. 2 (17 September 1973): 240.

30. Musset, P., CERN technical memorandum TC–L/PA/PM/ju, 17 April 1973.

31. Musset, P., *Journal de Physique*, Colloque C3, 34, nos. 11/12 (supp.) (November–December 1973): C3–1, at C3–6 (trans. by authors).

32. Interview, Larry Sulak, an office at New York University, 28 November 1983; interview, Carlo Rubbia, his office, CERN, 11 February 1984; interview, David Cline, Weak Interaction Conference, Wingspread, Wisconsin, 31 May 1984; Benvenuti, A., et al., *PRL* 30, no. 21 (21 May 1973): 1084; Galison, *op. cit.*, p. 495.

33. Hasert, F. J., et al., *PL*, section *B*, 46, no. 1 (3 September 1973): 121.

34. Letter, Ettore Fiorini to Paul Musset, 11 July 1973; letter, J. Sacton to Paul Musset, 11 August 1973 (so dated, but internal evidence suggests the actual date is 11 July).

35. Both letters quoted in Galison, *op. cit.*, p. 495.

36. Hasert et al., *op. cit.*, p. 138.

37. Interview, Paul Musset, his office, CERN, 2 February 1984.

38. Letter, Patrick Coomey to Carlo Rubbia, 24 July 1973, Rubbia's files.

39. Benvenuti, A., et al., typescript. The published paper (see below) was "received 3 August 1973." See also, Rubbia, C., and Sulak, L., Harvard technical memorandum, 18 August 1973.

40. Interview, David Cline, Weak Interaction Conference, Wingspread, Wisconsin, 31 May 1984; Galison, *op. cit.*, pp. 497–98.

41. Moreover, the novel manner of presentation of the HPW results drew shouts of protest from other experimenters. "People didn't even understand the statistics," Sulak said later. "It took us a while to learn how to state the argument. Now it's really very common. But we just had to hammer away, and people sort of refused to understand that if you ask—take old physics. Old physics is a neutrino comes in and a muon goes out. Given old physics, how many would you expect with a neutrino in and nothing out, just because you miss [the muon]? And for that type of event, something being missed, we had way too many muonless events. It was like a five or six standard deviation effect, which is a really big effect in terms of statistical power." Most physicists were then used to the question, how well is the value of the ratio known? The certainty was then much less, which distressed some experimenters. They did not understand the significance, Sulak said, "of saying, how many muonless events do you *expect*—it turns out to be eight—and how many do you *see*—thirty-four. Well the probability that if you expect eight to all of a sudden see thirty-four, that's roughly the numbers, is just a far more significant thing. Somehow we could not convince people to understand that" (interview, Larry Sulak, a restaurant in Ohio, 6 June 1985). All quotes below by Sulak from same interview.

42. Interview, David Cline, Weak Interaction Conference, Wingspread, Wisconsin, 31 May 1984.

43. The actual angular difference between the first and second setup was small, as can be deduced from the description above. The difference between a "wide–angle" and "narrow–angle" muon was a matter of a few degrees, because the particles were traveling with sufficient forward momentum to swamp the effects of the sideways motion imparted by the interaction.

44. Actually, he provided others with descriptions of the HPW results. See Myatt, G., in Rollnik, H., and Pfeil, W., *Proceedings of the Sixth International Symposium on Electron and Photon Interactions at High Energies*, Physikalisches Institut, University of Bonn (27–31 August

1973): 389, 395, and 405; Musset, P., *Proceedings of the II International Conference on Elementary Particles*, Aix-en-Provence, 6-12 September 1973. Musset was informed by L. Sulak.

45. Lubkin, G., *PT* (November 1973): 17.

46. Barish, B. C., and Sciulli, F., "Neutral Current Investigations at NAL," Proposal 262, Fermilab Archives, Proposal Shelf, 24 October 1973. The original team proposal (Sciulli, F., et al., "Neutrino Physics at Very High Energies," Proposal 21, Fermilab Archives, Proposal Shelf, 10 June 1970) did not mention neutral currents.

47. Cline, D., Wisconsin technical memoranda, 1 and 11 October 1973, quoted in Galison, *op. cit.*, pp. 497-98.

48. The referees also criticized the way the statistics were handled, not understanding the arguments cited in note 41 above.

49. A copy of the first page of the typescript is in Galison, *op. cit.*, p. 500.

50. Musset, P., "Les reactions de courant neutre dans Gargamelle et le schema des particules elementaires," photocopied typescript, 6 November 1973, his files; interview, Paul Musset, his office, CERN, 2 February 1984.

51. The letter is reproduced in Galison, *op. cit.*, p. 501; account confirmed in interviews with Musset, Rubbia, and Sulak.

52. Imlay, R., Wisconsin technical memorandum, 29 November 1973, Imlay's files. The full handwritten calculation by Aubert, Imlay, and Ling was done by December 18.

53. Telephone interviews, Richard Imlay, 26 June 1985, Larry Sulak, 7 October 1985; Cline, D., Wisconsin technical memorandum, 13 December 1973, Imlay's files, includes the transparencies and Cline's quote below.

54. A follow-up HPW paper appeared eight weeks later (Aubert B., et al., *PRL* 32, no. 25 [24 June 1974]: 1454); Hasert, F. J., et al., *NP*, section *B*, 73, 1 (24 June 1974): 1. The Caltech group didn't publish until the next year; Barish, B. C., et al., *PRL* 34, no. 9 (3 March 1975): 538.

55. Telephone interview, Sheldon Glashow, 29 August 1983.

56. Glashow, S. L., in Garelick, D. A., ed., *Experimental Meson Spectroscopy—1974*, AIP Conference Proceedings, No. 21 (New York: American Institute of Physics, 1974): 392.

57. Iliopoulos, J., in Smith, J. R., ed., *Proceedings of the Seventeenth International Conference on High Energy Physics*, London, 1-10 July 1974 (London: Science Research Council, 1974): section III, pp. 97-100.

CHAPTER 18

1. The earliest piece of evidence for charm turned up earlier, when Kiyoshi Niu, a cosmic ray experimenter from the University of Tokyo, found what is likely a charmed particle while conducting experiments in the cargo hold of a Japan Air Lines flight with a new type of emulsion chamber designed to study high-energy showers. Niu and company had heard nothing of charm, but they realized that they had made an important discovery. They sent their result to *Physical Review Letters*, which rejected it because the journal editors had no confidence in cosmic ray results (Telephone interview, Kiyoshi Niu). It eventually appeared in a Japanese journal (Niu, K., Mikumo, E., and Naeda, Y., *PTP* 46, no. 5 [November 1971]: 1644). What might have been the last great cosmic ray particle discovery remained buried. Julian Schwinger used it as a bolster to his prediction of the J/psi (Schwinger, J., *PR*, section *D*, 8, no. 3 [1 August 1973]: 960-64).

2. Interview, Nicholas Samios and Robert Palmer, Brookhaven National Laboratory, 18 May 1983.

3. Anonymous, *Brookhaven Bulletin*, 25 January 1974, p. 1. The details of the run given below were provided for us from the Brookhaven archives by Neil Baggett and Michael Murtagh, whom we take this opportunity to thank.

4. Interview, Milda Vitols, Brookhaven scanning room, 3 May 1983.

5. Interview, Michael Murtagh, 17 May 1985.

6. We thank Milda Vitols for providing us with a copy of the sketch.

7. Interview, Helen LaSauce, May 1983.

8. Interview, Nicholas Samios and Robert Palmer, 18 May 1983.

9. Brookhaven Nuclear Laboratories (BNL) memorandum, R. B. Palmer and N. P. Samios to R. R. Rau, 21 May 1973; BNL memorandum, R. H. Phillips to R. B. Palmer, 22 December 1974, both from Samios's files.

10. Telephone interview, Samuel C. C. Ting, 6 July 1983.

11. The various pieces of biographical information were collected from prefatory information

in Ting's Nobel lecture and the interviews with Becker, Chen, Deutsch, Lederman, Pipkin, and Ting cited below.

12. See, for instance, Kroll, N. M., *NC*, section *A*, 45, no. 1 (1 September 1966): 65.

13. The CEA was completed in 1962.

14. All quotes from interview, Francis Pipkin, his office, Harvard, 7 May 1985.

15. Blumenthal, R. B., et al., *PR*, 144, no. 4 (29 April 1966): 1199.

16. Interviews, Samuel C. C. Ting, an office at CERN, 9 February 1984; his office, MIT, 15 October 1985.

17. Interview, Samuel C. C. Ting, an office at CERN, 9 February 1984.

18. The Cornell group had presented their first results at an American Physical Society meeting some months before; Talman, R., *Bulletin of the American Physical Society* 11, no. 3 (26 April 1966): 380. We were aided by a telephone interview with Richard Talman, 19 June 1985.

19. Another controversy involved something called the "Omega-Rho Interference Effect." A review of early experimental work on the effect is G. Goldhaber, in Baltay, C., and Rosenfeld, A. H., eds., *Experimental Meson Spectroscopy* (New York: Columbia University Press, 1970): 59–128. The Ting group's final results were presented in Alvensleben, H., et al., *PRL* 27, no. 13 (27 September 1971): 888. Quotes here from interview, Samuel Ting, his office, MIT, 15 October 1985.

20. Letter, Samuel Ting to Min Chen, 8 December 1971, Chen's files.

21. Becker, U. J., et al., "AGS Proposal," no. 598, Brookhaven archives, 11 January 1972; letter, Robert Phillips to Sam Ting, 29 February 1972, Brookhaven archives.

22. Letters, Robert Phillips to Sam Ting, 3 and 26 May 1972, Brookhaven archives.

23. Dates of AGS operation taken from the AGS operations journal, June to December 1974, AGS control room, Brookhaven. Dates confirmed by examination of Ting's experimental log books, Ting's files.

24. Interviews, Min Chen and Ulrich Becker, their offices, MIT, 1 March 1984.

25. Reproduction of printout taken from Becker's files; to fit the page, we have changed the number of asterisks representing events from four to three. A copy of this graph recently displayed in an MIT exhibit is incorrectly dated 2 September.

26. Account and quotes from interview, Ulrich Becker, his office, MIT, 2 March 1984; telephone interview, Wit Busza, May 1985.

27. Interview, Min Chen, his office, MIT, 1 March 1984.

28. Letter, Samuel Ting to Ronald Rau, 20 September 1974; letter, Ronald Rau to Samuel Ting, 26 September 1974, both in Brookhaven archives.

29. An amusing account of this and other experiments that did not observe charm can be found in Lederman, L., in Gaillard, M. K., and Stora, R., eds., *Théories de jauge en physique des hautes énergies*, Les Houches summer school, session 37, 1981 (New York: North Holland, 1983): especially 837–47.

30. SPEAR history from Richter, B., *Slac Beam Line*, Special Issue No. 7, November 1984; quotes below from interview, Burton Richter, his office, SLAC, 20 February 1985. Description of ADA from visit to Frascati in February 1984.

31. This, and much of what follows, is from G. Goldhaber, in *Adventures in Experimental Physics* 5: 131–40. Richter's quote from letter, Burton Richter to authors, 12 August 1985.

32. The magnet, used to bend the beam path 8°, was a landmark in accelerator engineering, the first superconducting magnet used as a standard operating part of an accelerator. Its liquid helium coolant had malfunctioned.

33. Interview, Martin Deutsch, his home, Cambridge, 7 May 1985.

34. Interview, Samuel Ting, an office at CERN, 9 February 1984.

35. Interviews, Bill Taylor and AGS control staff, AGS control room, 20 May 1985. Ting also visited the office of *Physical Review Letters* to learn about the rules for unrefereed publication.

36. We thank Dr. Schwartz for checking his travel records to provide us with exact dates.

37. Telephone interview, Jayashree Toraskar, 12 April 1985.

38. This account and subsequent quote from interview, Mel Schwartz, his home, California, 19 February 1985; see also, Ting, S. C. C., *Adventures in Experimental Physics* Vol. Epsilon (1978): 115.

39. Interviews cited above with Deutsch and Chen.

40. Interview, Gerson Goldhaber, his office, Berkeley, 19 February 1985 (all quotes).

41. Interview, Min Chen, his office, MIT, 1 March 1984.

42. The shoulder is described in Christenson, J. H., et al., *PRL* 25, no. 21 (23 November 1970): 1523. There was another, more theoretical reason: The ratio of the number of muons to

pions was a mysterious number that theorists thought should be explained by the number of heavy photons, but wasn't by the three known. Ting redid the calculations adding in the *J*, and found that that didn't explain it, either. This led him to hope that there might be others just around the corner (Ting, S., *RMP* 49, no. 2 [April 1977]: 244).

43. Interviews with Richter and Goldhaber cited above, supplemented by notes and photographs from Goldhaber's files.

44. AGS Utilization Daily Report, Brookhaven Archives; interview, Austin McGeary, 20 May 1985.

45. Telephone interview, Burton Richter, 5 July 1983.

46. Aubert, J., et al., *PRL* 33, no. 23 (2 December 1974): 1404; Augustin, J., et al., *PRL* 33:23:1406; Bacci, C., et al., *PRL* 33:23:1408.

47. Abrams, et al., *PRL* 33, no. 24 (17 December 1974): 1453.

48. The two physicists quoted here, like many of their colleagues, talked freely on the subject but requested anonymity.

49. Interview, Martin Deutsch, his home, Cambridge, 7 May 1985.

50. The work of two groups of theorists, one at Harvard, the other at Cornell, was instrumental in the events that followed, helping to transform the quark model, in the words of Cornell physicist Kurt Gottfried, from "a mysteriously helpful Rube Goldberg into a serious intellectual edifice" (Letter, Kurt Gottfried to Gerson Goldhaber, 6 October 1994). Important papers include: Appelquist, T., et al., *PRL* 34, no. 6 (10 February 1975): 365–369; Eichten, E., et al., *ibid.*:369–372; Eichten, E., et al., *PR* D, 17, no. 11 (1 June 1978): 3090.

51. Telephone interview, Burton Richter, 5 July 1983.

52. Cazzoli, E. G., et al., *PRL* 34, no. 17 (28 April 1975): 1125. Nguyen–Khac, L., in "*La Physique du Neutrino à Haute Énergie*," *Proceedings*, (Paris: 18–20 March, 1975): 173.

53. Interview, Sheldon Glashow, 6 June 1983.

54. Interview, Gerson Goldhaber, Summer 1983.

55. Telephone interview, Roy Weinstein, Summer 1983.

56. The standard model faced two other experimental obstacles in these years, both erected by the HPW group. In 1974 and again in 1976 the HPW group reported an inexplicably high value of a number called Y, while in March 1977 they announced six events with three muons; these findings, they said, could not be explained by the standard model. Both the "High-Y" and "trimuon" anomalies, however, were spurious, as a CERN team showed in July and August 1977 to the considerable embarrassment of the Fermilab group. The first HPW papers were: Aubert, B., et al., *PRL* 33, no. 16 (14 October 1974): 984; Benvenuti, A., et al., *PRL* 38, no. 20 (16 May 1977): 1110. The CERN group papers were Holder, M., et al., *PRL* 39, no. 8 (22 August 1977): 433–36; see also Barish, B., et al., *PRL* 38, no. 11 (14 March 1977): 577–80.

57. Bouchiat, M. A., and Bouchiat, C., *PL*, section *B*, 48, no. 2 (21 January 1974): 111.

58. Telephone interview, Edward Fortson, 24 June 1985.

59. *Ibid.*; Feinberg, G., *Nature* 271, no. 5645 (9 February 1978): 509.

60. Interview, Patrick Sandars, his office, Oxford, 9 October 1984.

61. Baird, P. E. G. et al., *Nature* 264, no. 5585 (9 December 1976): 528.

62. Close, F. E., *Nature* 264, 5585 (9 December 1976): 505–6.

63. Interview, Steven Weinberg, outside Harvard, 7 May 1985.

64. Walgate, R., *New Scientist*, 31 March 1977, p. 766.

65. Barkov, L. M. and Zolotorev, M. S., *JETP Letters* 27, no. 6 (20 March 1978): 357.

66. Fortson and Sandars had performed calculations based on an experimentally determined number of 0.35 for $\sin^2 \theta$. In the summer of 1977, this was revised to 0.25 and some months later it fell to 0.21. Moreover, atomic physicists had discovered other factors that affected the way the Weinberg–Salam model was applied to atoms. The effect of the two changes was that the number predicted for the line studied by the Seattle experiment was now 10.5×10^{-8}— within the margin of error of the original experiment! The team had then released more results in contradiction with the new prediction. The Oxford experiment, in contradiction from the start, remained so. The Russian result, originally seeming to confirm the SU(2) × U(1) prediction, did so no longer.

67. Interview, Charles Prescott, SLAC, 20 February 1985.

68. Proposals E94, E122, SLAC proposal file, SLAC archives. Interview, Richard Taylor, his office, SLAC, 20 February 1985. We have been assisted further by letter, Richard Taylor to authors, 6 August 1985.

69. The results were first published as Prescott, C. Y. et al., *PL*, section *B*, 77, no. 3 (14 August 1978): 347–52.

CHAPTER 19

1. Interview, Frank Wilczek, his office, University of Santa Barbara, 22 February 1985.

2. The W and Z were discovered by two CERN collaborations with a new accelerator that collided protons and antiprotons. For their role in the construction of the machine and the subsequent discoveries, Carlo Rubbia and Simon van der Meer, both of CERN, were awarded the 1984 Nobel Prize for physics. A lengthy popular account of the finding is given in the final chapters of Sutton, C., *The Particle Connection* (New York: Simon and Schuster, 1984); considerably shorter and less technical is Crease, R., and Mann, C., *Science Digest*, September 1984, 51; see also the forthcoming Taube, G., *Nobel Dreams* (New York: Simon & Schuster, 1986). References to original papers are in Sutton, *op. cit.*, p. 174.

3. Bjorken, J., in *Neutrino '79: Proceedings of the International Conference on Neutrinos, Weak Interactions, and Cosmology*, 18–22 June 1979, Bergen, Norway (Bergen: University of Bergen and Nordita, 1980): 9. Bjorken's talk carried the very good point to his fellow theorists that most physical theories have been disproven, and all are at the least incomplete; he warned against the "dangers common to any orthodoxy."

4. These came from an SU(4) model done by Wigner, which treated spin and isotopic spin identically. The fundamental representation consisted of four particles: neutron, spin up and spin down; and proton, spin up and spin down. When the SU(3) eightfold way was invented, that was combined with the SU(2) spin to produce SU(6); quarks were incorporated to produce another version. Salam and two other physicists produced a relativistic version, which was called U(6,6) as a relativistic theory of Zerseth order Lagrangian vertices. It was mistaken for an S-matrix theory. Then a variety of "no–go" theorems were produced that said no fundamental theory could accommodate particles of different spins. Nowadays the arrival of supersymmetry has pointed out a flaw in the various no–go theorems. Salam "demolished"—interview, Abdus Salam, an office at CERN, 19 September 1983.

5. Interview, Abdus Salam, a hotel in New York, 27 May 1984.

6. Berlin, I., *Russian Thinkers*, Hardy, H., and Kelly, A., eds. (New York: Penguin Books, 1978): 22.

7. Lecture, Pati, J., Conference on the History of the Weak Interactions, Wingspread, Racine, Wisconsin, 31 May 1984. More technically, why must the proton and electron in beta decay have the same spin? It would be perfectly possible to give the neutron and electron the same spin in the theory by rewriting the currents.

8. Bjorken, J. D., in *Proceedings of the XVI International Conference on High Energy Physics*, vol. 2 (Batavia, Illinois: Fermilab, 1972): 304. Pati and Salam came up with four models, of which this is the one they most extensively discussed. We have used somewhat more modern notation.

9. Pati, J., and Salam, A., *PR*, section *D*, 8, no. 4 (15 August 1973): 1249. Worth mentioning here is an earlier, more confused attempt at unification: Bars, I., Halpern, M. B., and Yoshimura, M., *PR* 7, no. 4 (15 February 1973): 1233.

10. Pati, J. C. and Salam, A., *PRL* 31, no. 10 (3 September 1973): 661–64.

11. Salam, A., *RMP* 52, no. 3 (July 1980): 530.

12. Weinberg, S., comment in *Second International Conference on Elementary Particles*, Aix–en–Provence, 6–12 September 1973, *Journal de Physique* 34, no. 10 (supp.), Colloque C1, p. 47.

13. Interview, Abdus Salam, an office at CERN, 19 September 1983.

14. Interview, Howard M. Georgi III, his office, Harvard, 14 June 1983.

15. Telephone interview, Murray Gell–Mann, 22 May 1984.

16. Interview, Howard M. Georgi III, a restaurant in Manhattan, 29 January 1985.

17. The proton decays in Georgi and Glashow's SU(5) for entirely different reasons than in Pati and Salam's model. The proton decay of the Pati and Salam model does not arise from the single coupling constant, but stems instead from their special Higgs bosons. Sometimes the same phenomenon can be predicted for different theoretical reasons.

18. Quotes below from interview, Howard M. Georgi III, his office, Harvard, 14 June 1983; interview, Sheldon Glashow, his office, Harvard, 2 December 1982.

19. In the same experiment that discovered the neutrino, Reines, Cowan, and Maurice Goldhaber used the data to set a limit on proton lifetime (Reines, F., Cowan, C., Jr., Goldhaber, M., *PR* 96, no. 4 [15 November 1954]: 1157). Georgi and Glashow used the limit from Gurr, H. S., et al., *PR*, 158, no. 5 (25 June 1967): 1321. For an historical review, see Goldhaber, M., Langacker, P., and Slansky, R., *Science* 210 (21 November 1980): 851.

20. Georgi, H., and Glashow, S. L., *PRL* 32, no. 8 (25 February 1974): 438.

21. Telephone interview, James Bjorken, 5 July 1983.

22. Telephone interview, Howard M. Georgi III, 11 May 1984.

23. Interview, Steven Weinberg, outside Harvard, 7 May 1985. We have condensed the discussion slightly. It is worth noting here that the paper contains a prediction of the weak mixing angle that was initially quite low. During the 1970s, the value of the angle drifted down, to where it is now in good agreement.

24. A good nontechnical introduction to cosmology with some historical overtones is Weinberg, S., *The First Three Minutes* (New York: Basic Books, 1977).

25. Penzias, A., in Reines, F., ed., *Cosmology, Fusion, and Other Matters: George Gamow Memorial Volume* (Boulder, Co.: Colorado Associated University Press, 1972).

26. Interview, Sheldon Glashow, his office, Harvard, 13 June 1983.

27. If the proton does decay, it may solve a long–standing puzzle about the relative proportions of matter and antimatter in the Universe, namely the preponderance of matter. The exact mechanism is rather complicated, involving parity–violating effects. See Yoshimura, M., *PRL* 41, no. 5 (31 July 1978): 281; errata, 42, no. 11 (12 March 1978): 746; and Sakharov, A., *JETP Letters* 5, no. 1 (1 January 1967): 24.

28. Biographical information from interview, M. G. K. Menon, his office, Planning Commission, New Delhi, 21 March 1985. A history of the Tata Institute can be found in Lala, R. M., *The Heartbeat of a Trust* (New Delhi: Tata-MacGraw Hill, 1984), chaps. 8–10.

29. Krishnaswamy, M., et al., *PL*, section *B*, 106, no. 4 (12 November 1981): 339; interview, V. S. Narasimham, Kolar Gold Fields, 14 March 1985 (all quotes).

CHAPTER 20

1. Hawking, S., "Is the End in Sight for Theoretical Physics?" in Boslough, J., *Stephen Hawking's Universe* (New York: Morrow, 1985).

2. The exact source of this often-quoted remark is difficult to discover (telephone interview, Freeman Dyson, 1 December 1985).

3. Einstein, A., in Lorentz, H. A., et al., *The Principle of Relativity* (New York: Dover Publications 1952): 191.

4. Wolfgang Pauli to J. R. Oppenheimer, no date (but certainly in early 1958), Library of Congress, Manuscript Division, General Case File, Box 56, Pauli file. George Gamow reproduces a similar letter in Gamow, G., *Thirty Years That Shook Physics* (New York: Doubleday, 1966): 162.

5. Interview, Gerard 't Hooft, his office, University of Utrecht, 26 September 1984.

6. Telephone interview, Murray Gell–Mann, 5 February 1985.

7. Kaluza, T. F. E., *Sitzungsberichte der Berliner Akademie* 54 (1921): 966. The article was communicated by Einstein on 8 December. At first Einstein had said that he would be pleased to submit the article. (Letter, Albert Einstein to Theodor Kaluza, 21 April 1919, quoted in Freedman, D., and van Nieuwenhuizen, P., *Scientific American*, March 1985, p. 78.) Then, a week later, Einstein wrote to Kaluza with the thought that although he had found nothing wrong in the paper, "the arguments brought forward so far do not appear convincing enough," and asked for more detailed calculations (letter, Albert Einstein to Theodor Kaluza, 28 April 1919). Kaluza apparently did. Einstein didn't respond. Two and a half years later he unexpectedly relented (Letter, Albert Einstein to Theodor Kaluza, 14 October 1921).

8. Klein, O., *ZfP* 37, no. 12 (10 July 1926): 895.

9. Veneziano, G., *NC*, section A, 57, no. 1 (1 September 1968): 190; interview, Yoichiro Nambu, his office, University of Chicago, 13 February 1985. Nambu's first string article is Nambu, Y., in Chand, R., ed., *Symmetries and Quark Models*, Proceedings of an International Conference at Wayne State University, 18–20 June 1969 (New York: Gordon & Breach, 1970): 269. Later, independent discoveries of the string are listed in Scherk, J., *RMP* 47, no. 1 (January 1975): 123.

10. First described by Lovelace, C., *PL*, section B, 34, no. 6 (29 March 1971): 500 (which on p. 502 dismisses the twenty-six dimensions in passing as "obviously unphysical"); proven by Brower, R.C., *PR*, section D, 6, no. 6, 15 September 1972.

11. Neveu, A., and Schwarz, J., *NP*, section B, 31, no. 1 (1971): 56. A related model for spin-½ particles was proposed subsequently by Pierre Ramond, then incorporated in the following: Neveu, A., and Schwarz, J., *PR*, section D, 4, no. 4 (15 August 1971): 1109; Thorn, C.B., *ibid.*, p. 1112; Schwarz, J., *NP*, section B, 46, no. 1 (1972): 61.

12. Schwarz, J., and Scherk, J., *NP*, section B, 81, no. 1 (1974): 118.

13. The superstring theories in note 11 are historically the first example of supersymmetry.

14. Green, M., and Schwarz, J., *PL*, section B, 148, no. 1 (1984): 117; *ibid.*, 151, no. 1 (1985): 21.

15. As of the beginning of 1986, the most favored candidate was proposed by Gross, D., Harvey, J., Martinec, E., and Rohm, R., *PRL* 54 (1985): 502.

16. Princeton theorist Ed Witten has flatly predicted that superstring theory "will dominate the next half century, just as quantum field theory has dominated the previous half century" (quoted in *PT*, July 1985, p. 20).

17. Interview, Howard Georgi, a restaurant in New York City, 29 January 1985.

18. Bridgman, P., *The Logic of Modern Physics* (New York: Macmillan, 1946): 207.

19. Interview, Sheldon Glashow, his office, Harvard, 10 November 1982.

20. Interview, Julian Schwinger, a restaurant in Los Angeles, 4 March 1983.

21. Interview, Richard Feynman, his office, 23 February 1985.

22. Interviews, Steven Weinberg, 28 November 1984 and 7 May 1985.

23. Interviews, Toichiro Kinoshita, Cornell, 10 February 1995; Gerry Bunce, Brookhaven, 23 March 1995.

INTERVIEWS

Although we have based our work upon traditional historical sources, this book would not have been possible if many physicists had not been willing to set aside their own work for a while and talk to us. Most sessions were in person, at length, and taped, although there are exceptions; telephone interviews are indicated below with an asterisk (*).

STEPHEN ADLER   30 May 1985*

P. W. ANDERSON   28 January 1985*

PIERRE AUGER   3 October 1984

ULRICH BECKER   1 and 2 March 1984

JOHN BELL   6 February 1984

HANS BETHE   8 August 1984*, 4 December 1984*

JAMES BJORKEN   10 May 1983*, 5 July 1983*, 13 February 1985, 3 May 1985, 23 November 1985*

MARTIN BLOCK   30 May 1984

CLAUDE BOUCHIAT   4 October 1984

MARIE-ANNE BOUCHIAT   4 October 1984

GIORGIO BRIANTI   7 February 1984

GERRY BUNCE   26 January 1995, 23 March 1995

WIT BUSZA   5 June 1985*

NICOLA CABIBBO   21 February 1984

MIN CHEN   1 and 2 March 1984, 6 September 1985*

DAVID CLINE   31 May 1984

MARCELLO CONVERSI   16 February 1984

RICHARD DALITZ   11 October 1984

ALVARO DE RUJULA   9 February 1984

MARTIN DEUTSCH   31 May 1985, 7 May 1985

SAMUEL DEVONS   4 January 1984*, 23 May 1984, 4 October 1985

J. W. DRINKWATER   19 October 1984*

GERALD FEINBERG   27 January 1983, 11 May 1984, 9 January 1985, 1 May 1985, 25 June 1985*

MARKUS FIERZ   21 September 1984*

ETTORE FIORINI   8 February 1984

RICHARD FEYNMAN   23 February 1985

E. NORVAL FORTSON   1 July 1985*

J. BRUCE FRENCH   11 July 1984*

RAOUL GATTO   29 May 1984

MURRAY GELL-MANN   3 March 1983, 3* and 27* June 1983, 13 July 1983*, 10 November 1983*, 18* and 22* May 1984, 5* and 21 February 1985, 24 April 1985, 28* and 29* May 1985

HOWARD GEORGI   14 June 1983, 11 May 1984*, 29 January 1985, 4 September 1985*

SHELDON GLASHOW   9 and 10 November 1982, 2 December 1982, 5*, 8* and 25 April 1983, 12 May 1983*, 6*, 8* and 13 June 1983, 1 July 1983*, 16 January 1984, 8 and 15 May 1984*, 21 May 1984*, 8 May 1985

WOODY GLENN   20 May 1985

GERSON GOLDHABER   11 July 1983*, 1 June 1984, 19 February 1985

MAURICE GOLDHABER   4 February 1983, 2 April 1983, 18 May 1983, 21 December 1983

JEFFREY GOLDSTONE 26 November 1984, 23 January 1985*, 7 May 1985
KURT GOTTFRIED 9 February 1995
OSCAR GREENBURG 4 February 1984*
DANIEL GREENBERGER 16 February 1983*
DAVID GROSS 2 April 1985
PETER HIGGS 29 and 30 May 1984
CHARLES HILL 7 February 1984
GERARD 't HOOFT 26 September 1984
JOHN ILIOPOULOS 8 May 1984
RICHARD IMLAY 25 July 1985*, 16 October 1985*
HENRY KENDALL 26 February 1985*
THOMAS KIBBLE 5 October 1984
TOICHIRO KINOSHITA 10 February 1995
MICHIJI KONUMA 31 May 1984
MASATOSHI KOSHIBA 6 March 1985
NORMAN KROLL 11 July 1984*
HELEN LASAUCE May 1983
WILLIS LAMB 9 August 1984*
RICHARD LEARNER 5 October 1984
LEON LEDERMAN 14 February 1984, 24 May 1984*
Y. Y. LEE 7 June 1983*, 11 January 1984
LUCIANO MAIANI 17 February 1984
WILLIAM MARCIANO 2 April 1983
ROBERT MARSHAK 30 May 1984, 14 January 1985*
AUSTIN MCGEARY 20 May 1985
SIMON VAN DER MEER 7 February 1984
M. G. K. MENON 21 and 22 March 1985
PHILIP MORRISON 19 June 1984
LLOYD MOTZ 19 December 1984*
MICHAEL MURTAGH 3 May 1983, 17 May 1985*
PAUL MUSSET 2 and 6 February 1984
YOICHIRO NAMBU 30 and 31 May 1984, 13 February 1985, 30 May 1985*, 5 June 1985*
V. S. NARASIMHAM 15 and 16 March 1985
YUVAL NE'EMAN 3 May 1985
ABRAHAM PAIS 5 February 1985*, 30 March 1985
ROBERT PALMER 4 February 1983, 18 May 1983
RUDOLF PEIERLS 9 October 1984
ANDRÉ PETERMANN 10 February 1984
ORESTE PICCIONI 3 May 1985
FRANCIS PIPKIN 7 May 1985
DAVID POLITZER 21 February 1985
JACQUES PRENTKI 7 February 1984
CHARLES PRESCOTT 20 February 1985
I. I. RABI 9 May 1984
BURTON RICHTER 5 July 1983*, 24 February 1985
ALAŃ RITTENBERG 7 June 1983*
GEORGE ROCHESTER 30 May 1984, 8 October 1984
BRUNO ROSSI 1 November 1984
CARLO RUBBIA 11 February 1984, 17 May 1984*
ABDUS SALAM 19 September 1983, 23 February 1984, 27 May 1984, 2 May 1985
NICHOLAS SAMIOS 4 February 1983, 18 May 1983, 12 February 1985
PATRICK SANDARS 9 October 1984
HERWIG SCHOPPER 1 February 1984
MEL SCHWARTZ 16 June 1983*, 19 February 1985
JOHN SCHWARZ 22 February 1985
SILVAN SCHWEBER 19 March 1984
JULIAN SCHWINGER 4 March 1983
HERWIG SCHOPPER 1 February 1984
ROBERT SERBER 22 November 1983, 18 July 1984, 22 and 29 October 1984, 19 and 26 November 1984, 28 March 1985, 2, 8 and 26 April 1985, 6, 15, 23, and 29 May 1985, 5 June 1985

R. F. SHUTT   May 1983*
B. V. SREEKANTAN   14 March 1985
ERNST STUECKELBERG   10 February 1984
LAWRENCE SULAK   14 July 1983*, 28 October 1983, 14 February 1985*, 3 March 1985*, 5 and 6 June 1985
RICHARD TAYLOR   20 February 1985, 26 February 1985*
SAMUEL TING   6 July 1983*, 9 February 1984, 15 October 1985
ED UEHLING   20 July 1984*, 25 October 1984*
GEORGE UHLENBECK   10 May 1984
MILDA VITOLS   May 1985
MARTINUS VELTMAN   20 July 1983*, 13 December 1983, 26 February 1985*
STEVEN WEINBERG   28 November 1984, 7 May 1985, 30 September 1985*
ROY WEINSTEIN   Summer 1983*
VICTOR WEISSKOPF   24 September 1984
JOHN WHEELER   14 December 1983
FRANK WILCZEK   22 February 1985
ROBLEY WILLIAMS   20 and 21 November 1984*
C. N. YANG   3 April 1983*, 10 October 1985*
GAURANG YODH   11 January 1984*
TONY ZEE   5 October 1985*
ANTONINO ZICHICHI   3 and 15 February 1984

Index

475

INDEX

INDEX